Organisms Amplify Diversity

This book presents a hypothesis and evidence that organisms promote and ecosystems maximize biodiversity. All species have a net positive effect on their environment, other species, and diversity. The sun is 30% hotter than when life began, but the temperature has been kept moderate by life. Life created high oxygen, the ozone layer, and fertile soil, a diverse, living system. No species evolves in isolation, and most evolution is coevolution. The nature and number of links between species are as important as species number. Eukaryotes coevolve with complex ecosystems of microbes with which they exchange genes. Genomes and intraspecific interactions both act to promote evolution and diversification. Viruses increase diversity of their hosts and cause macroevolutionary transitions.

Key Features

- Life alters the Earth in ways that increase biodiversity
- All species make their environment better for other species and promote diversity
- Life created the life-friendly atmosphere, temperature, and soil of today

Organisms Amplify Diversity

An Autocatalytic Hypothesis

David Seaborg

CRC Press
Taylor & Francis Group
Boca Raton London New York

CRC Press is an imprint of the
Taylor & Francis Group, an **informa** business

First edition published 2023
by CRC Press
6000 Broken Sound Parkway NW, Suite 300, Boca Raton, FL 33487-2742

and by CRC Press
4 Park Square, Milton Park, Abingdon, Oxon, OX14 4RN

CRC Press is an imprint of Taylor & Francis Group, LLC

Library of Congress Cataloging-in-Publication Data
Names: Seaborg, David, author.
Title: Organisms amplify diversity : an autocatalytic hypothesis / David Seaborg.
Description: First edition. | Boca Raton : CRC Press, 2023. |
Includes bibliographical references and index.
Identifiers: LCCN 2022035944 (print) | LCCN 2022035945 (ebook) |
ISBN 9781032158020 (hardback) | ISBN 9781032159935 (paperback) |
ISBN 9781003246640 (ebook)
Subjects: LCSH: Biodiversity. | Coevolution. | Macroevolution. |
Symbiosis. | Biotic communities. | Evolution (Biology)
Classification: LCC QH541.15.B56 S464 2023 (print) |
LCC QH541.15.B56 (ebook) | DDC 333.95–dc23/eng/20221230
LC record available at https://lccn.loc.gov/2022035944
LC ebook record available at https://lccn.loc.gov/2022035945

ISBN: 9781032158020 (hbk)
ISBN: 9781032159935 (pbk)
ISBN: 9781003246640 (ebk)

DOI: 10.1201/9781003246640

Typeset in Times
by Newgen Publishing UK

Dedication

I dedicate this work to the following special people, who I admire.

Sumer and Jake Barcelon; Connor Cazelot Bloyd; Matthew Cook; David Hofstetter; Morgan Hurst; Drew and Denise Inman; Roger Jakobson; Dallas, Matthew, and Elias Martin; Ruth, Michael, and Grace Sheridan; Benjamin Tubbs

"Your charm is slow silver
tempered with turquoise."
"all that's in nature by
nature is good."

Charles Darwin, Alfred Russell Wallace, Albert Einstein, Isaac Newton

"To see a World in a Grain of Sand
And a Heaven in a Wild Flower

Hold Infinity in the palm of your hand
And Eternity in an hour"

Edward O. Wilson, David Sloan Wilson
"The force that through the green fuse drives the flower
Drives my green age"

Orlando Cepeda
"Time let me hail and climb
Golden in the heydays of his eyes"

Elvis Presley
Buddy Holly
John Lennon, Paul McCartney, George Harrison, Ringo Starr
Brian, Carl, and Dennis Wilson, Al Jardine, Mike Love
Frankie Vali, Bob Gaudio, Tommy DeVito, Nick Massi
"The wind the wind had meant to be—
A little through the lips and throat.
The aim was song—the wind could see."

The Scarlet King Snake, *Lampropeltis elapsoides*
"He prayeth well, who loveth well
Both man and bird and beast."

- The first quotation is from the poem, "You", in *Honor Thy Sow Bug*, by David Seaborg
- The second quotation is from the poem, "Fingers of Nightshade", in *Honor Thy Sow Bug*, by David Seaborg
- The third quotation is from the poem, "Auguries of Innocence", by William Blake
- The fourth quotation is from the poem, "The Force That through the Green Fuse Drives the Flower", by Dylan Thomas
- The fifth quotation is from the poem, "Fern Hill", by Dylan Thomas
- The sixth quotation is from the poem, "The Aim Was Song", by Robert Frost
- The seventh quotation is from the poem, "The Rhyme of the Ancient Mariner", by Samuel Taylor Coleridge

Contents

Acknowledgments

Sterling Bunnell encouraged and inspired me and had several valuable discussions with me about the Autocatalytic Biodiversity Hypothesis/Pachamama Hypothesis and my ideas before I began writing this book or gave the hypothesis its name. He had similar ideas to mine. He seems to have thought about a hypothesis similar to mine independently, although he never published it or anything concerning it.

The following scientists discussed the Autocatalytic Biodiversity Hypothesis with me, encouraged me, and gave me confidence and inspiration, the latter three before I began writing the book:

James Lovelock, Lynn Margulis, Eugene Odum, and Edward O. Wilson.

The following scientists helped me with ensuring that the science and logic are correct:

Robert Burner, Donald E. Canfield, David Catling, John Coffin, Bruce Damer, Romain Darnajoux, David Deamer, Paul G. Falkowski, Scott F. Gilbert, Graham F. Hatful, Robert Hazen, Isaac Hilburn, Jaime Iranzo Sanz, James Kasting, Joseph Kirshrink, Eugene V. Koonin, Lee R. Kump, Mark Leckie, Timothy M. Lenton, Graham A. Logan, Clenton Owensby, Marisa L. Pedulla, Forest Rohwer, Matthew Salzman, Alex L. Sessions, Norman H. Sleep, Curtis Suttle, Luis Villarreal, Yuk L. Yung, and Frank Zindler.

Of those above, the following scientists gave extraordinary help and a great deal of their time, discussing ideas and/or reading, commenting on, and/or suggesting changes to parts of the manuscript, and giving invaluable feedback to assure scientific accuracy:

Robert Burner, John Coffin, Romain Darnajoux, David Deamer, Graham F. Hatfull, Jaime Iranzo Sanz, Eugene V. Koonin, Clenton Owensby, Norman H. Sleep, Curtis Suttle, Luis Villarreal, Yuk L. Yung, and Frank Zindler.

The following laypeople read chapters of the manuscript, doing valuable editing, improving the English and presentation of my ideas, including making the wording more succinct, comprehensible, interesting, and exciting:

Hovig Bayandorian, Lynne Cobb, Kevin Langdon, Steve Seaborg, and Reid Stuart. Lynne Cobb was especially helpful, editing several chapters, with exceptionally helpful suggestions.

The following laypeople sent me articles supporting the hypothesis that were helpful because they addressed aspects I was at least partially unaware of:

Lynne Cobb, Bob Jansen, Adele Seaborg, and Steve Seaborg.

Lynne Cobb sent me the book, *The Medea Hypothesis*, by Peter Ward, making me aware of a hypothesis that is in opposition to the Autocatalytic Biodiversity Hypothesis.

Mamade Kadreebux encouraged me constantly and provided me forums to speak to audiences about my hypothesis a number of times.

Hovig Bayandorian and Adele Seaborg solved computer issues that were beyond my limited knowledge in that area.

Chuck Crumly and Kara Roberts of CRC Press/Taylor & Francis, the publisher of this book, kindly provided me with guidance through the publishing process.

Countless teachers, extended family members, friends, and scientists I knew, but did not speak with specifically about this book, made this book better. So, too, did innumerable living organisms, especially snakes; museums; and beautiful ecosystems in nature. It would not be in keeping with the spirit of this book to fail to acknowledge these.

Any errors of any sort in this book are solely my fault and not the fault of my commensal helpers.

About the Author

David Seaborg is an evolutionary biologist. His undergraduate degree is from the University of California at Davis in zoology, where he learned from Theodosius Dobzhansky and Ledyard Stebbins. His graduate degree is from the University of California at Berkeley, also in zoology. He originated the concept that organisms act as feedback systems in their evolution, and that they thus play an important role in their evolution. This concept is a mechanism for punctuated equilibrium. He showed that the standard genetic code is on an adaptive peak, and how populations cross over maladaptive valleys from one adaptive peak to another. He published a hypothesis to explain how homosexuality evolved even though it theoretically reduces the number of offspring produced. He has written two books on his Autocatalytic Biodiversity Hypothesis, which states that all species in natural ecosystems are ecosystem engineers that make their environment better for life, other species, and biodiversity. It states that life increases biodiversity and ecosystems maximize biodiversity. This is the second of the two books he wrote on this important hypothesis.

He has taught various biology courses from evolution to entomology at all levels from kindergarten to the university level. He taught the basic biology course at the University of California at Berkeley, University of California extension courses, courses at museums, and courses at the Fromm Institute for Lifelong Learning at the University of San Francisco.

David is an environmental leader. He founded and is President of the World Rainforest Fund (worldrainforest.org), a nonprofit, tax exempt foundation dedicated to saving the Earth's tropical rainforests and biodiversity by empowering the indigenous people who live in rainforests. This organization has saved rainforests in Brazil, Ecuador, Peru, Columbia, the Democratic Republic of the Congo, the Philippine Islands, and Borneo. It set the record for the most species saved per dollar when it helped stop a road that if built would have resulted in the destruction of a 10,000-acre rainforest in Ecuador that has the highest biodiversity of any ecosystem on Earth, according to scientists at the Missouri Botanical Garden. Had the road been built, exploiters would have used it to access the rainforest and destroy it. The World Rainforest Fund spent only $3,500.00 to stop the road and save this rainforest.

He raised $20,000 in less than a year to successfully help save Acalanes Ridge, a pristine oak and grassland habitat in the hills in Lafayette, California.

He wrote an article that is a summary of the scientific research on the effects of high atmospheric levels of carbon dioxide other than global warming. Unlike the climatic effects, these effects are not well known to the general public. They are very serious, and have the potential to cause high levels of extinction of species and greatly disrupt ecosystems and our food supply.

He was on the city of Lafayette's General Plan Advisory Committee, which he guided to producing a ten-year General Plan for that city that emphasized environmental sustainability, preserving open space, combating global warming, and energy conservation.

He is currently on the nominating committee for the Goldman Environmental Prize, the most prestigious grassroots environmental prize in the world.

He conceived, and helped secured passage by the Berkeley City Council, an ordinance banning the use of old growth rainforest and redwood in all products used by the city of Berkeley. This ordinance also required all businesses contracting with Berkeley to stop using old growth rainforest and redwood in any products or services Berkeley hires them to use or perform, or in any product they sell to this city.

David carried the Ten Commandments for the Earth, a version of the original Ten Commandments re-written to focus on saving the Earth's environment, while riding a camel down Mount Sinai in Egypt. Then, in a brief ceremony, he presented these Ten Commandments to a Bedouin youth, who represented the indigenous people and the youth of the planet, the generation inheriting the Earth for its stewardship.

David conceived the idea for and was an organizer for a press conference of Nobel Prize winners on global environmental and poverty issues that was held at the time of the 100th Nobel Prize ceremonies in Stockholm, Sweden, in December, 2001.

David is an award-winning nature and wildlife photographer and an award-winning poet. He wrote a popular and acclaimed poetry book called *Honor Thy Sow Bug*. He is listed is in *Who's Who in America*. An excellent public speaker, he lectures to various scientific, environmental, civic, business, and other organizations on evolutionary biology, the philosophical implications of science, and environmental issues.

1 Introduction

This is the second book I have written on my hypothesis that organisms create biodiversity, called the Autocatalytic Biodiversity Hypothesis (ABH), or the Pachamama Hypothesis. I will generally refer to it as the Autocatalytic Biodiversity Hypothesis, or the ABH, rather than use both names every time I refer to it. Both books are about, propose, and present evidence for, the same hypothesis, but the first book (Seaborg, 2022) covers different aspects and topics concerning the hypothesis than this second book. The first book is a companion volume to this second book on my hypothesis.

In some cases in this book, I cite an article only once or twice when I list a number of different points it makes over several sentences, rather than repeating the same citation several times in a short space of text. This avoids redundancy. When I do this, the reader should assume that all of my statements refer to the same work

In this book, I departed from the first book in two ways. In the first book, I capitalized the common names of species, thinking it would provide clarity. In this book, I did not. This is because the convention is to not capitalize the common names of species, and because I ran into some problems that were potentially confusing to the reader in the first book. I have learned a lesson and decided this change is for the better, and is in fact clearer.

The second change I made is the use of the term symbiosis. Symbiosis can be defined in two ways. One definition is that it is any interaction between two species in which one species receives a benefit. By this definition, there are three types of symbiosis. The first is mutualism, in which both species benefit, as in a fruit bat that eats a fruit and receives a nutritious meal, while benefiting the tree species that produced the fruit by dispersing its seeds. The second type of symbiosis by this definition is commensalism, in which one species benefits and the other is unaffected either positively or negatively, as in the hermit crab using the shell of a dead sea snail. The crab benefits by acquiring a protective home, while the snail is not affected because it is dead. The third type of symbiosis by this definition of it is parasitism, wherein one species benefits and the other is harmed, as in a tape worm that receives a habitat and food from a host such as a human. The host is harmed, becoming weakened or sick. The other definition of symbiosis is the same as the definition of mutualism, in which both species benefit. In the first book, I used the term mutualism for symbiosis in which two species benefit each other, to avoid confusion with the other definitions of symbiosis. I now think this was not the best decision. In this book, I will use the term symbiosis for any relationship between species in which both species benefit; that is, I use it as a synonym for mutualism. Therefore, I will avoid the term mutualism. I made this change because so many scientists now use symbiosis to mean mutualism, and because the educated lay public is more familiar with this definition of symbiosis. This also avoids the confusion that could be created in the minds of some readers when symbiosis is equated with commensalism and parasitism. The first definition of symbiosis I listed in this paragraph and the use of the term mutualism appear to have decreased in usage and popularity. In my view, the use of symbiosis to refer only to two species that benefit each other, and avoidance of the term mutualism, will provide more clarity and less ambiguity.

Ecology, population biology, and evolution have served us well, revealing many truths and providing many insights. However, they have given far too little emphasis to the role organisms played and still play in increasing biodiversity. Evolutionary theory is correct in stating that the environment affects organisms, including their structure, behavior, and genetics, and played an important role in creating the great diversity of life we observe today. Yet organisms affect the environment, and not randomly. They affect it in a way that aids life and increases biodiversity. There is feedback between life and the environment, each affecting the other. There is also too little attention given to this. This book addresses these issues and attempts to rectify these shortcomings.

In this second volume, I made some modifications to the mechanism(s) by which the ABH works. These will be explained below.

The Autocatalytic Biodiversity Hypothesis/Pachamama Hypothesis can be stated as follows: *organisms create biodiversity; life makes biodiversity significantly higher than would be the case if solely nonbiological factors such as physical, chemical, and geological forces acted in the absence of life; and ecosystems maximize biodiversity.* A simplified way to express this hypothesis in layman's terms is that life creates more species, although this does not fully capture the meaning of the idea because biodiversity is more than simply the number of species.

The two books on the ABH discuss several apparently unrelated mechanisms by which life increases biodiversity; that is, by which the ABH works and how ecosystems maximize biodiverstiy: natural selection; speciation promoted

DOI: 10.1201/9781003246640-1

by life; the actions of plants, herbivores, predators, symbiotic species, decomposers, viruses, and microbes; production of oxygen; burial of greenhouse gases; altruism; building of the soil; behavior of DNA; and so on. In science, it is desirable to have one or a few unifying mechanism(s) to explain phenomena. Fortunately, I can propose that there are three unifying mechanisms by which the ABH works, and this may even be reducible to one mechanism.

The first and principal mechanism by which life increases biodiversity and ecosystems maximize biodiversity is*: every species* (except *Homo sapiens*) *is an ecosystem engineer that has a net positive effect on biodiversity, other species, and its ecosystem, under natural conditions over time periods long enough to fairly represent its impact*. This literally means *all* species (except humans): all viruses, bacteria, archaea, fungi, plants, animals, and so on—*all* species from *every* group of organism. Beavers are well recognized as ecosystem engineers because they create pond habitat for species that live in ponds. But species of bacteria that live in the human digestive tract are ecosystem engineers that help with digestion, are involved in the development of the immune and nervous systems, and so on. The companion book to this one discusses how plants, herbivores, predators, decomposers, symbiotic species, and others are ecosystem engineers that increase diversity. *Natural conditions* are conditions without significant human effects. *Sufficient time* to fairly represent the impact of a species can best be explained by an example. Migratory locusts (*Locusta migratoria*), which occur in Africa, Asia, Australia, and New Zealand, periodically denude large areas of natural vegetation. However, these events are short-lived, and the ecosystem recovers in time. The overall net effect of the migratory locust on its ecosystem over long time periods is positive and increases diversity in natural conditions. When the first bacteria to evolve photosynthesis did so, the oxygen they produced was toxic and combined with methane, removing this powerful greenhouse gas from the atmosphere, causing the earth to become very cold. Both of these effects reduced biodiversity for a period of time. But in the long run, the effects of these first photosynthesizers was a great increase in biodiversity, since higher life needed the oxygen that they produced as an energy source to evolve and diversify.

The second unifying mechanism by which life increases biodiversity is that *all species are genetic engineers that promote evolution and increase biodiversity in natural conditions over sufficiently long time periods*. Natural conditions and sufficiently long time periods have the same meaning as for the statement above that all species are ecosystem engineers. By genetic engineers, I mean they alter genetic material through sexual recombination, transposable elements (segments of DNA that move from one area of the DNA to another in the same organism, between organisms, or even between species), duplication of genes, shuffling of segments of DNA, doubling of entire genomes, and other mechanisms. (The genome is all of the DNA—all of the genetic material—of an organism.) These mechanisms increase the genetic variability upon which natural selection can act, and can result in large evolutionary breakthroughs that are followed by great diversification into many new species. Different species can "share" their DNA; that is, DNA can move between species, promoting evolution and diversification. This is discussed in Chapter 9. Microbes in the microbiome are genetic engineers that donate DNA to their hosts (Chapter 7). Viruses act as genetic engineers in promoting evolutionary breakthroughs and diversification in their hosts (Chapter 8).

The third unifying mechanism by which life increases biodiversity is that *the organisms of some species are "social engineers" that cooperate with and act altruistically toward with each other, maintaining the population, its variability, and biodiversity* (see Chapter 10).

If one wishes to have the ABH simplified to one unifying principle instead of three, one could expand the definition of ecosystem engineer to include genetic engineering and "social engineering." One could consider the movement of DNA segments, duplication of genes and genomes, shuffling of DNA segments, and so on as a form of ecosystem engineering. Additionally, one could view altruistic behavior and cooperation—that is, "social engineering"—as a type of ecosystem engineering. If one accepts this expanded definition of ecosystem engineering, then the first mechanism that all species are ecosystem engineers in a broad sense is the only mechanism needed to explain how the ABH works, how organisms increase biodiversity. Some may object that this definition of ecosystem engineer is too broad, in which case three mechanisms are needed. Either interpretation is acceptable to me. One has the advantage that it reduces the number of mechanisms by which species increase biodiversity to one; the other has the advantage that it uses a more widely accepted definition of ecosystem engineer. Readers can decide for themselves which interpretation they prefer.

It is acceptable to state the ABH in terms of its principal mechanism. Thus, the ABH could be stated as follows: *Every species* (except *Homo sapiens*) *is an ecosystem engineer that has a net positive effect on biodiversity, other species, and its ecosystem under natural conditions over time periods long enough to fairly represent its impact*. This way of stating the ABH assumes ecosystem engineering includes genetic engineering and "social engineering." Again, if one views this as too broad a definition of ecosystem engineering, one can modify how one states the ABH to reflect this.

One can eliminate the term "ecosystem engineer" and state the ABH in a simple, direct way as follows: *All species have a net positive impact on their environment, other species, and biodiversity under natural conditions over sufficient time.*

A helpful way to view the ABH/Pachamama Hypothesis is its emphasis of life's effect on the environment and the Earth. We know that the environment shapes organisms and their genes through natural selection, and does so in a way that increases their fitness, and, as you will see in this book and its companion volume (Seaborg, 2022), increases biodiversity. *The ABH postulates that this process also works in the other direction, that organisms profoundly affect their environment and the Earth in a way that helps life and increases biodiversity.*

Although not necessarily unifying, another mechanism by which the ABH works is the idea that symbiosis and commensalism are very common, pervasive, fundamental, important, help structure ecosystems, affect evolution, and are very important in maintaining and increasing biodiversity. They are far more important than commonly recognized. This is shown in the microbiome (see Chapter 7), but also in many other areas of study. The importance of symbiosis (mutualism) is discussed in detail in Chapter 4, and of commmensalism in Chapter 5, of the companion book to this one (Seaborg, 2022). Both of these relationships between species are tied to the ABH because the principal mechanism by which the ABH works is that all species are ecosystem engineers that have a positive effect on diversity and their ecosystem. When a butterfly pollinates a flower, obtaining nectar, it is doing ecosystem engineering in a symbiotic relationship. Commensalism is a relationship between two species in which one species receives a benefit from the other, and the other species is neither helped nor hurt by the relationship. When mites are transported between flowers in the bills of hummingbirds, the bird is doing ecosystem engineering. When an elephant digs a water hole to obtain water and other animals use it, this is ecosystem engineering. In fact, all symbiosis and commensalism involve ecosystem engineering by at least one of the species involved.

Another idea related to the ABH is that all genes are connected to other genes, the genetic system of the organism they are in, the gene pool of the population they are in, and the genes of the ecosystem they dwell in. No gene exists or evolves autonomously. Genes coevolve with each other. Organisms are systems of coadapted genes. Similarly, all organisms of a population are inter-related and no organism exists or evolves in isolation. Likewise, all species are connected and no species exists or evolves in isolation. Each species is connected to other species in its community and to its ecosystem. All animals, plants, and other eukaryotes are connected to their microbiomes, and other species in their community. Coevolution between is fundamental and common. All species that strongly interact coevolve with each other. Finally, ecosystems are connected to and interact with other; concerning this last point, see Chapter 4 of the companion book to this one (Seaborg, 2022).

One major reason why species are ecosystem engineers that improve their ecosystem for other species and maintain and increase biodiversity is that their numbers are regulated. For example, mice help their ecosystem by aerating the soil with tunnels, fertilizing the soil with heir defecation and death, controlling populations of what they consume, and providing food for their predators. But if their populations were not controlled and their numbers became too high, they would deplete their food supply, devastating other species dependent on their food; pollute their environment with too much feces and to many corpses; and make many more tunnels in the soil than is optimal. These actions would harm them and many other species tremendously, greatly reducing diversity. This is true for all species; every species would negatively impact its environment if it were not regulated. This is one reason why humans are an exception to the ABH, acting as ecosystem engineers with a negative impact on their environment and biodiversity. Hence, predators and pathogens, herbivores and pathogens that control populations of photosynthesizers, and consumers and pathogens of fungi, are key in causing the ABH to work, for allowing species to increase diversity, and for helping ecosystems maximize biodiversity. This is especially true because these population regulators control populations by negative feedback. Competition between and within species also regulates populations, causing the ABH to work and diversity to increase.

The ABH is an original, radical, testable, qualitative (not mathematically expressed as yet) hypothesis supported by the evidence. This book, like the first book, presents evidence and arguments supporting the ABH. I recommend that the reader read the first volume before reading this second volume for a better understanding of my hypothesis and for a full comprehension of all the topics pertinent to my hypothesis. The first book covers topics different from this one that are relevant to the ABH. These topics include what biodiversity is and how organisms create it; how natural selection is one mechanism by which the ABH works; the relevance of symbiosis and commensalism to the ABH; interspecific competition and the ABH; how plants, herbivores, predators, parasites, and decomposers are ecosystem engineers that create high biodiversity; and eight new proposed principles of ecology and evolution relevant to the ABH. However, it is not necessary to read the first book to understand this second book. This chapter provides a summary of this second book.

I call my hypothesis the Autocatalytic Biodiversity Hypothesis because it proposes that life causes biodiversity to increase; hence, life promotes life and its own diversification, so is autocatalytic. This is the main name for the hypothesis. However, for those who desire a name in keeping with historic precedent concerning hypotheses of this type, I have given the ABH an alternative, second name, the Pachamama Hypothesis, taking the name from a goddess. This is consistent with two other hypotheses about life's effect on life that are named after goddesses, the Gaia Hypothesis and the Medea Hypothesis. These are both named for Greek goddesses. The Gaia Hypothesis, proposed by James Lovelock (1972, 1979) and Lovelock and Lynn Margolis (1974), postulates that organisms made the atmosphere favorable to life, and regulate it, keeping it favorable to life, by negative feedback. Gaia is one of the Greek primordial deities, the personification of the Earth and ancestral mother of all life. The Gaia Hypothesis is similar to and influenced the ABH/Pachamama Hypothesis. The Medea Hypothesis, proposed by Peter Ward (2009), postulates that life is self-destructive and acts to decrease diversity or biomass. This hypothesis is counter and in opposition to both the Gaia Hypothesis and the ABH, proposing essentially the opposite of the latter two hypotheses. I address it in Chapter 14, which covers arguments against the ABH. Medea is a destructive Greek goddess. Pachamama is the ever-present mother goddess of Earth and time who has her own self-sufficient and creative power to sustain life on Earth. She is the prime origin of her four cosmological Quechua principles: Sun, Moon, Water, and Earth. She is a goddess worshipped by indigenous peoples of the Andes, such as the Quechua and Aymara. Although Pachamama is usually translated as Mother Earth, a more literal translation from Quechua and Aymara is World Mother. Although I am departing from the use of a Greek goddess, I am continuing the use of the names of goddesses that are appropriate metaphors for hypotheses concerning life's effects on life. And, like Gaia, Pachamama is a mother goddess of the Earth.

As I pointed out in the first volume, although the ABH has areas of agreement with the Gaia Hypothesis, it is not identical to it. The Gaia hypothesis influenced and was inspirational to the ABH, and I appreciate and am indebted to James Lovelock and Lynn Margolis for this important work. However, there are three major differences between the two hypotheses. The first is that the Gaia Hypothesis is a subset of the ABH. The ABH is more general than the Gaia hypothesis, and incorporates it. The Gaia Hypothesis is principally concerned with the fact that life created an atmosphere favorable to life, although it does address other factors, such as oceanic salinity. Margulis did emphasize the importance of symbiosis, especially with respect to the origin of the mitochondrion and chloroplast (see, for example, Margulis, 1998), but symbiosis is not emphasized in, nor a major part of the Gaia Hypothesis. A hypothesis related to and generated by the ABH is that symbiosis is very important in structuring ecosystems and causing diversity to substantially increase. The ABH is much more comprehensive and general than the Gaia Hypothesis. It recognizes that life created an atmosphere favorable to life. But it also includes key innovations leading to diversification; natural selection as a mechanism for maintaining and increasing diversity; and the roles of symbiosis, commensalism, interspecific competition, plants, herbivores, predators, prokaryotes, the microbiome, viruses, decomposers, the behavior of the genome, cooperation between organisms of the same population, soil organisms, and other mechanisms, in increasing biodiversity.

Second, by far the main, and perhaps the only, mechanism for life maintaining a planet favorable to biology proposed by the Gaia Hypothesis is biologically mediated negative feedback on the nonbiological aspects of the Earth system, especially atmospheric gases. The Gaia Hypothesis emphasizes this as a major mechanism for keeping the Earth, especially the atmosphere, favorable to life. Life-induced negative feedback is central to the Gaia Hypothesis. The ABH agrees with the idea of life-induced negative feedback as a mechanism for maintaining atmospheric gases at levels favorable to life, but sees it as something that can be over-ridden by catastrophic events, and does not emphasize it as much as or give it as much importance as the Gaia Hypothesis does. According to the ABH, negative feedback is most important not in regulating atmospheric gases, but in predator-prey and parasite-host systems, which the Gaia Hypothesis does not address. The ABH proposes many other mechanisms than life-mediated negative feedback on nonbiological aspects of the Earth such as the atmosphere, with emphasis on the unifying one stated above, that all species are ecosystem engineers that net have a positive effect on diversity in natural ecosystems over time. Also, the ABH states that positive feedback can generate diversity in some cases, such as when coevolution between two species leads to symbiosis. The Gaia Hypothesis does not propose or mention any of these mechanisms.

Finally, the Gaia hypothesis does not propose that any variable in particular in an ecosystem is maximized by life. The ABH postulates that life's activities maximize an important variable of the ecosystem: biodiversity. The ABH postulates that ecosystems maximize biodiversity. I hope to demonstrate these claims in this book.

Additionally, I do not think Daisyworld (Watson and Lovelock, 1983), a computer simulation introduced to illustrate the plausibility of the Gaia Hypothesis, has any representation in the real world, in nature. Daisyworld is seeded with black daisies and white daisies as its only life. White daisies reflect light and cool the planet, while black daisies absorb light and heat it. The surface temperature of Daisyworld remains almost constant over a broad range of solar output by negative feedback. I do not think there could be a planet anything like this. It should not be considered part of the ABH. If anything, it represents a difference between the two hypotheses, in that the Gaia Hypothesis posits that Daisyworld could be representative of a real planet, and the ABH claims it could not.

For two reasons, it is necessary to briefly define terms and explain concepts that would not need definition or explanation if this book were written solely for scientists in one field. The first reason is that this book is written for two audiences. It is written principally for scientists because it presents an original, profound, admittedly controversial, scientific hypothesis. It also presents mechanisms of how life increases biodiversity and evidence to support the hypothesis. Yet, it is also written for the educated layperson because many such readers will be interested in this hypothesis and the evidence for it. The second reason is that this book covers a large variety of topics in evolutionary biology, ecology, and animal behavior ranging from how life regulates temperature and produced Earth's oxygen to the origin of life. There will be scientists who read this book who are familiar with the terminology and concepts in some chapters and not familiar with the jargon and concepts in others. I sincerely hope that specialists familiar with a topic being discussed will patiently bear with these brief definitions and explanations, and understand that I am not intending to talk down to them. And bear in mind that it is possible to skip these definitions and explanations.

Evolution, ecology, and population biology have given us valuable insights and understanding, but have had far too little emphasis on and recognition of the fact that life created most of the biodiversity on Earth, and this book is about that insight. The ABH includes the ideas that symbiosis and commensalism are fundamentally important, very common, help structure ecosystems, and create high diversity. Symbiosis often comes about via coevolution between two or more species, and when this is the case, it results in a tremendous increase in diversity.

The ABH is scientific, naturalistic, and mechanistic. Any attempt to interpret it as teleological, pseudoscientific, or

"New Age," as some incorrectly did for the Gaia Hypothesis, is incorrect.

The presentation of the hypothesis and the evidence for it are the subject of both this book and its companion volume. Many supporting concepts, observations, facts, and collections of data are themselves not original. What is original about the ABH is the integration of the facts and evidence into a comprehensive hypothesis that life generates biodiversity. Hence, when I provide an explanation of the microbiome—that animals and plants are ecosystems that have tens of thousands of species of microbes on and in them—I do not claim that this is an original idea. Rather, it is a piece of evidence showing that organisms create and maintain biodiversity. To my knowledge, *this book and its companion volume are the first attempt in the modern era at a comprehensive compilation of evidence that life itself creates biodiversity.*

Following is a summary of this book. The first life on Earth consisted of *prokaryotes*, simple one-celled organisms. Prokaryotes have no nucleus in their cell, and their DNA is scattered throughout their cell. They have no special organelles with membranes around them. *Organelles* are parts of cells with specific functions, analogous to organs in animals. The *nucleus* (structure that has most of the cell's DNA) and *chloroplast* (structure that carries out photosynthesis) of the cell are organelles. Prokaryotes consist of *bacteria* and simple organisms called *archaea*. Some archaea can live in extreme conditions, such as environments with very high temperatures or extremely acidic conditions. Prokaryotes contrast with *eukaryotes*, whose cells have a nucleus and other organelles with membranes around them. Everything from one-celled protozoans, such as amoebae, to humans are eukaryotes.

Following is a summary of this book. The sun is 30% hotter today than when life began. Prokaryotes called *methanogens* produced methane when life had not existed for very long and the Earth was very cold because the sun produced very little heat. Methane is a greenhouse gas that is about 30 times as powerful, molecule for molecule, as carbon dioxide, the main greenhouse gas. (This number is very complex to calculate and will yield different results depending on how it is calculated. I will use 30 times as powerful because it is a reasonable and fairly conservative number, and I will consistently stick with it throughout this book.) Thus, methanogens warmed the Earth and made it habitable for life in Earth's early days when the sun produced much less heat for Earth. As the sun became hotter as time passed, and volcanoes added carbon dioxide and methane to the atmosphere, organisms buried the carbon used to make these two main greenhouse gases, lowering their atmospheric concentrations, balancing out the effects of the sun and volcanoes, cooling the Earth, making the temperature more favorable for life. The burial of carbon by organisms also prevented nonclimatic effects of excess carbon dioxide in the biosphere that are deleterious to life, such as excessively high acidity in the sea and freshwater ecosystems.

However, this burial of carbon would in the distant future cause the amount of carbon available to life to be too little to allow most life on Earth to survive, if life did not provide a solution to this problem. This is because carbon is the most important element in biology. It is in DNA, RNA, protein, and carbohydrates. Carbon dioxide is needed by plants to carry out photosynthesis, which produces oxygen and their carbohydrates. However, life has mechanisms that conserve carbon and keep it available to the biosphere, even as organisms bury carbon-containing greenhouse gases and lower their concentration in the atmosphere. Among these mechanisms are storage of carbon in trees, viruses, and the soil, and the evolution of more efficient photosynthesis. This conservation of carbon by life will prolong the time that there will be sufficient carbon in the biosphere for life to survive until the sun makes temperatures too hot for life to survive on the Earth's surface and in the seas. Hence, the removal of carbon from the biosphere by organisms will not have a significant negative impact on life, even in the distant future.

Organisms have transformed Earth's atmosphere from negligible oxygen levels of less than 1% to the 21% level of today. These high oxygen levels are necessary for higher life to exist. By increasing atmospheric oxygen, life also indirectly created the ozone layer, which protects organisms from harmful ultraviolet radiation. Without the ozone layer, higher life on land and in much of the oceans could not exist. Organisms may help maintain close to optimal atmospheric levels of carbon dioxide and oxygen.

Species influence the evolution of other species. This often occurs by coevolution, which is two species influencing each other's evolution, with positive feedback involved. An example is the evolution of a natural insecticide in a species of plant to repel a species of insect that eats it. The insect evolves resistance to the pesticide. The plant in turn evolves a more potent pesticide. The insect responds with resistance to the stronger pesticide, and so on. Coevolution is fundamental in interactions between species. In fact, all evolution between interacting species involving feedback is coevolution. All interspecific relationships except commensalism and decomposition involve coevolution. The vast majority of evolution is coevolution.

Key innovations are important evolutionary adaptations that help species survive and thrive. Wings and flight in birds are key innovations. Many macroevolutionary transitions and key innovations are the result of coevolution or at least the influence of species on each other. Examples of large adaptive evolutionary changes that occurred as a result of coevolution or species influencing the evolution of other species include: antibiotics from fungi and bacteria; the evolution of multicellular organisms; the alliance of plants, root fungi, their helper bacteria, and rodents; flowers and their pollinators; fruits and animals that eat fruits and thereby disperse plant seeds; the immune system; and extremely toxic snake venom. Organisms also influence the evolution of other members of their species. Examples of this are elaborate feathers, songs, and dances in male birds in order to compete with other males of their species and attract females, and the influence of social systems on the evolution of intelligence in primates.. Life also coevolved with the nonbiological environment, which includes the soil, atmosphere, oceans, rivers, minerals, chemicals, and so on. In other words, life affected the nonbiological environment, which affected life, and this went back and forth for ages.

In ecological succession, plant communities are replaced by succeeding plant communities. Communities occur in stages, one following another. Each stage is discrete, and has its own characteristic species. Simpler, less diverse, earlier stages change the environment, often to their detriment, making it more favorable for the succeeding stage. Each stage is more diverse and complex than the preceding one, except that the climax community (the final stage) is less diverse and perhaps no more complex than the subclimax stage (the stage just before the climax community). Fires and herbivores cause disturbances, creating a mosaic of all successional stages, maximizing diversity. Herbivores cause a tendency for the most diverse stage, the subclimax community, to predominate. Since each stage changes the environment by ecosystem engineering to make it better a more diverse, complex stage that replaces it (except for the subclimax community to the climax community), ecological succession is a mechanism by which life increases biodiversity.

For the most part, life created the soil, which is an immensely diverse ecosystem. Soil is fascinating, and analogous to an organism. Soil helps life and life helps soil. Soil both aids and benefits from terrestrial ecosystems above it, primarily through connections provided by plants. Soil's properties that benefit life—such as friability and tunnels for the transport of nutrients, air, and water—are created by life by ecosystem engineering. The bacteria in the nitrogen cycle keep the levels of different nitrogen compounds and numbers of bacteria of each species in the cycle, as well as overall nitrogen, at levels favorable for life, in the soil, by negative feedback. Life breaks down rocks and organic matter, such as dead organisms and feces, building and adding nutrients to the soil. These nutrients are spread throughout the soil by fungi, and worms, insects, and other animals. Soil microbes and animals build the soil and help other species in the soil. Earthworms are keystone species that help the soil and other organisms in many ways. Keystone species are species that help an unusually large number of other species. Nitrogen gas in the air cannot be used by life. Some soil bacteria alter it chemically to make it useable by life. Soil life is a mechanism by which life increases diversity that exemplifies the ABH particularly well.

All eukaryotes are complex, diverse ecosystems, each with a diverse community of viruses, archaea, bacteria, fungi, and invertebrates in and on them called the *microbiome*. The organism that has the microbiome is called the *host* of the microbiome. Humans, lions, and redwood trees are all hosts, each with microbiomes of tens of thousands of species. Even eukaryotes that consist of only one cell, such as the amoeba, have large, diverse microbiomes. The species of the microbiome benefit because the host provides them with a home—a habitat. Some species in the microbiome are symbiotic with the host and help it through building its immune system, protecting it from disease, helping it with digestion, helping make vitamins for the host, aiding the host's absorption, helping with development of the host's nervous system, and providing many other functions. Others are commensal with the host. Only a very small percentage of microbes cause disease. The microbiome was involved in the splitting of one wasp species into two. The microorganisms in the microbiome can transfer their genes to their host, and this can promote host evolution, potentially creating evolutionary breakthroughs and key innovations in the host. The microbiome challenges the concept of the individual, shows the fundamental importance of symbiosis and commensalism, and provides a powerful example of evolution at the level of the entire group, called group selection.

There are more species of virus than any other group of organisms by far. They promote an increase in the number of species of the organisms they infect. They influence nutrient cycles, including the carbon cycle and nitrogen cycle, and regulate the climate. Marine viruses kill 20 to 40% of marine bacteria daily. This fertilizes the sea and helps life in the sea thrive. In addition, marine viruses increase the diversity of marine prokaryotes and phytoplankton by selectively killing the more abundant species—the species that are the superior competitors—allowing the species that are poorer competitors to coexist with the better competitors. Viruses may have helped in the replacing of RNA with DNA as the genetic material in organisms with cells, and in the evolution of: sexual reproduction, cells, the nucleus of the cell, the mitochondrion (the part of the cell that carries out cellular respiration and produces the cell's energy), the eukaryotic cell, multicellular organisms, and the three domains of life. (A domain is the highest group or category—called a taxon[1]—in the classification of organisms; bacteria, archaea, and eukaryotes are the three domains of life.) Amazingly, viruses move segments of DNA from one place to another within the DNA of a single organism, and between organisms of the same and different species! Viruses are sometimes coopted by their hosts to become part of their host's DNA. These last two phenomena can cause major macroevolutionary transitions and key innovations in their hosts. Certain viruses may have aided the evolution of the regulation of genes, the immune system in humans, and the ability to give live births instead of the laying of eggs in mammals. A virus was captured and repurposed to become a gene necessary for long-term memory and learning in mammals. Thirty percent of all protein adaptations since the divergence of humans from chimpanzees have been driven by viruses. About 8% of the human genome originates from viruses, and another 40% is made up of repeating segments of DNA thought to have a viral origin. For comparison, only about 2% of the human genome guides the production of proteins.

Genomes and their behaviors promote variation within populations and species, evolutionary change and even evolutionary innovations, and diversification. This is accomplished by such mechanisms as the combining and exchange of DNA segments via sexual reproduction, transposable elements (jumping genes), mutations in genes that control other genes, duplication of genes, and doubling the entire genome in one generation. The nature of genetic variation itself is subject to natural selection, and this "secondary selection" is an important mechanism by which major macroevolutionary innovations take place. For example, when segments of DNA jump or are exchanged or multiplied in number, the organism acquires DNA segments that have already undergone the test of natural selection. And the segments of DNA that are moved, exchanged, or multiplied in number are more likely than random mutations to be compatible with the DNA of the organism acquiring them. As a result of these two characteristics of DNA segments that are transported or multiplied in number, they are much more likely to be helpful to the organism that acquires them than random mutations, which are likely to be incompatible with the well-adapted genome of the organism. This is one reason why the mechanisms listed in the second sentence of this paragraph have persisted.

A simple, naïve look at standard Darwinian theory would lead one to think that selfishness would be favored by natural selection. However, there is a tremendous amount of altruism and cooperation in nature. Social ants and bees, certain birds, dolphins, and primates, including humans, display a great amount of altruism and cooperation. Worker honey bees are females that give up reproduction to help the queen reproduce, and sometimes sacrifice their lives for the hive. In fact, altruism and cooperation by individuals within populations are often favored by natural selection. Mechanisms for this are selection that favors the altruist because it passes on its genes because it is related to the organism(s) it is helping (kin selection), alliances whereby individuals help each other (reciprocal altruism), and perhaps natural selection between groups whereby groups with more altruistic and cooperating individuals outcompete groups of uncooperating and selfish individuals (group selection). Cooperative behaviors include working together to raise young, and alarm calls that warn others of the same species of the existence of predators. Many animals could kill or seriously injure each other in fights over territory and mates. But they usually do not do so. They fight according to rules, and determine the stronger animal without either party being injured. This helps the group. But game theory indicates that it is explained by standard Darwinian selection on the individual, not the group. This "social engineering" discussed in this paragraph is a mechanism by which the ABH works.

Network theory is relevant to and supports the ABH/Pachamama Hypothesis. It supports the idea that ecosystems with more biodiversity are more stable and resilient than ones with less biodiversity, if the species in the ecosystem coevolved together over a long time period.

The laws of physics and chemistry indicate that the evolution of chemicals into life is inevitable on any planet or moon with the right conditions, such as was the case on Earth when the first cells evolved there. Thus, the chemical evolution of life was inevitable on Earth, may have occurred on other planets or moons in our solar system, and is highly likely on some planets circling other stars. The molecules that are the building blocks for life that are needed to make proteins, DNA, and RNA have been found in space and were delivered to Earth by meteorites and dust from space. Experiments indicate life likely started in thermal hot springs such as those found in Yellowstone National Park in the United States. Chemicals that are used to make proteins and DNA were mixed with other chemicals in a small, boiling spring, and produced primitive "cells" spontaneously. These precursors to cells (protocells) can divide spontaneously. Natural selection might have started when the more robust protocells survived long enough to be transported to other hot springs. Fossils of microbes show chemical evolution started at a geologically short time after conditions permitted it, and life emerged and diversified quickly, with complex cells appearing as early as about 3.465 billion years ago (bya). A general hypothesis that the Universe is primed to increase its order, information content, diversity, and complexity in local areas with the right conditions incorporates the ABH as a subset of it. It states that there is a tendency toward increased order, information, diversity, and complexity as a result of scientific laws acting on systems with the right conditions, such as exist on Earth. Note that biodiversity is a measure of information content.

The solar system, sun, Jupiter, moon, and nonbiological Earth all help life, evolution, and biodiversity. The sun provides heat and light for life on Earth. The Earth's distance from the sun and tilt are close to optimal for life on Earth. The Earth's magnetic field shields it from the solar wind, which is harmful to life and can strip a planet of its atmosphere. Jupiter has protected the Earth from meteor strikes. This shows nonbiological factors aid life, so the ABH is not a comprehensive explanation for Earth's great biodiversity. But it suggests the ABH is a subset of a larger hypothesis that the entire solar system promotes life and biodiversity.

There are counterexamples to and arguments against the ABH/Pachamama Hypothesis. These are presented in Chapter 14, along with answers to them. They include the fact that when photosynthesis evolved and the first oxygen-producing microorganisms caused atmospheric oxygen levels to rise, some species of microbes decreased in number or went extinct because oxygen was toxic to them. The oxygen produced by life also combined with the powerful greenhouse gas methane, reducing its concentration in the atmosphere and causing the Earth's temperature to drop tremendously, possibly causing much of the Earth to be covered in ice. However, only prokaryotes existed at the time. They evolved resistance to the toxic oxygen, and then evolved the ability to use it as an energy source, through cellular respiration. This allowed the evolution and diversification of higher plants and animals. Life in time stabilized the temperature. In the long run, the production of oxygen by life led to a great net increase in diversity, in spite of the problems it caused for life. Sulfate-reducing microbes produce the toxic gas, hydrogen sulfide, harming life. This seems to have played a role in the Permian-Triassic mass extinction that killed off the vast majority of species about 251.9 million years ago (mya). But these sulfate-reducing microbes cannot grow in the presence of oxygen, and the production of oxygen by life has limited them ever since photosynthesis evolved.

Also, sulfate-reducing microbes have many positive effects on life, including having symbiotic relationships with many animal species. For example, they live in worms in deep-sea vents and help them obtain nutrients. They have over time had a net positive effect on diversity. Humans have caused a great deal of environmental destruction and extinction of species, and seem to contradict the ABH. However, they are not natural and will not survive long if they do not stop reducing diversity and destroying the Earth. There are other answers to this counter argument that humans contradict the ABH covered in Chapter 14.

The final chapter of this book discusses the implications of the ABH. These include the following:

> (1) Biodiversity should be valued and conserved. (2) Many human activities would benefit if they were modified to conserve biodiversity. These include agriculture (e.g., soil is a diverse ecosystem that has lost a great amount of diversity because of pesticides) and forestry (Earth's forests absorb about 16 billion tons of CO_2 per year, and about one-half of this is lost to the cutting and burning of forests and other disturbances). (3) Deep ecology states that nature has inherent value, while pragmatic environmentalism states that we should preserve nature because we are dependent on it for our survival and well-being. The ABH is relevant to both of these. (4) Humans are unwittingly performing a scientific test of the ABH, causing extinction rates 1,000 times higher than natural background rates of extinction. As we cause species to become extinct, additional species go extinct because they benefit from or depend on the ecosystem engineering performed by the species we exterminate. This is strong evidence for the ABH. This also shows that symbiosis and commensalism cause diversity to increase, and these observations are corollaries to the ABH. (5) Biodiversity of a taxon is positively correlated with its ecological importance, and negatively correlated with the size of the organisms in the taxon. (6) Coevolution is extraordinarily important. All evolution between interacting species except commensal species is coevolution. (7) Ecosystems are holistic, interacting systems, much like "super-organisms." The number and nature of relationships between species are at least as important as the number of species in determining the stability and resilience of ecosystems. (8) The concept of the individual is only partly correct. No gene, organism, or species evolves or exists in isolation. They are all connected to other genes, organisms, and species, and to their ecosystems. (9) The Universe is primed for life, scientific laws are precisely right for the appearance of life in limited and isolated areas with favorable conditions, and chemical evolution of life is inevitable under the right conditions. (10) *Symbiosis and commensalism are pervasive, of fundamental importance, structure ecosystems, and greatly increase biodiversity.*

It is time for an entirely new way of thinking and new model and synthesis for ecology, evolution, and population biology. It is time for a paradigm shift to recognizing that life and organisms are the main generators of biodiversity, that ecosystems maximize biodiversity, and that all species are ecosystem engineers that have a net positive effect on other species, their ecosystems, and biodiversity, in natural ecosystems over sufficiently long time periods.

NOTE

1 A taxon (plural taxa) is a group of organisms of any rank, such as a species, genus, family, phylum, or kingdom, used for biological classification.

REFERENCES

Lovelock, J. E. (1972). Gaia as seen through the atmosphere. *Atmospheric Environment* 6 (8): 579–80. Bibcode: 1972 AtmEn...6..579L. doi: 10.1016/0004-6981(72)90076-5.

Lovelock, J. E. (1979). *Gaia. A New Look at Life on Earth.* Oxford Univ. Press, Oxford, London.

Lovelock, J. E. & Margulis, L. (1974). Atmospheric homeostasis by and for the biosphere: the Gaia hypothesis. *Tellus. Series A. Stockholm: Internat. Meteorol. Inst.* 26 (1–2): 2–10. Bibcode: 1974Tell...26....2L. doi: 10.1111/j.2153-3490.1974.tb01946.x. ISSN 1600-0870.

Margulis, L. (1998). *Symbiotic Planet: A New Look at Evolution.* Basic Books. A Member of the Perseus Group. New York, NY.

Seaborg, D. (2022). *How Life Increases Biodiversity: An Autocatalytic Hypothesis. CRC Press/Taylor & Francis.* Boca Raton, FL; London, UK; New York, NY.

Ward, P. (2009). *The Medea Hypothesis. Is Life on Earth Ultimately Self-Destructive?* Princeton Univ. Press, Princeton, NJ, and Oxford, UK.

Watson, A. J. & Lovelock, J. E. (1983). Biological homeostasis of the global environment: the parable of Daisyworld. *Tellus B.* 35 (4): 286–9. Bibcode:1983TellB..35..284W. doi:10.1111/j.1600-0889.1983.tb00031.x.

2 Life Regulates the Atmosphere's Greenhouse Gas Levels and the Earth's Temperature

2.1 TERMINOLOGY

I shall start with three introductory paragraphs following this one that define some terms necessary in order for the layperson to understand this and subsequent chapters. I recommend that the reader who is not familiar with food webs and the carbon cycle read these paragraphs carefully at least once, and refer back to it as necessary, to be able to follow the discussion without difficulty. There is also a glossary at the end of this book that will be helpful with unfamiliar terms. The scientist or layperson familiar with the carbon cycle and food webs can skip these paragraphs.

Primary producers are organisms that acquire their energy from sunlight and nonliving sources, such as nutrients in the soil. Plants are primary producers. *Phytoplankton* are small, one-cell organisms that live in the shallower portions of the ocean. They are primary producers. They are at the base of most ocean food webs. A *food web* is a system of nutrient flow that starts with primary producers that are eaten by herbivores that are in turn eaten by predators. All of these organisms are broken down by *decomposers* when they die. Decomposers are part of the food web. They recycle the nutrients in the organisms they decompose. A food web is not called a food chain because there are many species of primary producers, herbivores, predators, and decomposers, so it is a branched web, not a linear chain. Nature has food webs, not food chains. An example of a food web is as follows. Phytoplankton produce food (carbohydrate) by photosynthesis, and the phytoplankton are eaten by various species of small organisms called zooplankton, which are eaten by many small species of invertebrate, which are eaten by many species of small fish, which are eaten by a few species of large fish, and all of these species are decomposed by bacteria and fungi after their death, in the sea.

Some phytoplankton are *prokaryotes*, some are *eukaryotes*. A *prokaryote* is a simple organism that has no nucleus enclosing its DNA, so its DNA is spread throughout its cell. The cells of prokaryotes have no internal membranes or organelles. Almost all prokaryotes are organisms with just one cell. They are microbes. Bacteria constitute one of the two major groups of prokaryotes. The other is *archaea* (singular is *archaeon*), which sometimes live in extreme conditions, such as in hot springs that have very high temperatures. A *eukaryote* is an organism with a nucleus in each of its cells. Most of the DNA of the cell is in this nucleus, which is surrounded by a membrane. Eukaryotes also have organelles that are bound by membranes. *Organelles* are specialized structures within cells with specific functions. An example of an organelle is the *mitochondrion*, which carries out cellular respiration, using oxygen to produce energy for the cell. The *chloroplast* is an organelle in cells of eukaryotes that can carry out photosynthesis. Land plants, seaweeds, phytoplankton, and bacteria called cyanobacteria are examples of organisms capable of photosynthesis. I said *most* of the DNA is in the eukaryote's nucleus because mitochondria and chloroplasts have their own DNA; they have their own genomes. The *genome* is the complete set of genes or genetic material present in a cell or organism.

Cyanobacteria are bacteria, and hence are prokaryotes, so they do not have chloroplasts. But they do carry out photosynthesis. Prokaryotes that carry out photosynthesis do so without chloroplasts. *Oxygenic photosynthesis* is photosynthesis that produces oxygen. Oxygenic photosynthesis is the chemical reaction that uses the energy of sunlight to convert carbon dioxide (CO_2) and water into carbohydrate (food) and oxygen. Plants, phytoplankton, and cyanobacteria carry out oxygenic photosynthesis. It is important because it produces carbohydrates, which are a major food source for organisms, and oxygen. Many organisms, including plants, use the pigment *chlorophyll* to carry out oxygenic photosynthesis. Chlorophyll is in the chloroplasts of the cell. Some bacteria carry out *anoxygenic photosynthesis*, which uses the energy of sunlight to produce food, but uses hydrogen sulfide instead of water, and produces sulfur instead of oxygen. Anoxygenic photosynthesis is photosynthesis that does not produce oxygen. *Respiration* is the opposite or reverse reaction of photosynthesis, and is carried out by the vast majority of organisms. It uses carbohydrates and oxygen to produce carbon dioxide, water, and energy. It is important because it is a source of energy that allows organisms to survive and thrive, and makes higher life possible. Oxygen is important because it is used to produce energy used by many organisms, including higher organisms, via respiration. Without respiration and sufficiently high atmospheric levels of oxygen to allow organisms to carry it out, higher life would not be possible.

There is a glossary at the end of this book that I encourage the reader to use if he or she is uncertain of the meaning of any term I use.

DOI: 10.1201/9781003246640-2

2.2 THE CARBON CYCLE AND NEGATIVE FEEDBACK LOOPS

After the sun heats the Earth, some of the heat radiates back into space, cooling the Earth. Greenhouse gases trap the heat leaving the Earth, and so raise its temperature. Carbon dioxide and methane are the main greenhouse gases, and methane is about 30 times as powerful as CO_2, molecule for molecule, at warming the Earth. Yet CO_2 is the major greenhouse gas because it is a powerful one, and there is much more of it in the atmosphere than methane. Both methane and CO_2 contain carbon. Therefore, if carbon is buried, the amount of CO_2 and/or methane will be reduced in both the atmosphere and in rivers, lakes, and the sea. If carbon is added to the system, there will be more CO_2 and/or methane in the air and in water. Carbon naturally moves through the air, water, organisms, and ground. It gets buried and released back into the atmosphere. This constant movement of carbon through all of these places is called the *carbon cycle*.

The carbon cycle has two aspects, the long-term and short-term carbon cycles. The *long-term carbon cycle* generally occurs over millions of years. It has profound effects on atmospheric carbon. It is wholly nonbiological. It includes *carbon sources*, which add carbon to the biosphere and atmosphere. (The *biosphere* is the part of the Earth occupied by life; it is the sum of all Earth's ecosystems.) They do this by such mechanisms as volcanoes, movement of the continents over the seafloor's calcium carbonate (which contains carbon, is important in the carbon cycle, and is in many kinds of organisms) and organic sediments, and spreading of the seafloor. The long-term carbon cycle also includes carbon sinks. A *carbon sink* is anything that removes more carbon from the atmosphere than it releases to it. In the case of the long-term carbon cycle, this carbon is stored in the solid Earth, including the seafloor. This process of removal of carbon from the air and storing it elsewhere is called *carbon sequestration*, and includes weathering and burial. *Weathering* is the breaking down of rocks, soil, minerals, wood, or even artificial materials through contact with the Earth's atmosphere, water, and biological organisms. In it, CO_2 is combined with minerals in chemical reactions. The resulting material is then carried via rivers and creeks to the sea, where it falls to the ocean bottom and is buried. Weathering is a key step in sequestering carbon, removing it from the biosphere. Weathering is the main way carbon is removed from the atmosphere nonbiologically. Weathering can be done biologically too, which also sequesters carbon, and I will discuss this later.

A common way weathering removes CO_2 from the air is silicate weathering through the carbonate-silicate geochemical cycle. This cycle favors removal of CO_2 under the normal temperatures at Earth's surface. The reaction shifts to the other direction, with the release of CO_2 into the atmosphere, at temperatures above 300°C (572°F). On Venus, with its very high temperatures, the process was important in causing the high atmospheric CO_2 and thus even higher temperatures, favoring yet more CO_2 release via this reaction, in a positive feedback loop.

The *short-term carbon cycle* ranges from days to tens of thousands of years, and includes all the biological processes affecting atmospheric carbon, such as photosynthesis and burial of dead plants. It can have geologically rapid effects on carbon in the system and climate. Nonbiological factors can affect the short-term carbon cycle too.

Nonbiological effects on atmospheric carbon can be profound. The Earth's continental plates move in the process of continental drift. Sometimes there is plate subduction, where one plate slides under another. This buries carbon, sometimes in large quantities. This often buries carbon locked in living or dead organisms or parts of organisms, such as in seashells and microbes. So nonbiological factors and life can work together to sequester carbon. Volcanoes return this carbon to the atmosphere in the carbon cycle. When the Indian subcontinent crashed into Asia as a result of continental drift 50 million years ago, the collision of the two land masses caused the Himalayan Mountains to form, exposing new rocks to weathering. This sequestered CO_2 from the atmosphere in very large amounts, and greatly lowered Earth's temperature.

Nonbiological processes can also send large amounts of greenhouse gases into the atmosphere. Carbon dioxide could potentially build up to high levels in the atmosphere if chemical and physical processes were the only carbon sources and sinks. It is added in great quantities annually through gases leaving the Earth's interior, such as in volcanic eruptions. Also, water vapor is the gas that volcanoes emit the most, and it is a greenhouse gas. They also sometimes methane. Some volcanoes emit mainly particles that block solar radiation and thus cool the planet, but most emit more greenhouse gases than particles, and so warm it. There have been times when massive volcanism spewed tremendous quantities of greenhouse gases into the air and greatly heated the Earth.

Over time, scales of tens to hundreds of millions of years, CO_2 levels in the air may vary due to natural perturbations. Yet there have been billions of years of life-friendly temperatures that have stayed within a fairly confined range, which required a tight balance between carbon sources and sinks over these timescales. Zeebe and Caldeira (2008) presented direct evidence for a close mass balance of surface carbon on shorter timescales than these. They concluded that, over the last 610,000 years, the net imbalance between carbon inputs and outputs from the surface environment was less than 1–2%. This means that carbon on Earth's surface has been kept at a very constant level for over the last 610,000 years.

This is likely largely because under normal circumstances, both nonbiological and biological factors keep the carbon levels stable by negative feedback. *Feedback* is a process whereby a cause is modified by its effect. *Negative feedback* is a process whereby one factor, call it factor A, causes another factor B to increase, and factor B causes factor A to decrease. There can be more factors in the loop. So if A increases B, which increases C, which comes back and decreases A, there is negative feedback. Since it makes a circle, one often uses the term negative feedback loop to refer to the process. Negative feedback loops keep the system stable, and keep the level of CO_2 (and methane) in the air and the temperature stable, and at intermediate levels beneficial to life. Normally, nonbiological factors stabilize CO_2 levels and temperature by negative feedback, by the following mechanism. When there is more CO_2 in the atmosphere, the temperature increases by the greenhouse effect, air can hold more water vapor at higher temperatures, more water will evaporate and eventually fall somewhere as rain, which increases weathering, which will remove CO_2 from the atmosphere. When there is less CO_2 in the air, temperature decreases, so the atmosphere can hold less water, so rainfall decreases, weathering that removes CO_2 from the air decreases, carbon sequestration decreases, and so CO_2 level and temperature increase. This mechanism stabilizes CO_2 levels and temperature.

Positive feedback or a *positive feedback loop* is destabilizing, and occurs when A causes B to increase and B causes A to increase. Again, there can be more than two agents. If A increases B, which increases C, which loops back and increases A, there is a positive feedback loop. If CO_2 levels in the air become too high, positive feedback can replace the negative feedback. For example, humans have caused increased CO_2 in the air, warming the temperature, and causing the ice over the sea on the coast of Greenland to melt. This exposes the sea, which is darker than the ice that has melted and disappeared. Darker surfaces absorb heat, while lighter surfaces reflect heat. So melting the ice and exposing the darker sea causes more heat to be absorbed, temperatures get hotter, and more ice melts. This causes the sea to be darker, causing it to absorb more heat, leading to higher temperatures, causing more sea ice to melt. The positive feedback loop could keep going until all the sea ice is gone. Positive feedback can lead to catastrophe. Although nonbiological factors in normal circumstances stabilize greenhouse gas levels and temperature via negative feedback, they can cause positive feedback and destabilize the system in rare circumstances. This is because large, uncommon perturbations caused by catastrophic nonbiological events such as massive volcanism or large meteor strikes can throw the temperature to extremes. Then there can be destabilizing positive feedback that also involves biological factors for a while. The system might then take a very long time to return to negative feedback and stable conditions with moderate, life-friendly greenhouse gas levels and temperatures. But it eventually succeeds in doing so.

Both the Gaia Hypothesis and the ABH (Pachamama Hypothesis) would be best supported if there were stabilizing negative feedback of either or both greenhouse gas levels and temperature by life. In fact, life has a homeostatic (stabilizing) mechanism involving negative feedback that regulates both carbon and temperature (Lovelock and Whitfield, 1982).

Here is how life employs negative feedback to keep greenhouse gases levels and temperature stable, and at life-favorable, intermediate levels. Bear in mind in this discussion that burial of carbon removes the two main greenhouse gases, CO_2 and methane, from the air, since carbon makes up a large part of both of these gases. If CO_2 rises, plants and phytoplankton carry out photosynthesis faster. Photosynthesis consumes CO_2, removing it from the atmosphere. Laboratory experiments have demonstrated that if there are higher atmospheric levels of CO_2, plants and phytoplankton take up more of it for photosynthesis if their growth is not limited by light, water, or nutrients [e.g., Bazzaz, 1990]. They increase in number, and so more die. About the same percent of plants are buried and about the same percent of phytoplankton sink to the seafloor. But because there are more of both when the air has more CO_2, more plants are buried and more phytoplankton sink to the bottom of the sea. So the dead plants and phytoplankton sequester more carbon when they are buried. Also, more phytoplankton and plants support more animals at all levels of the food web. There are thus more zooplankton, shrimp, fish, and so on, in the sea. Thus, more organisms die, sink, and are buried in the seafloor. And the organisms, being more numerous, produce more feces that sink and are buried in the seafloor. The same principle applies to life in the food webs on land. The dead organisms and the feces of living ones contain carbon, so carbon is buried with both the carcasses and feces. By this mechanism, increasing the population size of the phytoplankton increases the sequestration of carbon. Also, when it is warmer, phytoplankton and plants grow faster (up to a point), so more biomass is buried when they die, burying more carbon. The atmosphere thus has less of the main two greenhouse gases, CO_2 and methane. This lowers the temperature. On the other hand, when CO_2 levels are low, plants and phytoplankton absorb less of it, and its atmospheric level increases. And when there is less CO_2 and it is cooler, plants and phytoplankton grow more slowly and become less abundant, so less of their biomass and carbon is buried, and CO_2 levels and temperatures rise. Thus, life acts as a thermostat that regulates CO_2 and methane levels and temperatures to within a range favorable to it, by negative feedback. And as will be discussed later, animals called salps are, like phytoplankton, important in negative feedback and regulation of carbon in the sea. In a book, Lenton and Watson (2011) discussed various mechanisms by which negative feedback has stabilized the Earth's systems, including its temperature, throughout its history, many of these mechanisms driven by life.

Both these feedback systems have limits, and are far from perfect. In biologically driven negative feedback, if greenhouses gas levels or temperature increase excessively, plants and phytoplankton grow more slowly or even do not do well, and the negative feedback breaks down. And nonbiological factors can destabilize the system for long periods. For example, volcanism could greatly increase, adding tremendous amounts of CO_2 to the air and driving the temperature up enormously. It is thought that the Permian-Triassic mass extinction of approximately 251.9 mya, the largest extinction in Earth's history, resulted from volcanism belching so much CO_2 into the air that temperatures soared so high that the frozen methane in the sea in the form of what is called methane hydrates were melted, releasing great quantities of methane into the air, making the temperature exceedingly hot. This melted more methane hydrates, adding yet more methane to the atmosphere, making the Earth even hotter, in a positive feedback loop, until the planet was too hot for most life to exist. Meteorite strikes can also throw the system into dark conditions and extremely cold temperatures that last a long time, until negative feedback eventually restores favorable temperatures and stable conditions. A large meteor hitting the Earth caused the Cretaceous-Paleogene mass extinction of about 66 mya that caused the extinction of the dinosaurs.

Additionally, the stabilizing biological negative feedback mechanism can be disrupted by large external inputs that throw the system to an extreme. In such cases, life takes part in a destabilizing positive feedback loop. For example, when CO_2 was doubled experimentally in a wetland, plants increased their productivity, but also pumped more organic matter underground, which stimulated microbes that produce methane, increasing methane emissions significantly over control sites (Dacey et al., 1994). And ice core data indicate that, if the climate warms enough, although the picture is not fully clear, nitrous oxide (N_2O), which is about 2,000 times more powerful a greenhouse gas than CO_2, increases in the atmosphere. This is likely because the relevant microbes emit more of this gas at higher temperatures. And human-induced global warming has led to melting of the permafrost, resulting in the release of greenhouse gases by organic matter and microbes, in a positive feedback loop.

The positive feedbacks are not permanent, and throughout life's history, negative feedback and stable, favorable conditions have always eventually been restored. The general rule is that the system tends to bring greenhouse gas levels and temperatures back to favorable levels for life after a period of time. Negative feedback gets restored. As a whole, the system works at stabilizing greenhouse gas levels and temperature. And this is accomplished at least partly by life. There can be little doubt that, although they can be over-ridden for finite periods by large perturbations, the nonbiological and biological negative feedback systems are both involved in and effective at maintaining, and when necessary, restoring, an atmosphere and temperature regime beneficial to life, as the usual state.

Other nonbiological effects on climate include the tilt of the Earth and changes in Earth's orbit.

With this background, this chapter will make the case for the importance of life in regulating carbon (and hence CO_2 and methane) in the atmosphere and stabilizing Earth's temperature, major tenets of both the Gaia Hypothesis and the ABH. Although Lovelock (1979) argued that life has regulated and does regulate Earth's greenhouse gases and temperature, the arguments presented for the ABH here are more specific, broad, and thorough, partly because we know more about this topic now than when he wrote about it.

2.3 LIFE IS THE MAIN REGULATOR OF ATMOSPHERIC CARBON

Figure 2.1 shows atmospheric CO_2 levels for about the last 570 million years. The fluctuations in CO_2 levels show that the negative feedback loops that stabilize them involving both life and nonbiological factors are imperfect. The figure shows that CO_2 peaked about 520 mya, and, although there have been both large fluctuations and periods of increase, CO_2 levels have decreased to preindustrial levels since then, in a somewhat steady downward trend. It is also true that that the average global temperature has fluctuated a good deal, but is lower than it was 600 mya, and has been decreasing steadily from about 60 mya until preindustrial times. As will be discussed, these two downward trends are mainly due to life's sequestration of carbon, and are highly beneficial to life.

The amount of heat the sun gives off has increased about 30% since its origin 4.5 billion years ago (bya) (ibid.). Thus, the heat the sun has provided the Earth has increased gradually and steadily, except for minor variations, until the present, and will continue to do so into the future, with its brightness increasing by 1% every 100 million years. There is also a continuous input of carbon into the air from mid-ocean ridges, volcanoes, and other sources, although this is small compared to the transfer of carbon from the air to the living organisms via photosynthesis (Frings, 2019). For life to thrive, something had to keep the temperature warm when life had just begun and the sun produced much less heat. As time went on, there had to be a mechanism to keep cooling the Earth to counter the continual heat increase from the sun and the warming greenhouse gas input to the atmosphere from the Earth via volcanoes and other sources. Both of these were accomplished mainly by life.

With the sun producing such a small amount of heat at life's beginning, when the only life were prokaryotes, Earth would have been an icy, cold wasteland with its seas frozen, unless a heat-generating mechanism existed. Ueno et al. (2006) found evidence that the heat needed to keep Earth favorable to life was provided by methanogens. Methanogens are archaea that produce the powerful greenhouse gas, methane. These researchers found the carbon-13 isotope signature produced by methanogens in deposits that bear methane in about 3.5 bya structures in Australia, showing that methanogens had arisen by 3.46 bya, 0.7 billion years earlier than previously thought, warming the Earth by producing large quantities of methane over a 100-year period, acting as ecosystem engineers. Sediments laid down by water show the Earth and its seas were not frozen, and liquid water existed 3.8 to 2.5 bya. In fact, geological and biological evidence suggests that Earth was warm during most of its early history, despite the fainter young sun (Haqq-Misra et al., 2009). Something provided a source of heat to allow early evolution to occur, and it was life in the form of methanogens. Haqq-Misra et al. (2009) pointed out that CO_2 levels 40 times higher than preindustrial levels because of volcanism and reduced weathering were present then as well. These high CO_2 levels were also needed to maintain sufficient temperatures for life to thrive. In fact, the high levels of methane would have blocked solar radiation with a thick haze, keeping the atmospheric temperatures close to freezing levels inhospitable to life, had the CO_2 not diluted it. Methane is a powerful greenhouse gas that generally warms the Earth, but at high levels, it can block out solar radiation and lower the temperature considerably. So nonbiological factors in the form of CO_2 not produced by life also played

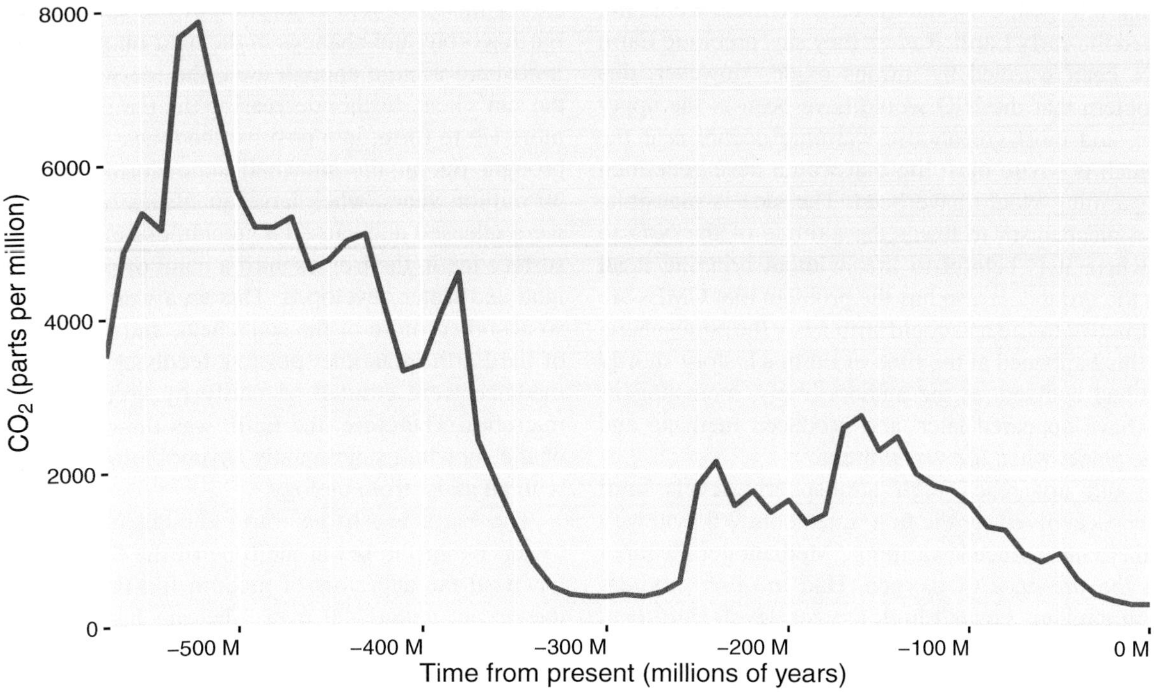

FIGURE 2.1 Carbon dioxide concentration over the past 570 million years.

IGBP Pages/World Data Center for Paleoclimatology
Data Contribution Series # 2002-051
NOAA/NGDC Paleoclimatology Program, Boulder CO, USA
Name of Data Set: GEOCARB III: A Revised Model of Atmospheric CO_2 over Phanerozoic Time
Last Update: 8/2002 (Original Receipt by WDC Paleo)
Contributor: Robert A. Berner, Yale University
Data from Berner and Kothavala (2001)

a role in heating Earth. However, without the methanogen-generated methane warming the air, the atmosphere would have been too cold for life to thrive and diversify, even with the high CO_2 levels, because the sun provided the Earth with much less heat. Atmospheric methane levels would have had to be 1,000 parts per million by volume, combined with levels of CO_2 100 times preindustrial levels, in the middle Archean eon (3.8 to 2.5 bya), to reach temperatures comparable to those of today. The right combination of CO_2 and methane concentrations were needed to keep the planet warm enough for the prokaryotic life at the time to thrive. And although methane can be produced from nonbiological reactions of simple inorganic compounds, and some of this early methane may have been produced this way, the finding of Ueno et al. shows that the bulk of the methane of about 3.46 bya was almost surely produced by methanogens. The model of Haqq-Misra et al. (2009) shows other greenhouse gases, especially ethane, also contributed to the warming. But the ethane would have been made by the combining of the methane molecules made by the methanogens, so ultimately it was biological. Further evidence that atmospheric methane levels were elevated is that temperatures dropped and the Earth was glaciated in the Paleoproterozoic, which started 2.5 bya, corresponding with the rise in oxygen. Falling temperatures correlated with high oxygen levels suggests high levels of atmospheric methane because oxygen causes the level of methane to drop by combining with it, causing the temperature to drop (ibid.). Thus, biological and nonbiological factors, with the biological factors being more important, combined to create a climate favorable to life, allowing diverse, actively evolving communities of prokaryotes to thrive. There are other hypotheses that attempt to account for the heat required for life to thrive at this time, but in my view, they lack evidence as convincing as that of the work by Ueno et al. and Haqq-Misra et al. Thus, life was the primary provider of heat to the early Earth, with help from a nonbiological factor.

However, Airapetian et al. (2016) used simulations combined with Kepler Space Telescope observations to conclude that, in the early Earth, Coronal Mass Ejections (CMEs) initiated reactions converting molecular nitrogen, CO_2, and methane to the potent greenhouse gas nitrous oxide, which would have greatly warmed the planet, and to hydrogen cyanide (HCN), a compound likely important in the chemical evolution of life. CMEs emit billions of tons of charged particles at temperatures of 10s of millions of degrees F that travel millions of miles per hour (see Chapter 13). These authors argue that the reactions would have destroyed the molecular nitrogen, CO_2, and methane, so these would not have warmed the early Earth. Rather, they say, prebiotic Earth would have been warmed by nitrous oxide. However, this has the problem that the N_2O would have been in the upper atmosphere, and most greenhouse warming occurs near the surface, which is where most life that would have benefitted from this warming would have been. The idea is plausible, but needs a mechanism to lower the altitude of the N_2O to the point where it is helpful to life, without bringing it all the way to the ground. It also has the problem that CMEs are highly destructive to life and could strip away the atmosphere. Finally, if this happened at the time in Earth's history shortly before life had evolved, as the authors imply, methanogens could still have appeared later, and produced methane and warmed the planet when life was present.

Oxygen did not reach high atmospheric levels until photosynthesis evolved for the first time about 2.5 bya, well after the methane-induced warming. Methanogens cannot survive in the presence of oxygen. Had the first oxygen-producing organisms created high oxygen levels 3.46 bya, the methanogens would have been greatly limited in their numbers by the high oxygen levels, and the vast majority of the methane they produced would have reacted with oxygen and been destroyed. This would have prevented the life-induced warming of Earth by methane. So, methanogens evolving and thriving long *before* oxygen-producing organisms became abundant was key to making possible the warming produced by life. This warming of the planet allowed prokaryotes on the young Earth to diversify into many different species. This is an intriguing example in support of the ABH.

Ueno et al. account for the heat to keep the sea liquid starting 3.46 bya, but not starting when we know the seas were not frozen, 3.8 bya, or when life may have started, 4.1 bya. Possibly methanogens arose by 3.8 or 4.1 bya, since Ueno at al. found they had arisen *at least* by 3.46 bya, thus perhaps earlier. We cannot currently account for the heat source from about 4.1 to 3.46 bya, and more research is needed to solve this problem.

Pavlov et al. (2000) found evidence that methane in the atmosphere 2.8 bya, presumably produced by methanogens, could have warmed Earth at this later time, and that its elimination with the rise of oxygen about 2.5 bya could have triggered the Earth's first widespread glaciation. So methanogens and the heat they provided were fundamental to the thriving and diversity of prokaryotes starting at least 3.46 bya and possibly as late as about 2.5 bya, although we do not know if this occurred continuously for this entire time.

It is hypothesized that there was a Snowball Earth, when much of the Earth was frozen, postulated to have lasted from about 750 to about 660 mya. This is not universally accepted. If it happened, life helped bring Earth out of this frozen state. At this time, the sun's heat output was considerably less than it is today. If it occurred, it was triggered by a nonbiological cause or combination of causes. Possible causes are the eruption of a supervolcano or many volcanoes that exuded particles that blocked sunlight, a reduction in the atmospheric levels of CO_2 and/or methane, less absorption of the sun's heat, changes in Earth's orbit, and changes in the heat output of the sun. Once it covered a large enough area, the ice would have reflected the sun's heat, further decreasing the temperature and causing more ice to form, in a positive feedback loop. The Earth was brought out of the snowball state during a period of 4 to 30 million years, when large quantities of CO_2 and methane were released and caused a greenhouse effect that melted the surface ice in the tropics until a band of permanently ice-free land and water developed. This area was darker than the ice, so absorbed more of the sun's heat, and helped thaw the rest of the Earth in another positive feedback loop. The CO_2 and methane were emitted primarily by volcanoes, but also by microbes. Therefore, the Earth was thawed and brought out of the snowball state mainly by nonbiological processes, but with an assist from biology.

The Earth had to be warm enough for complex animals to survive in the sea at the time of the Cambrian explosion, when all the ancestors of modern forms of life appeared in the ocean, about 550 mya. The sun has increased its heat output by about 6% since this time. And mid-ocean ridges, volcanoes, and other sources have added greenhouse gases since this time. The surface of the Earth would have become torrid, far too hot for higher life, including mammals, birds, reptiles, plants, and fungi, unless something removed greenhouse gases from the atmosphere, to counter these two sources of heat. *Organisms sequestered carbon for the last 550 mya, regulating the temperature and the acidity of the seas and freshwater ecosystems, allowing life to thrive.*

It is not known what percentage of carbon is sequestered because of biological as opposed to nonbiological causes, and the relative importance of the two to the carbon cycle and climate has varied at different times in geologic history. But biological processes have been and are of crucial importance and are the major mechanism of reducing greenhouse gases in the atmosphere, seas, and freshwater ecosystems, sequestering significantly more carbon than nonbiological forces have done and do. Life has buried immense quantities of carbon, thereby removing it from the atmosphere, maintaining the Earth at temperatures favorable to life.

Life regularly adds carbon to the atmosphere when organisms respire and decompose. Life warmed the planet by adding methane when life had not existed for very long and methanogens were dominant, as discussed above. But since those times, the average amount of carbon sequestered by life has far exceeded the average amount that it has been added by respiration and decomposition.

We know that life, with some help from nonbiological weathering, took CO_2 levels from a peak of about 7,000 parts per million in the Cambrian period (541 to 485.4 mya), to below 500 parts per million in about the middle Carboniferous period, about 330 mya (see Figure 2.1). This was correlated with a drop in global mean temperature from about 21.5°C (70.7°F), which, for a *global average*, is absolutely scorching and inhospitable to life, about 550 mya, to about 17°C (62.6° F), which is still hot for a global average, but much more favorable to life and biodiversity, about 330 mya. For perspective, the World Meteorological Organization estimates the Earth's annual mean temperature was about 14°C (57.2°F) between 1961 and 1990. Tremendous volcanism shot CO_2 levels to 3,000 ppm at the end of the Permian, about 251.9 mya. Global mean temperature spiked to a torrid 22°C (71.6°F). Since then, life has brought atmospheric CO_2 and methane down consistently and gradually over a period of almost 250 million years, to the preindustrial levels of what are now thought to be about 260–270 parts per million, resulting in a mean temperature of about 17°C (62.6°F) again, by about 2 mya. By this time, our Hominid ancestors were present. The changes in greenhouse gas levels and temperature were not straight lines, since nonbiological factors and variations in the amount that life sequestered carbon through time influenced these variables. Still, the profound effect of life in molding the atmosphere and temperature to favor life and biodiversity is clear and of great magnitude. Let us explore how this feat of ecosystem engineering by organisms was done.

It must be borne in mind in this discussion that it is imperative that life did not remove carbon in such great quantities or so fast that it made Earth too cold or temperatures to fluctuate too much for higher life and high diversity. For the ABH/Pachamama Hypothesis to be correct, it is crucial that life removed the right amount of carbon (within a certain range), and at a sufficiently gradual rate. If life sequestered too much carbon too quickly, Earth would become too cold to be optimal for life. And since carbon is necessary for photosynthesis and the survival of all living organisms, organisms would not thrive if there were too little of it in the biosphere.

The biological pump is the sea's biologically driven sequestration of carbon to the ocean's interior and seafloor sediments. It removes carbon from the atmosphere. The sea receives the carbon mainly from water running from the land into the sea via rivers. It removes CO_2 and methane from the system through the burial of carbon, thus regulating atmospheric temperature. Every year, the biological pump transfers 11 billion tons of carbon to the ocean's interior. This is equivalent to more than 66 million blue whales (*Balaenoptera musculus*), or about the mass of all nonhuman land mammals on Earth, or approximately twice the mass of all of the people in the world. This takes carbon out of contact with the atmosphere for several thousand years or longer. If it were not for the biological pump, atmospheric CO_2 levels would be about 400 parts per million higher than they are today. This would make the Earth so hot that diversity would be very low.

Biology accounts for most carbon sequestration in the ocean (Morse and Mackenzie, 1990). A great deal of carbon is buried in the seafloor as a result of a continuous shower of debris falling through the sea. This debris originates in the upper layers of the sea where plankton carry out photosynthesis. It can be seen by scuba divers. It formed the White Cliffs of Dover. It looks somewhat like falling snow, so is called marine snow. It is made up of parts or all of dead or dying animals, seaweeds, plankton, bacteria, archaea, fecal matter, sand, soot, and other inorganic dust. Most of it consists of aggregates of small particles bound together by a sugary mucus exuded as a waste product, primarily by bacteria and phytoplankton. And mucus secreted by zooplankton contributes to marine snow aggregates (Miller, 2004). Thus, marine snow is mostly, but not exclusively, made up of and generated by organisms.

The aggregates sometimes increase in size, sometimes reaching several centimeters in diameter, and travel for weeks before reaching the seafloor. Marine snow often forms as a result of phytoplankton blooms, which occur when phytoplankton in the sea reproduce profusely and bloom into great numbers. The phytoplankton from these blooms aggregate, which speeds up their sinking rate. This is a big source of loss of phytoplankton from the surface (Lazier, 2006). Most organic components of marine snow are eaten by organisms, from bacteria to filter-feeding animals, within about the first 1,000 meters (about 3,280.8 feet) of their downward journey. There is an entire group of animals at depths between about 100 (about 328.1 feet) and 1,000 meters that eat marine snow. Most of what is eaten gets sent back to the air in a relatively short time, because organisms expel CO_2 when they breathe and because if organisms decompose without sinking, this emits CO_2 into the air. However, a significant amount is never consumed, and stays in the sea's circulation for a long time, and shuttles carbon into the sea's depths; this can keep it out of contact with the atmosphere for more than 1,000 years. This is long enough that marine snow certainly qualifies as an effective, natural carbon sequestration mechanism by life. Marine organisms sequester about 10^{15} grams of carbon per year as marine snow (Boyd et al., 2019, and references therein). Marine snow has buried huge quantities of carbon in the seafloor throughout the history of life.

Marine viruses are likely major players in the biogeochemical cycles that run the planet (Danovaro et al., 2011). The viral shunt is the process by which marine viruses burst open the cells of one-celled organisms such as bacteria, releasing nutrients from the cells and fertilizing the ocean with them. A liter of seawater collected in marine surface waters typically contains at least 100 billion viruses. Viruses of bacteria are called bacteriophage or simply phage. Marine bacteriophages kill 20 to 40% of ocean bacteria every day (Fuhrman, 1999; Weinbauer, 2004, Suttle, 2005; Proctor and Fuhrman, 1990). Mojica and Brussard (2015) showed that as much as 80% of the carbon made into carbohydrates by photosynthesis moves through the viral shunt into the pool of dissolved matter in the ocean, while Wilhelm and Suttle (1999) stated that as much as a quarter of the carbon in complex compounds used by life in the sea flows through the viral shunt. This means viral infection is a very important mechanism for recycling carbon in the marine environment. Every second, approximately 10^{23} viral infections occur in the ocean (Suttle, 2007). This releases vast amounts of nutrients, including nitrogen, phosphorus, iron, and carbon compounds, into the sea. The nutrients released cause the growth of tremendous amounts of archaea, bacteria, and phytoplankton. The archaea, bacteria, and phytoplankton that receive nutrients as a result of the viral shunt support the major food webs in the ocean. Since the viral shunt provides nutrients to the ocean's main food webs, it greatly increases productivity, biomass, and biodiversity in the sea.

We do not know for certain if the effect of the viral shunt is to slow down or speed up carbon sequestration (Weitz and Wilhelm, 2012). Mojica and Brussard (2015) think the viral shunt decreases carbon sequestration. But Suttle (2007) pointed out that the shunt very likely increases carbon sequestration. The carbon-rich compounds are among the most complex compounds released by the shunt, and are not easily ingested, and so tend to sink with greater probability than the other compounds, selectively sequestering carbon (ibid.). In addition to bacteriophage, viruses of eukaryotic plankton appear to increase carbon sequestration (Blanc Mathieu et al., 2019). In fact, some virus-infected phytoplankton sink more rapidly than uninfected ones (Lawrence and Suttle, 2004), making them more likely to be buried at the seafloor (Lawrence et al., 2002), although this has only been examined in a few plankton. Considering all this, it is likely that the net direct effect of the viral shunt is increased carbon sequestration.

However, the major effect of marine viruses on carbon burial is likely indirect. They liberate nutrients for the growth of prokaryotes and phytoplankton, fueling the microbial carbon pump, which is proposed as a major mechanism for carbon sequestration in the ocean, through the production of *dissolved organic carbon, which comprises the same amount of carbon as is in the atmosphere* (Suttle, 2007; Zhang et al., 2018). Some of this sinks to the seafloor, sequestering carbon. In addition, the nutrients that are made available by the viral shunt allow the rapid growth of prokaryotes and phytoplankton. In fact, the shunt causes the release of sufficient iron to supply the needs of phytoplankton (Poorvin et al., 2004). Adding iron and other nutrients to the sea increases phytoplankton and prokaryote numbers. The increase in the populations of these organisms means more eventually sink to the seafloor, which sequesters carbon. They are also eaten, and more dead bodies and feces of organisms that eat them sink, burying carbon. Also, organisms at all levels of the food webs supported by these prokaryotes and phytoplankton (including predators and decomposers that do not directly consume these small primary producers) defecate, and some sink after death, with both of these processes burying carbon. Since the increased iron and other nutrients from the shunt enhances prokaryote growth and thus the number of prokaryotes that sink, the shunt probably has increased carbon burial since essentially the origin of life, because prokaryotes were the first life. So the bulk of evidence is that the viral shunt increases the biological pump and carbon burial. And the amount of carbon sequestered by the viral shunt as well as its effect on the carbon cycle is likely large. And viruses cause the formation of much of the marine snow by causing the cells they kill to release components that bind to each other and sink, sequestering carbon. Hence, viruses sequester great quantities of carbon, much of which they keep available to the biosphere by storing it in their DNA and proteins.

Stromatolites are structures that reach heights of up to one meter (about 3.3 feet) or more and are formed mainly by cyanobacteria and layers of sandy and pebbly sediments. Cyanobacteria are probably the first organisms to evolve the type of photosynthesis that produces oxygen. One of the places they grow is in shallow seas. Cyanobacteria produce adhesive compounds that cement sand and other pebbles to form microbial mats. Another layer of cyanobacteria grows over the old one, cementing more sand and pebbles. By this process, they gradually build up in layers over time. The lower layers store carbon that is kept out of the biosphere. Living stromatolites exist, but are rare today. But they are a major constituent of the fossil record of the first forms of life on Earth, so might date to over 4 bya. They peaked about 1.25 bya and subsequently declined, so that by the start of the Cambrian period about 541 mya, they had fallen to 20% of their peak. But they were abundant before the Cambrian. Some fossils that are called stromatolites are not biological, so cannot be used as evidence that life sequestered carbon. But biological stromatolite fossils are abundant enough to support the thesis that they sequestered a good deal of carbon. These fossils store carbon that cyanobacteria removed from the biosphere. As cyanobacteria photosynthesize, a layer of mucus often forms over mats of their cells. This mucus can trap debris from the surrounding habitat. This debris can be cemented together by calcium carbonate to grow thin laminations of limestone. Limestone is in fact mostly calcium carbonate and hence rich in carbon. So this process sequesters carbon. This process happened for long time periods and over great areas in the geologic past. Similarly, bacterial mats in freshwater, such as those in Yellowstone National Park, are layered, with top layers growing over and burying carbon-rich bottom layers. So stromatolites and bacterial mats sequestered a large amount of carbon in the past.

Life started as prokaryotes at least 3.5 bya. One-celled eukaryotes first appeared between 2.1 and 1.6 bya, and the first multicellular organisms appeared between 1.6 and 1 bya. So, from 3.5 bya or earlier to between 1.6 and 1 bya, over half of life's history, one-celled forms and viruses were the only life and hence the only biological mechanism of carbon sequestration. These life forms sequestered a good deal of carbon. This was largely accomplished in the sea, where it happened via the viral shunt, by prokaryotes and phytoplankton sinking, by layering of bacteria in stromatolites, by zooplankton defecation, and death and sinking of these small organisms.

Carbon fixation is the process by which simple carbon molecules, mainly CO_2, are converted to more complex molecules by living organisms. The molecules are then used to store energy and as structure for other biological molecules, including DNA. Today, most planetary carbon cycling actually occurs in the deep sea where it is too dark for photosynthesis. Pachiadaki et al. (2017) showed that certain nitrite-oxidizing bacteria are the main organisms that do carbon fixation between approximately 220 to 1,100 meters (721.8 to 3,608.9 feet) below the ocean surface, fixing 15 to 45% of inorganic carbon in this zone, in the western North Atlantic. They are the most abundant and globally distributed nitrite-oxidizing bacteria in the ocean, and the primary producers and base of the deep-sea food webs, meaning they support countless species that eat them and are eaten by other species, which are eaten by still others, and so on up to the top of the food webs. The researchers identified over 30 species of nitrite-oxidizing bacteria in this dark part of the sea. They sequester large amounts of carbon when they sink and are buried, although much less than that sequestered by surface phytoplankton. More carbon is sequestered by animals that eat them directly or are higher up the food web than by the sinking of these bacteria. That is because some of these animals, which include crabs and fish, die and sink to the sea bottom, and this results in the burial of a great deal of carbon there. Carbon is also sequestered when these animals defecate and their feces sink and get buried in the seafloor. The amount of carbon these bacteria sequester, directly and indirectly, is currently unknown, but the researchers say that the bacteria in this part of the sea that carry out nitrite oxidation may have a greater impact on the carbon cycle than previously assumed.

Ever since the first appearance of photosynthetic eukaryotes about 1.5 bya, eukaryotic phytoplankton in the oceans have been among the most important groups of organisms at sequestering carbon. Marine phytoplankton carry out about half of the photosynthesis that occurs on the planet (Baumert and Petzoldt, 2008; Simon et al., 2009), and they do this at an astonishingly rapid rate.

Phytoplankton, which in this discussion includes photosynthetic prokaryotes such as cyanobacteria, sequester carbon by three mechanisms. First, they bloom in great numbers, followed eventually by large numbers of them dying, sinking, and being buried as sediments, mainly in shallow seas. The second mechanism is aggregation and sinking of cells, which is more important for small cells, such as cyanobacteria. Finally, some are eaten by zooplankton and plankton feeders, which starts carbon's ascent up the food web. Then some of the organisms at every level in the food web, from zooplankton to whales, defecate or die, and some of the carcasses and feces sink to the seafloor, where some of the carbon is buried. *A staggering 5 to 12 billion tons of carbon per year is sequestered directly or indirectly in the sea by phytoplankton* (Siegel et al., 2016). *This is a larger quantity than the amount of carbon consumed every year by all the ecosystems on land combined* (Boyd et al, 2019, and references therein). *It is over half of the CO_2 produced each year by burning fossil fuels.*

A population of phytoplankton can double its numbers on the order of once per day. Large populations of phytoplankton, sustained over long periods of time, can significantly lower atmospheric CO_2 levels. However, note that photosynthesis by phytoplankton or plants alone would not remove CO_2 from the biosphere, because the photosynthesizers respire, and decompose after death. Both processes add CO_2 back to the biosphere. These organisms must be buried in the sea bottom or underground on land, or be eaten by organisms that then acquire their carbon and die and get buried or defecate with their feces getting buried, in order for carbon to be removed from the biosphere. Of course, they can also keep carbon out of the biosphere by storing it in their cells and molecules until they die and decompose. This can be a long time and involve a great deal of carbon in large organisms, such as whales, and a tremendous amount of carbon in organisms with extraordinarily high population numbers, such as bacteria and viruses.

There are four major groups of phytoplankton today, and they are all ecosystem engineers that have been sequestering carbon for over 185 million years. They all can carry out photosynthesis. Diatoms comprise one group of these phytoplankton. They are one of the major groups of photosynthetic eukaryotes. There are probably well over 10,000 species of them (Mann and Droop, 1996). Their cell walls have silica, not carbon. The silica acts as a shield for defense against predators. Even so, they sequester large amounts of carbon, for the following reasons. Diatoms contain carbon as well as silica. The silica cell walls act as ballast when the diatoms die, so they have more of a tendency to sink and thus get buried than other phytoplankton. There are many of them. Living diatoms make up a significant portion of the Earth's biomass, and make up almost half of the organic material found in the oceans. The shells of dead diatoms can reach a thickness of over 800 meters (2,624.7 feet) deep on the seafloor. This is about a half-mile. Also, being silicon-rich, they cause great quantities of silicon to be buried in the seafloor, and this also helps life considerably. Too much silicon in the environment can harm organisms. This latter benefit to life, the burial of silicon, will be discussed later in this chapter.

Dinoflagellates are another type of marine phytoplankton. There are about 1,555 described species of free-living marine dinoflagellates. There are estimated to be about 2,300 total living dinoflagellate species, if you count the estimated number of species not yet discovered or described. There are marine, freshwater, and parasitic dinoflagellates. About half of dinoflagellate species are photosynthetic. Dinoflagellates can form protective shells called cysts that lie dormant when conditions are unfavorable, and these contain carbon and sometimes get buried and fossilized, sequestering carbon. Their cysts are found as certain microfossils from the Triassic period (about 252 to 201 mya), with a few probable records of them from the Permian period (298.9 to 251.9 mya) (Fensome et al., 1996). They form a major part of the marine phytoplankton fossils from the middle Jurassic, about 73 mya, to the present day, showing how long dinoflagellates have existed, and that they sequestered large amounts of carbon because of the great numbers of them buried over large time spans. The primary time of diversification when dinoflagellates evolved into many species was in the Jurassic period (about 201 to 145 mya) and Cretaceous period (about 145 to 66 mya) (Tappan, 1980; Falkowski et al., 2004). Though they sequestered carbon from their first appearance until today, they may have done this the most during their primary diversification period.

Coccolithophores are almost exclusively single-celled marine phytoplankton that are found in large numbers throughout the ocean's photic zone, the shallowest part of the ocean, where enough light penetrates to allow photosynthesis. Many have a broad distribution. When they die and get buried, coccolithophores can take a great deal of carbon out of the biosphere, largely due to their carbon-rich coccoliths, special calcium carbonate plates that cover and protect them and that are important very small fossils (Prothero, 2004). Iglesias-Rodriguez et al. (2008) found that in at least some circumstances, the species *Emiliania huxleyi* becomes 40% heavier—so acquiring a greater tendency to sink—and also more abundant, in waters with higher CO_2 concentrations. With more of them, more die and sink, sequestering more carbon. Heavier ones contain more carbon, so sequester more carbon when they die and sink. Significantly, this would allow more carbon sequestration in the presence of high CO_2, regulating both atmospheric CO_2 and the Earth's temperature like a thermostat by negative feedback. *E. huxleyi* has population explosions that can be seen from orbit. The blooms are killed in great numbers by viruses, sequestering large amounts of carbon. Coccolithophores have both short-term and long-term effects on the carbon cycle. Short term, the production of their coccoliths actually adds CO_2. They take up some of this CO_2 through photosynthesis, but blooms release and hence increase CO_2 short term. But long term, their coccoliths containing calcium carbonate sink and become part of the seafloor sediment (Marsh, 2003). The long-term effect greatly outweighs the short-term one, so the net effect of the two is a big decrease in atmospheric carbon and cooling of the Earth.

The 10,000 living and 40,000 fossil species of foraminifera make up the other major group of phytoplankton that has sequestered significant amounts of carbon for long periods of Earth's history. They have a test, or shell, commonly made of calcium carbonate or agglutinated sediment particles. The ones with tests of calcium carbonate sequester significant amounts of carbon when foraminifera die, sink, and get buried in the seafloor. The principal source of seafloor carbonate from approximately 150 mya to the present is from calcareous phytoplankton, mainly coccolithophores and foraminiferans. Foraminiferans arose about 150 mya, and have grown in abundance since then.

The remains of calcareous phytoplankton, such as coccolithophores and foraminifera, fall to the seafloor, accumulate, and form carbonate-rich sediment there. Sedimentation of carbonates has increased since the appearance of calcareous phytoplankton, largely due to them. Presently more than half the global carbonate deposits are in the seafloor (Milliman, 1974). Much of these are from phytoplankton.

It turns out that the great majority of plankton are mixotrophs, which are hybrid organisms that are both predators and photosynthesizers. A system with mixotrophs interacting in a food web with bacteria and phytoplankton removed 65 grams of carbon per square meter of seawater, compared to 30 grams of it per square meter if only microzooplankton that are exclusively predators interacted with bacteria and phytoplankton (Mitra et al., 2016; Mitra, 2018). Thus, combining photosynthesis with predation in one organism increases carbon burial significantly. Diatoms, foraminiferans, and radiolarians are mixotrophs (Flynn and Mitra, 2009; Stoecker et al., 2017). Radiolarians are a phylum of protozoa zooplankton found throughout the global ocean that primarily eat smaller plankton, but many have symbiotic photosynthesizers in them.

When the asteroid that caused the Cretaceous-Paleogene extinction event of about 66 mya hit, debris from its impact and wildfire ash blocked the sun, and the Earth was dark. Photosynthesis in the sea was greatly reduced, including by the phytoplankton at the base of the ocean's food webs. Gibbs et al. (2020) used marine plankton microfossils and ecoevolutionary modeling to find that coccolithophores saved the marine ecosystem by adapting, with some becoming mixotrophs. The system flipped to a bacteria-dominated quasi-stable state when sunlight was lacking. Coccolithophores became mixotrophs, and ate the bacteria. Some of the foraminifera that survived also apparently became mixotrophs, evolving spines that would have worked with miniature tenacles to capture prey. Mixotrophy allowed the coccolithophores and forminiferans to survive and balance the food webs when photosynthesis was difficult to carry out. These two groups may have kept the marine ecosystem from going extinct. Most of them became photosynthesizers once again when sunlight returned and the ecosystem was re-established two million years later.

When higher animals had just appeared, in the Cambrian period, which started about 541 mya, there were carbon-rich calcareous algae; animals called brachiopods that looked roughly like clams and had carbon-containing shells and consist of 12,000 fossil species; molluscs; and simple corals with carbon-containing skeletons. All of these removed a good deal of carbon from the biosphere in that period, many continuing to do so up until the present. This is because they contained a considerable amount of carbon and sometimes got buried in the seafloor after death.

Every autumn in the North Atlantic, billions to trillions of the copepod *Calanus finmarchicus* (no common name), which is about the size of a grain of rice, migrate down to hibernate in winter at ocean depths of about 1,400 to 1,600 meters (about 4,593.2 to 5,249.3 feet). Copepods are a group of small crustaceans found in great numbers in nearly every freshwater and saltwater habitat. They transport carbon in fatty molecules called lipids deep in the sea, where it is released and sequestered via their respiration and defecation, and the death and sinking of some of them. Unlike other components of the biological pump, this so-called "lipid pump" does not strip the ocean surface of a significant amount of nutrients. Copepods are especially important, then, since they are among the few organisms that sequester great amounts of CO_2 in the sea without removing very many other nutrients from the system. The amount of CO_2 sequestered by copepods equals the sinking of marine snow (Jónasdóttir, et al., 2015). This has occurred for sufficiently long geological time periods to have sequestered a tremendous amount of CO_2.

Salps are barrel-shaped, free-floating marine animals (phylum Chordata; subphylum Tunicata) that constantly swim and feed on phytoplankton by filtering them from seawater, in tropical, temperate, and polar seas. They are abundant enough to affect the biological pump, and they sequester tremendous quantities of carbon. They transport many tons of carbon per day from the ocean surface to the deep sea. They can bud off clones and can probably grow their populations faster than any other multicellular animal, producing swarms that number in the billions and last for months; they can be even more abundant than krill. (Krill are small crustaceans found in all the world's oceans that feed on plankton, are extremely abundant, and are a major food source for whales, seals, penguins, squid, and some fish). Salps have no known predators. If their phytoplankton food is too dense, they can clog and die. And when there is no longer a sufficient quantity of phytoplankton to sustain them, they die in great numbers. Their carcasses often sink and get buried, sequestering carbon. They produce a great deal of heavy fecal pellets that quickly sink, efficiently transporting more carbon to the seafloor than their carcasses. One species, *Salpa aspera*, in the Mid-Atlantic Bight region, consistently multiplies into dense swarms that last for months. One swarm covered 100,000 square kilometers (about 38,610 square miles) of sea surface, containing trillions of salps the size of a human thumb. The swarm consumed up to 74% of the phytoplankton from the sea surface per day, and the salps' sinking fecal pellets transported up to 4,000 tons of carbon a day to deep water. *Salpa aspera* swim down to and stay at depths of 600 to 2,800 meters (1,968.5 to 9,186.4 feet) during the day, probably to avoid damaging sunlight (Madin et al., 2006). Because of this behavior, these salps may release fecal pellets in deep water, where it is unlikely that feces-eaters would consume them. So the pellets reach the ocean floor with high probability. They feed and reproduce at the sea surface at night, so they transport enormous quantities of carbon from the sea surface to its bottom—between five and 91 milligrams/square centimeter/night (Madin et al., 2006). And when they die, their bodies also sink rapidly, at 475 meters (1.558.4 feet) per day. Salp fecal pellets sink as much as 1,000 meters (about 3,280.8 feet) a day, and far faster and with higher carbon content, making their efficiency of downward carbon movement much higher, than that of the two other major marine invertebrate grazers, which are krill and copepods.

Perissinotto and Pakhomov (1998) found that in the Antarctic Ocean's Lazarev Sea during the austral summer (December to January) of 1994 to 1995, the salp *Salpa thompsoni* had dense swarms throughout the marginal ice zone, consuming tremendous quantities of phytoplankton. The resulting carbon flux could have attained levels of up to 88 milligrams of carbon/meter/day, accounting for the bulk of the vertical transport of carbon in the Lazarev Sea, showing the spectacular amounts of carbon this species can transport from the photic zone to the seafloor. Yet in January, 1995, when phytoplankton concentrations exceeded a threshold level, these salps experienced a drastic reduction in their feeding efficiency, possibly as a result of clogging of their filtering apparatus. There was apparently a quasi-stable state (a state that is mainly stable, but can be changed to an alternate state in extreme circumstances) of higher salp and lower phytoplankton populations that was dramatically reversed to a higher phytoplankton and lower salp quasi-stable state, when a threshold of very high phytoplankton numbers was reached, during which a dense phytoplankton bloom developed resulting in the salp population's collapse. Significantly, there appears to be some stabilizing negative feedback in the carbon cycle (and hence temperature) with *Salpa thompsoni*. Increases in its biomass and geographic range have occurred in several areas of the Southern Ocean, often in parallel with a rise in sea-surface temperature during periods of warming anomalies. It thus likely sequesters more carbon when temperatures are high, stabilizing carbon levels and the temperature, at least locally.

Giant Larvaceans (phylum Chordata; subphylum Tunicata), which are free-swimming relatives of salps, filter water for feeding at 20 milliliters (about 0.04 pints) per second, a higher rate than salps or any other group (Katija, 2017). They shed their filtering structures of mucus, which have diameters over 1 meter (about 3.3 feet), and these sink to the seafloor. They sequester vast amounts of carbon this way.

Ware et al. (1991) showed that the precipitation of calcium carbonate in coral reefs is accompanied by a shift in the acidity of the water that actually releases CO_2, making them, surprisingly, sources, not sinks, of atmospheric carbon, in the short term, releasing 0.02 to 0.08 billion tons of carbon as CO_2 annually. This is about 0.4% to 1.4% as much as the annual worldwide CO_2 production due to fossil fuel combustion by humans. However, KInsey and Hopley (1991) stated that coral reefs at present act as a sink for 111 million metric tons of carbon per year, the equivalent of 2% of the present output of CO_2 by humans, although they agree with Ware et al. that the immediate effect of calcium carbonate precipitation is to encourage the release of CO_2 to the atmosphere, causing the reef to be a short-term carbon sink. In the long term, the calcium carbonate skeletons of hard corals have been converted to limestone by the deposition of dead corals during long periods in the geologic past, removing carbon from the atmosphere. This has resulted in extensive limestone deposits in various areas around the globe, such as in the Caribbean. Some fish, such as bumphead parrotfish (*Bolbometopon muricatum*), eat coral and excrete white coralline sand, which is full of carbon that does not easily return to the atmosphere. So coral reefs have played a large long-term role in carbon sequestration. Since a few thousand years after hard skeletons appeared, reefs have existed. They were most prominent in the Middle Cambrian period (513-501mya), Devonian period (416-359 mya), Carboniferous period (359-299 mya), Late Cretaceous period (100-66 mya), and all of the Neogene period (23 mya-present). Not all reefs in the past were formed by corals. In the Early Cambrian period (541-513 mya), they consisted of calcareous algae and archaeocyathids (small animals with a conical shape, likely related to sponges). These sequestered some carbon. In the Late Cretaceous period, the dominant reef-builders were a group of strange bivalve molluscs called rudists that went extinct at the Cretaceous-Paleogene mass extinction that killed the dinosaurs about 66 mya. The fossilized deposits of their carbon-containing shells in the Arabian Peninsula, North Africa and the Near East, the Balkans, the Caribbean, France, Spain, Greece, Italy, Turkey, Mexico, and the U.S. Gulf Coast represent the sequestration of a large amount of carbon. Phytoplankton, some worms with shells, other bivalve molluscs, and other organisms that make calcium carbonate have added to these deposits, sequestering a great amount of carbon. Also, less commonly, limestone can precipitate directly out of solution without life's aid. All told, the amount of carbon in rock far exceeds the amount of free carbon in the air, water, and soil. Limestone is the largest repository of sequestered CO_2 on the planet. It represents a huge amount of carbon life has removed from the atmosphere.

Echinoderms are a phylum of animals that include starfish and sea urchins. They incorporate carbon directly from seawater into their skeletons in the form of minerals, the main one being calcium carbonate. They also obtain a great deal of carbon from their food. Lebrato et al. (2010) showed that these animals remove surprisingly substantial amounts of carbon—over a hundred billion kilograms (about 110,231,131 tons) of it per year—from the biosphere, since many get buried in the ocean bottom. This is more carbon than foraminifera deliver to the seafloor sediments and sequester there. The amount of carbon that Echinoderms sequester is so immense that Lebrato et al. declared, "Climate models must take this carbon sink into account."

Wilmers et al. (2012) found that sea otters in the Pacific northwest seas allow kelp ecosystems to develop with a net primary productivity (NPP) of 313–900 grams of carbon per square meter per year (g C/m^2/yr), whereas in their absence, these areas would have an NPP of only 25–70 g C/m^2/yr. NPP in this case is a measure of the growth of kelp, so sea otters substantially increase the growth of kelp forests by eating sea urchins, which eat kelp. Over a substantial area, the effect of sea otter predation on sea urchins on living kelp biomass alone causes an increase in carbon storage of 4.4 to 8.7 trillion grams. At the 2012 price of US$47 per ton of carbon, this stored carbon value is US$205 million to US$408 million on the European Carbon Exchange. Here a predator indirectly helps a primary producer, the kelp, two steps lower on the food web, store carbon and keep it out of the atmosphere, by eating the animal that eats it.

A large amount of carbon that sinks to the deep sea is not supplied by particulate organic carbon (such as marine snow) alone, so comes from another source. One main such source is the sinking of the corpses of large animals. Hypothesized percentages of the contribution of the sinking of large dead animals to the total carbon that sinks to the deep ocean range from 0.3% (Smith, 2005) to 4% (Higgs et al., 2012). The large quantities of carbon that these dead animals bring to the deep sea is generally sequestered by being buried in the seafloor.

The largest animals that sink to the sea bottom upon death are whales. Whale carcasses sequester tremendous amounts of carbon. The quantity of carbon tied up in only one typical 40-ton whale carcass is about two metric tons, which is approximately the amount of carbon that reaches one hectare (10,000 square meters, or 107,639 square feet) of abyssal ocean floor in 100–200 years (Smith and Baco, 2003). This amount of organic material reaching the seafloor at one time creates a pulse equivalent to about 2,000 years of background carbon sinking in normal conditions to the 50 square meters (538 square feet) of sediment that a whale carcass would occupy (ibid.). This not only sequesters great quantities of carbon, but provides nutrient to deep-sea ecosystems.

Whales need to breathe at the surface, so they tend to defecate in shallow water. Their excrement tends to float, so is available to phytoplankton in the photic zone. This reverses the normal downward flow of the biological pump, sending carbon upward in the sea; this is called the whale pump (Roman and McCarthy, 2010). This increases phytoplankton growth, helping the food webs in the shallow sea, and sequestering large amounts of carbon when the phytoplankton die and sink. Yet some whale excrement sinks, causing large quantities of carbon to be buried in the deep sea. Defecation by the Southern Ocean's Sperm Whales alone sequesters approximately 400,000 tons of carbon annually. Before commercial whaling, these whales sequestered approximately as much carbon as an average coal-fired plant produces in an equivalent amount of time.

The feces of krill-eating whales are up to 10 million times as iron-rich as the surrounding seawater (Nicol et al., 2011). This is a major source of the iron needed for maintaining phytoplankton biomass on Earth, and aids their blooms. This is especially important in the Southern Ocean, which is rich in other nutrients, but iron-poor (ibid.). This allows plankton populations to explode. This causes phytoplankton to sequester large quantities of carbon when they die and sink. As discussed earlier in this chapter, phytoplankton are major players in carbon sequestration. The iron from the defecation of merely the approximately 12,000 sperm whales (*Physeter macrocephalus*) in the Southern Ocean helps sequester about 200,000 tons of carbon annually. Whales also indirectly help countless species in the ocean's major food webs, since phytoplankton are the sea's main primary producers. That is, whale excrement helps phytoplankton bloom, and the phytoplankton are eaten by zooplankton, which are eaten by krill, and so on. A whole food web of animals of many species is supported in this way. Of course, these animals in the food web also die and sink, and some of their feces sink when they defecate. Some of these corpses and feces get buried in the seafloor. This sequesters large quantities of carbon. So whale feces and phytoplankton play important roles in sequestering large amounts of carbon, both directly and indirectly through food webs.

There are now about 8,000 blue whales worldwide, the largest animal ever to exist. There were once more than 200,000 blue whales in the Antarctic Ocean alone. All giant whale species were much more abundant in the past. They sequester large amounts of carbon now because of their size. But they sequestered much more carbon in the past because of their greater numbers before humans hunted them to low numbers.

The shells of bivalve molluscs, such as clams, mussels, and scallops, are made mainly of calcium carbonate, and can be broken down to sand, which stays a long time on the seafloor, or they can be buried. The outer skeletons of crustaceans, such as crabs, shrimp, and lobsters, contain a great amount of carbon. When these animals molt (shed their outer skeletons) or die, the molted skeleton or animal is sometimes buried. They molt several times in their lives. Bivalve molluscs and crustaceans have sequestered a great deal of carbon. Precise figures are not known.

In rainforests, elephants are ecosystem engineers that selectively eat small, fast-growing trees with wood of low density, allowing large, slow-growing trees with wood of high density to dominate. The latter store carbon better than the former. Without elephant foraging, billions of tons of carbon would be released into the air annually. At a typical density of 0.5 to one animal per square kilometer, elephant disturbances increase aboveground biomass by 26 to 60 tons per hectare. The extinction of forest elephants would decrease aboveground biomass by 7% in central African rainforests. Biomass is the total mass of organisms, including plants, in a given area. Elephants represent a carbon storage service of $443 billion per year (Berzaghi et al., 2019).

Freshwater ferns can be buried and sequester carbon in great quantities. There was a period when global mean temperature increased by over 5 to 8°C (9 to 14.4°F) (McInherney and Wing, 2011), and was over 8°C (14.4°F) higher than today. There were turtles, crocodilians, and palm trees in the Arctic. Its precise time and duration are uncertain, but it is estimated to have occurred about 55.5 mya (Bowen et al., 2015), lasting for about 200,000 years. The warming probably was ended by the Azolla event. Blooms of a freshwater fern of the genus *Azolla*, related to the modern fern, *Azolla filiculoides*, were abundant near the Arctic Ocean. When they died, many were washed by rivers into the Arctic Ocean, sank, and were buried and incorporated into the sediment, sequestering carbon. It is postulated that these ferns were buried over time in such abundance as to throw the world's climate from hothouse conditions to a long-term cooling trend that led eventually to the Pleistocene epoch glaciation that began 2.58 mya. (The Azolla event did not cause the Pleistocene glaciation, but rather brought Earth to favorable temperatures for life.) Sedimentary layers throughout the Arctic basin contain fossilized *Azolla* (Waddell and Moore, 2006). Evidence indicates the Azolla event lasted an estimated 800,000 years (Brinkhuis and Schouten, 2006). And it coincides precisely with a dramatic decline in CO_2 levels, which fell from 3,500 parts per million in the early Eocene epoch (this epoch started about 55.8 mya) to 650 parts per million during the Azolla event (Pearson and Palmer, 2000). *Azolla* can draw down 1.5 kilograms of carbon per square meter per year (six tons of carbon per acre per year) (ibid.), which is a huge amount. In the warmth and 20 hours of sunlight during the Arctic summer, it can double its biomass over two to three days. Other factors may have contributed to the cooling, but *Azolla* was a major carbon burial agent and important factor that brought Earth out of the very hot conditions to a more life-favorable climate.

Rainforests sequester large amounts of carbon. Much of this is accomplished with the help of phytoplankton, working together in a complex system. Leaves and other organic matter full of nutrients from rainforests enter rivers, which transport the nutrients to plankton in the sea, which greatly increase their numbers as a result of the increased nutrients. Some of the plankton die and sink, burying carbon. Some are eaten and the carbon goes up the food web. When animals higher on the food web supported by the phytoplankton die and sink or defecate and their feces sink, carbon is sequestered in the seafloor. This results in the sequestration of great amounts of carbon. And as a result of the increased numbers of phytoplankton due to the nutrients from the rainforest, the phytoplankton produce a great quantity of oxygen. This is a major source of Earth's oxygen. Surface plumes transport freshwater discharged by large rivers hundreds to thousands of kilometers away from the coasts. The nutrients carried by the plumes from the rainforest contribute to enhanced production of plankton in the ocean. Near the mouth of a river, the waters are dominated by typical coastal diatoms that are sustained by the nitrates and silicates in the plume. Far from the river mouth, it was thought that there was insufficient nitrate to sustain phytoplankton. But experiments by Subramaniam et al. (2008) in the Northwestern Tropical Atlantic on water flowing from the Amazon River showed that the dominant species of phytoplankton change as the plume moves and the nitrate runs out, so beneficial nitrogen-fixing cyanobacteria provide the nitrate to sustain phytoplankton far from the coast, at least some of which are different species than the ones near the mouth of the Amazon River. This leads to more carbon sequestration. The scientists calculated the production of phytoplankton supported by nitrate coming from the Amazon into the sea to be about 7.2 million tons per year, and that supported by nitrogen fixation by cyanobacteria to be about 20.4 million tons per year for a total carbon sink of about 27.6 million tons of carbon converted to phytoplankton per year in the plume. The key groups in this great carbon sequestration by interspecific teamwork are interacting rainforest trees, phytoplankton, and cyanobacteria. The authors have preliminary evidence that the same processes are at work in other tropical rivers that flow into oceans, such as the Congo, the Orinoco in South America, and the Mekong. Rainforest trees also store tremendous quantities of carbon in their leaves, trunks, and roots.

Trees sequester great quantities of carbon when buried. The appearance of rooted plants during the Devonian period, 419 to 360 mya, almost certainly dramatically lowered atmospheric CO_2 levels because their roots weathered rocks. Well-differentiated forest soils had developed by the Devonian period, and deep-rooted vascular plants spread to upland areas during this time (Berner, 1997). Both of these developments resulted in the sequestration of great quantities of carbon by life, as tree roots and their symbiotic fungi buried carbon by weathering. Vascular plants are higher plants with true roots, leaves, and stems, although some groups have secondarily lost one or more of these traits. They include the most complex plants, and consist of the clubmosses, horsetails, ferns, conifers, and flowering plants.

The Carboniferous period is the period when the most carbon was buried by plants by far. Lower sea levels allowed forests and swamps to grow on lowlands, which would otherwise have been under the ocean. This allowed swamps with large trees to be prevalent, and fallen trees are more prone to be covered with sediment in swamps. So great masses of large trees were buried, with the organic matter containing carbon being converted to coal and oil over millions of years, forming much of today's fossil fuel deposits. The trees of the Carboniferous period sequestered more carbon from the atmosphere than was removed at any comparable time period in life's history. Trees are among the major mechanisms by which life controlled and still controls atmospheric carbon and global temperature.

Lignins evolved in the Ordovician period (485.4 and 443.8 mya). They are a component of wood. Lignin in wood gave a decisive evolutionary advantage for many trees, protecting them from physical assaults, such as fire, and making them less digestible to animals. It also allowed trees to have bark, which gave them strength, structure, and protection from fire, herbivores, and other challenges. It plays a crucial role in conducting water and aqueous nutrients in plant stems. It aids disease resistance by accumulating at the site of pathogen infiltration, making the plant cell less accessible to cell wall degradation. Lignin is unique to woody plants. It took many millions of years for bacteria and fungi to evolve the ability to break lignin down. Even today, it is only broken down by a limited number of organisms, such as white rot fungi. Before organisms could break lignin down, lignin increased in the biosphere, and thus large amounts of organic matter accumulated. Trees sometimes attained heights in excess of two meters (about 6.6 feet) at this time. So plant debris with large amounts of undecomposed lignin was deposited and buried in coal swamps on land, and also in the sea, to which it was transported by rivers, and where it was buried with other ocean-deposited carbon-bearing compounds. Vast coal deposits from the Carboniferous and Permian periods attest to the great amount of lignin that was buried during these periods. Coal is formed mainly from lignin. Lignin makes up one-quarter to one-half of wood when wood is dry. The evolution of this protective substance and the slow response of organisms to evolve a mechanism to digest it for food caused the removal of vast quantities of carbon from the atmosphere. Hence, biological processes—an evolutionary breakthrough and a slow evolutionary response to it—contributed substantially to large reductions of carbon in the atmosphere and hence a lowering of temperature from the end of the Ordovician period to the Carboniferous period and beyond it to the Permian period. Since much of the carbon buried was not bound to oxygen; this also caused atmospheric oxygen levels to rise significantly. This is discussed in Chapter 3. The evolution of lignin also was a key innovation that allowed trees to greatly diversify, creating many new tree species.

From the early Earth of over 4 bya to the first appearance of vascular plants on land about 430 mya, weathering happened mostly nonbiologically, and with negative feedback (Berner, 1992). Since vascular plants appeared, they have sequestered, and still sequester, large quantities of carbon by weathering via their roots and the symbiotic fungi associated with their roots, and by holding soil in place and preventing erosion, thus enhancing the rates of chemical weathering of silicate minerals (for a summary, see Berner, Berner, and Moulton, 2003). The plants give the fungi carbohydrates and the fungi help obtain minerals and water for the plants. Plants and their associated fungi have increased the rate of weathering of land surfaces (Quirk et al., 2015). The mineralogy of highly weathered soils cannot be understood in the absence of plant activity (Lucas, 2001). Roots and their fungal partners secrete organic acids and other chemicals that break down mineral particles in the soil to free nutrients plants need, principally calcium, potassium, iron, phosphorus, and magnesium. The soil minerals combine with carbon, and some of the resulting compounds go into streams and are transported to the sea, where some are buried, sequestering carbon. This is weathering. And some of these compounds with carbon in them provided by plant and fungal weathering are used by fish, bivalve molluscs, coral, some phytoplankton, and others, to build shells, bones, coral skeletons, and so forth. This takes carbon out of the biosphere and stores it in these structures. This of course reduces the amount of carbon in the air, lowering temperature to a level more favorable to life. Plants and their symbiotic fungi are thus ecosystem engineers that regulate temperature and ocean acidity, and provide nutrients to a great number of species of sea life. Plant root fungi are also important in sequestering carbon in their own right. Careful studies have shown that plants with their fungi dissolve rocks five times faster than would be the case without the fungi, in all major forest types (Berner, 1998). Roots with their symbiotic fungi can penetrate rocks and pry them apart, exposing fresh surfaces, and completely dissolve minerals in them by secreting organic acids (April and Keller, 1990; Griffiths et al., 1996; Berner and Cochran, 1998; van Breeman et al., 2000). The actions mentioned in the last two sentences are forms of weathering and cause a good deal of carbon to be sequestered. And litter from dropped leaves and dead plants decomposes into acids that help chemically break down minerals, sequestering carbon by weathering. Vascular plants accelerate weathering and hence carbon burial by about four to 10 times (Ward, 2009).

Large terrestrial vascular plants are the most effective at weathering. Vascular plants first appeared during the Silurian period (about 443 to 419 mya), but large land plants with deep roots, such as trees, did not become dominant until the Devonian period (about 419 to 340 mya). These plants increased weathering and carbon sequestration immensely compared to the algae, lichens, and mosses that preceded them (Berner, 1998). Certainly lichens, and for the most part mosses, are tiny and slow-growing, and have a small interface with rocks. Therefore, though there is evidence that they weather minerals, they do it very slowly. In contrast, trees have root systems that divide into vast numbers of rootlets and work with extensive symbiotic fungal networks associated with their roots and that act as extensions of these roots, causing a large interface with minerals. They take up nutrients quickly and grow quickly. They break up rocks and soil, liberating carbon. We know trees have built much more soil than lichens, demonstrating that trees accelerate weathering much more than lichens. Three different field studies showed trees accelerate weathering rates over lichens and mosses by factors of eight (Drever and Zobrist, 1992), four (Arthur and Fahey, 1993), and three to four (Moulton et al., 2000).

So weathering and hence removal of CO_2 from the air was tremendously increased by large vascular plants. Today, about one-tenth of the carbon fixed each year is buried with plants and algae that are washed or blown down from land surfaces into seas, rivers, swamps, and other bodies of water. Most weathering by plant roots and their symbiotic fungi results in carbon becoming bound to minerals, being washed into the sea, and getting buried in the seafloor. But also, much of the carbon from plant weathering builds up high levels of CO_2 in the soil (Berner, 1992), where much of it is stored and used by life. Water is recycled by plant weathering as well (ibid.).

The increase in weathering after the spread of land plants drove a 40-million-year decline in atmospheric CO_2, which fell from around 4,000 parts per million by volume (ppmv) to about 1,000 ppmv (Beerling and Berner 2005), during the Carboniferous period. This rate is as fast as any in the deep geologic past, although it is much slower than the changes during the industrial era, which are an increase of over 2 ppm/year over the past decade (IPCC, 2013). Without plants and their fungal allies accelerating the chemical weathering of rocks ever since this symbiosis first appeared, the atmosphere would have a CO_2 content 15 times as high as today's (Berner, 2013). That would make the average temperature of our planet a torrid 10°C (50°F) hotter!

When plants drop their leaves, they form a leaf litter layer above the soil. This forms a continuous, acidic, moist environment that breaks down minerals, causing yet more weathering and hence carbon sequestration. Organic plant litter accumulates in the soil too, where bacteria help the process by decomposing the litter to organic and carbonic acids, providing chemicals that dissolve minerals that can combine with carbon, making compounds that sometimes get buried. Trees actually create rain, which provides more water to dissolve minerals. Much of the rain from the Earth's heavily forested areas is created by trees (Shukla and Mintz, 1982). The recycling of water as rain by trees causes repeated flushing of the water through the soil, increasing weathering (Berner, 1998). The Amazon rainforest recycles rain that comes from the Atlantic Ocean numerous times before it returns to the sea via the Amazon River and its tributaries. About half of the Amazon's rainfall is from rainforest trees. However, some rainforests, such as those of the Amazon lowlands, are flat and have been exposed to so much rain for so long that their soils are thoroughly leached and their roots no longer weather bedrock. So, they do not help sequester carbon. The soils of these forests have very little nutrients.

Another mechanism by which plants sequester carbon is that some plants form microscopic silica grains in their leaves and stems called phytoliths. This includes all grasses, and some herbs, shrubs, and trees. A small amount of carbon becomes encapsulated in each grain. This encapsulated carbon is very resistant to decomposition, even if the plant burns or dies. Unlike most plant matter, which readily decomposes in soil and returns CO_2 to the air, the carbon in phytoliths effectively removes CO_2 from the atmosphere for millennia (Parr and Sullivan). The carbon can accumulate in the soil (ibid.). Phytolith carbon is estimated to currently extract 300 million tons of CO_2 per year from the atmosphere, storing it in the soil for thousands of years. Carbon sequestration by phytoliths varies with the plant species and subspecies, type of biological community (forest, grassland, etc.), climate, and soil conditions. Since this process has likely been going on since the origin of higher plants, the quantity of carbon sequestered by this mechanism since it started has probably been considerable.

The grasslands of the world are a major carbon sink. A school of thought that challenges conventional wisdom and cannot be discounted, given the evidence, argues that grasslands have persistently contributed to keeping the planet comfortably cool for life since the end of the Eocene epoch, 34 mya (Retallack, 2001). When grasses die, they decompose, and a large portion of their nutrients are returned to the soil. Carbon from grasses is sequestered and stored underground in this way as organic matter. This organic matter held in the soil is a little-known resource of grasslands. It makes the soil fertile and nutrient-rich, and helps it with water retention. Grasslands thus remove large amounts of CO_2 from the atmosphere, and store and conserve it as organic soil nutrient for plants and soil microorganisms. Grasses also cause weathering, freeing minerals from soil. The minerals combine with carbon, and the carbon combined with minerals is taken to the sea via rivers, and some of this sinks to and is buried in the seafloor, resulting in yet more sequestered carbon. This also supplies nutrients to phytoplankton, making marine ecosystems more productive. And as previously discussed, more phytoplankton means more carbon sequestration.

The mycelium (plural mycelia) is the vegetative part of a fungus that does not make spores for reproduction, and consists of a mass of branching, thread-like, usually white filaments that grow underground. Mycelia of soil fungi make oxalic acid and other acids, which, with the enzymes of the fungi, break down rocks in the first step in soil formation. (Enzymes are proteins that aid chemical reactions) They pull two CO_2 molecules from the air to make oxalic acid, adding calcium, to make calcium oxalate, sequestering CO_2. They may make other chemicals called oxalates that combine with CO_2, keeping it in the soil. Fungi also change the soil solution chemistry to dissolve minerals, causing weathering. Fungi bury a great amount of CO_2 by these mechanisms.

The above list of organisms that sequester carbon is not comprehensive. Skeletons and shells, which have large quantities of carbon, first appeared in the fossil record about 550 mya and have been getting buried with their carbon ever since. Eighteen evolutionary lines of animals have independently evolved exoskeletons with carbon as a major component. All life has carbon in it. Any organism that gets buried on land or in the sea removes carbon from the ecosystem. All groups of organisms sequester carbon because some individuals in each group are buried when they die, and some of the time their feces are buried when they defecate (for those species that defecate).

2.4 LIFE CONSERVES CARBON FOR THE BIOSPHERE, EVEN AS IT REGULATES IT

Life's sequestering of carbon, though beneficial to life, creates a dilemma. It regulates temperature and keeps acidity in the sea and freshwater systems at levels that are beneficial for life. But it depletes the biosphere of carbon, the most important element to life. Carbon is needed by all life. It is in all important biological molecules, including DNA, RNA, proteins, and carbohydrates. It is used by plants and phytoplankton to carry out photosynthesis, which supports most ecosystems on land and in freshwater and the sea. About 99.6% of all carbon on Earth is stored in rock, and only 0.002% of carbon exists in the biosphere (Berner, 1992, and references therein). When carbon is in rocks, it is not accessible to life. The only nonbiological sources of carbon to life are mid-ocean ridges, volcanoes, and other sources of that nature. There is so little CO_2 in the atmosphere-ocean system that if CO_2 were removed even a little faster than it is replaced, all CO_2 in the air could disappear in less than a million years (Berner and Caldeira, 1997)!

But life even has mechanisms to alleviate this problem, to slow down the loss of carbon, to conserve it and keep it available for future use, even as it keeps it out of the atmosphere, regulating Earth's temperature and the acidity of the ocean and freshwater ecosystems. *One of the primary mechanisms of accomplishing this is keeping carbon stored in the system in living organisms themselves.* Of course, the organisms are using the carbon as they do this. Large amounts of carbon are stored in viruses, prokaryotes, phytoplankton, giant fungi, trees, coral reefs, the shells of bivalve molluscs (such as clams), the external skeletons of arthropods, large whales, other large mammals, and other organisms. This keeps the carbon out of the air, regulating temperature, without losing it from the biosphere, keeping it available to life. All living organisms are involved in this solution, some more than others.

Bacteria, archaea, and viruses store great amounts of carbon because of their high numbers. Since there are at least 10 billion microbes and 100 billion viruses in a liter of seawater, tremendous amounts of carbon are kept out of the atmosphere while being conserved for use by the biosphere by these organisms. Viruses are among the world's most important organisms in terms of amount of carbon stored, because of their large numbers. Although viruses are very small, they contain and store a large amount of carbon because of their great numbers. If all the approximately 10^{31} viruses on Earth were laid end to end, they would stretch for 100 million light years. Suttle (2005) estimates that marine viruses collectively contain 200 million grams of carbon. The average marine virus has only 2×10^{-16} grams of carbon, but collectively they lock up the same amount of carbon as in 200 million beluga whales (*Delphinapterus leucas*)! Since viruses are short-lived and turn over quickly, this carbon is relatively quickly available to the biosphere. Thus, viruses are one of the most effective mechanisms for keeping large amounts of carbon out of the atmosphere, while storing it and keeping it available to the biosphere. Prokaryotes on the land, in the sea, in freshwater ecosystems, and underground also store spectacularly large amounts of carbon. *Terrestrial plants store by far the most carbon of any life form.* Trees store tremendous amounts of carbon that is not lost from the ecosystem. They store enormous amounts of it in their trunks, leaves, and roots. Today, the transfer of carbon from the air to the biosphere via photosynthesis, including trees, other land plants, and phytoplankton, is about 1.2×10^{17} grams of carbon per year, which is a great deal of carbon removed from the air in a small amount of time (Frings, 2019). The amount that terrestrial plants contribute to carbon storage and temperature regulation while conserving carbon, and hence the aid they give to the biosphere and biodiversity, is astounding. The biomass of the biosphere is dominated by land plants. *An estimated 99.9% of the global standing crop, the total quantity of biomass in the biosphere, is classified as vegetation* (Lieth and Whittaker, 1975). Marine photosynthesizers are only 0.2% of the global standing crop (Woodwell et al., 1978). Thus, *terrestrial vegetation constitutes almost all of the biomass on Earth.* Plant biomass is mainly cell wall materials (Duchesne, 1989). Between 40 and 60% of the total mass of plant cell walls is cellulose (ibid.), which is rich in carbon. Therefore, terrestrial plants store almost all of the biosphere's carbon, most of it in their cell walls, and about half of this is cellulose. This is a vast amount of carbon being kept out of the atmosphere and water, regulating temperature and aquatic acidity. All of this carbon is kept available for use by the biosphere. Forests of the coast redwood (*Sequoia sempervirans*), the tallest tree species on Earth, store more carbon than any other forest on Earth. Old-growth coast redwood forests produce more aboveground biomass annually, storing more carbon, than forests dominated by any other species (Sillett et al., 2020).

A surprising 50% of carbohydrate production from photosynthesis by plants is stored underground, where carbon is stored and used in the root system, nourishes the symbiotic root fungi, and feeds soil organisms. Typically, 10% of the remaining 50% that is aboveground is dropped to the soil as leaves. This leaf litter is decomposed by fungi and bacteria, contributing nutrients to soil life, while returning nutrients to the plants, and keeping some carbon out of the air. Rainforests absorbed great quantities of carbon before humans destroyed large portions of them. Before these human impacts, the Amazonian rainforest stored an estimated 1.5 billion tons of CO_2 per year, cooling the Earth considerably, while keeping the carbon available to the biosphere. Marine photosynthesizers, a good percentage of which are phytoplankton, make up 32% of the total annual primary production, even though only 0.2% of the standing crop (ibid.). Thus, they sequester a great deal of carbon, but do not store very much of it.

Bello et al. (2015) found that the loss of large animals that eat the fruit and disperse the seeds of trees in Atlantic tropical forests led to a great loss of the capacity for carbon storage. This demonstrates that large seed-dispersing animals are very important in keeping carbon out of the atmosphere while ensuring that it is stored, conserved, and available to the ecosystem. This is because they help trees by dispersing their seeds. They eat the fruit, then move far away from the tree, and defecate or drop the seeds a great distance from the tree. Trees that grow far from their parent tree survive much better than those close to their parent tree because they are less subject to herbivores and disease. Large animals that disperse plant seeds are thus important ecosystem engineers whose actions remove greenhouse gases from the atmosphere.

Forests cool the Earth by other means than sequestering CO_2 (Lawrence et al., 2022). Tropical forests cool Earth by about 1.5°C (2.7°F.). About two-thirds of this is from the capture and storage of atmospheric carbon. The other one-third is the following three factors, in order of decreasing importance: the uneven canopy of the forest has a cooling effect because it provides an undulating surface that can bump overpassing fronts of air up and away from the forest; the release of water vapor by trees via evapotranspiration; and the release of aerosols that lower temperatures by reflecting sunlight and seeding clouds. All these mechanisms conserve carbon. Boreal forests in Alaska, Canada, Scandinavia, and Russia increase temperature because they temper the reflectivity of the snow, which would decrease temperature. Clearing these forests exposes more snow cover in the winter, decreasing temperature because the white snow reflects heat. But considering the collective effects of the Earth's forests, their net effect is to lower the average global temperature by about 0.5°C.

Marine sediments at the bottom of the ocean are important carbon stores that can hold carbon for millennia. Much of the carbon is delivered to them by marine snow and the sinking of dead organisms and feces. Bottom trawling (fishing by dragging heavy nets across the seafloor) can release tremendous quantities of carbon into the ocean.

Another mechanism of keeping carbon out of the air while conserving it concerns the fact that bacteriophages recycle the carbon in deep-sea sediments, making it available to the marine portion of the biosphere (Danovaro et al., 2005). Phages infect, burst open, and kill huge numbers of prokaryotic cells in the ocean depths, liberating and recycling nutrients, and making them available to other prokaryotes. This is vital to the health and diversity of life in the deep sea (ibid.). The researchers also found that a great portion of deep-sea nutrients liberated by the viruses are carried by animals from the ocean depths to the shallower sea, and contribute substantially to the ecosystems in the photic zone. The photic zone sends nutrients, including carbon, to the deep sea as marine snow and from the sinking of carcasses and feces. Thus, the photic zone near the sea surface and the deep sea exchange nutrients, in a mutually beneficial recycling system. These two spatially separated ecosystems are interrelated and aid each other. Viral attack of prokaryotes in the deep ocean closes the loop in the carbon cycle, helping to recycle carbon back to the photosynthetic zone in the shallow sea. This is of crucial importance, because it keeps a substantial amount of carbon from being buried under the seafloor, keeping it available to the oceanic life, without adding it to the atmosphere. However, it does increase carbon in the sea, so the acidity of the ocean increases. But under natural circumstances, this is not enough of an increase in acidity to be a problem.

SAR11 is a bacterium with the smallest known genetic structure of any independent cell. It is the most abundant nonviral organism in the sea, making its effect tremendous. Globally, there are 2.4×10^{28} SAR11 cells in the seas, half of which are located in the photic zone. A milliliter of seawater has 500,000 SAR11 cells. Their combined weight exceeds that of all the fish in the sea. On average, the SAR11 clade makes up a third of all the cells present in surface waters and nearly a fifth of the cells in the area between about 200 and 1,000 meters (about 656 and 3,281 feet) deep. In some regions, members of the SAR11 clade make up 50% of the total surface microbial community (Morris et al., 2002)! Thus, they store huge amounts of carbon, keeping it out of the air and sea, but available to the biosphere. Recycling organic matter, when they die and decompose, they provide nutrients needed by phytoplankton. SAR11 is very important in the Earth's carbon cycle. But it produces CO_2 and H_2O as end products of its metabolism. And when starved for phosphorus, SAR11 generates methane by breaking down a chemical called methylphosphonic acid to obtain phosphorus (Carini et al., 2014). The amount of methane produced is not known. Pelagiphages are marine bacteriophages that greatly slow the addition of these two greenhouse gases to the air by killing millions of SAR11 cells per second. This also adds crucial nutrients for phytoplankton growth. The result is more phytoplankton, hence more phytoplankton that die, sink, and get buried in the seafloor. This sequesters a great carbon of carbon. It is likely the SAR11 and phage taken together remove more carbon than they add, but more study is needed to know this for certain.

Soil stores great amounts of carbon. Plant roots, soil microorganisms, and soil animals store large amounts of carbon in soil for varying time periods, and the amount of carbon and length of time it remains in the soil varies with climate, vegetation, soil texture, and how easily water drains from the soil. Carbon is the main element in soil organic matter. Carbon can be released into the atmosphere in days, or remain in the soil for millennia. Burial and decomposition of plants and animals, animal feces, and falling leaves add carbon to the soil. Most of the carbon stored in the soil is soil organic matter, a complex mixture of carbon compounds, consisting of decomposing plant and animal tissue, roundworms, fungi, protozoa (one-celled organisms, which include the *Euglena*, *Paramecium*, and some amoebas), bacteria, archaea, and viruses, as well as carbon associated with soil minerals. All of these organisms and minerals contribute to carbon sequestration and storage in soil. Of course, the ultimate source of most soil carbon is photosynthesis by plants. Many of the soil organisms benefit each other in a symbiotic web. Plants exude carbon they do not require for growth through their roots, feeding soil fungi, prokaryotes, and other soil organisms, which render the carbon stable, keeping it in the soil, and creating humus, the organic component of soil, itself a carbon and energy source for soil microbes and plants. Most of the soil organisms are symbiotic with plants, since they help plants obtain nutrients.

Some of the carbon used by plants is not sequestered or stored in the soil, but returned to the atmosphere. Carbon also helps give soil its water-retention capacity, structure, and fertility. There is a "symbiotic feedback" whereby the species of the soil community benefit the soil, which in turn benefits them.

Plants and soil organisms store carbon in the soil that can be used later by the ecosystem. Soil bacteria and archaea, and soil itself, are major carbon reservoirs. Bacteria play a huge but poorly understood role in the carbon cycle. The planet's soils contain more than 2.5 trillion tons of carbon, four times as much as aboveground in all organisms and the atmosphere combined (Hathaway, 2001), much of it stored in soil prokaryotes. Photosynthetic bacteria store a good deal of carbon. There is no doubt that soil microbes are a reservoir that stores carbon, keeping it out of the atmosphere and conserving it. Yet much is not known about their net effect on the carbon cycle and climate. It is crucial that we study this important carbon sink and storage bank further. There are a number of other ways that carbon can be stored in the ground by life while still being kept available to life, including organic matter recycling; the formation of humus; plant sugars and substances secreted from roots; previously discussed phytoliths; and carbonates.

Special soil habitats also store and conserve carbon. Organisms store a considerable quantity of carbon by creating peatlands, which cover only 3 to 5% of Earth's land surface, but store a quarter of all soil carbon. And when grasslands bury carbon, it is removed from the atmosphere, but is still available for later use. Grasslands store a great quantity of carbon in the roots of grasses and in soil microbes. And when grasslands cool the planet because of their relatively high reflectivity, they do so without burying carbon and removing it from the biosphere—another mechanism by which life creates temperatures favorable to life while conserving carbon.

There is evidence that coastal habitats, such as mangroves, sea grasses, and salt marshes, store 10 times more carbon in their soils per hectare than temperate forests and 50 times more than tropical forests (Pidgeon, 2009). The long-term carbon storage capacity of 1 square kilometer (0.39 square mile) of mangrove forest is equal to 50 square kilometers (19.3 square miles) of tropical forest. Mangroves, sea grasses, and salt marshes are very efficient at burying carbon in sediment, where it can stay for centuries or even millennia. Most of the carbon stays in the soil rather than the plant, so little is released when the plant dies. (Rainforests store much of their carbon in the trees. Of course, much of this carbon is recycled into organisms and not sent into the air when the trees die.) Seagrass beds are among the most efficient carbon sinks. Ricart et al. (2015) found that seagrass can absorb carbon to the point that it can reduce ocean acidification at local scales by sequestering carbon. Over six years, they found that areas with seagrass meadows along more than 600 miles of California coastline were on average less acidic with about 20% more carbon stocks than patches at seagrass-sand edges and bare sediments, even at night, when the plants are not performing photosynthesis. Coastal vegetation accounts for about half of the total carbon sequestration in sea sediments even though it accounts for less than 2% of the ocean surface, largely because it usually grows deeper below ground or seafloor than it grows above, with some plants going as deep as eight meters (about 26.2 feet). The carbon is stored mainly in the root. Mangroves, sea grasses, and salt marshes also purify water, provide habitat for fish and invertebrates to breed, and protect coral reef and coastal habitats from storms.

Another mechanism of carbon conservation is by decreasing atmospheric methane. Recall that methane is 30 times as powerful at heating Earth as CO_2. Some methane is produced by nonbiological processes, but admittedly most is made by life. Biological production on land is by methanogens, mainly in wetlands, and by microbes in the digestive tracts of termites and ruminants, which release some of it into the atmosphere. (Ruminants are mammals that include cattle, sheep, antelopes, deer, and giraffes.) In the ocean, methane is produced biologically when microorganisms break down organic matter that settles to the seafloor, including dead fish, invertebrates, and bacteria. However, life strongly limits methane production and accumulation. The oxygen in the atmosphere that was produced by photosynthesis reacted in the past, as it does today, with methane to produce CO_2 and water vapor, removing methane from the air. Water vapor is a greenhouse gas, but like CO_2, a much less potent one than methane. The net result of this reaction is a lowering of atmospheric temperature. Since one molecule of methane is lost per molecule of CO_2 gained, there is no net loss of carbon. Today, each year, O_2 from photosynthesis removes much of the between 531 and 792 million tons of methane produced by methanogens, without losing any carbon from the system.

Methanogens, the archaea that produce methane, can only grow in soil in the absence of oxygen. They cannot grow in soil that is well aerated. Oxygen made by photosynthesis limits methanogens. Plant roots and their symbiotic fungi break up and aerate soil. Burrowing animals aerate soil. These include gophers, moles, prairie dogs, ground hogs, badgers, marmots, mice, desert and gopher tortoises, earthworms, and several other animal species. Thus, methanogen populations are kept low in soil by photosynthesizing plants that produce oxygen and organisms that aerate the soil with oxygen.

In marine sediments, methanogen numbers are kept low by other microbes that use sulfate, not oxygen, as an energy source. These microbes outcompete methanogens for nutrients. Marine sediments are sulfate-rich. Thus, methane from methanogens accumulates in the ocean for the most part in the rare areas where sulfate is depleted and oxygen is very low or absent. Dissolved oxygen in the sea from photosynthesis also limits methanogens.

Bacteria of the genus *Beggiatoa* grow near deep-sea hydrothermal vents, in cold seeps in the sea, in sulfur springs, in mud layers of lakes, and in the root area of swamp plants. Some species oxidize hydrogen sulfide to elemental sulfur for energy, reducing the amount of this toxic gas. Some form colonies, or mats, that support microbes that consume methane. Working together, the species of these communities act as biological filters over methane seeps, limiting the amount of methane that enters the water column and the atmosphere.

In addition to this, methanotrophs are prokaryotes that consume methane as their source of carbon to unlock the energy in oxidized chemicals. Some methanotrophs are bacteria, some are archaea. They play a major role in the reduction of methane released into the air from bogs and swamps, where methane production is relatively high.

In some areas in the deep sea, a great amount of methane is released from fossilized reservoirs. Spectacularly, an estimated 90% of this methane is converted to other compounds by groups of various species of methane-oxidizing archaea and sulfate-reducing bacteria, mainly in marine sediments that have no oxygen. Here, methanotrophs that are archaea that do grow in environments that lack oxygen, unrelated to the methanotrophs that require oxygen, consume 90% of the methane formed, as their food source, preventing its build-up. This produces toxic hydrogen sulfide. But this does not persist. Filamentous sulfur bacteria consume it as a food. So a group of methane-consuming archaea and sulfur bacteria work together, reducing the amount of methane released into the air, without increasing toxic hydrogen sulfide. They can form small aggregates or even voluminous mats. And there is another microbe in the archaea that can use nitrate to consume methane (Haroon et al., 2013).

There are many species of dung beetles (superfamily Scarabaeoidea); Australia alone has 400 species. They decompose dung by eating it. Dung is a source of methane. A study showed that six-day-old cow pats (feces) without dung beetles produced five times higher methane emissions per day than those with the beetles. Dung beetles tunnel through dung, allowing oxygen to enter it. This kills the methanogens because, recall, they cannot grow in the presence of oxygen (Penttilä et al, 2013). And of course, the oxygen combines with the methane, reducing the amount of it. Reducing methane and the methanogens that produce it controls temperature while conserving carbon. The beetles also disperse grass seeds, and fertilize and aerate the soil, and make channels that allow water to soak into it more easily. And soil aeration, just as aeration of dung does, reduces the amount of methane by destroying it and preventing the growth of methanogens. Their tunneling in the soil also gives plant roots more room to grow. Without dung beetles, animal feces would reduce grazing land, and would provide ideal breeding habitat for disease-spreading flies and parasites, including worms and bacteria. These could greatly reduce animal populations, potentially decreasing diversity. Hardened cow pats can last years. Dung that is not decomposed can cover grasslands or pastures, and convert them from carbon sinks to carbon emitters. So dung beetles greatly reduce atmospheric carbon while conserving it.

Also, since methane depletes oxygen, the reduction of the release of methane into the air by life helps maintain higher atmospheric oxygen levels.

Methane is also kept out of the atmosphere nonbiologically by the formation of methane hydrates at low temperatures and high pressures in the seafloor. These are frozen forms of methane, most of which was generated by methanogens. They are also found frozen in Antarctic ice cores. If they are heated too much, the methane can escape and warm the Earth, potentially to temperatures catastrophic for biodiversity. If organisms did not sequester carbon through the ages, the Earth would be so hot that the methane hydrates would melt, releasing huge quantities of methane into the air, making Earth too hot for most life, in a positive feedback loop. Methane hydrates are also in continental rocks in beds of sandstone or siltstone at depths of less than 800 meters (about 2624.7 feet), in Alaska, Siberia, and Northern Canada, where they are formed from a mix of thermally and microbially derived gas.

Life evolved another way to conserve carbon while regulating temperature, which is the evolution of more efficient photosynthesis. Photosynthesis converts CO_2 into carbohydrate through a pathway with several chemical intermediates along the way, catalyzed by the enzyme Rubisco (ribulose-1,5-bisphosphate carboxylase/oxygenase). In most photosynthetic organisms, the first compound created from CO_2 in the photosynthetic pathway has a backbone of three carbon atoms, so this is called C3 photosynthesis, and the plants are called C3 plants. But certain plants, such as tropical grasses, create a molecule with a backbone of four carbon atoms as the first chemical created from CO_2 in the pathway of photosynthesis. This process is called C4 photosynthesis, and the plants are called C4 plants. C4 plants evolved a revolutionary upgrade of photosynthesis well after photosynthesis originally evolved. The upgrade comes in the form of a solar-powered CO_2 pump that increases the concentration of CO_2 around Rubisco. The result is a concentration of CO_2 in the plant's cells that is 10 times that in the air, allowing Rubisco in C4 plants to convert CO_2 to carbohydrates with an incredible efficiency, much better than in C3 plants. Tropical grasses evolved this much more efficient photosynthetic system about 30 mya, although there is a good deal of uncertainty about this date. C4 grasses expanded tremendously throughout most of the world 6 to 8 mya, with C4 grasslands and savannas replacing forests in many areas. C4 photosynthesis thrived and spread, and today there are about 7,500 species of C4 plants, occupying one-fifth of the vegetated land surface of Earth, and accounting for 30% of the primary productivity of the terrestrial biosphere (Lloyd and Farquhar, 1994), although they are only 3% of plant species. Most C4 plants are confined to tropical and subtropical climates, because they function best in bright sunlight at high temperatures. By far the majority are subtropical grasses. They dominate subtropical grasslands and savannas. *The ascendency of C4 plants is one of the most profound transitions of global ecosystems in the Earth's history. The evolution and spread of a photosynthetic system that uses carbon more efficiently conserves carbon and prolongs the time that there will be enough carbon for photosynthesis and life on Earth's land, in the sea, and freshwater to continue.*

Studies show that the evolution of C4 plants was facilitated, even largely caused, by life (Beerling, 2007). Millions of years before C4 plants appeared, life—mainly large land plants—caused CO_2 levels to fall precipitously. Many were buried, and their roots and their symbiotic fungi caused weathering. Both of these occurences caused CO_2 to drop to extremely low levels. The resultant low levels of CO_2 caused natural selection to favor plants that used CO_2 efficiently when carrying out photosynthesis. Hence, some plants evolved the more efficient C4 system of photosynthesis. It took C4 plants millions of years to evolve after low atmospheric CO_2 was the norm. It is unclear why.

Finally, Earth's atmosphere is about 78% nitrogen, 21% oxygen, and the rest CO_2 and trace gases. The high concentration of nitrogen causes nitrogen to collide often with CO_2 molecules (Li et al.,2009). This broadens the infrared (heat) absorption band of the CO_2 molecule via an effect of quantum physics, increasing the strength of CO_2's greenhouse effect. Thus, a high concentration of nitrogen in the atmosphere increases the greenhouse warming effect of CO_2. Nitrogen-fixing bacteria remove nitrogen from the air, while denitrifying bacteria add it back. This has no long-term net effect on nitrogen in the air, because it is being added and removed about equally. But the actions of some bacteria add nitrogen to the sea. Some of this nitrogen makes it to the seafloor. Some types of bacteria help bury nitrogen in sediments in the sea bottom, and this becomes hard rock over a long time period. This is the major mechanism of removing and sequestering nitrogen from the system. Nitrogen is removed from the system whenever an organism is buried, since all organisms have nitrogen in their biomolecules. However, bacteria are likely the organisms that bury the most nitrogen, although this is not known for sure. We do know that a significant amount of nitrogen is sequestered by life. This weakens the greenhouse effect of CO_2 and cools Earth without losing CO_2 from the system.

Li et al. (2009) state that, by this process, life can extend the time of oxygen-generating photosynthesis and hence higher life and high biodiversity at least 2.3 billion years into the future, more than doubling previous estimates. Thus, it appears that life regulated temperature and cooled the planet as solar output increased throughout life's history by burying nitrogen as well as carbon. Also, if life did not bury nitrogen throughout its history, there would be an excess of it. This would cause the problems humans cause today when they add too much nitrogen to ecosystems. This includes phytoplankton and bacteria growing in excessively large numbers due to the abundance of nitrogen, which is a nutrient they need. When they die, other bacteria decompose them, and this requires tremendous amounts of oxygen. Thus, oxygen gets depleted, and animals from fish to crabs die off in great numbers. This process, called *eutrophication*, can occur in both freshwater and marine ecosystems. There are dead zones with essentially no higher life in several areas in the seas as a result of this today. These dead zones were caused by humans adding too much nitrogen to the oceans. A well-known dead zone in the Gulf of Mexico results from run-off of agricultural and other wastes and delivery of nitrogen loads into the Gulf of Mexico. Freshwater ecosystems also have die-offs from eutrophication due to excess nitrogen as a result of human activity.

This conservation of carbon by life ensures that life's beneficial depletion of carbon will not cause a catastrophe. Life is sequestering carbon and hence keeping the temperature and acidity in aquatic ecosystems favorably low for itself, but keeping the rate of carbon loss from the biosphere so slow that nonbiological factors will kill off life before life's burial of carbon removes so much carbon from the biosphere that life is significantly harmed. The catastrophe of the distant future will be caused by a nonbiological force, the increased heat output of the sun, not life removing carbon from the biosphere. As the sun continues to increase its heat output, the weathering rate of silicate minerals will increase, resulting in a decrease in the atmospheric CO_2 necessary for photosynthesis. As the increased heat causes water to evaporate from the Earth's surface, rocks will harden, causing plate tectonics to slow and eventually stop. This will cause all volcanoes to become inactive and stop adding carbon to the atmosphere and minerals to the soil. In about 600 million years, the level of CO_2 will be too low for C3 plants to photosynthesize. Plants that use C4 photosynthesis can survive at CO_2 concentrations as low as 10 parts per million, but in time, CO_2 will be too low for them to carry out photosynthesis. All life on Earth's surface and in fresh or salt water will suffer the same fate, for all life needs carbon. Thus, the sun's increased heat output will cause the biosphere to have too little carbon for any life to survive except possibly that deep underground by shortly after 600 million years from now, a time period about 1.7 billion years less than the 2.3 billion years that biology is extending the time life will take to significantly deplete carbon.

2.5 LIFE LIKELY REGULATES TEMPERATURE WITH NEGATIVE FEEDBACK AND DIMETHYL SULFIDE

There is another important homeostatic, negative feedback process that regulates temperature, also driven by biology. This one is independent of CO_2 and methane. The CLAW Hypothesis, which takes its acronym from the first letter of the surnames of its four originators (Charlson, Lovelock, Andreae and Warren), proposes that particular phytoplankton that produce the chemical called dimethyl sulfide [DMS; chemical formula CH_3-S-CH_3] are responsive to variations in temperature, and that these responses lead to a negative feedback loop that acts to stabilize the temperature of the Earth's atmosphere (Charlson et al., 1987). Here is how the negative feedback loop works. If the sun increases its output, phytoplankton increase their growth rates, by a physiological response to higher temperature, and by increasing their photosynthetic rate due to enhanced solar radiance. Thus, there will be more phytoplankton. Some phytoplankton make a chemical called dimethylsulfoniopropionate [DMSP; chemical formula $(CH_3)_2S^+CH_2CH_2COO^-$]. Their increased growth increases their DMSP production. When these phytoplankton are damaged or killed, for example by grazing zooplankton, a virus bursting them open, or turbulence, their DMSP is released. Then bacteria and other phytoplankton convert the DMSP to DMS, a gas essential to Earth's biogeochemical cycles that easily leaves the ocean's waters. Some of it goes into the air. DMS is also released from some phytoplankton when grazed by herbivores, such as tiny fish, copepods, invertebrate larvae, and small protozoa called ciliates. DMS represents 95% of the sulfur that travels from the sea to the air, and supplies about 50% of global sulfur to the air from life. DMS is oxidized with the sun's aid to sulfur dioxide in the atmosphere, leading to the synthesis of sulfate aerosols, which attract molecules of water to form water droplets that make up clouds. This increases cloud droplet number, elevating cloud water content and cloud area. The effect of clouds on weather depends on the cloud type. Some clouds reflect heat exiting the Earth back to the ground, enhancing the greenhouse effect and increasing air temperature. However, the clouds produced by these sulfate aerosols increase cloud reflectivity, so the clouds reflect incoming sunlight, and thus lower the temperature, which is what most clouds do. About a third of solar radiation that reaches Earth's atmosphere is radiated back by clouds and surfaces like snow and ice.

On the other hand, less solar input and heat from the sun leads to less DMS, less cloud cover, and an increase in atmospheric temperature. The CLAW hypothesis states that DMS is a mechanism by which life regulates and stabilizes climate significantly by negative feedback, on a large scale. Therefore, it is compatible and supportive of both the Gaia Hypothesis and the ABH. The effect is significant, since marine phytoplankton that synthesize DMSP cover a large area of the sea at any given time. Some seaweeds, including some red and some green algal species, also produce DMSP, and have the enzyme to convert it to DMS. How much they affect cloud formation is not known.

There is empirical evidence supporting the CLAW Hypothesis (Andreae et al., 1995; Cropp et al., 2005), although it is equivocal (Vallina et al., 2007), and some have suggested that CLAW-like negative feedback may operate in the planet's sulfur cycle without an active biological component (Shaw et al., 1998).

It is likely that the negative feedback of the CLAW Hypothesis can be thrown into positive feedback with a sufficiently strong disruption, such as tremendous volcanism that emits great quantities of CO_2. In fact, Lovelock (2007) suggested that if humans throw the system beyond a tipping point, positive feedback can replace negative feedback, destabilizing the system. In his so-called anti-CLAW Hypothesis, human-induced global greenhouse warming could heat the oceans to the point that nutrients no longer get delivered from the deep sea to the shallow sea. This would mean phytoplankton would not receive much nutrient. This will make the number of phytoplankton and their activity decrease, so phytoplankton will produce less DMS. Thus, there will be fewer particles that cause cloud formation, and so a decrease in cloud cover. This will cause a lower reflectivity in the sky, leading to a further increase in atmospheric temperature, leading to fewer nutrients ascending to the photic zone, so fewer phytoplankton, and thus even less DMS production, and hence further warming of the atmosphere, and so on, in a positive feedback loop. Studies that simulate future oceanic plankton growth have shown evidence supporting this idea (Cox et al., 2000). Volcanism emitting tremendous amounts of CO_2 could cause the same destabilizing positive feedback by the same mechanism of shutting down the flow of nutrients from the deep to shallow sea.

Phytoplankton release DMS when they are eaten by krill, which are small, tremendously numerous crustaceans that feed on phytoplankton as one of their primary food sources. Savoca and Nevitt (2014) showed this is beneficial to the phytoplankton because far-ranging seabirds such as petrels and albatross use the smell of DMS as a cue to locate the prey they eat, which includes krill. The sea is so uniform that it is hard for seabirds to locate their prey by sight. So the phytoplankton attract predators of the herbivores that eat them. These seabirds have excellent senses of smell. Seabirds also fertilize the sea, aiding plankton with iron from their feces, in this symbiosis between phytoplankton and seabirds. In the vast Southern Ocean, iron is a limiting resource that phytoplankton need. Thus, these predatory seabirds are tied to both global climate regulation and the health of the ocean ecosystem through a symbiotic relationship with phytoplankton. This symbiosis of phytoplankton with seabirds has also been shown with various other animals that use the smell of DMS as a cue to find krill to eat, including penguins, seals, sharks, sea turtles, coral reef fishes, and possibly whales. Plastic that has been soaking in the sea emits DMS. This attracts some birds, some fish, and loggerhead sea turtles (*Caretta caretta*) (Pfaller et al., 2020). All of these animals then mistakenly eat the plastic and are harmed by it.

Viruses are also involved in this system, since DMSO is released when they burst open and kill the cells of one-celled organisms (Fuhrman, 1999). Thus, the DMS negative feedback loops that act as thermostats and regulate and stabilize the atmosphere's temperature involve viruses, phytoplankton, krill, red and green algae, fish, sea turtles, sea birds, penguins, and marine mammals. All of these organisms are thus ecosystem engineers that help regulate temperature.

In reviewing aspects of the influence of phytoplankton on climate, Watson and Liss (1998) speculated that the overall effect of phytoplankton today is to cool the planet by about 6°C (about 10.8°F), with one-third of this figure being due to carbon burial and two-thirds due to cooling by DMS.

Regulating temperature by negative feedback using DMS is another way life controls temperature while conserving carbon in the biosphere.

In addition, a key process in the entire planetary sulfur cycle is the transfer of sulfur compounds from the ocean to the land via the atmosphere, and DMS is the dominant marine sulfur compound generated by life in this process. DMS is therefore essential to the global sulfur cycle and in transferring sulfur from the sea to the air and from there to the soil. DMS is transferred from the air to the soil by rain, and is easily incorporated into soil, elevating its sulfur content. So biology helps add sulfur to the soil from the sea. This is beneficial to soil life, and the process does not tend to add too much of this nutrient. Since sulfur compounds significantly alter temperatures, cloud formation, and hence rainfall, they also significantly influence the global water cycle. So the sulfur and water cycles influence each other via biology, and are both of crucial importance to life and profoundly influenced by it in feedback loops wherein life helps life.

2.6 COCCOLITHOPHORES MAY REGULATE TEMPERATURE THROUGH NEGATIVE FEEDBACK

In another mechanism, phytoplankton, mainly coccolithophores, release particles that act as seeds for cloud formation. Phytoplankton blooms in the Southern Ocean increase the number of cloud-forming droplets over the sea by about 60% annually. Cocclithophores shed some of their coccoliths (the tiny plates that make up their shells), and these form calcium nitrate particles in the air, which are magnets for water particles, aiding cloud formation. Coccolithophores infected with viruses shed three times as many coccoliths as uninfected ones, and ocean spray above virally infected phytoplankton has about 10 times as many coccoliths as that above uninfected plankton. But the situation is complicated. Coccoliths also hinder cloud formation by removing other potential cloud-forming particles from the air. The net effect of phytoplankton and their viruses seems to be to enhance the formation of clouds that reflect light, hence lowering temperature, but this is not certain. They clearly have a profound effect on cloud formation. Further study is needed to determine whether they are involved in negative feedback similar to DMS. The negative feedback loop would be: more heat leads to more coccolithophores because they grow faster in warm water, and this means more coccoliths are sent into the air, which produces more clouds of the heat-reflecting type, which lowers the temperature. It would be another way life regulates temperature without removing carbon from the biosphere.

2.7 ORGANISMS REGULATE SILICON, HELPING LIFE

In addition to regulating carbon, organisms regulate silicon, which has similar chemical properties to carbon. Silicon is abundant in the ocean, with 80% coming from rivers. It originates from weathering, volcanoes under the sea, and sea sediments. All naturally occurring types of silicon are nontoxic and it is even required by life. But high concentrations of soluble silicon compounds in water may impede the use of the very important element, phosphorus, by organisms, and limit algal growth. Some of it had to be removed to bring its levels closer to what is optimal for life. Organisms played and still play a large role in this. Plankton ingest it and use it for various processes. This is a major mechanism of removal, with diatoms removing over 6.7 billion metric tons of silicon each year from the ocean (Treguer, 1995). Diatoms have cell walls made of silica, which is silicon dioxide combined with water molecules. Freshwater diatoms also remove silicon from rivers (Roubiex et al., 2008). Sponges remove large amounts for their skeletons. Various plant species remove it. Plants contain about 200–62,000 parts per million by dry weight of silicon. Dandelions and bamboo contain silicon in their stems and leaves, increasing their stability. In all these groups of organisms, silicon is sequestered by burial upon death, the same as the major mechanism by which organisms sequester carbon. Yet, also, by storing silicon in their cell walls, skeletons, stems, and leaves, organisms keep a large quantity of silicon available to the biosphere, while preventing it from doing harm as a result of being too abundant. Silicon is also removed nonbiologically by reactions of dissolved silicon with clay minerals and by its settling into sediments such as the seafloor and muddy lake bottoms.

2.8 BIOLOGY MIGHT AID LIFE BY REGULATING THE OCEAN'S SALT CONTENT

Organisms might also contribute to regulating the salinity of the ocean, which has been constant at about 3.5% for a very long time (Segar, 2012). This is important for sea life because the cells of most organisms need a stable salinity, and cannot survive at levels above 5%. Sea organisms have adaptations to regulate the amount of salt they take in from seawater, but this ability is not unlimited and takes a good deal of energy. Salinity that is too high is detrimental to life. Salts from the weathering, erosion, and dissolution of rocks have been transported by rivers and deposited into the oceans through the ages, and this would increase the salinity in the sea if there were no counterbalancing force, and none is known. Salinity in the sea might be greatly influenced by seawater circulation through hot volcanic rocks at the bottom of the ocean, and its emergence from hot water vents on mid-ocean ridges. However, this might not sufficiently regulate the sea's salt concentration, and seawater's salinity is far from what it would be if no outside forces were affecting it, suggesting life may have made it favorable to life. Segar (2012) hypothesized that bacterial colonies reduce the sea's salinity by removing salt from seawater and forming salt plains on the seafloor. If so, bacteria are key to making the sea's salinity favorable to life, keeping seawater from being too salty for organisms to thrive. But more research needs to be done before we can reach any definitive conclusions.

2.9 VENUS AND HUMAN IMPACTS SHOW THAT ORGANISMS HAVE PROFOUNDLY HELPED LIFE BY REGULATING ATMOSPHERIC GREENHOUSE GASES AND TEMPERATURE

A null hypothesis is one that is the alternative to a proposed hypothesis. The null hypothesis with respect to the ABH (Pachamama Hypothesis) is that life does nothing to increase biodiversity, help life, or make Earth better for life. The planet Venus serves as a good control to test the ABH against the null hypothesis, showing what happens when there is no life to sequester carbon from the atmosphere. Venus is closer to the sun than is Earth, but this is a small percentage of the cause of Venus being well over 800°F (about 427°C) hotter than Earth.

Outgassing of CO_2 and other gases from more than 1,000 volcanoes or volcanic centers larger than 12 miles in diameter on its surface has caused surface temperatures to reach 900°F (about 482°C). There is no life to sequester the carbon, and hence nothing to prevent a runaway greenhouse effect. On Earth, without the sequestering of carbon by life, there would be about 1,000 parts per million of CO_2 in the atmosphere. The Earth would be torrid, and inhospitable to all but possibly some simple forms underground and deep in the sea.

One part of the null hypothesis is that if life did not sequester greenhouse gases such as CO_2 and methane, their higher levels would make no difference in diversity. Interestingly, but unfortunately, the effects of increased atmospheric CO_2 and methane are currently being tested in an unintended experiment by the addition of these and other greenhouse gases to the atmosphere by humans. Some of the results of this test of the ABH vs. the null hypothesis—of the effects of this increased carbon on biodiversity—are discussed in the following. Preindustrial levels of atmospheric CO_2 were about 260 to 270 parts per million (Wigley, 1998). In early 2015, the seasonally adjusted concentration was about 400 parts per million, with a recent growth rate of between two and four parts per million per year.

The purpose of the following discussion is not to convince the reader of the seriousness of human-induced climate change, but to explore it as a test of the ABH against the null hypothesis. Hence, the discussion will focus on that. This chapter will discuss only effects on biodiversity, since effects exclusively on humanity are not relevant to this test of the ABH.

Humans have warmed the Earth by an average of 1.1°C (2°F) since the Industrial Revolution.

The hotter water is, the less oxygen it can hold, and a rapid increase in sea temperatures can create areas in the ocean that have very low oxygen content, which occurred in some areas during the last glacial melting, 18,000 to 11,000 years ago (Moffitt et al., 2015). This harms sea life. So the carbon and oxygen cycles are linked by temperature, and life helped maintain healthy oxygen levels in the ocean and freshwater ecosystems by sequestering carbon. Some rivers have lost fish that need high oxygen because of their lower oxygen content due to the fact that they are hotter because of human-induced climate change. Salmon and trout need water with high oxygen content, and have been impacted considerably in some areas. Some invertebrate species have also been negatively impacted.

Ocean currents are favorable to life partly because of carbon sequestration by life, although the position of continents is also important. If organisms did not sequester the amount of carbon they did, the Greenland ice sheet would have melted. This would have caused the ocean current system to largely collapse. The Gulf Stream would have ceased. The increased temperatures due to human-induced climate change are melting Greenland's ice sheet, adding freshwater to the sea near Greenland. This lowers the salt concentration of the water that normally sinks off Greenland. The sinking of this water powers the Gulf Stream. With a lower salt concentration, the water does not sink. There is danger the Gulf Stream will cease if this melting of the ice sheet continues. The Gulf Stream warms Europe, preventing it from being very cold. Rome is at about Chicago's latitude, but has a Mediterranean climate because of the Gulf Stream. Biodiversity in Europe would decrease tremendously if temperatures there went down a great amount because of the loss of the Gulf Stream.

In fact, all ocean currents would be weakened or cease if the melting of ice at Greenland and Antarctica continues for too long as a result of human-induced warming. The ocean currents keep the temperature of the ocean evenly distributed. If the currents were to slow considerably or shut down, there would be extreme temperature differences between different parts of the ocean, causing many marine species of all types to decrease or die off. This would also cause climate effects on land, negatively impacting land species. And the currents aid the transport of many species, including sea turtles, fish and their larvae, invertebrate larvae, jellyfish, and many other groups. The weakening or loss of the sea currents would cause the reduction and extinction of many of these species. Nonbiological factors also make the currents favorable to life today. This is discussed in Chapter 13.

Climate change is causing animals that live in mountains to move to higher altitudes, and others to move toward the poles. About 80% of the species studied Root (2003) have migrated to a higher elevation or toward the poles. In the ocean, animal species have traveled an average of 59 kilometers (37 miles) toward the poles per decade. This is disrupting ecosystems, causing species that are not adapted to each other to live together, thus decreasing population sizes, sometimes to zero. Some species are running out of space to move higher on the mountains they inhabit. Pikas (genus *Ochotona*) are small, rabbit-like mammals that live in mountains. Pikas of the Great Basin have disappeared from 40% of the places where they occurred in the first half of the twentieth century because they have no livable habitat above them on their mountain homes (Root, 2003; Parmesam and Yohe, 2003).

Thousands to millions of king crabs of the family Lithodidae are moving south to invade the seas around the Antarctic as a result of global warming. Their homes to the north have become too hot for them to thrive in. Antarctic clams, snails, brittle stars, sea squirts, and other thin-shelled and soft-bodied animals are not adapted to the crushing pincers of these predatory crabs, which can easily eat them. Scientists are worried that the crabs could wipe out the vulnerable Antarctic ecosystem. This shows that species of ecosystems adapt and keep diversity high. Prey species evolve hard shells and external skeletons as defense against predators with crushing jaws and claws, but do not waste energy evolving these defenses when such predators do not live with them. When an unnatural situation arises wherein a predator that the species are not adapted to appears because humans change the climate, there can be a catastrophic loss of diversity. Under normal circumstances, it is cold temperatures that kept the king crab out of the Antarctic ecosystem, preserving its diversity.

Animals and plants are changing the timing of their life cycles. Species that depend on other species are becoming out of synchronization with them. Dates on which caribou give birth have remained steady, because their schedule is set by day length. The plants they eat are emerging earlier because they respond to temperature, causing caribou to starve. Birds are no longer nesting when their insect prey is at its peak, so they do not have enough food for their young. Insects that eat the nectar and pollen of flowers and pollinate them are finding their host flowers are no longer blooming when the adult insect populations are peaking. Thus, the insects have too little food and the plants cannot reproduce as effectively because the insects do not pollinate their flowers. This lack of synchronicity of flowers, insects, and birds is due to the fact that these groups respond to climate change differently. It is causing large declines in several species in all of these groups, and could result in the extinctions of some of these species.

Bates et al. (2022) found that many species of birds are nesting and laying eggs nearly a month earlier than they did a century ago. They assessed changes in nesting dates of 72 bird species, including blue jays (*Cyanocitta cristata*), yellow warblers (*Setophaga petechia*), and field sparrows (*Spizella pusilla*), in the Upper Midwest of the United States, by comparing egg-laying dates of currently nesting birds with those obtained from archived, historical nest records in museums over a 143-year period (1872–2015). They found that overall, egg-laying dates advanced about 10 days over the 143-year period. Twenty-four species, exactly one-third of them, significantly advanced egg-laying dates; of these, the overall average advance in first date of egg laying was 25.1 days, with the minimum advance being 10.7 days, and the maximum 49.9. Other factors in addition to temperature affect when birds nest in spring, such as the budding of plants and increase in insect numbers. But increased temperatures are a major factor in driving earlier nesting times. Spring temperatures are occurring earlier due to human-induced climate change. Other factors such as habitat loss and the indiscriminate use of pesticides are also contributing to the decline of birds. But early nesting is a major factor, mainly because it causes bird nesting season to no longer coincide with when the populations of their insect prey are at their peaks. Also, earlier arrival and nesting time subjects birds to early spring cold spells, which can kill large numbers of them.

Broad-tailed hummingbirds (*Selasphorus platycercus*) migrate north to high altitude breeding sites in the western U.S. each spring. Males arrive to set up territories before the first flowers bloom, but global warming has narrowed the gap between first arrival of the birds and first floral bloom by thirteen days over the last four decades in the Rocky Mountains. This decreases the already short time that the males have to set up territories and breed. Now the birds have time to raise only one brood of at most two chicks per summer, which is a major decrease in the young they raise (McKinney, 2012). And European pied flycatchers (*Ficedula hypoleuca*), migratory birds, now have much less food for their young because the caterpillars they once fed their chicks are now peaking well before the birds arrive and nest (Both et al., 2006).

Bears in the United States are emerging from hibernation earlier, making it harder for them to obtain food and more prone to starvation. Cherry, peach, pear, apple, and plum trees are blossoming weeks earlier than they once did, and in the UK, plants flowered a full month earlier between 1987 and 2019 than they did before 1986, putting all of these trees and plants out of sync with their pollinators.

Lian et al. (2020) found the fact that spring is coming earlier due to climate change is increasing the rate that plants release water into the air more than it increases rainfall, resulting in the drying of soil. This amplifies the frequency and intensity of summer heatwaves in a positive feedback loop. This effect is strong in Europe. The local soil moisture loss in Siberia is offset by moisture from Europe, showing a connection of different parts of the planetary system.

Life's regulation of climate has kept moisture levels unfavorably low for some fungal species in many areas, and human-induced increased temperatures have increased moisture levels in many of these areas, creating favorable conditions for certain fungal pathogens. Many suspect such climatic change played a role in the following fungal diseases of animal species, because they only became serious epidemics after climate change increased humidity in the disease areas.

Chytrid fungi such as *Batrachochytrium dendrobatidis* and *B. salamandrivorans* cause a disease called chytridiomycosis in amphibians that has caused dramatic population declines and even extinctions of amphibian species in western North America, Central and South America, the Caribbean, and eastern Australia. It has been proposed as a contributing factor to a global decline in amphibian populations that has affected *30% of the Earth's amphibian species* (Stuart and Chanson, 2004). Many scientists think global warming is a key facilitator of the fungus, while others do not. The fact that the problem is correlated with rising temperatures and humidity in certain areas suggests that climate change is at least part of the cause.

As of 2012, white-nose syndrome (WNS), caused by the fungus *Pseudogymnoascus destructans*, was estimated to have killed 5.7 million to 6.7 million bats in North America (Froschauer and Coleman, 2012). As of 2017, 15 bat species, had been impacted by WNS. Four species have had great declines, and one is in danger of extinction. WNS started in a New York cave, and has been found in 25 states and 5 Canadian provinces as of 2014.

The often-lethal fungus, *Ophidiomyces ophiodiicola*, has infected 23 unrelated species of snakes from the eastern to midwestern United States. It first received widespread attention in 2008.

It is not caused by a fungus, but since white band disease was first reported in the 1970s, the disease has led to the devastation of approximately 95% of the elkhorn coral (*Acropora palmata*) and staghorn coral (*Acropora cervicornis*) in the Caribbean. Warmer water due to global warming causes the bacteria to shift from a species of nonpathogenic bacteria of the genus *Pseudmonas* being dominant to a disease-causing species of bacteria of the genus *Vibrio* becoming dominant. This is an example of two quasi-stable states, with the natural state of high-*Pseudomonas*, low-*Vibrio* causing a much more diverse and complex system than the unnatural, human-caused state.

Since the 1970s, droughts have become longer, more intense, and have covered larger areas, notably in the tropics and subtropics. Higher temperatures allow air to hold more moisture, so it does not fall as rain. Moisture-laden tropical air often moves to higher latitudes, where lower temperatures allow it to fall as rain. This causes droughts in the tropics and subtropics, which reduces the number of trees and diversity of tree species in these ecosystems, and weakens those that manage to live. It makes them more vulnerable to insect, bacterial, viral, and fungal attacks. Drought dries the soil. This increases the number, intensity, and size of forest fires, making the amount of fire far above the optimum for life. Rainforests have high rainfall and moisture under normal conditions, and so do not have large or frequent fires. However, they are now so dry that large fires have burned rainforests throughout the world: in Brazil, Mexico, Africa, southeast Asia, and other places. Plants other than trees also have reduced diversity due to the drought and increased fires. With fewer plant species, animal diversity from worms to insects to vertebrates decreases. Dry conditions cause rainforests to dry out on the edges, with the trees on the edges dying. The area that becomes the new edge then dries out and dies, the process continues, and the rainforest continuously shrinks. These effects on diversity are profound because the tropics and subtropics are among the most diverse terrestrial ecosystems. Tropical rainforests are the most diverse ecosystems on land.

Climate change is causing droughts in regions outside the tropics and subtropics as well, including the western United States. It causes lakes, ponds, wetlands, and rivers to dry up, causing aquatic life to die off, and animals on land to die of thirst. Increased temperature and the decreased precipitation that goes with it in boreal forest, the Earth's largest biome, have caused a large increase in fires that burn this forest type (Veraverbeke et al., 2017). Drought is also causing increased fires and insect outbreaks in temperate forests. As a result of drought, fires are becoming more frequent and larger in the western United States. This releases more CO_2 into the air in a positive feedback loop. Between 1997 and 2010, over 5 million hectares (19,305 square miles) were affected by bark beetles in the western United States, mainly the mountain pine beetle (*Dendroctonus ponderosae*), spruce beetle (*Dendroctonus rufipennis*), and pinyon ips beetle (*Ips confusus*). The amount of carbon in trees killed by these insects exceeds that of the carbon in trees killed by fire, so insect attacks release tremendous quantities of CO_2 in another destabilizing positive feedback loop. In the southeastern and northeastern United States, the southern pine beetle (*Dendroctonus frontalis*) has affected more than 14,000 hectares (54 square miles) since 2008, especially in New Jersey and Mississippi.

In many areas, a threshold has been crossed, and forests will not grow back after being burned, because of drought. Conditions in some forests have become too dry or hot to support them. A firefighter analyzed sites across the Rocky Mountains, and found that almost one-third of places that had burned since 2000 had no trees regrowing. Instead of tree seedlings, there were shrubs and flowers. In the Rocky Mountains, it is estimated that about 15% of the forests would not grow back if burned because today's climate will not support them. Forests are being replaced with shrub ecosystems. In the southern Canadian Rockies, 72% of the trees of the whitebark pine (*Pinus albicaulis*) were found dead. Climatic variation caused drought stress that may have predisposed the trees to blister rust, a fungal disease. Subsequent decline in snowpack and increased heat likely incited further climatic stress, with blister rust reducing tree resistance to bark beetles, which killed yet more trees. And when trees do not receive sufficient water, they cannot produce enough pitch to protect themselves from bark beetles. Thus, an interaction of warming, drought, blister rust, and beetles caused the tremendous tree mortality (Wong and Daniels, 2017). The southwestern United States is undergoing a megadrought. In this area, as much as 30% of forests could be converted to shrubland or another ecosystem. This is an example of two quasi-stable states, with the natural one being more complex and diverse, which is the rule. The natural quasi-stable state of forest is more complex and diverse, but is being replaced by the unnatural, human-caused quasi-stable state of shrub systems, which is less complex and less diverse. In the Amazon, a forest mortality threshold is looming, and a good deal of rainforest could be converted to savanna. Here, climate change has lengthened the dry season and caused the rainfall to decline in various areas. This favors savanna and hurts forests. This is another case of quasi-stable states, with the natural quasi-stable state of rainforest having higher complexity and diversity than the unnatural, human-caused quasi-stable state of savanna. Around the world, research has suggested that the tree mortality rate in some temperate and tropical forests has doubled or more in recent decades. As forests die back, they switch from storing carbon to emitting it, exacerbating global warming, in a maladaptive, destabilizing positive feedback loop.

The coral in coral reefs, the most diverse marine ecosystems there are, are dying due to coral bleaching, a process whereby the symbiotic alga leaves the coral when the sea's temperature rises beyond a threshold temperature. The alga carries out photosynthesis, providing carbohydrate to itself and the animal part of the coral, which is a distant relative of sea anemones. Each species of coral has its own unique algal species, and great numbers of coral species are dying off as they lose their algal allies. If coral bleaching is not stopped, only a few species of algae that are not symbiotic with the coral animal will grow on the dead coral, supporting much fewer species of fish and invertebrates than the healthy coral reef. The result is a reef system of incomparably lower diversity. The high-diversity reef is replaced with a new, depauperate quasi-stable state.

Upwelling is the movement of water that is rich in nutrients from deeper water in the ocean to surface waters, bringing nutrients to phytoplankton, which are in water shallow enough for sunlight to penetrate. This way they can carry out photosynthesis. They need both nutrients and light to produce large numbers of themselves. Upwelling requires the right ambient air and water temperatures. When temperatures heat up due to global warming, upwelling occurs less often or ceases completely. Thus, warmer seas are decreasing upwelling. Phytoplankton populations become lower or nonexistent where upwelling decreases or ceases. Phytoplankton are the base of most food webs in the ocean. So when phytoplankton populations go down, there are fewer animals at all levels of the major food webs of the sea. This means there are decreases in populations, in some areas possibly to zero, of zooplankton, krill, crabs, lobsters, shrimp, sea stars, small fish, large fish like sailfish, sharks, sea turtles, seabirds, seals, whales, and many other kinds of animals.

Greater sea temperatures reduced upwelling off the coast of Alaska between the late 1970s and middle 1980s, decreasing phytoplankton growth. This caused decreases in crab, shrimp, and forage fish such as capelin (*Mallotus villosus*), and increases in groundfish high up on the food web, such as flatfish, which are very flat and include flounders, soles, and halibut. These lower-quality food sources caused large declines of Steller sea lions (*Eumetopias jubatus*), northern fur seals (*Callorhinus ursinus*), and harbor seals (*Phoca vitulina*). So killer whales (*Orcinus orca*) had fewer sea lions and seals, their preferred prey, to eat. Thus, they ate more sea otters (*Enhydra lutris*), causing them to decline. This allowed an explosion of one the otters' main prey species, the green sea urchin (*Strongylocentrotus droebachiensis*). The urchin populations skyrocketed to hundreds per square meter in some areas. They destroyed kelp forests along hundreds of miles of coastal areas. Many species of kelp are keystone species. Kelp provides habitat, nurseries, and food to an ecosystem of invertebrate and fish species. Thus, all the species dependent on kelp declined or died off. This included many fish and invertebrate species. There was a decline of species like bald eagles, which eat the fish of kelp forests. So there was a cascade of inter-related species declines. The ecosystem underwent a tremendous loss of diversity.

Due to rising sea level, Vanuatu and the Marshall Islands in the Pacific Ocean, much of Bangladesh, and coastal areas worldwide, are being submerged. This will destroy some oceanic island habitats and many coastal habitats. It will flood and destroy coastal forests, including mangrove forests, which clean toxic chemicals from ecosystems and are breeding habitat for many invertebrate and fish species. Intertidal habitats will be permanently submerged. Saltwater will intrude into freshwater wetlands, degrading these habitats and causing the local loss of many species. All of this will result in a tremendous loss of biodiversity.

Warmer winters would cause many reptile species to decrease in numbers of individuals or perish because they need cold to lower their metabolism and prevent starvation during hibernation. If it is too warm in winter when snakes, lizards, turtles, and tortoises hibernate, they will burn too much energy and starve to death before it is warm enough for them to hunt for food. The sex of crocodiles, alligators, and turtles is determined by the temperature the eggs are incubated at. If the temperature is raised even a few degrees, most or all offspring would be the same sex, causing population declines and/or extinctions.

Glaciers in mountains supply rivers with water all over the world. As the planet gets hotter, glaciers worldwide are melting and not being replenished as fast as they are being lost. When the glaciers are melted, there will be no source of water for rivers, and they will dry up. Melting of the Himalayan glaciers is removing the water supply for seven rivers in Asia that support many freshwater species. Drying out of the rivers will result in the total destruction of riparian, wetland, forest, and other habitats, and the loss of essentially all the species in all of these habitats. Entire forests will die from desiccation. This is a huge area, including Tibet, Nepal, China, India, Pakistan, Bangladesh, Cambodia, Laos, Myanmar (Burma), Thailand, Vietnam, and more. The Earth is also losing glaciers in the Sierra Nevada, which supply the water for major rivers in California; the Rocky Mountains, which supply the Colorado River; the Andes, which supply the Amazon River system; the Alps; and other mountain ranges. This will result in the drying of all the rivers these glaciers supply water to. The Amazon rainforest, the most diverse terrestrial ecosystem on Earth, will have very little or no water, resulting in the extinction of millions of species of animals and plants.

The Amazon is the world's largest tropical rainforest. It has played a key role in absorbing and storing carbon. Deforestation has had a great impact on it. Gatti et al. (2021) found that the southeastern Amazon now is a net carbon source, and no longer a sink. It released about 20% more CO_2 into the atmosphere than it took in from 2010–2019. The Amazon's ability to act as a major carbon sink has declined as a result of deforestation and climate change. Increased temperatures are promoting changes in dry season conditions and stress to trees, causing them to emit more carbon. The authors found that an increase in deforestation and intensification of the dry season that results from this and global climate change promote increases in fire occurrence, ecosystem stress, and carbon emissions in the eastern Amazon. Temperatures there have risen by three times the global average during the hottest months, and increased in the two hottest months of the year by 3.07°C (about 5.53°F), which is about the same increase seen in the Arctic. The southeast is the worst affected, with higher rates of tree loss and an increasing number of fires. The eastern part of the Amazon, which is about 30% deforested, emitted 10 times more carbon than in the west, which is about 11% deforested. The carbon is mainly emitted from the decomposition of trees as they die in greater numbers than usual and from the increased occurrence of fire. The researchers are worried that the changing climate is also interfering with rainfall. Human impacts have changed the Amazon from a quasi-stable state that absorbs carbon to a quasi-stable state that emits it.

Drying of cloud forests has driven the golden toad (*Incilius periglenes*) of Costa Rica extinct. It was last seen in 1989 and is the first species in the scientific literature to go extinct because of climate change. Warming has reduced mist and made conditions drier in the cloud forest of Monteverde National Park in Costa Rica, where the average altitude of clouds has been rising since the 1970s. This caused this toad to die off because it needs moist conditions. It also allowed the keel-billed toucan (*Ramphastos sulfuratus*), a bird which normally breeds at lower, drier altitudes to invade and nest at higher elevations on the mountains. Normally a dweller of lowland rainforests, it now nests in the cloud forest because of the drier conditions. This causes it to compete with the resplendent quetzal (*Pharomachrus mocinno*) for nesting sites and foraging territories. This is reducing the population of quetzals with the danger that they will be eliminated from their cloud forest habitat. Lizards, snakes, amphibians, and other types of animals are being directly negatively impacted by the drying of cloud forest. In addition, some animals, such as snakes, are being indirectly impacted because species they rely on, such as their prey, including frogs, are declining.

Many polar species are losing habitats. Polar bears (*Ursus maritimus*), seals, penguins, and other Arctic and Antarctic wildlife need sea ice and are experiencing population declines as their sea ice habitat decreases. Polar bears need sea ice for hunting. Seals are the primary prey for polar bears. Seals are very agile in the water but slow on top of the sea ice. Hence, sea ice is the habitat where the bears have the best chance of catching seals. As a result of the loss of sea ice, many polar bears are starving and switching from seals to other prey.

Seibold et al. (2021) found a positive feedback loop in human-induced climate change involving biology. They looked at 140 types of trees in 55 forest areas on six continents. The dead parts of a tree, bush, or plant are called deadwood. Deadwood stores about 8% of the carbon in trees and other plants in forests (in forests, it is mostly in trees). Deadwood's decomposition is largely governed by climate, with decomposer groups such as microorganisms and insects contributing to its rate of decomposition. They found that decomposition rates increase with temperature, with the largest temperature effect at high precipitation levels. Precipitation affects these rates negatively at low temperatures and slightly positively at high temperatures. They discovered that wood-boring insects such as termites and wood-boring longhorn beetles (family Cerambycidae), along with indirect effects through interactions with microorganisms, greatly accelerate the deadwood decomposition in tropical forests, at a median mass loss of 3.9% per year. These insects cause 29% of the carbon released from deadwood every year, and it is higher in the tropics. Tropical forests contribute 93% of all carbon released via deadwood, since they have a high mass and decompose rapidly. Globally, decaying wood releases about 10.3 ± 3.2 billion tons of carbon per year, or 115% of the annual carbon emissions from fossil fuels. Globally, the net effect of insects may account for 29% of the carbon transfer from deadwood to the atmosphere. This is not from the study, but it should be noted that more trees die from drought as temperatures increase, creating more deadwood. This shows that biology's negative feedback that stabilizes atmospheric CO_2 and methane under normal circumstances can be thrown into a destabilizing positive feedback loop by adding significant quantities of greenhouse gases to the system. The natural quasi-stable state of moderate greenhouse gas levels in the air, moderate temperatures, a small amount of deadwood, and low numbers of wood-boring insects in a stabilizing negative feedback system can be thrown into an unnatural (human-caused) quasi-stable state with a destabilizing positive feedback loop with high greenhouse gas levels, temperatures, deadwood, and wood-boring insects. The former system is much more complex and diverse than the latter one.

Methane hydrates are found in certain areas in the sea, some lakes, and in Alaska and Siberia. They consist of the powerful greenhouse gas methane under low temperatures, forming a solid similar to ice. If the Earth were heated to high enough temperatures by greenhouse gases, methane hydrates would release large quantities of methane into the air in a devastating positive feedback loop, and would make Earth extremely hot and unfavorable for life, causing a great decrease in diversity, even a probable mass extinction. The end-Permian mass extinction, the largest extinction in Earth's history, probably resulted from a rise in temperature from volcanoes that led to the release of large quantities of methane from methane hydrates. Methane hydrates are widespread in the deep sea and contain as much as 20% of all carbon on Earth. All the living organisms on the planet together have less carbon in them than there is in Earth's methane hydrates.

A report from the UN's Intergovernmental Panel on Climate Change by researchers from 67 countries warned that human-induced global warming is threatening Earth's biodiversity, threatening large numbers of species with extinction, even under relatively conservative estimates (O'Neill et al., 2017). If Earth warms by 1.5°C (2.7°F), up to 14% of all plants and animals on land will face a high risk of extinction. This temperature rise is almost certain to occur. With a 3°C (5.4°F) temperature rise, up to 29% of species on land could face extinction. A shocking 47% of species have already lost some of their populations because of climate change. For example, in 2005, heatwaves decimated a subspecies of the rare ringtail possum in Australia. In the summer of 2021, high temperatures killed hundreds of millions of marine animals in the Pacific Northwest, from sea stars to mussels. Warming of only 1.5° C has the potential kill up to 90% of tropical coral reefs, which support over 25% of all marine species.

The above list of ways that climate change from greenhouse gases can devastate biodiversity, and indeed is doing so, is not comprehensive. It is only meant to give an idea of the great extent to which life has increased biodiversity by sequestering carbon and hence regulating Earth's temperature. It shows that we can reject the null hypothesis that the sequestering of carbon by life has no effect or a negligible effect on biodiversity.

2.10 THE DETRIMENTAL IMPACTS OF HIGH ATMOSPHERIC CARBON DIOXIDE LEVELS ON LIFE ARE NOT LIMITED TO TEMPERATURE EFFECTS

There are several extremely severe negative impacts on biodiversity that high CO_2 levels produce, quite apart from temperature effects.

The higher the atmospheric concentration of CO_2, the greater the amount of it absorbed by the oceans. This is in fact happening today, and it acts to alleviate global warming to some extent. However, the excess CO_2 in the seas reacts with water to form carbonic acid, raising the acidity of the oceans. The higher the acidity of water, the less calcium can come out of it as a solid, so more of it stays in solution. The less calcium is able to come out of solution, the less organisms can use it to form their protective calcium shells. Groups of animals with protective calcium shells include various kinds of phytoplankton. Others that use calcium to build protective shells or skeletons are the myriad species of bivalve molluscs, such as clams and scallops; sea snails; barnacles; crustaceans from crabs to shrimp to the tiny krill that feed many species of great whales and several other species; sea urchins; sand dollars; and coral. Thus, if organisms did not sequester carbon, acidity in the seas would cause great decreases or extinctions of a number of species in all of these animal groups. This would severely impact all groups that depend on these animals, directly or indirectly. Entire food webs would be disrupted. Any species directly or indirectly dependent on coral reefs for food or shelter would decline or become extinct. This includes the approximately 25% of all marine fish species that live in coral reefs. Coral are the most diverse ecosystems in the sea. Their loss would result in the decline or extinction of countless species of invertebrates, including sea stars, sea urchins, crabs, snails, clams, and roundworms. And the high acidity would also disrupt the reproduction of many species, alter biogeochemical cycles, disrupt other physiological processes of marine organisms, and damage marine ecosystems in other ways. Increased temperature and acidity would likely act together, with negative effects on ecosystems (see Doney, 2006; Fabry et al., 2008). There is also a positive feedback loop in that diatoms build their silica cell walls at a slower rate in higher acidity. They therefore have on average thinner, lighter cell walls, so do not as readily sink and sequester carbon. So temperatures and the acidity both become even higher. This means diatoms build their cells walls even more slowly and sink even less. The loop would continue and the ecosystem would deteriorate. Animals that build shells and live in freshwater rivers, ponds, and lakes would also be impacted by increased acidity of their water. This includes freshwater clams, mussels, snails, shrimp, and so on. So increased CO_2 in aquatic ecosystems causes great decreases in diversity. This effect has already started to manifest itself in several ecosystems.

A recent study of soil microbes found high soil CO_2 causes lower bacteria population numbers and lower numbers of bacterial species (Zhili et al., 2012). Reduced microbial diversity would reduce diversity of many eukaryote species, including soil algae, earthworms, plants, and other groups, since bacteria are a food source and carry out processes vital to these organisms. The reduction of soil algae, earthworms, plants, and other groups would cause the decline of the myriad species that benefit from them.

Plants that grow in very high atmospheric CO_2 grow faster, but have a low protein to carbohydrate ratio. In general, insect larvae feeding on plants grown in high CO_2 conditions eat more in response to the lack of sufficient protein, but still become protein deficient. They are more subject to predation, malnutrition, starvation, and overall death rates, and smaller if and when they do emerge as adults. Adult spittlebugs (family Aphrophoridae), which suck sap from plants, had their survival rates decrease 27% in high CO_2 (Brooks and Whittaker, 2001). Four times as many leaf miners (insects whose larvae live in and eat leaves; the vast majority are moth larvae) grown in high CO_2 were attacked by parasitoid wasps as controls that were grown in normal CO_2 levels (Stiling et al., 1999). Parasitoid wasps are small wasps of several species that lay their eggs on other insects; most are in the suborder Apocrita. After the larvae hatch from the eggs, they eat the insect, killing it. In contrast, aphids reproduce 10 to 15% faster in elevated CO_2 conditions. Infested bean plants grown in high CO_2 were so overwhelmed by the aphids' rapid population growth that they could not grow flowers or new shoots (Awmack et al., 2003).

A decrease of insect herbivores would have serious negative effects on ecosystems, because most insect species are herbivorous, and there are about 900,000 known species of insects, which is about 80% of all animal species. Most insect species have not been described; it is estimated there are a total of 20 to 30 million insect species, and there are about 10^{19} insects alive at any given time. If there were too few plant-eating insects, the plants they eat would not be regulated, so they would become so dense that many animal species would no longer be able to live in them, and they would deplete soil nutrients, leading to the death of great numbers of plants, and the animals that eat them. Also, plants would crowd each other and compete for space, water, and nutrients. The poorer competitors would die off, so there would be a much lower number of species of plants. Plant-eating insects keep plants that are better competitors from driving plant species that they compete with locally extinct. A decrease in herbivorous insects would also tremendously negatively impact all species they aid, such as their predators, the plants they pollinate, and plants whose seeds they disperse.

Herbivorous mammals such as deer also suffer from lower protein if they consume plants grown in high CO_2. Their loss would negatively impact the plants they eat, their predators, and many other species that they benefit. The fact that plants grown in high CO_2 have a low protein-to-carbohydrate ratio has the potential to cause tremendous reductions in diversity in ecosystems, even causing them to crash.

Some plant species, such as loblolly pines, do better than other plant species in high CO_2 environments (Tangley, 2001), out-competing them, and causing their population to decline or go locally extinct. This causes a drop in plant diversity (diversity here is number of species), which causes the diversity of herbivores to fall, and hence predator diversity to decline. Poison ivy of the eastern United States is an example of a plant that does very well and increases its growth in high CO_2. Vines in the rainforest that grow on trees often do not harm them, but in CO_2-enriched air, some of these vines grow much bigger, smothering and killing their host trees with a much greater likelihood.

2.11 DID BIOLOGY CREATE OPTIMAL CONDITIONS FOR LIFE?

Methanogens kept the early Earth warm when the sun was weak. Life continuously removed greenhouse gases from the air as the sun increased its heat output through time. So life regulated the temperature, keeping Earth in a favorable, cool temperature range.

It is possible that organisms come close to optimizing the levels of oxygen, water vapor, CO_2, and methane in the atmosphere for life. Biology seems to regulate the two major greenhouse gases, CO_2 and methane, by negative feedback, keeping them at favorable levels to create temperatures beneficial to life. Organisms seem to regulate the amount of oxygen in the atmosphere, keeping it at near-optimal levels for life. Life increased the atmospheric oxygen level from less than 1% to today's life-favorable 21% by producing oxygen through photosynthesis and burying reduced carbon. *Reduced carbon* is carbon that is not combined with oxygen. Because of this, when it is buried, there is a net gain of oxygen in the air. This contrasts with carbon that is combined with oxygen, the burial of which does not necessarily result in an increase in atmospheric oxygen. The current 21% oxygen in the air is beneficial to life. Oxygen levels are sufficient to provide the energy needed for higher organisms. But at excessive levels, oxygen would be harmful to cells and create damaging chemicals called free radicals. In regulating the level of atmospheric oxygen, biology may also come close to optimizing the amount and intensity of fires for life to thrive. In regulating the amount of CO_2 and methane in the air, life influences the temperature and hence the amount of water vapor in the air. The higher the temperature, the more water evaporates and rises into the air. The amount of water vapor influences the amount of cloud formation and hence the amount of lightning, which is generated by clouds. The amount of lightning influences the amount of fire. Thus, life may cause a favorable amount of water vapor in the air to help cause a life-favorable amount of fire. The atmospheric oxygen level is also important in determining the number and intensity of fires. The more oxygen there is in the air, the more easily fires burn, because fire uses oxygen as a fuel to burn. If life did not produce the intermediate atmospheric oxygen levels we see today, there would not be the favorable, intermediate amount of fire for life to thrive. Life may even keep the atmospheric oxygen level at a biology-favorable 21% by negative feedback (see Chapter 3). This optimizes the amount and intensity of fires. Too much fire is bad for forests, but so is too little. Forests need fires to recycle nutrients and create fertile soil. Trees grow vigorously after fires. This is discussed in more detail in Chapter 3. Life produces almost all of the oxygen and most of the methane, which combine with and destroy each other. Given its depletion by oxygen, to sustain the amount of methane in the Earth's atmosphere, at least a billion tons of it must be added to the air annually. Most of this is made by life. It was needed to maintain a warm enough temperature before humans started heating the planet. Twice as much oxygen as methane must be added back to the air because methane destroys it, and this is almost entirely done by life. Life produces just enough of each that neither depletes the other or becomes too abundant. Also, the level of ozone in the ozone layer might be close to optimal to protect life and yet cause a favorable amount of mutations for evolution and adaptation. This level is due to the level of oxygen created mainly by life.

It is also possible that the actions of organisms resulted in greenhouse gas levels nearly optimal for the best number and intensity of hurricanes for the intermediate disturbance of coral reefs, maximizing their diversity. Higher temperatures cause conditions for more hurricanes, and more intense ones. Too many hurricanes and ones that are too intense would be destructive to coral reefs, but too few would allow the better competitors to drive less competitive coral species locally extinct, reducing the number of species of coral and hence of fish and invertebrate species dependent on coral. An intermediate number and intensity of hurricanes results in the highest biodiversity on coral reefs. This underscores the important role of life in regulating atmospheric greenhouse gas levels with the result that storm activity is close to optimal for coral reef biodiversity.

These ideas of life creating optimal levels of greenhouse gases, oxygen, water vapor, and temperatures for life are hypotheses I am proposing that need testing. I am not suggesting they are to be accepted as established truths at this point.

Interestingly, Earth's temperature and CO_2 levels have remained remarkably stable from about 10,000 years ago to just before industrial times. There have been periods when temperature departed considerably from the mean, such as the Little Ice Age (not a real ice age; it was a period of lower temperatures that lasted from about the 16th to the 19th centuries), but these have been small fluctuations compared to some of the large fluctuations throughout the last 500 million years. Also, during this recent 10,000-year period, the temperature and greenhouse gas levels have been close to optimal for life. This is interesting. It raises the possibility that life brought global temperatures and greenhouse gas levels to near-optimal levels by about 10,000 years ago, and kept them there, until humans intervened. It is striking how preindustrial atmospheric greenhouse gas levels were so close to optimal. At these levels, droughts and floods were minimized, glaciers did not disappear and kept replenishing rivers with water, rainforests did not suffer droughts and major fires, forests of all types had intermediate and diversity-maximizing levels of fire and moisture and insect attacks, coral reefs thrived in favorable water temperatures, the size and frequency of hurricanes were near-optimal for coral reef diversity, trees caused life-favorable amounts of local rainfall and atmospheric moisture, the sea and freshwater ecosystems were at beneficial levels of acidity for life, there were near-optimal carbohydrate levels in plants for herbivores, ocean currents were favorable to life, there were near-optimal temperatures for ocean upwelling and hence phytoplankton blooms and healthy ocean food webs and healthy kelp forests, and many more favorable conditions for biology.

But we must be aware that the continents are approximately at their maximum possible distances apart, which may be optimal for life by allowing favorable sea currents, winds, phytoplankton blooms, and temperatures. The continents were clumped together as one land mass more than once in the geologic past, and will in time move back together. When the continents are together as one continent, oceans currents are different and probably less favorable for keeping temperature differences between different areas in the sea low. Today's currents make temperatures in the sea more uniform from one place to another, which is good for life. If the continents were currently clumped together, perhaps the current greenhouse gas levels would not be so close to optimal for currents. We do not know what greenhouse gas levels and temperatures will be when the continents coalesce again. Nor do we know to what extent life and perhaps nonbiological forces caused the highly life-favorable greenhouse gas levels and temperatures for the last 10,000 years. The ABH, or Pachamama Hypothesis, is carrying out one of the functions of a hypothesis, which is revealing potentially fruitful areas of research. This question of whether life and perhaps nonbiological factors created nearly optimal conditions for various parameters is a particularly important and interesting area of research that should be pursued.

2.12 CONCLUSION

The sun is 30% hotter today than when life began. Life warmed the young Earth by producing methane when the sun was cooler, then cooled it by sequestering carbon, to life's benefit. Biology has also regulated other aspects of the physical-chemical environment to the benefit of life, and it still does. If organisms did not sequester carbon to the extent that they have, it is highly probable that the Earth would be so hot that at best, only ecosystems underground and deep in the sea, and no life at all or only simple ecosystems with only one-celled organisms on the Earth's surface, would exist.

REFERENCES

Airapetian, V. S., et al. (2016). Prebiotic chemistry and atmospheric warming of early Earth by an active young Sun. *Nature Geoscience* 9: 452–5.

Algeo, T.J. & Scheckler, S. E. (1998). Terrestrial-marine teleconnections in the Devonian: links between the evolution of land plants, weathering processes, and marine anoxic events. *Philosophical Transactions of the Royal Soc. B. Biological Sciences* 353 (1365): 113–30. doi: 10.1098/rstb.1998.0195 PMC 1692181.

Andreae, M. O., et al. (1995). Biogenic sulfur emissions and aerosols over the tropical South Atlantic, 3. Atmospheric dimethylsulfide, aerosols and cloud condensation nuclei. *Journal of Geophysical Research* 100: 11335–56. doi: 10.1029/94JD02828 www.agu.org/pubs/crossref/1995/94JD02828.shtml.

April, R. & Keller, D. (1990). Mineralogy of the rhizosphere in forest soils of the eastern United States. *Biogeochemistry* 9: 1–18.

Arthur, M. A. & Fahey, T. J. (1993). Controls on soil solution chemistry in a subalpine forest in north-central Colorado. *Soil Science Society of America Journal* 57: 1122–30.

Awmack, C. (30 Oct., 2003). Host plant effects on the performance of the aphid *Aulacorthum solani* (Kalt.) (Homoptera: Aphididae) at ambient and elevated CO_2. *Global Change Biology* 3 (6): 545–9. https://doi.org/10.1046/j.1365-2486.1997.t01-1-00087.x.

Bates, J. M., Fidino, M., Nowak-Boyd, L., Strausberger, B. M., et al. (March, 2022). Climate change affects bird nesting phenology: Comparing contemporary field and historical museum nesting records. *Journal of Animal Ecology*. doi: 10.1111/1365-2656.13683.

Baumert, H. Z. & Petzoldt, T. (22 Oct., 2008). The role of temperature, cellular quota and nutrient concentrations for photosynthesis, growth and light-dark acclimation in phytoplankton. *Limnologica* 38 (3-4): 313–26.

Bazzaz, F. A. (1990). The response of natural ecosystems to the rising global CO_2 levels. *Annual Review of Ecology, Evolution, and Systematics* 21: 167–96.

Beerling, D. (2007). *The Emerald Planet: How Plants Changed Earth's History*. Oxford Univ. Press, Oxford, UK.

Beerling, D. J. & Berner, R. A. (2005). Feedbacks and the coevolution of plants and atmospheric CO_2. *PNAS USA* 102: 1302–5.

Bello, C., et al. (18 Dec., 2015). Defaunation affects carbon storage in tropical forests. *Science Advances* 1: e1501105: 1–10.

Berner, E. K., Berner, R. A., & Moulton, K. L. (2003). Plants and mineral weathering: Present and past. *Treatise in Geochemisty* 5: 169–88.

Berner, R. A. & Caldeira, K. (1997). The need for mass balance and feedback in the geochemical carbon cycle. *Geology* 25: 955–6.

Berner, R. A. & Cochran, M. F. (1998). Plant-induced weathering of Hawaiian basalts. *Journal of Sedimentary Research* 68: 723–6.

Berner, R. A. (1998). The carbon cycle and CO_2 over Phanerozoic time: the role of land plants. *Philosophical Transactions of the Royal Society B: Biological Sciences* 353: 75–82.

Berner, R. A. (2013). Personal communication.

Berner, R. A. (25 April, 1997). The rise of plants and their effect on weathering and atmospheric CO_2. *Science* 276 (5312): 544–6. doi: 10.1126/science.276.5312.544.

Berner, R. A. (Aug., 1992). Weathering, plants, and the long-term carbon cycle. *Geochimica et Cosmochimica Acta* 56 (8): 3225–31.

Berner, R.A. & Kothavala, Z. (Feb., 2001). GEOCARB III: A revised model of atmospheric CO2 over Phanerozoic time. *American Journal of Science* 301: 182–204.

Berzaghi, F., et al. (2019). Carbon stocks in central African forests enhanced by elephant disturbance. *Nature Geoscience* 12: 725–9.

Blanc Mathieu, R., et al. (22 July, 2019). Viruses of the eukaryotic plankton are predicted to increase carbon export efficiency in the global sunlit ocean. *BioRxiv*. doi: https://doi.org/10.1101/710228.

Both, C., et al. (4 May, 2006). Climate change and population declines in a long-distance migratory bird. *Nature* 441: 81–3. doi: 10.1038/nature04539.

Bowen et al. (2015). Two massive, rapid releases of carbon during the onset of the Palaeocene–Eocene thermal maximum. *Nature* 8 (1): 44–7. doi:10.1038/ngeo2316.

Boyd, P.W., Claustre, H., Levy, M., et al. (2019). Multi-faceted particle pumps drive carbon sequestration in the ocean. *Nature* 568: 327–35. https://doi.org/10.1038/s41586-019-1098-2.

Brezinski, D.K., et al. (2009). Evidence for long-term climate change in Upper Devonian strata of the central Appalachians.

Palaeogeography, Palaeoclimatology, Palaeoecology 284 (3–4): 315–25. doi: 1 0.1016/j.palaeo.2009.10.010

Brinkhuis, H. & Schouten, S. (2006). Episodic fresh surface waters in the Eocene Arctic Ocean. *Nature* 441 (7093): 606–9. doi: 10.1038/nature04692.

Brooks, G. L. & Whittaker, J. B. (24 Dec., 2001). Responses of three generations of a xylem-feeding insect, *Neophilaenus lineatus* (Homoptera), to elevated CO_2. *Global Change Biology* 5 (4): 395–401. https://doi.org/10.1046/j.1365-2486.1999.00239.x.

Carini, P., White, A. E., Campbell, E. O., & Giovannoni, S. J. (7 July, 2014). Methane production by phosphate-starved SAR11 chemoheterotrophic marine bacteria. *Nature Communications* 5, Article number 4346. doi: 10.1038/ncomms5346.

Charlson, R. J., Lovelock, J. E., Andreae, M. O., & Warren, S. G. (16 April, 1987). Oceanic phytoplankton, atmospheric sulphur, cloud albedo and climate. *Nature* 326: 655–61. doi: 10.1038/326655a0. www.nature.com/nature/journal/v326/n6114/abs/326655a0.html.

Cox, P. M., Betts, R. A., Jones, C. D., Spall, S. A., & Totterdell, I. J. (2000). Acceleration of global warming due to carbon-cycle feedbacks in a coupled climate model. *Nature* 408 (6809): 184–7. doi: 10.1038/35041539. PMID 11089968. www.nature.com/nature/journal/v408/n6809/abs/408184a0.html.

Cropp, R. A., Gabric, A. J., McTainsh, G. H., Braddock, R. D., & Tindale, N. (2005). Coupling between ocean biota and atmospheric aerosols: Dust, dimethylsulphide, or artifact?. *Global Biogeochemical Cycles* 19: GB4002. doi: 10.1029/2004GB002436. www.agu.org/pubs/crossref/2005/2004GB002436.shtml.

Dacey, J., Drake, B. & Klug, M. (1994). Stimulation of methane emission by carbon dioxide enrichment of marsh vegetation. *Nature* 370: 47–9. https://doi.org/10.1038/370047a0.

Danovaro, R., Corinaldesi, C., Dell'anno, A., Fabiano, M. & Corselli, C. (2005). Viruses, prokaryotes and DNA in the sediments of a deep-hypersaline anoxic basin (DHAB) of the Mediterranean Sea. *Environmental Microbiology* 7: 586–92.

Danovaro, R., Corinaldesi, C., Dell'anno, A., Fuhrman, J. A., Middelburg, J. J., Noble, R. T., & Suttle, C. A. (Nov., 2011). Marine viruses and global climate change. *FEMS Microbiology Reviews* 35 (6): 993–1034. https://doi.org/10.1111/j.1574-6976.2010.00258.x.

Doney, S. C. (March, 2006). The dangers of ocean acidification. *Scientific American* 3: 58–65. (secondary literature).

Drever, J. I. & Zobrist, J. (1992). Chemical weathering of silicate rocks as a function of elevation in the southern Swiss Alps. *Geochim. Cosmochim. Acta* 56: 3209–16.

Duchesne, L. C. & Larson, D. W. (Apr., 1989). Cellulose and the evolution of plant life. *BioScience* 39 (4): 238–41.

Emilson, E. J. S., et al. (4 May, 2018). Climate-driven shifts in sediment chemistry enhance methane production in northern lakes. *Nature Communications* 9, Article number 1801. doi: 10.1038/s41467-018-04236-2.1089969.

Fabry, V. J., et al. (April, 2008). Impacts of ocean acidification on marine fauna and ecosystem processes. *ICES Journal of Marine Science* 65 (3): 414–32. https://doi.org/10.1093/icesjms/fsn048.

Falkowski, P. G., Katz, M. E., Knoll, A. H., Quigg, A., et al. (2004). The evolution of modern eukaryotic phytoplankton. *Science* 305: 355–60.

Fensome, R. A., MacRae, R. A., Moldowan, J. M., Taylor, F. J. R., & Williams, J. L. (1996). The early Mesozoic radiation of dinoflagellates. *Paleobiology* 22: 329–38.

Flynn, K. J. & Mitra, A. (1 Sept., 2009). Building the "perfect beast": Modeling mixotrophic plankton. *Journal of Plankton Research* 31 (9): 965–92.

Frings, P. J. (Aug., 2019). Palaeoweathering: How do weathering rates vary with climate? *Elements* 15: 259–65. doi: 10.2138/gselements.15.4.259.

Froschauer, A. & Coleman, J. (17 Jan., 2012). North American bat death toll exceeds 5.5 million from white-nose syndrome. White-Nose syndrome.org. U.S. Fish & Wildlife Service.

Fuhrman, J. A. (10 June, 1999). Marine viruses and their biogeochemical and ecological effects. *Nature* 399 (6736): 541–8.

Fuhrman, J. A. (10 June, 1999). Marine viruses and their biogeochemical and ecological effects. *Nature* 399: 541–8.

Gatti, L. V., Basso, L. S., Miller, J. B., et al. (2021). Amazonia as a carbon source linked to deforestation and climate change. *Nature* 595: 388–93. https://doi.org/10.1038/s41586-021-03629-6.

Gibbs, S. J., et al. (30 Oct., 2020). Algal plankton turn to hunting to survive and recover from end-Cretaceous impact darkness. *Science Advances* 6 (44). doi: 10.1126/sciadv.abc9123.

Griffiths, R. P., Baham, J. E., & Caldwell, B. A. (1996). Soil solution chemistry of ectomycorrhizal mats in forest soil. *Soil Biology and Biochemistry* 26: 331–7.

Haqq-Misra, J. D., Domagal-Goldman, S. D., Kasting, P. J., & Kasting, J. F. (3 Feb., 2009). A revised, hazy methane greenhouse for the Archean Earth. *Astrobiology* 8 (6): 1127–37. https://doi.org/10.1089/ast.2007.0197.

Haroon, M. F., et al. (2013) Anaerobic oxidation of methane coupled to nitrate reduction in a novel archaeal lineage. *Nature* 500 (7464): 567–70. 10.1038/nature12375.

Hathaway, W. (June, 2001). The microbe factor and its role in our climate future. *Journal of Bacteriology* 183 (12): 3770–83. doi: 10.1128/JB.183.12.3770-3783.2001.

Higgs, N. D., et al. (1 June, 2012). Evidence of *Osedax* worm borings in Pliocene (~3 Ma) whale bone from the Mediterranean. *Historical Biology* 24 (3): 269–77. doi:10.1080/08912963.2011.621167. ISSN 0891-2963. S2CID 85170976.

Iglesias-Rodriguez, M. D. Halloran, P. R., Rickaby, R. E. M., Hall, I. R., et al. (2008). Phytoplankton calcification in a high-CO_2 world. *Science* 320 (5874): 336–40.

IPCC (2013). *Climate Change 2013: The Physical Science Basis. Contribution of Working Group I to the Fifth Assessment Report of the Intergovernmental Panel on Climate Change.* Stocker, T. F. and 9 coeditors. 1535 pp. Cambridge Univ. Press, Cambridge, UK.

Jónasdóttir, S. H., et al. (29 Sept., 2015). Seasonal copepod lipid pump promotes carbon sequestration in the deep North Atlantic. *PNAS USA* 112 (39): 12122–6, doi: 10.1073/pnas.1512110112.

Katija, K., et al. (3 May 2017). New technology reveals the role of giant larvaceans in oceanic carbon cycling. *Science Advances* 3 (5). e1602374. doi: 10.1126/sciadv.1602374.

Kinsey, D. W. & Hopley, D. (March, 1991). The significance of coral reefs as global carbon sinks—response to Greenhouse. *Palaeogeography, Palaeoclimatology, Palaeoecology* 89 (4): 363–77.

Lawrence, D., et al. (24 March, 2022). The unseen effects of deforestation: Biophysical effects on climate. *Frontiers in*

Forests and Global Change. https://doi.org/10.3389/ffgc.2022.756115.

Lawrence, J. E., Chan, A. M., & Suttle, C. A. (March, 2002). Viruses causing lysis of the toxic bloom-forming alga, *Heterosigma akashiwo* (Raphidophyceae), are widespread in coastal sediments of British Columbia, Canada. *Limnology and Oceanography* 47 (2): 545–50. https://doi.org/10.4319/lo.2002.47.2.0545.

Lawrence, J. E. & Suttle, C. A. (2004). Effect of viral infection on sinking rates of *Heterosigma akashiwo* and its implications for bloom termination. *Aquatic Microbial Ecology* 37: 1–7.

Lazier, J. R. N. (2006). *Dynamics of Marine Ecosystems*, p. 35. Wiley-Blackwell, Hoboken, NJ.

Lebrato, M., Iglesias-Rodriguez, D., Feely, R. A., Greeley, et el. (2010). Global contribution of echinoderms to the marine carbon cycle: a reassessment of the oceanic $CaCO_3$ budget and the benthic compartments. *Ecological Monographs* 80 (3): 441–67. doi: 10.1890/09-0553. www.esajournals.org/doi/abs/10.1890/09-0553.

Lenton, T. & Watson, A. (2011). *Revolutions that Made the Earth*. Oxford Univ. Press, Oxford, UK.

Li, K.-F., et al. (16 June, 2009). Atmospheric pressure as a natural climate regulator for a terrestrial planet with a biosphere. *PNAS USA* 106 (24): 9576–9. https://doi.org/10.1073/pnas.0809436106.

Lian, X. (3 Jan., 2020). Summer soil drying exacerbated by earlier spring greening of northern vegetation. *Science Advances* 6 (1): eaax0255. doi: 10.1126/sciadv.aax0255.

Lieth H. & Whittaker, R.H. (eds.). (1975). *Primary Productivity of the Biosphere. Ecological Studies (Analysis and Synthesis)*, vol. 14. Springer, Berlin, Heidelberg, Germany. https://doi.org/10.1007/978-3-642-80913-2_1.

Lloyd, J. & Farquhar, J. D. (1994). ^{13}C atom discrimination during CO2 assimilation by the terrestrial biosphere. *Oecologia* 99: 201–15.

Lovelock (1979). *Gaia. A New Look at Life on Earth*. Oxford University Press, Oxford, New York, Toronto, Melbourne.

Lovelock, J. (2007). *The Revenge of Gaia*. Penguin, London, UK. ISBN 0141025972.

Lovelock, J. E. & Whitfield, M. (8 April, 1982). The lifespan of the biosphere. *Nature* 296: 561–3.

Lucas, Y. (2001). The role of plants in controlling rates and products of weathering: importance of biological pumping. *Annual Review of Earth and Planetary Sciences* 29: 135–63.

Madin, L. P., et al. (May, 2006). Periodic swarms of the salp *Salpa aspera* in the slope water off the NE United States: Biovolume, vertical migration, grazing, and vertical flux. *Deep Sea Research Part I: Oceanographic Research Papers* 53 (5): 804–19. https://doi.org/10.1016/j.dsr.2005.12.01.

Mann, D. G. & Droop, S. J. M. (1996). Biodiversity, biogeography and conservation of diatoms. Biogeography of freshwater algae. *Hydrobiologia* 336: 19–32.

Marsh, M.E. (2003). Regulation of $CaCO_3$ formation in coccolithophores. *Comparative Biochemistry and Physiology Part B: Biochemistry and Molecular Biology* 136 (4): 743–54. doi:10.1016/s1096-4959(03)00180-5. PMID 14662299.

McInherney, F. A. & Wing, S. (2011). A perturbation of carbon cycle, climate, and biosphere with implications for the future. *Annual Review of Earth and Planetary Sciences* 39: 489–516. Bibcode: 2011AREPS..39..489M. doi:10.1146/annurev-earth-040610-133431.

McKinney, A. M., et al. (1 Sept., 2012). Asynchronous changes in phenology of migrating Broad-tailed hummingbirds and their early-season nectar resources. *Ecology* 93 (9): 1987–93. https://doi.org/10.1890/12-0255.1.

Miller, C. B. (2004). *Biological Oceanography*, pp. 94–5, 266–7. Blackwell Science Ltd.. Oxford, UK.

Milliman, J. D. (1974). *Marine Carbonates*. Springer-Verlag, New York, NY.

Mitra, A., et al. (April, 2016). Defining planktonic protist functional groups on mechanisms for energy and nutrient acquisition: Incorporation of diverse mixotrophic strategies. *Protists* 167: 106–20. http://dx.doi.org/10.1016/j.protis.2016.01.003.

Mitra, A. (April, 2018). Tiny creatures, part plant and part animal, may control the fate of the planet. *Scientific American*. www.scientificamerican.com/article/tiny-creatures-part-plant-and-partanimal-may-control-the-fate-of-the-planet/. (secondary literature).

Moffitt, S. E., et al. (2015). Paleoceanographic insights on recent oxygen minimum zone expansion: Lessons for modern oceanography. *PLoS ONE* 10 (1): e0115246. https://doi.org/10.1371/journal.pone.0115246.

Mojica, K. D. A. & Brussaard, C. P. D. (2015). The viral shunt in a stratified Northeast Atlantic Ocean. In Mojica, K. D. A., *Viral Lysis of Marine Microbes in Relation to Vertical Stratification*. UvA-DARE (Digital Academic Repository); Ch. 7. Univ. of Amsterdam, Amsterdam, the Netherlands.

Morris, R. M., et al. (13 Dec., 2002). SAR11 clade dominates ocean surface bacterioplankton communities. *Nature* 420: 806–10. doi: 10.1038/nature01240.

Morse, J. W. & Mackenzie, F. T. (1990). *Geochemistry of Sedimentary Carbonates*. Elsevier, Amsterdam, the Netherlands.

Moulton, K. L, West, J., & Berner, R. A. (2000). Solute flux and mineral mass balance approaches to quantification of plant effects on silicate weathering. *American Journal of Science* 300: 539–70.

Nicol, S., et al. (June, 2011). Southern Ocean iron fertilization by baleen whales and Antarctic krill. *Fish and Fisheries* 11 (2): 203–9. doi: 10.1111/j.1467-2979.2010.00356.x.

O'Neill, B., Oppenheimer, M., Warren, R., et al. (4 Jan., 2017). IPCC reasons for concern regarding climate change risks. *Nature Climate Change* 7: 28–37. https://doi.org/10.1038/nclimate3179.

Pachiadaki, M. G., et al. (24 Nov., 2017). Major role of nitrite-oxidizing bacteria in dark ocean carbon fixation. *Science* 358 (6366): 1046–51. doi: 10.1126/science.aan8260.

Parmesan, C. & Yohe, G. (2 Jan., 2003). A globally coherent fingerprint of climate change impacts across natural systems. *Nature* 421: 37–42. doi: 10.1038/nature01286

Parr, J. & Sullivan, L. (2005). Soil carbon sequestration in phytoliths. *Soil Biology and Biochemistry* 37: 117–24. doi: 10.1016/j.soilbio.2004.06.013.

Pavlov, A. A., et al. (5 May, 2000). Greenhouse warming by CH_4 in the atmosphere of early Earth. *Journal of Geophysics Research; Planets* 105 (E5): 11981–90. https://doi.org/10.1029/1999JE001134.

Pearson, P.N. & Palmer, M.R. (2000). Atmospheric carbon dioxide concentrations over the past 60 million years. *Nature* 406 (6797): 695–9. doi: 10.1038/35021000. www.nature.com/nature/journal/v406/n6797/pdf/406695a0.pdf.

Penttilä, A., Slade, E. M., Simojoki, A., Riutta, T., Minkkinen, K., & Roslin, T. (Aug., 2013). Quantifying beetle-mediated effects on gas fluxes from dung pats. *PLoS ONE* 8 (8): e71454. doi: 10.1371/journal.pone.0071454.

Perissinotto, R. & Pakhomov, E. A. (1998). Contribution of salps to carbon flux of marginal ice zone of the Lazarev Sea, southern ocean. *Marine Biology* 131: 25–32.

Pfaller, J. B., et al. (9 March, 2020). Odors from marine plastic debris elicit foraging behavior in sea turtles. *Current Biology* 30 (5): PR213–R214. doi: https://doi.org/10.1016/j.cub.2020.01.071.

Pidgeon, E. (2009). Carbon sequestration by coastal marine habitats: Important missing sinks. *The Management of Natural Coastal Carbon Sinks*. IUCN 47–53.

Poorvin, L., Rinta-Kanto, J. M., Hutchins, D. A., & Wilhelm, S. W. (2004). Viral release of iron and its bioavailability to marine plankton. *Limnology and Oceanography* 49: 1734–41.

Proctor, L. M. & Fuhrman, J. A. (1990). Viral mortality of marine bacteria and cyanobacteria. *Nature* 343: 60–2.

Prothero, D. R. (2004). *Bringing Fossils to Life: An Introduction to Paleobiology* (2nd ed.), pp. 210–13. McGraw-Hill, Boston, MA.

Quirk, J., et al. (22 Aug., 2015). Constraining the role of early land plants in Palaeozoic weathering and global cooling. *Proc. of the Royal Soc. B. Biol. Sciences* 282 (1813). doi: 10.1098/rspb.2015.1115.

Retallack, G. J. (2001). Cenozoic expansion of grasslands and climatic cooling. *Journal of Geology* 109: 407–26.

Ricart, A. M., York, P. H., Rasheed, M. A., et al. (15 Nov., 2015). Variability of sedimentary organic carbon in patchy seagrass landscapes. *Marine Pollution Bulletin* 100 (1): 476–482. https://doi.org/10.1016/j.marpolbul.2015.09.032.

Roman, J. & McCarthy, J. J. (11 Oct., 2010). The whale pump: Marine mammals enhance primary productivity in a coastal basin. *PLoS ONE* 5 (10): e13255. doi: 10.1371/journal.pone.0013255.

Root, T. R., et al. (2 Jan., 2003). Fingerprints of global warming on wild animals and plants. *Nature* 421: 57–60. doi: 10.1038/nature01333.

Roubiex, V., et al. (June, 2008). Diatom succession and silicon removal from freshwater in estuarine mixing zones: From experiment to modelling. *Estuarine Coastal and Shelf Science* 18 (1): 14–26. doi: 10.1016/j.ecss.2007.11.007.

Savoca, M. S. & Nevitt, G. A. (18 March, 2014). Evidence that dimethyl sulfide facilitates a tritrophic mutualism between marine primary producers and top predators. *PNAS USA* 111 (11): 4157–61. https://doi.org/10.1073/pnas.1317120111.

Segar, D. (2012). *Introduction to Ocean Sciences*, 3rd Ed., Ch. 5. www.reefimages.com/oceans/SegarOcean3Chap05. ISBN 978-0-9857859-0-1. Corpus ID: 129389136.

Seibold, S., Rammer, W., Hothorn, T., et al. (1 Sept., 2021). The contribution of insects to global forest deadwood decomposition. *Nature* 597: 77–81. https://doi.org/10.1038/s41586-021-03740-8.

Shaw, G. E., Benner, R. L., Cantrell, W., & Clarke, A. D. (1998). The regulation of climate: A sulfate particle feedback loop involving deep convection—An editorial essay. *Climate Change* 39: 23–33. doi: 10.1023/A:1005341506115. www.ingentaconnect.com/content/klu/clim/1998/00000039/00000001/00164320.

Shukla, J. & Mintz, Y. (1982). Influence of land-surface evapotranspiration on the Earth's climate. *Science* 215: 1498–501.

Siegel, D. A., et al. (8 March, 2016). Prediction of the export and fate of global ocean net primary production: The EXPORTS Science Plan. *Frontiers in Marine Science. Ocean Observation*. doi: 10.3389/fmars.2016.00022.

Sillett, S. C., et al. (15 Feb., 2020). Aboveground biomass dynamics and growth efficiency of *Sequoia sempervirens* forests. *Forest Ecology and Management* 458: 117740. https://doi.org/10.1016/j.foreco.2019.117740.

Smith, C. (2005). Bigger is better: The role of whales as detritus in marine ecosystems. *Whales, Whaling, and Ocean Ecosystems* 12 (3): 1–46.

Smith, C. R. & Baco, A. R. (2003). Ecology of whale falls at the deep-sea floor. *Oceanography and Marine Biology: an Annual Review* 41: 311–54.

Simon, N., et al. (Feb.-Mar., 2009). Diversity and evolution of marine phytoplankton. *Comptes Rendus Biologies* 332 (2–3): 159–70.

Stiling, P., et al. (1 Feb., 1999). Decreased leaf miner abundance in elevated CO_2: Reduced leaf quality and increased parasitoid attack. *Ecological Applications*. https://doi.org/10.1890/1051-0761(1999)009[0240:DLMAIE]2.0.CO;2.

Stoecker, D. K., et al. (Jan., 2017). Mixotrophy in the marine plankton. *Annual Review of Marine Science* 9: 311–35.

Stuart, S. N. & Chanson, J. S. (2004). Status and trends of amphibian declines and extinctions worldwide. *Science* 306: 1783–6.

Subramaniam, A., et al. (29 July, 2008). Amazon River enhances diazotrophy and carbon sequestration in the tropical North Atlantic Ocean. *PNAS USA* 105 (30): 10460–5. https://doi.org/10.1073/pnas.0710279105.

Suttle, C. A. (15 Sept., 2005). Viruses in the sea. *Nature* 437: 356–61. doi: 10.1038/nature04160.

Suttle, C. A. (Oct., 2007). Marine viruses—major players in the global ecosystem. *Nature Reviews Microbiology* 5 (10): 801–12. doi: 10.1038/nrmicro1750. PMID 17853907.

Tangley, L. (6 April, 2001). High CO_2 levels may give fast-growing trees an edge. *Science* 292 (5514): 36–7. doi: 10.1126/science.292.5514.36.

Tappan, H. (1980). *The Paleobiology of Plant Protists*. W. H. Freeman and Co., San Francisco, CA.

Treguer, P., et al. (1995). The silica balance in the world ocean: A reestimate. *Science* 268 (5209): 375–9. doi: 10.1126/science.268.5209.375. PMID 17746543.

Ueno, Y., et al. (23 March, 2006). Evidence from fluid inclusions for microbial methanogenesis in the early Archaean era. *Nature* 440: 516–9. doi: 10.1038/nature04584.

Vallina, S. M., Simo, R., Gasso, S., De Boyer-Montegut, C., del Rio, E., Jurado, E., & Dachs, J. (2007). Analysis of a potential "solar radiation dose-dimethylsulfide-cloud condensation nuclei" link from globally mapped seasonal correlations. *Global Biogeochemical Cycles* 21: GB2004. doi: 10.1029/2006GB002787. www.agu.org/pubs/crossref/2007/2006GB002787.shtml.

van Breeman, N., Finlay, R., Lundstrom, U., Jongmans, A. G., Giesler, R., & Olsson, M. (2000). Mycorrhizal weathering: a true case of plant nutrition? *Biogeochemisty* 49: 53–67.

Veraverbeke, S., et al. (26 June, 2017). Lightning as a major driver of recent large fire years in North American boreal forests. *Nature Climate Change* 7: 529–34.

Waddell, L. M. & Moore, T. C. (2006). Salinity of the early and middle Eocene Arctic Ocean from oxygen isotope analysis of fish bone carbonate. *American Geophysics Union, Fall Meeting 2006, abstract# OS53B-1097.* www.agu.org/cgi-bin/wais?hh=OS53B-1097.

Walsh, K. J. E., McBride, J. L., Klotzbach, P. J., et al. (Jan./Feb., 2016). Tropical cyclones and climate change. *WIREs Climate Change* 7 (1): 65–89. https://doi.org/10.1002/wcc.371.

Ward, P. (2009). *The Medea Hypothesis. Is Life on Earth Ultimately Self-Destructive?* Princeton Univ. Press, Princeton, NJ, Oxford, UK, p. 86.

Ware, J. R., Smith, S. V., & Reaka-Kudla, M. L. (1991). Coral reefs: Sources or sinks of atmospheric CO_2? *Coral Reefs* 11 (3): 127–30.

Watson, A. J. & Liss, P. S. (29 Jan., 1998). Marine biological controls on climate via the carbon and sulphur geochemical cycles. *Philosophical Transactions of the Royal Society B: Biological Sciences* 353 (1365). https://doi.org/10.1098/rstb.1998.0189. Print ISSN:0962-8436. Online ISSN:1471-2970.

Weinbauer, M.G. (2004). Ecology of prokaryotic viruses. 28: 127–81.

Weitz, J. S. & Wilhelm, S. W. (5 Sept., 2012). Ocean viruses and their effects on microbial communities and biogeochemical cycles. *F1000Reports/Biol*. 4: 17. doi: 10.3410/B4-17.

Wigley, T. M. L. (1983). The preindustrial carbon dioxide level. 5: 315–20. https://doi.org/10.1007/BF02423528.

Wilhelm, S. W. & Suttle, C. A. (Oct., 1999). Viruses and nutrient cycles in the sea: viruses play critical roles in the structure and function of aquatic food webs. *Bioscience* 49:781–8. https://doi.org/10.2307/1313569.

Wilmers, C. C., et al. (2012). Do trophic cascades affect the storage and flux of atmospheric carbon? An analysis of sea otters and kelp forests. *Frontiers in Ecology and the Environment* 10: 409–15.

Wong, C. M. & Daniels, L. D. (May, 2017). Novel forest decline triggered by multiple interactions among climate, an introduced pathogen and bark beetles. *Global Change Biology* 23 (5): 1926–4. https://doi.org/10.1111/gcb.13554.

Woodwell, G. M. (Jan., 1978). The biota and the world carbon budget. The terrestrial biomass appears to be a net source of carbon dioxide for the atmosphere. *Science* 199 (13): 141–6.

Zeebe, R. E. & Caldeira, K. (2008). Close mass balance of long-term carbon fluxes from ice-core CO_2 and ocean chemistry records. *Nature Geosci* 1: 312–5. https://doi.org/10.1038/ngeo185.

Zhang, C., et al. (July, 2018). Evolving paradigms in biological carbon cycling in the ocean. *National Science Review* 5 (4): 481–99. doi: 10.1093/nsr/nwy074 (2018). https://academic.oup.com/nsr/article/5/4/481/5053813.

Zhili, H., et al. (2012). The phylogenetic composition and structure of soil microbial communities shifts in response to elevated carbon dioxide. *ISME Journal* 6: 259–72. doi: 10.1038/ismej.2011.99.

3 Organisms Created High Oxygen Levels, Which Allowed Complex Life to Evolve and Diversify

3.1 LIFE INCREASED EARTH'S OXYGEN LEVEL IN THE GREAT OXIDATION EVENT

It is clear from Chapter 2 that life profoundly changed the chemistry of the atmosphere in ways favorable to life and biodiversity. One of the most important such changes is life's creation of an atmosphere high in oxygen. Oxygen concentrations in the Earth's atmosphere rose from negligible levels of much less than 1% in the Archaean Eon (4 to 2.5 bya) to about 21% in the present day. This also oxygenated the soil and both fresh- and saltwater ecosystems. It is undeniable that life, through ecosystem engineering by both photosynthesis and the removal of hydrogen and carbon, produced the high level of oxygen present in today's atmosphere, and needed for higher life. Prior to the origin of life, atmospheric oxygen was negligible near Earth's surface (Kasting, 1993). Atmospheric oxygen levels remained negligible for almost the entire first half of Earth's 4.6 billion-year history. Thus, complex life could not evolve, and only prokaryotes existed. Then oxygen levels increased, in an event called the Great Oxidation Event, or GOE (Sessions et al., 2009, and references therein). This occurred about 2.3 to 2.45 bya (ibid.). Before the GOE, atmospheric oxygen was likely less than 10^{-5} of the level that is present today (ibid.). Oxygen levels rose a great deal during the GOE, but, by the end of it, were still very low compared to today. During the GOE, atmospheric oxygen levels rose greatly from likely less than 10^{-5} of today's level to a few percent of today's level of about 21%.

Recall from the previous chapter that oxygenic photosynthesis produces oxygen. It is carried out primarily by plants, algae, phytoplankton, and cyanobacteria. It is by far the main source of oxygen in the atmosphere, and was so in the past. Oxygenic photosynthesis produces 98 to 99% of the atmosphere's oxygen, while the breakup of water molecules by ultraviolet radiation from the sun, a nonbiological process, makes the other 1 to 2%. Oxygen production by photosynthesis is necessary but not sufficient for the accumulation of oxygen in the biosphere. Respiration, the decay of dead organisms, and fire consume oxygen, with the result that there is no net gain of it from photosynthesis alone. Carbon combines with oxygen and so depletes it during respiration, the decomposition of dead organisms, and combustion. Carbon and hydrogen react with oxygen and consume it in other reactions as well. Burial of CO_2 regulates temperature, but does not increase oxygen, since one oxygen molecule (O_2) is buried with each carbon atom when CO_2 is buried. To increase free atmospheric oxygen, the system must remove hydrogen or carbon that are not attached to oxygen, in addition to producing oxygen by photosynthesis. This is mostly done by removing reduced carbon, rather than hydrogen. Reduced carbon is usually combined with hydrogen, so burying it generally helps increase oxygen in the air in two ways: burial of carbon and burial of hydrogen. When carbon or hydrogen that are not combined with oxygen are buried, there is less of them to combine with and deplete oxygen, so there are more free oxygen molecules that are not combined with either of these elements in the biosphere. Hence, for example, burying methane not only lowers temperature, it also increases oxygen in the biosphere because one atom of carbon and four atoms of hydrogen are buried without burying any oxygen every time one molecule of methane is buried. Most reduced carbon is recycled back into the biosphere. And some is sequestered by nonbiological processes, principally movement of the continents and nonbiological weathering. However, life's sequestration of reduced carbon is the main mechanism that works in conjunction with photosynthesis to increase oxygen. The many kinds of organisms that bury carbon discussed in the last chapter bury both reduced carbon and carbon that is combined with oxygen.

As discussed in the previous chapter, there is another form of photosynthesis, called anoxygenic photosynthesis, which also uses sunlight as an energy source. It uses hydrogen sulfide (H_2S) instead of water and produces molecular sulfur instead of molecular oxygen. It also makes carbohydrates, a food source for the photosynthesizer. This process is carried out by several groups of bacteria, such as green sulfur bacteria, purple bacteria, heliobacteria, and acidobacteria. Anoxygenic photosynthesis produced many new niches, leading to prokaryotic diversification into many species, showing that life promotes diversification and supporting the ABH. But it does not produce oxygen, so is not relevant to this discussion. Oxygenic photosynthesis is the type of photosynthesis that produces oxygen, and in this chapter, the word "photosynthesis" will refer to oxygenic photosynthesis, unless otherwise indicated, since that is the type of photosynthesis this discussion is primarily concerned with.

It had been generally accepted, based on the fossil record, that cyanobacteria were the first organisms to evolve oxygenic photosynthesis, and that several species of them in the ocean produced the oxygen responsible for the GOE.

DOI: 10.1201/9781003246640-3

They often occurred in stromatolites, microbial mats that consist of several species of microbes in layers. There has been a vigorous debate as to when cyanobacteria evolved, as Anbar et al. (2007) and Kaufman et al. (2007) found evidence that small, but significant amounts of oxygen appeared in the Earth's oceans and atmosphere 2.5 bya, 50 to 200 million years prior to the GOE, and others presented evidence in molecular fossils that oxygenic photosynthesis originated no later than 2.7 or even 2.75 bya (Brocks et al., 1999, 2003; Summons et al., 2006; Planavsky et al., 2014), and therefore at least 250 million years before the GOE. Others question the authenticity of the early molecular fossils and other evidence, asserting that there was an abrupt, catastrophic oxygenation almost as soon as cyanobacteria evolved, and this was at the time of the GOE (Brasier et al., 2006; Kopp et al., 2005; Johnson et al., 2013). Oxygen was toxic to many prokaryotes when photosynthesis first appeared, for many were not yet adapted to it, and it could kill them. There are microbes today that can only live in habitats where there is no oxygen. So a sudden rise in oxygen certainly caused a decrease in the diversity of microbes. In this chapter, by diversity, I mean number of species. It could have potentially been catastrophic to diversity at that time, and some think the GOE caused a mass extinction of prokaryotes, providing a counterexample to the ABH. Prokaryotes (bacteria and archaea) and viruses were the only life at the time of the GOE.

But Cardona (2018) presented evidence that oxygenic photosynthesis evolved 3.6 bya, a billion years earlier than previously thought. This suggests a tendency of evolution toward complexity, innovation, and to new adaptations. It suggests cyanobacteria were not the first to evolve oxygenic photosynthesis. However, dating of the GOE likely will not change as a result of this, for there was a time lag between the first appearance of oxygenic photosynthesis and the GOE. In addition, cyanobacteria were still likely among the main organisms causing the GOE. Above all, Cardona's findings suggest prokaryotes were exposed to oxygen for ample time to adapt to its toxic effects before the GOE, and that die-offs because of high toxic oxygen were not severe, if oxygen was kept at low levels.

This apparently was the case, for oxygen could not have accumulated abruptly after the advent of photosynthesis. There were a great deal of compounds of sulfur, iron, manganese, nitrogen (mainly as ammonia), and other elements that combined with the oxygen, keeping it at low levels in the atmosphere for a long time period (Kump and Barley, 2007, and references therein). Oxygen could accumulate only after it fully oxidized these compounds, making it so that they could no longer combine with oxygen. This accounts for the lag between the first appearance of oxygenic photosynthesis and the GOE. Also, weathering of rocks removes a large quantity of oxygen from the air today, and it would likely have done so in these early times. This would have kept oxygen levels low, but would not have depleted it, because photosynthesis produced more oxygen, and organisms and geological processes buried reduced carbon.

Therefore, the transformation to an atmosphere high in oxygen was a gradual process that took several hundreds of millions of years, and there were continual small rises in oxygen levels followed by small decreases as oxygen combined with the reduced chemicals mentioned above. This gave microbes sufficient time to adapt to the oxygen increases. Microbes reproduce and mutate very rapidly, and so could have adapted rapidly to the oxygen increases. We know enzymes that protect cells from damage done by oxygen evolved, for they are common today. So it appears that microbes that oxygen was initially lethally toxic to and that were adversely affected by the GOE adapted to the increased oxygen levels. Furthermore, even in today's high-oxygen Earth and seas, the organic-rich marine sediments have no oxygen below a surface film of oxygenated sediment that has depths ranging from one millimeter to a few centimeters. These habitats shielding microbes from oxygen that is lethal to them surely existed during the GOE, and would have been a refuge from toxic oxygen. Some microbes that were sensitive to oxygen may very well have gone extinct, because of an inability to adapt rapidly enough to high oxygen or to settle in a habitat shielded from it. We do not know for certain that the GOE did not cause a great number of extinctions, but a mass extinction is very unlikely. And *the oxygen increase caused by photosynthesis and the burial of reduced carbon by life ultimately greatly increased biodiversity and complexity. It made the evolution of large, complex organisms possible.*

In fact, even the short-term effects of the GOE promoted diversity. Some prokaryotes suddenly had a way to take advantage of the many niches created by sulfides, ammonia, iron in its form that readily combines with oxygen, and other compounds that easily combine with oxygen that were abundant at the dawn of photosynthesis. These prokaryotes therefore diversified: some species evolved that oxidized sulfides, others oxidized ammonia, others oxidized iron, and others evolved to oxidize yet other compounds. These bacteria created several new compounds that represented new niches, such as sulfate, nitrate, and iron and manganese oxides, that several species of microbes evolved to exploit, and these latter forms thus also diversified. Still other new niches opened up with the sudden abundance of readily available reduced carbon compounds produced by oxygenic photosynthesis that were available at a low energy cost, leading to the evolution and diversification of species of microbes that could exploit these niches by consuming these compounds. Then other bacteria evolved that could eat these microbes, and diversified. Thus, photosynthesis led to the creation of many new niches in the form of several new chemical compounds that were consumed by microbes that evolved to exploit them. The result was a great deal of diversification of prokaryotes, and thus a great increase in the number of species of them, due to the evolution of photosynthesis. Said another way, photosynthesis created many new niches that microbes evolved to utilize, resulting in a great diversification of microbes, tremendously increasing their species number. This happened numerous times, and in several different groups of microbes. The reason all the different groups of microbes that evolved mentioned in this paragraph greatly diversified is that they had many new niches in the form of new chemical food sources and new prokaryotic species available to them to consume. Each of these new species also made possible the evolution of many new species of virus that preyed on them, so virus diversity greatly increased. These evolutionary innovations and adaptive radiations occurred on both land and in the sea. Thus, oxygenic photosynthesis directly and indirectly created several new niches for many species of prokaryotes and viruses, allowing both to diversify greatly.

Further diversification of the microbes occurred as a result of the fact that oxygen was toxic to many of them, and they had to adapt to it. They evolved chemical systems that protected them from its toxic effects. And oxygen is a great energy source. Some of the microbes evolved the ability to use oxygen to produce energy, and in the process, the chemicals that protected the microbes against oxygen were converted to enzymes that use oxygen to produce energy; that is, to the enzymes that carry out cellular respiration. *The very chemicals that protected microbes against the toxic effects of oxygen evolved into the enzymes that are used in cellular respiration, which is the process that uses oxygen as an energy source.* This converted oxygen from an enemy to an ally. Now the microbes could carry out cellular respiration—they could "breathe." Molecular oxygen is a very good molecule for producing energy. The production of higher oxygen levels by life provided the selective pressure that aided the evolution of the use of oxygen to produce the energy needed to support higher life. So because of life, life now had access to the energy needed to evolve complex, many-celled organisms, such as birds and mammals. Thus, *life sparked the evolution of the complex cellular respiration system of enzymes that uses oxygen to produce the energy source for all higher life.* Oxygenic photosynthesis selected for and led to the evolution of cellular respiration, with its utilization of oxygen as an energy source, a key innovation that allowed microbes to diversify into many new species. The high oxygen in the air and life's ability to use it for energy, both created by life, led to the evolution of complex organisms with many cells, such as redwood trees and humans.

Microbial mats flourished during the GOE. A microbial mat is a multilayered sheet of microorganisms of a number of species, usually bacteria and archaea, but sometimes just bacteria, in close symbiotic association. In moist conditions, mats are generally held together by slimy substances secreted by the microorganisms. Mats of coexisting bacteria and archaea date back to the early Archaean Eon and may have had a nearly modern structure as early as 3.5 bya (Nisbet and Fowler, 1999). The fossil record of early photosynthesis shows that three distinct prokaryote communities lived in vertical layers along a gradient of oxygen availability, within a few centimeters of each other (ibid. and references therein). In the top layer, bacteria produced organic carbon and oxygen through photosynthesis. Immediately below them, a middle layer dominated by bacteria that utilized oxygen and that could produce their own energy chemically (not through sunlight, as is the case in photosynthesis) took advantage of the oxygen produced by the top layer to get energy through combining oxygen with such chemicals such as sulfur, ammonia, iron, manganese, and others. The bacteria of this layer benefitted from the top layer. Slightly further below and at the very bottom, where there was no oxygen, a third community of bacteria lived off nutrients that seeped down from the two communities above it. Sometimes dead microbes seeped down and the microbes of this community ate them. These bacteria did not use oxygen, but rather used sulfate, nitrate, oxidized iron, and other oxidized by-products from the community above them for oxidation of the reduced carbon produced by photosynthesis by the bacteria in the top layer, as some of the oxidized compounds and reduced carbon trickled down to them. This layer produced sulfide and other compounds that were used by the middle layer. So the middle and lowest layers were symbiotic, benefiting each other. The bottom layer produced CO_2, used in photosynthesis by the top layer, making the bottom and top layers symbiotic as well, since the top layer also helped the bottom layer as explained above. Thus, a stable coexistence based on symbiosis was established early on in evolution. *The symbiotic evolution of chloroplasts* (which carry out photosynthesis for the cell) *and mitochondria* (which carry out respiration for the cell) *could have occurred in the mats, because of the close proximity of the species involved.* The chloroplast was originally a bacterium that could perform photosynthesis that was eaten by a microbe that did not digest it. It slowly evolved over many generations from a parasite to a symbiotic resident of the cell to the organelle that carries out photosynthesis for the cell. The mitochondrion evolved by the same mechanism of ingestion by a microbe and coevolution with it until it became the organelle of the cell that performs cellular respiration. These two cases of gradual coevolution were key to the evolution of the eukaryotic cell. Thus, *the evolution of the eukaryotic cell could have happened in microbial mats.* The close proximity of the layers to each other could have facilitated the ingestion of the bacteria that evolved into the chloroplasts and the mitochondria.

Poulton et al. (2021) found that oxygen rose and fell three times between 2.43 and 2.22 bya. After these events, oxygen increased until a tipping point was reached and it did not go down again. Permanent oxygenation of the atmosphere had arrived, and oxygen levels were high enough to combine with and deplete a good deal of the methane in the air. At that point, oxygen was able to end the time when the powerful greenhouse gas methane was the major influence on the climate. Henceforth, CO_2 from volcanic activity and other sources was the dominant greenhouse gas for keeping the planet warm. The scientists found that this permanent oxygenation of the atmosphere did not occur until about 2.22 bya, almost 100 million years later than thought until they did their work. This extends the GOE by almost 100 million years and means it took almost 100 million years longer than thought to achieve permanent oxygenation of the atmosphere. This means the GOE was more gradual than previously thought. Thus, microbes had even longer to adapt to the increasing oxygen levels. So a mass extinction of prokaryotes due to the GOE is even less believable.

Sulfate-reducing bacteria, which are bacteria that get their energy from sulfur, helped with the GOE (Halevy et al., 2012). Sulfate-reducing bacteria on the seafloor used the sulfate dissolved in seawater to obtain energy, producing oxygen and sulfide. The sulfide combined with iron, making the iron unable to combine with oxygen, and thus decereasing the amount of it. So sulfate-reducing bacteria increased the amount of oxygen in two ways: producing it and causing less depletion of it by iron. This aided the start of the GOE. Therefore, *sulfate-reducing bacteria had a major role in setting the stage for the GOE.*

It has been proposed (Catling et al., 2001) that hydrogen was removed from the biosphere in the early period of the GOE. Life played the major role. The very cyanobacteria that produced oxygen produced carbon compounds in so doing. Other bacteria consumed these compounds, yielding molecular hydrogen and acetate. Methanogens consumed this hydrogen and made methane. Hydrogen cannot easily escape into the air, because it readily combines with other chemicals in the soil and sea. But methane does not easily combine with most chemicals and is volatile. It is so light that much of it would have ascended to the stratosphere, where the ultraviolet light from the sun would have access to it, and break it into hydrogen and carbon. Since hydrogen is light, much of it would escape into space, while much of the heavier carbon would stay in Earth's atmosphere. Pope et al. (2012) provided evidence that this occurred during the GOE. Oxygen levels were too low at the start of the GOE to deplete the methane or for the methane to deplete it. And scientists have measured massive hydrogen production by today's microbial mats that include cyanobacteria on the coast of Mexico's Baja Peninsula (Hoehler et al., 2001), finding they pumped a thousand times more hydrogen into the air than typical volcanoes and hydrothermal vents, the two nonbiological hydrogen sources. It is reasonable to assume early microbial mats behaved similarly. At the time of the GOE, methanogens would have converted much of this hydrogen to methane, some of which would have escaped into space by the mechanism just described. Catling's group thinks the excess methane accumulated in the atmosphere to concentrations a few hundred to a few thousand times greater than modern levels, so copious quantities of hydrogen would have escaped into space by this mechanism. Thus, although not fully proven, *it appears most probable that microbial mats and methanogens were primarily responsible for a good deal of the early oxygen build up via hydrogen removal. This is another way that life helped oxygen increase in the atmosphere during the GOE.*

However, Kump and Barley (2007) proposed that nonbiological factors also contributed to the early rise in oxygen about 2.5 bya. They say the major oxygen sink in the Archaean Eon, which was enhanced submarine volcanism, was abruptly and permanently diminished about 2.5 bya. Land-based volcanism became widespread at the beginning of the Proterozoic eon. Volcanoes under the sea consume more oxygen than ones on land, so the shift from mainly volcanoes under the ocean to a mix of the two would have led to the rise of atmospheric oxygen.

3.2 HIGH OXYGEN LEVELS DIVERSIFIED MINERALS AND INCREASED THE NUMBER OF AMINO ACIDS USED BY LIFE. BOTH OF THESE INCREASED BIODIVERSITY

The GOE also caused a tremendous increase in the variety of minerals on Earth. It is estimated that the GOE was directly responsible for about 2,900 new minerals on Earth, by combining already existing minerals with oxygen (Hazen et al., 2008). There is a total of about 4,400 minerals on Earth today. After the GOE, chemical elements, such as uranium, vanadium, manganese, copper, selenium, tellurium, arsenic, antimony, bismuth, silver, and mercury could be present in one or more forms combined with oxygen in minerals near Earth's surface. This development resulted in an explosive growth in the diversification of minerals in the near-surface environment (Sverjensky et al., 2010). Also, the relative abundances of clay minerals were enhanced by major biological events, including the rise of deep-rooted vascular plants and their associated mycorrhizal fungi.

It is not just photosynthesis that increased the number of minerals. Bacteria interact with metal ions more than any other type of life. They have played and are continuing to play a big role in the minerals of most soils and sediments, with widespread importance on shaping the Earth and the recycling of many mineral elements. *Many of life's most important essential elements (carbon, sulfur, nitrogen, phosphorous) are abundant in the outer part of the Earth, where they are largely unavailable to many prokaryotes and eukaryotes. Some bacteria make these elements available to many other species today by processes such as weathering, and use them to make a great range of minerals. Their ability to accumulate metal ions has led to suggestions that bacteria clean natural environments of metals that are toxic when in concentrations that are too high, such as cadmium, nickel, and copper. Microbial communities work together to accomplish a great deal of these functions* (Douglas and Beveridge, 1998).

The evolution of the composition of Earth's atmosphere and minerals are intimately linked to the evolution of life throughout much of Earth's history. Life caused an increase in mineral diversity as well as biological diversity, especially during the GOE.

The life-induced increase in mineral diversity, like an increase in biodiversity, is an increase in information content on Earth caused by life. We know that bacteria and archaea often partition mineral niches, with different species specializing on different mineral compounds. *It is thus reasonable to assume that the increase in minerals led to an increase in prokaryote diversity, as prokaryotes diversified and filled the new mineral niches created by life, in a life-induced positive feedback loop between life and minerals. This is another example of the coevolution of the biosphere and geosphere.*

Amino acids are the building block molecules used to make proteins in all life. Granold et al. (2018) showed that high atmospheric O_2 was the main reason why the number of amino acids used to make proteins by life was increased from 13 to

20, even though functional proteins can be assembled by the original 13 amino acids alone. Oxygen caused the formation of free radicals, which the 7 new amino acids, which had greater reactivity than the original 13, reacted with. Obviously, the increase in the number of amino acids employed by life from 13 to 20 greatly increased the variety of proteins available. This in turn profoundly increased evolutionary innovation and diversification into new life forms, and hence biodiversity.

3.3 LIFE CAUSED FURTHER RISES IN OXYGEN LEVELS AFTER THE GOE

In the time since the GOE, reduced carbon was sequestered by phytoplankton, trees, echinoderms, salps, and other organisms, by the mechanisms described in Chapter 2. This happened while plants, phytoplankton, cyanobacteria, and other photosynthesizers added oxygen to the air by photosynthesis. Life performed the two necessary actions needed to increase oxygen levels: production of oxygen and sequestration of reduced carbon. Thus, life was almost the sole cause in the rise of oxygen levels after the end of the GOE. Of course, organisms sequestered oxidized carbon too, but this did not increase atmospheric oxygen levels. Nonbiological forces also removed some reduced carbon and hydrogen from the system, but much less than life did. And nonbiological factors have removed a great deal of oxygen from the atmosphere in the past because substances like iron, sulfur, manganese, and so on combined with it.

It is estimated that 1% of present-day oxygen levels were likely reached during the GOE, by 2.5 to 2.3 bya (Canfield, 2005; Holland, 2002). And it is estimated that atmospheric oxygen may have approached today's levels 2.22 to 2.06 bya, although these levels were maintained for only a few hundred million years (Bekker and Holland, 2012; Canfield et al., 2014; Lyons et al., 2014). The rise was accomplished, in addition to photosynthesis, partly by phosphate providing nutrient for prokaryote blooms, resulting in great numbers of prokaryotes. When some of the prokaryotes died, they sank, and this buried both reduced and carbon combined with oxygen. Of course, burying the reduced carbon increased oxygen in the air and sea. Thus, life was involved in this oxygen increase. But this ended and oxygen fell to low levels again, and oxygen levels did not rise back until about 580 mya. The oxygen content of the early Earth's surface environment is thought to have permanently increased in two broad steps: the previously discussed GOE and the Neoproterozoic Oxidation Event (NOE), during which oxygen possibly accumulated to the levels required to support animal life and oxygenate the deep oceans. The Neoproterozoic era was from one bya to 541 mya. The GOE is well accepted. The timing and extent of the NOE is not well agreed upon. Och and Shields-Zhou (2012) reviewed the evidence for the NOE and presented convincing evidence that it occurred and was important and was instrumental in major biological innovations and an abrupt diversification of the Ediacarian biota. The Ediacaran biota are the life of the Ediacaran period (about 635–541 mya). They were the first complex multicellular organisms, although many of the animals could not move, and were attached to rocks or the seafloor. Many of the animals were frond-shaped and tubular. They included trilobites and jellyfish. They were not as complex on the average as the animals that followed them. They first appeared about 580 mya and diversified into many forms about 575 mya. Canfield et al. (2014) also presented convincing evidence that oxygen levels rose about 580 mya to 15% of today's levels, or about 3% of the atmosphere. This corresponds well in time with the proposed event called the Avalon explosion, when the Ediacaran biota are thought to have undergone a tremendous diversification into many new species 575 million years ago (Shen et al., 2008). These researchers made a convincing case for a rise in oxygen on a global scale during the Late Neoproterozoic era, which would be the NOE. Therefore, *this oxygen increase is significant because it made possible the rise and diversification of the first complex multicellular animals*. It is not certain what caused this rise in oxygen, but there are some good hypotheses. Below are what I consider to be the two best ones. In both hypotheses, life played a major role in increasing oxygen levels.

Lichens and bryophytes (mosses and their relatives) likely aided this rise in oxygen by increasing weathering and thus burial of reduced carbon. Porada et al. (2014) showed that lichens and bryophytes have the ability to cause weathering, and these groups were abundant at this time.

The second mechanism is that there is convincing evidence that sponges were a key factor in this rise of oxygen. They first appear in the fossil record about 650 mya. Mills et al. (2014) and Mills and Canfield (2014) provided experimental evidence from studies with the crumb sponge (*Halichondria panacea*) that sponges could have survived very well in atmospheric oxygen levels as low as 0.5% to 4% of present ones, which probably existed well before higher animals evolved. Some modern sponges can live under oxygen levels 200 times less than present in the air today, which is about 0.105% of present levels. Sponges filter hundreds of liters of water per day, removing carbon, some of which is reduced and gets sequestered when the sponge (after its death) or its waste products are buried. A great number of sponges doing this would have cooled the planet and increased oxygen substantially. Sponges would have been abundant enough to increase oxygen levels in the sea substantially by the time of the appearance of the Ediacaran fauna.

It is not necessarily the case that only one mechanism is correct to the exclusion of the others. The increase in atmospheric oxygen preceding the rise of the Ediacaran animals could have resulted from a combination of different mechanisms. Life played a major role in all of the possible mechanisms, including ones that I did not discuss.

The Cambrian explosion was a huge diversification of animals that started about 541 mya, and brought about the origin of all modern animals. It is when higher animals appeared, when shells, boney skeletons, complex nervous systems, and sensory organs appeared. There are several hypotheses as to why life suddenly became more complex and diverse at this time. Not all of the hypotheses emphasize a rise in oxygen, although oxygen had to rise sufficiently to support this great increase in complexity and diversity. Chen et al. (2015) presented data demonstrating that there was an increase in oxygen to modern-like levels in oceanic bottom waters that coincided with the Cambrian diversification of both animals and eukaryotic phytoplankton, which he dates to 540–520 mya. They argued that animals benefitted from and contributed to the keeping of oxygen levels in the sea stable. Lenton et al. (2014) and Butterfield (2007) pointed out that animal and algal ecosystem engineers may have contributed to ocean oxygenation. This includes sponges, as previously discussed. They say small, planktonic animals and algae diversified nearly simultaneously in the early Cambrian. These groups sequestered reduced carbon because large numbers were buried in the seafloor when they died.

Logan et al. (1995) may have the best mechanism for the oxygen increase that precipitated the Cambrian explosion. In it, too, the driver is life. It was accomplished by the feces of small organisms. They showed from hydrocarbons in sediments that bacteria thrived in the Proterozoic eon, and they propose a significant portion of their energy was produced from sulfate, rather than using sunlight and photosynthesis. They consumed oxygen near the sea surface, inhibiting transport of oxygen to the deep ocean. These researchers found evidence suggesting that zooplankton (one-celled organisms that lived in the shallow waters of the sea) outcompeted the bacteria and greatly reduced their numbers, greatly reducing their consumption and depletion of oxygen near the ocean's surface. Importantly, the increase and diversification of zooplankton capable of forming large fecal pellets when they excreted their solid waste resulted in great quantities of these pellets sinking to the seafloor. This occurred at the beginning of the Cambrian, just before the Cambrian explosion. This would have buried a great deal of reduced carbon, and caused a large increase in oxygen in surface waters, leading in time to tremendous oxygenation of deep waters when waters at different depths were mixed. They suggest this provided the needed oxygen for the Cambrian explosion. In other words, zooplankton helped in two ways to produce the oxygen necessary for the increase in biodiversity and complexity in the Cambrian explosion, acting as ecosystem engineers with a positive effect of diversity, just as the ABH predicts.

The presence of organisms on exposed land accelerates the burial of carbon, some of it reduced, by weathering through physical and chemical processes (Schwartzman, 1999). Lenton and Watson (2004) pointed out that it has been proposed that colonization of land by eukaryotes occurred in the Neoproterozoic era, which ended just before the Cambrian explosion. This would include algae, fungi, and other small organisms. They hypothesized that this colonization involved selective weathering of phosphorous from rocks and an increase in overall weathering rates. They presented two models that predict that an increase in weathering of phosphorus to the sea would have caused an increase in oxygen in the air and sea in the Neoproterozoic. Increased phosphorous increases phytoplankton numbers, so they would have produced more oxygen through photosynthesis, and more of them would have sunk to the seafloor upon death, burying carbon, some of it reduced. This created the high oxygen levels in the sea necessary for the Cambrian explosion.

Heckman et al. (2021) analyzed protein sequences to provide evidence that green algae and fungi were present on land 1 bya and that land plants appeared by 700 mya. This is much earlier than previous estimates of 480 to 460 mya based on the age of the first fossil land plants and fungi. The authors say the land plants and fungi were present early enough to have produced the oxygen needed for the Cambrian explosion by weathering and burying reduced carbon. They also say the timing of the appearance of these organisms implicates them in at least one of the Snowball Earth events of about 750 to 580 mya. Lichens produced acids strong enough to dissolve rocks, releasing calcium that combined with carbon and was carried to the sea in rivers. Calcium carbonate limestone formed in the ocean, locking up carbon, keeping greenhouse gases made from carbon out of the atmosphere, and causing Snowball Earth. The key point of their research is that the early appearance of lichens and early land plants made it possible that they had key roles in the Cambrian explosion and Snowball Earth, acting as ecosystem engineers.

Cyanobacteria remained the principal oxygen producers throughout almost all of the Proterozoic eon. It was cyanobacteria that evolved into the chloroplast used by green plants for photosynthesis, by entering a cell from the outside. Green algae, which are eukaryotes, joined cyanobacteria as major photosynthesizers and producers of oxygen on the continental shelves of the seas near the end of this eon. It was not until the Mesozoic era, 252 to 66 mya, that the waters of the continental shelf took on their modern form in terms of oxygen production, with the diversification of the four groups of phytoplankton at that time: dinoflagellates, coccolithophorids, foraminiferans, and diatoms. These one-celled phytoplankton became important oxygen generators from the time of their evolution until the present. They are the major phytoplankton that produce most of the oxygen and support the major food webs in today's oceans. Cyanobacteria are still important oxygen generators in the ocean. Of course, land plants, especially trees, have been important producers of oxygen ever since they first appeared. Today, cyanobacteria, phytoplankton, and land plants produce almost all of Earth's oxygen. Land and sea plants produce 105 billion tons of biomass per year from CO_2 (Beerling, 2007), indicating they have a profound effect on regulating CO_2, reduced carbon, oxygen, and climate. Scientists are debating the relative importance of each group of photosynthetic species, and what follows are approximations. Freshwater and marine phytoplankton are responsible for over half of the Earth's total productivity and oxygen production, although they are less than 1% of the Earth's photosynthetic biomass (ibid.). Land plants account for 20 to 45% of Earth's oxygen, and constitute 90% of the photosynthetic biomass (ibid.). Cyanobacteria account for most of the remainder of the Earth's oxygen production. Cyanobacteria, photosynthesis, respiration, nitrogen fixation by bacteria symbiotic with plants that live in their roots, the symbiotic evolution of chloroplasts from cyanobacteria inside microbial cells, and the invasion of land by fungi were all necessary for the colonization of land by plants, which led to spectacular diversification of plants, making possible the colonization of land by animals, then the diversification of land animals. So the colonization of land by plants occurred as a result of many different kinds of organisms working together symbiotically. It then allowed animals to colonize land and greatly diversify into countless species on land.

The increase in atmospheric oxygen, eventually to 21%, created by life, provided the energy source necessary for all higher life, which could not exist without it. Without high atmospheric oxygen levels, all life would be simple. It would be unicellular or a few cells. High oxygen in both the air and water is necessary for all vertebrates, invertebrates, higher plants, and fungi. And these higher organisms provide habitat for over tens of millions of species of fungi, bacteria, archaea, and viruses that live in or on these animals and plants.

Lenton (2012) pointed out that the origin and spread of primitive plants such as algae and moss progressively increased weathering of phosphorus from rocks. This released phosphorus into the sea. Phosphorus is a nutrient for phytoplankton. Its release into the ocean caused an increase in phytoplankton. Since phytoplankton produce oxygen through photosynthesis, this resulted in more oxygen. And more phytoplankton means more phytoplankton died and were buried, increasing the burial of reduced carbon into the seafloor, leading to a rise in atmospheric oxygen from about 12% by volume 570 mya, plus or minus 40 mya, to its present level of about 21% by volume by about 340 mya. His experiments reveal that spreading-leaved earth moss (*Physcomitrella patens*) amplifies phosphorus weathering by a factor of up to 60! Lenton et al. (2016) updated this to assert that land plants increased oxygen to today's levels by as early as 420 to 400 mya, and with this came the regulation and stabilization of atmospheric oxygen levels by fires, which continued ever since, shaping subsequent evolution (this will be discussed later in this chapter).

Dahl et al. (2010), looking at the isotopic composition and concentration of molybdenum in sedimentary rocks, found two episodes of global ocean oxygenation. The first coincides with the latter part of the existence of the Ediacaran fauna, about 550 to 560 mya. The authors offer no mechanism for this first rise in oxygen. The second is of more interest. It was likely larger, and about 400 mya, well after Cambrian explosion, indicating that this event occurred in relatively low oxygen levels, to an extent contradicting the view expressed by researchers earlier in this chapter that oxygen levels had to be high for the explosion to occur. This second rise in oxygen in the air is correlated with and likely facilitated by the rise and diversification of higher vascular land plants, which likely contributed to it through the enhanced burial of reduced carbon. It was done by the burial of these plants when they died, and by weathering by their roots and their fungal allies. Moreover and of great interest, *it is correlated with and facilitated the rise and diversification of large predatory fish, which required high oxygen*. Some were up to about 9 meters (about 30 feet) long. The research shows that *biologically induced increases in oxygen caused major breakthroughs in animal evolution. It also further illustrates evolutionary innovation (the origin of higher plants) can drive environmental change (rise in oxygen), which can then drive further biological innovation (the rise and diversification of predatory fish), and it thus provides another example of the coevolution of life and the physical/chemical environment.* The plants acted as ecosystem engineers with a positive effect on their ecosystem, other species, and biodiversity, supporting the ABH.

To give an idea of the importance of the burial of plants to oxygen levels, if the only portion of oxygen that was supplied to the atmosphere by the gradual burial of plant organic matter over the eons were removed, the remaining oxygen content would be an asphyxiating 10% today. This is equivalent to the thin atmosphere at about 5.5 km (about 3.4 miles) above sea level!

Atmospheric oxygen skyrocketed in the Carboniferous period (about 358.9 to 298.9 mya) because of the great number of very large plants buried in swamps, making the coal beds that provided much of the fuel for power generation during the Industrial Revolution and that is still used, exacerbating climate change today. The 35% oxygen content of the air at oxygen's peak levels during this period greatly increased the frequency and intensity of fires. So there was strong selection for defenses against fires at that time. The rise in atmospheric oxygen was slow and gradual, allowing time for the evolution of plant defenses against fire. Lignin is one such defense, and it also is likely that the abundance of thick, bark-like, corky layers on the outside of Carboniferous period plants evolved for fire resistance. These thick layers allowed trees and tree-like plants to become very tall. Plants that evolved this defense included trees distantly related to modern club mosses, which grew to over 40 meters (about 131 feet) tall. Club mosses today are small. Tall plants also included species of the genus *Calamites*, which were tree-like horsetails reaching over 30 meters (about 100 feet) in height. Horsetails today might reach only 1.5 meters (about 5 feet). These tall forms provided new habitats for insects and other animals to use for food and shelter. Even the spatial structure, methods of reproduction, and energy partitioning of ecosystems of the Carboniferous period seemed to have evolved as resistance to fires (Robinson, 1989, 1991). There was a sequence of change and diversification of life driven by life, as follows: plants were buried in swamps in great numbers; this buried tremendous amounts of reduced carbon; oxygen gradually increased to levels that peaked at 35%; fires gradually increased to great sizes and frequencies; lignin and thick, bark-like, corky layers evolved for fire resistance, which gave support and allowed some plants to get extremely tall, increasing plant diversity; and animals diversified as they took advantage of the towering three-dimensional habitat the tall plants provided. Plants thus drove their evolution and diversification and that of several animal species by ecosystem engineering.

The sequence I proposed above is supported by the fact that Carboniferous coals are very rich in fossil charcoal, a product of fire. This suggests a high frequency of Carboniferous period fires, presumably due to high oxygen levels. It has also been shown that fire-resistant trees produce more charcoal when they burn (Nelson, 2001). So the high fossil charcoal content of coals from the Carboniferous period indicates that there were many fire-resistant plants and many fires at that time. Large land plants and trees provided fuel for fires, causing more fires. So large plants indirectly selected for fire resistance in themselves and other plants. It is also noteworthy that charcoal is resistant to combining with oxygen, and tends to be buried as reduced carbon, not as carbon combined with oxygen. Thus, the formation and burial of charcoal was a biologically induced mechanism that increased atmospheric oxygen in a positive feedback loop (Berner et al., 2003) until a new stable state was reached. The feedback loop is as follows. More atmospheric oxygen causes more fires, so selects for fire resistance, including the ability to be charred and not burn. This produces charcoal, which tends to be buried as reduced carbon, increasing atmospheric oxygen, and hence more fires. The existence of more fires selects for more fire resistance, producing more charcoal, and so on. This stopped when a stable state was reached at about 35% oxygen in the air. The 35% oxygen was also caused by increased burial of reduced carbon because trees were buried in great numbers.

This high oxygen level did not lead to tragedy or a great loss of biodiversity. The 35% level of oxygen in the atmosphere allowed arthropods to become gigantic. The respiratory system of insects and many other arthropods functions by passive diffusion. They have no active pumps, such as lungs, to breathe with. So oxygen content of the atmosphere limits their size, which can get very large at levels of 35%. So the high oxygen level and allowed arthropods to become giants, and spectacular forms appeared. Huge insects evolved. The largest insect ever to evolve, an ancestor of the dragonfly, with a wingspan of 73 centimeters (almost 2 feet, 5 inches), appeared at this time. There were huge spiders and scorpions; centipedes and millipedes over 1 meter (3 feet, 3 inches) long, including a millipede called *Arthropleura*, which measured 2 meters long (about 6 feet, 7 inches) and a half meter (about 1 foot, 8 inches) wide. Since small arthropods existed as well, the diversity of this phylum likely increased a great deal. Some amphibians also became huge at this time—up to 5 meters (about 16 feet, 5 inches), with amphibian diversity likely increasing for the same reason. Although amphibians breathe actively with lungs, they also frequently breathe passively through the skin, so high oxygen levels could allow them to become large. And the presence of very large arthropods as potential prey could have selected for giant amphibians. So the increased oxygen caused by life—by plants and their burial—increased arthropod and amphibian diversity. By the end of the Permian, when oxygen levels were down to 15%, the age of global giants was over. Extreme fluctuations in oxygen need not lead to catastrophe, and extreme levels returned to intermediate ones in time. Indeed, the increase in oxygen in the Carboniferous period caused by plant burial is an example of life probably causing an increase in diversity of insects, amphibians, and some plants by ecosystem engineering, supporting the ABH.

Some scientists think the coming together of all continents into the one supercontinent called Pangaea, about 300 mya, caused a great amount of weathering, and flushed nutrients into the sea, causing more phytoplankton to grow and produce more oxygen. More phytoplankton also meant more were buried at sea, sequestering carbon, some of which was reduced, further increasing oxygen. This is a nonbiological mechanism working in concert with a biological one causing an increase in oxygen toward the end of the Carboniferous period.

On the basis of the carbon isotope record, Falkowski et al. (2005) modeled oxygen concentrations, finding that oxygen levels rose gradually from about 10% to today's 21% level, from 205 mya onward, with pronounced rises in the early Jurassic period, about 200 mya, and Eocene epoch, about 55 mya. The increase was not continuous, and had large fluctuations. They suggested that the overall increase in oxygen, mediated by the formation of continental margins along the Atlantic Ocean when the continents around this ocean drifted apart, causing weathering, and hence sequestering reduced carbon, was critical in the evolution, subsequent increase in average size, diversification, and rise to prominence of large mammals. In fact, the authors proposed that these changes in large mammals were possible only because oxygen levels increased. Large mammals have high metabolisms, so require high atmospheric oxygen. These events happened to mammals at about the time of the Eocene epoch increase in oxygen levels and immediately after it. This is also when placental mammals became prominent. Of course, life also buried reduced carbon and supplied the oxygen through photosynthesis. But the sequestration of reduced carbon was mainly nonbiological, according to the authors. So mainly nonbiological factors, with some help from life, produced the oxygen for the rise and diversification of placental mammals[1], in a co-evolution of geology and biology. This only partially supports the ABH.

Chennu et al. (2021) used modeling and microbial mats growing under extreme conditions at the bottom of a submerged Lake Huron sinkhole, to show that the increased speed of rotation of the Earth allowed more oxygen production. Daylength was as short as six hours in life's early days and has been increasing ever since because of the gravitational pull of the moon. At the sinkhole, sulfur-eating bacteria that did not produce oxygen covered cyanobacteria in the morning and evening, blocking their access to sunlight and preventing them from carrying out oxygen-producing photosynthesis. When there is sunlight, the sulfur-oxidizing bacteria migrate back down below the photosynthetic cyanobacteria in the Lake Huron sinkhole, enabling them to start producing oxygen. Presumably this happened with similar microbes in the early seas. When the length of time sunlight was available increased to a critical threshold, cyanobacteria were able to produce more oxygen, accounting for the GOE and Neoproterozoic Oxidation Event, according to the researchers. These are nonbiological factors, the moon and Earth's rotation speed, interacting with life to make the planet more favorable to life and biodiversity.

3.4 THE AMAZON RIVER AND PHYTOPLANKTON TOGETHER PRODUCE GREAT QUANTITIES OF OXYGEN

An interaction of the Amazon rainforest and ocean phytoplankton produce a great portion of Earth's oxygen. The Amazon Basin is one of Earth's oldest and largest ecosystems, with 5.5 million square kilometers (about 2.124 million square miles) of rainforest, which is 0.56 times, or just over half, as big as United States. It has one-tenth of the world's species. It is the largest continuous natural preserve of life left on Earth. It is very dense, has trillions of leaves, and each tree releases hundreds of thousands of cubic meters of oxygen in its life. It produces a fifth of Earth's oxygen. But it is almost a closed system, so almost all the oxygen gets reabsorbed in the forest through respiration and decomposition, and never leaves it to oxygenate the rest of the planet. This is not well known, as many people think the Amazon exports great quantities of oxygen to Earth's atmosphere on its own. But another step is necessary to oxygenate the Earth. The trees of the Amazon rainforest create half of its rain, the other half coming from Atlantic Ocean water evaporated by the sun. Average annual rainfall is very high, over 36 inches, and it rains up to 250 days per year. The Amazon River system supplies 20% of global water runoff to the seas. This water carries soil with it, totaling two million tons of sediment per day. A great deal of organic material, such as leaves, branches, and nutrients, are carried in the sediment. The longest distance it travels is 6,437 kilometers (about 4,000 miles), from the headwaters of the Amazon River at the Andes Mountains. It travels east and goes through the Amazon delta into the Atlantic Ocean, where the organic material and nutrients the river carries allow phytoplankton to explode into an immense bloom near the sea's surface. The bloom spreads out to sea, doubling in one day, growing to cover a massive 25,000 square miles of sea surface, carrying out photosynthesis, absorbing CO_2, and producing billions of gallons of oxygen each day. In as little as three days, almost all of the plankton that are not eaten die, some sinking to the seafloor, where they get buried, sequestering carbon, some of it reduced. The phytoplankton stay at the seafloor for millennia, making a blanket a half mile thick in some places. This provides the whole Earth with a huge net gain in oxygen. The largest quantity of oxygen created by any river system is produced by the Amazon, but all over the Earth are similar interactions of forests, rain, rivers, and phytoplankton that produce large amounts of oxygen. This a major mechanism by which phytoplankton produce over half the planet's oxygen. Forests indirectly make large amounts of it as well by this mechanism. It is the interaction of forests, rainfall, rivers, sea, and phytoplankton that produce large quantities of oxygen, and the rainforest and phytoplankton together, not rainforests alone, that can be called the lungs of the Earth.

3.5 LIFE'S PRODUCTION OF HIGH OXYGEN LEVELS HELPS DIVERSITY IN MANY WAYS BESIDES PROVIDING AN ENERGY SOURCE FOR THE EVOLUTION AND DIVERSIFICATION OF COMPLEX LIFE

The production of high oxygen levels by life led to the formation of the ozone layer. Ozone, O_3, is three atoms of oxygen bonded together into one molecule, and is formed from molecular oxygen, O_2, which is two oxygen atoms bonded together, and how oxygen normally exists in the atmosphere. The ozone layer was formed from the high atmospheric oxygen levels created by life. So life is the ultimate cause of the ozone layer. Over 90% of the planet's ozone is in the ozone layer, which is in the lower portion of the stratosphere, about 13 to 20 kilometers (about 8.1 to 12.4 miles) above Earth. This layer protects the Earth from high levels of ultraviolet (UV) radiation from the sun, absorbing 97–99% of the sun's damaging medium-frequency UV radiation. Without the ozone layer, UV radiation would prevent the evolution and existence of any life except the simplest forms, and those underground or deep enough in the sea to be shielded by water from UV. UV causes genetic damage by inducing mutations in the DNA; weakens the immune system; harms the eyes and can cause blindness; destroys vitamin A in the skin; can accelerate aging of the skin by damaging collagen fibers; is the main cause of sunburn; creates carcinogenic and cell-damaging free radicals; and can induce skin cancer, including the most deadly form of it, malignant melanoma. The protective effects of the ozone layer do not help humans alone, but many other types of animals, plants, fungi, one-celled eukaryotes, and prokaryotes. The formation of the ozone layer allowed phytoplankton that previously lived deeper in the ocean to move up to the surface, and hence better photosynthesize and increase their populations. They could thus support many more animals in the ocean's food web. They could also produce more oxygen, and sequester more carbon. So they increased atmospheric oxygen and hence enhanced the ozone layer, in a positive feedback loop that eventually reached an equilibrium and stabilized. The ozone layer also allowed plants to colonize land. Land plants photosynthesized and buried reduced carbon, increasing oxygen and ozone, in another positive feedback loop. Goldblatt et al. (2006) showed that once the ozone layer was produced, it shielded oxygen, causing a nonlinear increase in how long molecular oxygen can last in the atmosphere. This also allowed an increase in oxygen levels in the air. When oxygen rose to only 10^{-5} of the present atmospheric level, a threshold was reached and the ozone layer protected oxygen, allowing it to rise further, in this positive feedback loop. *Life is an ecosystem engineer that helped its own survival, evolution, and diversification, by producing oxygen and the protective ozone layer.*

High oxygen also causes some molecular nitrogen, N_2, which makes up about 80% of the atmosphere and which life cannot use, to form nitrate, NO_3-, which plants can use. This makes nitrogen available to life. This is important, because nitrogen is necessary for the survival of all forms of life. The nitrate then dissolves in water droplets in clouds, and falls to Earth with rain. Some of this water for clouds and rain is produced by trees, so life helps deliver the nitrate to soil. Of course, most nitrogen useable by life is produced by nitrogen-fixing bacteria.

Sulfate-reducing bacteria produce hydrogen sulfide, which is toxic to almost all life. These bacteria cannot live in the presence of oxygen. Life's production of a high-oxygen atmosphere limits these bacteria to small, limited habitats where there is no oxygen, greatly limiting toxic hydrogen sulfide production. There are also microbes that obtain their food by oxidizing hydrogen sulfide, destroying it; this is another way life limits the amount of hydrogen sulfide. Finally, oxygen destroys the potent greenhouse gas, methane, and limits the growth of methanogens, microbes that produce methane and cannot grow in oxygen's presence. This keeps the Earth at a cooler temperature more favorable to life.

3.6 OXYGEN AND FIRE ARE REGULATED BY NEGATIVE FEEDBACK MEDIATED BY LIFE

High oxygen, working in concert with lightning, also benefits life by the creation of fire. The atmospheric oxygen level must be at least 17% to ignite fires. For every increase in atmospheric oxygen of 1%, the probability of a fire resulting from a lightning strike increases by 70%. Oxygen, lightning, and fuel, such as trees, are the ingredients needed for natural fires. Lightning is constantly starting forest fires randomly around the world, and over 19 million square miles of fire burn on Earth each year (this figure applies to times before human-induced climate change). Within hours, a spark from one lightning bolt turns acres of forest to flames. This renews life. Forests are full of dead and diseased trees that lock up life's nutrients, such as carbon, nitrogen, sulfur, and phosphorous. Boreal forests make up the world's largest biome, are found in cold, high latitudes all around the northern hemisphere, and are characterized by conifers, mostly pines, spruces, and larches. They are vibrant, but in their cold environments, their trees can take decades to decompose and return their nutrients to the soil, because decomposing bacteria and fungi metabolize slower in cold environments. Fire can shorten the decomposition to hours, making nutrients available right away. It burns the trees and dead animals to the ash that goes into the soil and returns the nutrients to the soil, recycling them. It allows the biosphere to turn over more efficiently and effectively. A burst of new growth follows a fire almost immediately. Large burned areas are rejuvenated within months. Fire thus converts dead organisms to living ones. All this happens in temperate forests as well, although boreal forests, being colder, tie up nutrients longer in the absence of fire.

Tropical rainforests in their natural state are too wet and rainy for fires to be frequent or large. (In this discussion, I am referring to times before their destruction by industrialized societies.) They are hot and humid, ideal conditions for decomposers such as bacteria and fungi. So they recycle their nutrients quickly without fire. Much of their rain and hence moisture comes from the trees, and the vegetation holds in much of the humidity, so life creates conditions for both low fire and rapid recycling of nutrients. Forests at higher latitudes than the tropics create rain and hold humidity, but much less so than tropical forests, so they recycle nutrients much more slowly in the absence of fire, and burn more easily. Life sets the conditions for how readily fires occur and how quickly nutrients are recycled in the absence of fire in forests of various types. The less fire-prone a forest, the more life makes up for this by providing favorable conditions for rapid recycling of its nutrients by its decomposers. Tropical forests do not readily burn, but make up for this by recycling their nutrients readily without fire, and those forests in temperate and cooler regions recycle nutrients very slowly without fire, but make up for that by burning easily. Oxygen is involved here because it must be at favorable levels for a beneficial amount of fire for life. The near-optimal oxygen levels on Earth for life are largely maintained by biology, as will be discussed later in this chapter.

Fire also burns forests back to earlier stages of ecological succession, such as meadows. In many areas, only forests would exist without fire. Fire creates a mosaic of many different ecosystems because it burns some of the forest to the earliest stage of succession. Multiple fires can burn forests back to every stage of succession. This can result in many different stages of ecological succession existing simultaneously. In the greater Yellowstone ecosystem, for example, periodic forest fires ensure there are both forests and meadows, instead of only forests. Although forests are in general more diverse than meadows, the overall diversity is increased when both types of ecosystem are present, rather than just one. Over large areas, fire increases overall diversity by often causing all stages of succession to exist simultaneously.

Probert et al. (2019) showed that overgrazing by livestock in the Serengeti-Mara ecosystem from 2001 to 2014 decreased grass fuel, and damaged the ecosystem through a great reduction in fire. Herbivores such as wildebeest and gazelles in a natural ecosystem would not overgraze because predators would regulate their populations. This suggests predators can be a positive influence on ecosystems and increase their diversity by increasing fire toward an optimum level (in addition to all the other positive effects they provide).

Fire is such an integral part of ecosystems that many trees have adapted to it, changing their structure and chemistry, and even make use of it. Fire selects for adaptations to it. Pine trees drop their seeds after fires, taking advantage of the nutrients and open habitat. The seeds of some trees will not germinate and some pine cones will not open, unless stimulated by the heat of a fire. This is adaptive because it ensures germination only occurs in open habitats without competitors for sunlight and in soil that is nutrient-rich. Pine trees contain a great deal of resins, so burn easily. This creates fires that destroy their competitors, which are generally less fire-resistant than they are, and creates nutrient-rich, open, sun-lit habitats for their seedlings to thrive in. Trees of the genus *Eucalyptus* evolved in a fire-prone environment, and have many adaptations to fire. They even have adaptations to cause fires, which help *Eucalyptus* trees by burning out their less fire-adapted competitors, creating open areas where light can penetrate, enriching the soil, and recycling nutrients. *Eucalyptus* trees and their leaves and bark are rich in very flammable phenolic oils that burn easily and create high temperatures when they do. The trees' leaf litter and long peels of bark are excellent fire fuel. The bark peels off in long streamers that fall to the ground, and the fallen leaves create dense carpets. Both fallen bark and leaves provide potent fire fuel on the ground, drawing ground fires up, creating massive, fast-spreading fires at the top of the forest. Since only 10 to 40% of *Eucalyptus* leaves that fall decompose the year they drop in the absence of fire, fuel builds up. *Eucalyptus* tree oil, leaf litter, and peeled bark can turn a small ground fire into a huge, explosive fire storm in minutes. They have fire-adapted trunks, so can survive most fires. Burned trees can re-sprout from their stumps and roots. Their seeds are extremely resistant to heat. After a fire, trees that are not badly burned quickly drop their seeds. The seed capsules open when burned, and the seedlings thrive in the nutrient-rich soil created by fire. Fire gives the *Eucalyptus* an advantage over other plants. Still, fire recycles nutrients for the whole *Eucalyptus* forest, including its competitors. Fires in *Eucalyptus* forests cause a gain in diversity because far more plants and animals benefit than are hurt by fires there. Thus, the *Eucalyptus* has evolved to create periodic fires that benefit it and most of the other plants and animals of its forest. This is a form of ecosystem engineering. The *Eucalyptus* forest maintains itself by this mechanism.

Roche et al. (2019) found forest fires conserve water. In two California river basins, they found fires saved 153 to 218 millimeters of water per hectare of land, which equals about 77 billion liters of water per year in one of the river basins. Natural wildfires in the 5,310-square-kilometer (about 2,050-square-mile) American River basin save the forest about 773 billion liters of water per year. This is because trees emit water. With fewer trees losing water, trees can grow larger, the forest is healthier and wetter, and a greater diversity of animals can exist due to the larger, healthier trees and adequate water supply.

In the dry hills around Santa Fe, New Mexico, when fires burn off shrubby undergrowth, this makes space between trees for grasses to grow (Wine et al., 2016). This conserves water because grasses hold soil in place, allowing moisture to soak into the ground instead of running downhill and eroding the soil. This also prevents floods and mudslides.

Fires control understory growth, preventing hot, destructive fires that would occur if the understory became too dense. Fire sometimes increases diversity by preventing the plant community from reaching carrying capacity, a state where plants are crowding each other and competing for space, sunlight, and nutrients. Preventing them from reaching this state prevents the best competitors from eliminating the weaker competitors from the pant community. This results in more plant species living in the community. And that allows more animal species to exist there too. Fires are essential for the health of many terrestrial ecosystems.

Barkley et al. (2019) showed that fires in Africa supply up to half of the phosphorous annually to the Amazon basin, helping the productivity and diversity of the most biodiverse ecosystem on land. African fires are also an important source of soluble phosphorous to the Tropical Atlantic Ocean and oceans in the Southern Hemisphere and may be important for marine productivity, particularly in the northern summer and fall (ibid.). The increased phosphorous leads to phytoplankton blooms, feeding the countless animals in the marine food webs. The phytoplankton blooms also produce oxygen and sequester carbon, some of it reduced carbon, increasing oxygen and making the temperature lower and better for life. This also regulates acidity of the sea, preventing it from becoming too high for life to thrive. Thus, wildfires can help diversity on land and in the sea and increase carbon sequestration by fertilizing terrestrial and marine ecosystems far from them.

Wildfires that burned 21% of Australia's temperate and broadleaf forests in 2019 caused iron and phosphorous to be delivered to the southern Pacific Ocean thousands of miles away (Tang et al., 2021). Winds blew the smoke that had these nutrients. It settled in waters with relatively low iron levels. Some of the nutrients were from the trees that were burned, and others were from dust that the fires took in and lofted into the air. The nutrients spurred enormous phytoplankton blooms that grew over 2,000 miles wide, an area larger than Australia, concentrated in patches south of Australia and far off the western coast of South America, that peaked in January 2020, and lasted for approximately four months. This is surprising, since typically the abundance of phytoplankton in these regions is lower during this time of year.

The Thomas Fire burned from December 5, 2017 until January 12, 2018, burning almost 300,000 acres of land in Ventura and Santa Barbara counties, making it the largest wildfire in California history at the time. The nutrient-rich smoke fertilized the Santa Barbara Channel, resulting in an abnormally high number of dinoflagellates (Kramer et al., 2020), supporting the food web there. Wildfires also cause algal blooms in freshwater rivers and lakes, where the limiting nutrient is typically phosphorous.

However, too much fire can decrease diversity and be destructive to life in the sea. In 1997, wildfires in Indonesia fertilized a coral reef in Sumatra with too much nutrient, causing a red tide that killed the reef and its fish.

Wildfires that are not too large are beneficial to oceanic and freshwater ecosystems because of the nutrients they provide. But if they are too big, they deliver so much nutrient that the phytoplankton blooms can cause eutrophication in both oceanic and aquatic ecosystems. Under natural, preindustrial conditions, fires were generally of the beneficial type, and represent examples of life aiding ecosystems far away, increasing their diversity, and helping sequester carbon, some of which is reduced carbon. However, the large and frequent fires that result from human-induced heating can be so large that they harm far away aquatic ecosystems. In this unnatural case, terrestrial life actually can help harm distant aquatic ecosystems.

3.7 LIFE STABILIZED ATMOSPHERIC OXYGEN LEVELS FOR THE LAST 350 MILLION YEARS WITH IMPERFECT NEGATIVE FEEDBACK

The continuous charcoal record, interpreted with the help of the results of combustion experiments, indicates atmospheric oxygen has varied remarkably little over the past 350 million years (Lenton and Watson, 2000). Ignition experiments show that atmospheric oxygen must be at least 17% for fires to burn (Lenton, 2012). Charcoal first appears in the fossil record about 420 mya, showing fires could burn at this point. It remains in the fossil record from this date to the present. So atmospheric oxygen was at least 17% starting about 420 mya or earlier. And weathering experiments and modeling indicate the first lower (nonvascular) plants, which began colonizing the land surface in the Ordovician Period (488.3 to 443.7 mya), caused atmospheric oxygen to rise to at least 17% by this date (ibid.), by weathering and burial of reduced carbon. It requires over 35% atmospheric oxygen to cause so much fire that forests cannot exist, and the continued presence of forests for the last 350 million years indicates oxygen never exceeded this level during this time. So the range of atmospheric oxygen has been kept between 17 and 35%, just over a factor of two, during this 350 million-year period, during which time the entire oxygen reservoir has been replaced over 100 times. This is strong evidence for remarkable regulation of atmospheric oxygen at close to optimal levels for life by some mechanism. Too little oxygen would make complex, multicellular life impossible, while too much is toxic to cells. Too little oxygen would allow too few fires, and too much too many and too large fires, to maximally aid life. The 21% oxygen level of today's atmosphere is as near optimal for life as one could hope for, for both of these reasons. The regulation of atmospheric oxygen occurs on million-year time scales and is accomplished in large part by regulation of the burial of reduced carbon and weathering rates (ibid.), and life is involved in this.

If a negative feedback mechanism driven by life regulated atmospheric oxygen, keeping it at levels favorable to life, this would account for the relatively constant, life-favorable levels of it discussed in the above paragraph, and strongly support both the Gaia Hypothesis and the ABH/Pachamama Hypothesis.

Such stabilizing negative feedback mechanisms exist. One is that increased oxygen causes more fire, which reduces the oxygen level. Fire consumes and hence depletes oxygen. Fire also reduces the amount of forest, which means less weathering by tree roots and their symbiotic fungi, and hence less reduced carbon burial. This means that there is more reduced carbon in the air. The increased reduced carbon combines with and thus further decreases the amount of atmospheric oxygen. So there is a negative feedback loop that stabilizes atmospheric oxygen, and involves life in the form of forests. Another negative feedback loop, a more complete model, connects land and sea ecosystems and fire (Lenton and Watson, 2000; Lenton, 2012). Phosphorus is the ultimate limiting nutrient in the global ocean. It is supplied to the surface ocean partly by the upwelling of deeper water, but on geological time scales, the source of bioavailable phosphorus to the sea is the weathering of continental rocks, greatly accelerated primarily by land plants and their fungi. Less oxygen means less fire, and hence more plants, including trees. More plants mean more phosphorus is released from the soil. This is delivered to the sea via rivers. Phosphorus causes increased growth of phytoplankton, which produce more oxygen by their photosynthesis. Also, when there are more phytoplankton, more of them die and sink to the bottom of the sea. More carbon, some of it reduced, is buried in the seafloor as a result, so there is more oxygen in the biosphere. Also, more phytoplankton means more of them are eaten by small organisms that are eaten by larger ones. As these organisms defecate, their feces sinks, and as they die, they sink. This buries more reduced carbon in the seafloor. This means there is less reduced carbon to deplete atmospheric oxygen. So the result is an increase in atmospheric oxygen. On the other hand, an increase in oxygen means more fire and hence less vegetation, so less phosphorus is weathered. This causes there to be less phosphorous in the sea, thus less phytoplankton growth there, and hence less oxygen production via photosynthesis. And there is less burial of phytoplankton. So less reduced carbon is buried in the sea, so more reduced carbon is present to combine with and reduce the amount of oxygen in the biosphere. This is a negative feedback loop that stabilizes oxygen levels in the air and ocean, keeping them at levels favorable to life, over long time periods. It is accomplished by life, supporting the ABH.

There is a positive feedback loop, as follows: more oxygen means more fire; the burning of trees releases nutrients to the sea, causing more phytoplankton growth, burying reduced carbon and increasing oxygen. But the negative feedback is much stronger. However, the negative feedback is not foolproof and can be over-ridden by catastrophe, such as a large asteroid strike or massive volcanism. Nor does it keep atmospheric oxygen levels perfectly stable at all times. They reached 35% in the middle of the Carboniferous, and stayed this high for 50 million years. Furthermore, this was caused by life, when great masses of trees and plants were buried in swamps and covered by sediments, burying great quantities of reduced carbon. Yet, the biologically mediated negative feedback has always been restored, and, with the help of life, oxygen levels have remained relatively stable at between 17 and 35%, a range in which life can survive, during the last 350 million years, a long time.

Notably, this system also features a life-induced negative feedback loop that keeps fire at intermediate levels favorable to life. More fire reduces oxygen by consuming it, and reduces vegetation by burning it. Fire needs oxygen and fuel in the form of vegetation. With less of these, there is less fire. Conversely, less fire means more oxygen and fuel, so more fire. Fire and oxygen control each other in a stabilizing system utilizing biology—the trees and phytoplankton.

Some animals also help keep fire close to optimal levels. Hayward et al. (2016) showed that burrowing animals reintroduced to forests in Australia from which they had been driven locally extinct, such as small marsupials about the size of rats, acted as ecosystem engineers and reduced leaf litter by 24% by burrowing and turning over the dead leaves with soil, helping break the litter down. The species included the bilby (*Macrotis lagotis*), western ringtail possum (*Pseudocheirus occidentalis*), bridled nailtail wallaby (*Onychogalea fraenata*), greater stick-nest rat (*Leporillus conditor*), numbat (*Myrmecobius fasciatus*), southern brown bandicoot (*Isoodon obesulus*), and others. The reduction in leaf litter reduced the intensity and rate of spread of fires, so increased the number of species in the ecosystem. With too much fire due to the accumulation of leaf litter because of the killing off of burrowing species, diversity of Australian forests had decreased.

White rhinoceros (*Aceratherium simum*) are ecosystem engineers that maintain short grass patches by grazing (Waldram et al., 2008). This decreases the amount and intensity of fire, which benefits life. White rhino removals allowed grasses to grow higher, increasing the amount and continuousness of fuel. This resulted in larger, less patchy fires. The shorter grass height that they created also allowed shorter grazing animals to access food more easily, increasing the diversity of the grassland ecosystem. Similar effects of large herbivores on their ecosystems with respect to fire and hence diversity may have been much more widespread before the extinction of large grazers in the Pleistocene. The plain's viscacha (*Lagostomus maximus*), a rodent in the family Chinchillidae, reduces and increases fire intensity at different stages in its population cycle in the semiarid scrub of Argentina (Hierro et al., 2011). It creates natural fire-breaks through intense grazing, reducing fire, and generating over time patches of large unburned shrubs in grazed zones, when their colonies are active. They gather and place coarse wood on colonies during territorial displays. After colonies are abandoned, recovery of vegetation and the previous accumulation of this coarse wood increases fire intensity, creating patches where shrubs are burned. These opposing effects on fire created by these ecosystem engineers enhances the variation in amount of vegetation in their woody-dominated ecosystems. The researchers did not find or state this, but this variability in the environment is conducive to higher species diversity. Bower birds, which live in Australia and New Guinea, and lyrebirds, which dwell in Australia, alter litter volume and distribution, and thereby reduce fire likelihood (Mikami et al., 2010; Nugent et al., 2014).

Based on experiments and the fossil record, Watson et al. (1978) pointed out that methanogens also control oxygen, because the methane they produce combines with oxygen, removing it from the air. Today, methanogens play a key role in preventing oxygen from getting to levels that are higher than optimal for life. Each year, about 10^{14} moles of methane are produced, primarily by methanogens (ibid.). This is a substantial amount. Thus, methanogens also recycle a great deal of carbon, making a very significant contribution to the Earth's carbon cycle. Methane removes oxygen in the lower atmosphere at about 2,000 megatons per year (ibid.). Because of the great quantity of reduced carbon burial, without methanogens, atmospheric oxygen concentration would rise by 1% in as little as 12,000 years, a geologically short time. The authors suggest that oxygen has been regulated by methane, and perhaps by nitrous oxide and other chemicals, at levels of about 10–25% of the atmosphere for very long periods relative to the times given quantities of these gases stay in the atmosphere.

At the same time, photosynthesis produces oxygen, which removes the potent greenhouse gas methane when the oxygen combines with it. This regulates temperature while conserving carbon for the biosphere. The oxygen produced by photosynthesis also controls methane by controlling the growth of the methanogens that produce methane, which cannot grow in the presence of oxygen. So methanogens and photosynthetic organisms act as ecosystem engineers that help biodiversity by regulating both oxygen and methane, countering each other, and keeping these gases at levels favorable to life.

3.8 THE GOE MAY HAVE CAUSED A GLACIATION, BUT ALMOST SURELY NOT A MASS EXTINCTION

The rise of oxygen profoundly increased biodiversity and the complexity of life in the long run, but the GOE probably had a catastrophic consequence that decreased biodiversity for about 300 million years. Oxygen combined with methane, depleting it to the point that it became a trace gas. Methane had been helping keep Earth warm, and its loss made the planet so cold that ice formed over large areas. The large amount of ice coverage raised the reflectivity of the Earth, causing heat to be reflected from the planet into space, which caused further cooling and ice formation, in a positive feedback loop. This made Earth extremely cold. Some researchers think that a great deal of the Earth froze, triggering the Huronian glaciation, which occurred right after the GOE, lasting from about 2.4 to about 2.1 bya. Recall that the sun was giving off much less heat energy to Earth at this time. When most of the Earth is glaciated and covered in ice, including the upper layers of the seas, as some think was the case, this is called a Snowball Earth. There was doubtless some loss of diversity in the sea. It is thought that there was more than one Snowball Earth period in the planet's history, but some scientists think that this was the longest one ever (Kopp et al., 2005). However, this is controversial, and some think there was no Snowball Earth at all at this time. Peltier et al. (2007) coupled two models to show that the decline of ocean surface temperatures caused a good deal of oxygen to be dissolved into the sea, operating to increase the rate of remineralization of a massive pool of dissolved organic carbon. This led directly to an increase of atmospheric CO_2, increasing greenhouse warming, and preventing a snowball state. At any rate, the Earth was not entirely glaciated, because there was not enough water to form glaciers between 10° and 30° latitude, in both the northern and southern hemispheres. This allowed life to survive in these unglaciated areas.

Poulton et al. (2021) studied marine sedimentary rocks from the Transvaal Supergroup, South Africa. If there is no oxygen in the atmosphere, these rocks contain certain forms, or isotopes, of sulfur. When oxygen is high, these sulfur isotopes disappear because the chemical reactions that create them do not occur in the presence of oxygen. The researchers could therefore tell oxygen levels at various times in Earth's history by observing the quantities of these isotopes. They found that the initial oxygen level of 10^{-5} of today's level took until 2.22 bya to increase to the point that it reached permanent, irreversible, permanently high oxygen levels in the atmosphere, which is almost 100 million years later than previously estimated. This extends the GOE by almost 100 million years. Oxygen levels and temperatures fluctuated across the threshold of 10^{-5} of the present atmospheric level for about 200 million years by the following processes. Cyanobacteria produced oxygen by photosynthesis. This oxygen reacted with the methane, depleting it from the atmosphere. There was too little CO_2 to make up for the cooling that resulted from the loss of methane, so Earth became very cold and the glaciers expanded. Oxygen production was very low, since little sunlight reached the cyanobacteria because the seas were covered by ice, which blocked the sunlight from the cyanobacteria. Then subglacial volcanoes spewed out CO_2, warming the planet again. Also, volcanoes exuded methane, and microorganisms produced it, so it began to build up in the atmosphere again, heating the Earth yet more. So the glaciers melted, and cyanobacteria received more sunlight and increased in number. And volcanic activity sent nutrients to the sea, causing yet more cyanobacteria to bloom. Thus, the cyanobacteria produced more oxygen, and more of them sank to the seafloor and were buried, sequestering reduced carbon. So oxygen levels went up again, and this sent methane levels down again, causing another glaciation. This cycle occurred three or four times. Eventually, another geological change broke this oxygenation-glaciation cycle. The pattern seems to have ended about 2.2 billion years ago when the rock record indicates an increase in reduced carbon being buried, indicating increased oxygen production by cyanobacteria. It was at this point that the permanent high atmospheric oxygen levels started. It is not known what triggered this tipping point. Bekker and his colleagues hypothesize that there was volcanic activity at this time that was greater and perhaps more sustained than previous volcanisms, that provided abundant nutrients to the seas, allowing cyanobacteria to bloom and produce oxygen in great amounts. At this point, oxygen levels were high enough to permanently suppress methane's oversized influence on the climate, and CO_2 from volcanic activity and other sources became the dominant greenhouse gas for keeping the planet warm. The glaciations seem to have been within and part of the Huronian glaciation, suggesting this glaciation was interspersed with periods when there were large unglaciated areas on Earth. This work indicates that the rise of oxygen did indeed cause glaciations and reduced diversity, although the Earth recovered every time. It must be conceded that the glaciations were at least partly caused by ecosystem engineering by cyanobacteria when they photosynthesized and were buried in the seafloor, and that the warming after each glaciation was accomplished partly nonbiologically by volcanoes. But in the long run, the ecosystem engineering of the cyanobacteria greatly helped life and increased biodiversity because it significantly increased oxygen levels, and this was necessary for multicellular life to survive, thrive, evolve, and diversify in later times.

Only prokaryotes were alive at the time of the Huronian glaciation and proposed Snowball Earth (so eukaryotes were not affected), and they are very adaptable, and could exist under the ice, deep underground, in pockets of liquid water under ice caps as occurs in Lake Vostok in Antarctica, in geothermal hot spots on land such as in Yellowstone National Park and places in Iceland today, in tropical areas where the sun or volcanic heat heated bare rock sheltered from cold wind and made small temporary melt pools, in warmer waters in the deep sea, and at deep-sea vents. Prokaryotes also can become dormant and live for eons in freezing temperatures, becoming active when warmth returns. Microbes could also have existed in dirty ice covering shallow seas. This dirty ice forms small ponds and becomes small, diverse ecosystems full of many species of organisms (Barras et al., 2018); these environments may have covered about 12% of Earth's surface (ibid.). The Earth's temperature was lowered considerably, but how serious this was and how much ice cover was created is unclear. Microfossils such as oncolites and stromatolites prove that, at least in shallow marine environments, life survived without any significant loss of diversity, and developed complex food webs in spite of the glaciation (Corsetti et al., 2003). Oncolites are layered structures formed by cyanobacteria that are very similar to stromatolites, but that form approximately spherical structures instead of columns, as stromatolites do. A mass extinction from a Snowball Earth would noticeably reduce the diversity and change the composition of Earth's species. These changes have not been observed (ibid.). In fact, the groups most susceptible to climatic variation were largely unaffected (Grey et al., 2003). So there likely was not a mass extinction of the existing life, which at any rate was solely prokaryotes. Prokaryotes have very short generation times and can quickly evolve and adapt to environmental stresses such as cold. Thus, many species of them would have survived the glaciation.

Also, some think depletion of methane by oxygen was not the cause of the glaciation, and it is almost certain it was not the sole cause. Nonbiological causes were likely important. For example, silicates in fresh basaltic surfaces combined with CO_2, resulting in the burial of a great deal of carbon, taking great quantities of greenhouse gases out of the atmosphere (Melezhik, 2006). A lull of 250 million years without volcanic activity meant no CO_2 was generated nonbiologically. Placement of continental landmasses from continental drift in a position that greatly increased carbon sequestration removed more CO_2. Since the Earth had a much greater tilt on its axis at that time than today, more severe winters could have greatly lowered temperatures. And Earth may have orbited farther from the sun. There could have been changes in the sun's output. Any combination of these could have aided the cooling.

Crowley et al. (2001) estimated the CO_2 levels necessary to thaw the planet from a Snowball Earth by warming it up were 350 times today's levels, about 13% of the atmosphere. The thawing was accomplished by a combination of interacting biological and physical-chemical factors. The high percentage of the Earth covered by ice prevented the removal of CO_2 from the air by the weathering of silicon-rich rocks. Sufficient CO_2 and methane were produced over 4 to 30 million years, mainly by volcanoes, but also by microbes, to melt the ice in the tropics (Pierrehumbert, 2004). This melting produced an area darker than the ice, an area that absorbed more sunlight, causing a positive feedback loop that brought Earth out of the snowball phase (ibid.). This melted the ice in a relatively short period geologically, possibly less than 1,000 years. And it is possible that methane anhydrates (which are frozen and contain methane) in low-latitude permafrost were melted, adding a great deal more methane to the atmosphere, and triggering a strong positive feedback loop for deglaciation and warming as well (Kennedy et al., 2008). Life helped accelerate the thaw. During millions of years, cryoconite, which is made of small rock particles, soot, and microbes, would have accumulated on and inside the ice, darkening it. Microorganisms adapted to extreme cold, as well as volcanic ash and dust from ice-free locations, would settle on ice, covering several million square kilometers of it. These would darken the ice, making it absorb more heat, accelerating the melting (Hoffman, 2016). Thus, life in the form of microbes acting as ecosystem engineers combined with nonbiological substances to help warm the planet and melt the ice.

Significantly for the ABH, the Huronian glaciation may have helped the production of more oxygen in two ways. First, hydrogen peroxide, produced when ultraviolet light hits water, would be protected from being broken down by sunlight when trapped in the glaciers. When the glaciers melted, it would be split into water and oxygen, increasing oxygen tremendously, but gradually, in the seas and the air (Liang et al., 2006). The authors stated that the peroxides and molecular oxygen generated during Archean and earliest Proterozoic non-Snowball glacial intervals could have driven the evolution of enzymes that mediate and utilize oxygen, paving the way for the eventual appearance of oxygenic photosynthesis.

Second, the melting of the glaciers when the Huronian glaciation was coming to an end would have released tremendous amounts of glacial deposit. This would erode and weather, resulting in sediments washing to the ocean that were high in nutrients, such as phosphorus. These would cause a cyanobacteria population explosion. Of course, they would produce oxygen through photosynthesis. Many would sink upon death, sequestering tremendous quantities of reduced carbon, causing atmospheric oxygen to increase greatly. This would have taken millennia. This mechanism of increasing oxygen levels employed life.

Also, Snowball Earth helped life in that the high volcanism that released great quantities of greenhouse gases over 4 to 30 million years that would have caused extremely hot conditions on Earth was tempered by the fact that Earth was covered by ice and very cold at the start of the volcanism, at least partly because the oxygen life produced removed methane from the air. Otherwise, temperatures would have soared planet-wide.

There were possibly two or three other Snowball Earth periods between about 720 and about 580 mya, but, if they occurred, they were apparently caused by nonbiological factors.

3.9 ANOXYGENIC AND OXYGENIC PHOTOSYNTHESIS AND THE USE BY LIFE OF WHAT IS AVAILABLE

Anoxygenic photosynthesis is photosynthesis that does not produce oxygen. It is done by various bacteria, such as green sulfur bacteria; uses hydrogen sulfide instead of water; and produces sulfur instead of oxygen. Oxygenic photosynthesis produces oxygen. Both kinds of photosynthesis manufacture carbohydrates. Anoxygenic photosynthesis is more energetically efficient than oxygenic photosynthesis. Possibly oxygenic photosynthesis became the prevalent type of photosynthesis because water is much more abundant than hydrogen sulfide. If anoxygenic photosynthesis using hydrogen sulfide were dominant, the atmosphere would be high in sulfur instead of oxygen. Sulfur would likely not have evolved to be a key component of a respiration system that is as efficient as the one that uses oxygen. Therefore, higher, complex life would likely not have evolved, and biodiversity would likely be much lower. It is interesting that the fact that water is much more abundant than hydrogen sulfide could be a fundamental reason why higher life was able to evolve and diversify. This represents a possible increase in diversity and complexity not caused by life, but chemical conditions on Earth. It would support a more general hypothesis than the ABH, one that posits that the solar system is primed for higher life to evolve, which is discussed in Chapter 13. However, the idea that the abundance of water is the reason that the type of photosynthesis that uses it became the most common type of photosynthesis has yet to be demonstrated. This interesting question needs much research.

3.10 LIFE COEVOLVED WITH THE ATMOSPHERE

This chapter and the one before it demonstrated that life coevolved with Earth's atmosphere. Life reduced atmospheric CO_2 and methane levels, keeping temperatures low enough to favor life. Life also produced levels of acidity favorable to life in aquatic ecosystems. Life increased the levels of oxygen in the air and seas. These changes allowed life to increase in complexity, abundance, and diversity. They allowed the evolution of vascular plants, including trees. This allowed life to further regulate greenhouse gases and global temperatures, and to increase oxygen levels in the atmosphere and in rivers, lakes, and the sea. These favorable CO_2, methane, and oxygen levels, and favorable temperatures and levels of acidity in aquatic ecosystems, allowed life to further thrive, evolve complexity, and diversify, reaching spectacular levels of complexity and diversity. Thus, positive feedback was involved, but in this case, it did not destabilize the system. Rather, it helped the system increase its biodiversity, complexity, and information content.

NOTE

1 Placental mammals are mammals that carry the fetus in the uterus of the mother to a relatively late stage of development. They contain the vast majority of living mammals. Placental mammals are the most complex mammals, and include wolves, deer, lions, monkeys, and humans.

REFERENCES

Anbar, A. D., et al. (28 Sept., 2007). A whiff of oxygen before the great oxidation event? *Science* 317 (5846): 1903–6. doi: 10.1126/science.1140325

Barkley, A. E., et al. (29 July, 2019). African biomass burning is a substantial source of phosphorus deposition to the Amazon, Tropical Atlantic Ocean, and Southern Ocean. *PNAS USA* 116 (33) 16216–21. https://doi.org/10.1073/pnas.1906091116.

Barras, C. (March, 2018). Scott's dirty ice may solve mystery. *New Scientist* 237 (3171): 16. doi: 10.1016/S0262-4079(18)30558-X. Secondary literature.

Beerling, D. (2007). *The Emerald Planet: How Plants Changed Earth's History*. Oxford Univ. Press, Oxford, UK; New York, NY. First two citations, page 14.

Bekker, A. & Holland, H. D. (Feb., 2012). Oxygen overshoot and recovery during the early Paleoproterozoic. *Earth and Planet. Sci. Letters* 317–318: 295–304. https://doi.org/10.1016/j.epsl.2011.12.012.

Berner, E. K., Berner, R. A., & Moulton, K. L. (2003). Plants and mineral weathering: present and past. *Treatise in Geochem.* 5: 169–88.

Brasier, M., McLoughlin, N., Green, O. & Wacey, D. (2006). A fresh look at the fossil evidence for early Archaean cellular life. *Philos. Trans. of the Royal Soc. B—Biol. Sciences* 361: 887–902. https://doi.org/10.1098/rstb.2006.1835.

Brocks, J. J., Logan, G. A., Buick, R., & Summons, R .E. (1999). Archean molecular fossils and the early rise of eukaryotes. *Science* 285: 1033–6.

Butterfield, N. J. (18 Jan., 2007). Macroevolution and macroecology through deep time. *Palaeontology* 50 (1): 41–55. https://doi.org/10.1111/j.1475-4983.2006.00613.x.

Canfield, D. E. (2005). The early history of atmospheric oxygen: homage to Robert M. Garrels. *Annual Rev. Earth Pl. Sci.* 33: 1–36.

Canfield, D. E., et al. (2014). *Oxygen: A Four Billion Year History*. Princeton Univ. Press, Princeton, NJ.

Cardona, T. (2018). Early Archean origin of heterodimeric Photosystem I. *Heliyon*. doi: 10.1016/j.heliyon.2018.e00548.

Catling, D. C., Zahnle, K. J., & McKay, C. P. (2001). Biogenic methane, hydrogen escape, and the irreversible oxidation of early Earth. *Science* 293: 839–43.

Chen, X., et al. (2015). Rise to modern levels of ocean oxygenation coincided with the Cambrian radiation of animals. *Nature Communications* 6, Article no. 7142. doi: 10.1038/ncomms 8142

Chennu, A., Klatt, J. M., Arbic, B. K., Biddanda, B. A., & Dick, G. J. (2021). Possible link between Earth's rotation rate and oxygenation. Nature Geoscience. doi:10.1038/s41561-021-00784-3. http://hdl.handle.net/21.11116/0000-0009-4876-2.

Corsetti, F. A., et al. (7 April, 2003). A complex microbiota from snowball Earth times: Microfossils from the Neoproterozoic Kingston Peak Formation, Death Valley, USA. *PNAS USA* 100 (8): 4,399–404. Bibcode: 2003PNAS..100.4399C. doi: 10.1073/pnas.0730560100. PMC 153566. PMID 12682298.

Crowley, T. J., et al. (2001). CO_2 levels required for deglaciation of a "near-snowball" Earth. *Geophys. Res. Lett.* 28 (2): 283–6. doi: 10.1029/2000GL011836.

Dahl, T. W., et al. (19 Oct., 2010). Devonian rise in atmospheric oxygen correlated to the radiations of terrestrial plants and large predatory fish. *PNAS USA* 107 (42): 17911–5. doi:10.1073/pnas.1011287107

Douglas, S. & Beveridge, T. J. (1 June, 1998). Mineral formation by bacteria in natural microbial communities. *FEMS Microbiology Ecol.* 26 (2): 79–88. https://doi.org/10.1111/j.1574-6941.1998.tb00494.x.

Falkowski, P. G., et al. (30 Sept., 2005). The rise of oxygen over the past 205 million years and the evolution of large placental mammals. *Science* 309 (5744): 2202–4. doi: 10.1126/science.1116047.

Goldblatt, C., Lenton, T., & Watson, A. (2006). Bistability of atmospheric oxygen and the Great Oxidation. *Nature* 443 (7112): 683–6. Bibcode: 2006Natur.443..683G. doi: 10.1038/nature05169. PMID 17036001

Granold, M., et al. (2 Jan., 2018). Modern diversification of the amino acid repertoire driven by oxygen. *PNAS USA* 115 (1) 41–6. doi: 10.1073/pnas.1717100115.

Grey, K., Walter, M. R., & Calver, C. R. (1 May, 2003). Neoproterozoic biotic diversification: Snowball Earth or aftermath of the Acraman impact? *Geology* 31 (5): 459–62. Bibcode: 2003Geo....31..459G. doi: 10.1130/0091-7613(2003)031<0459:NBDSEO>2.0.CO;2

Halevy, S. E., et al. (20 July, 2012). Sulfate burial constraints on the Phanerozoic sulfur cycle. *Science* 337 (6092): 331–4. doi: 10.1126/science.1220224

Hayward, M. W., et al. (20 March, 2016). Could biodiversity loss have increased Australia's bushfire threat? *Animal Conservation* 19 (6): 490–7. https://doi.org/10.1111/acv.12269.

Hazen, R. M., et al. (2008). Mineral evolution. *Amer. Mineralogist* 93: 1693–720. doi: 10.2138/am.2008.2955 1693.

Heckman, D. S., et al. (10 Aug., 2021). Molecular evidence for the early colonization of land by fungi and plants. *Science* 293 (5532): 1129–33. doi: 10.1126/science.1061457

Hierro, J., Clark, K., Branch, L., & Villarreal, D. (2011). Native herbivore exerts contrasting effects on fire regime and vegetation structure. *Oecologia* 166: 1121–9.

Hoehler, T. M., Bebout, B. M., & Des Marais, D. J. (19 July, 2001). The role of microbial mats in the production of reduced gases on the early Earth. *Nature* 412: 324–7. doi: 10.1038/35085554.

Hoffman, P. F. (2016). Cryoconite pans on Snowball Earth: supraglacial oases for Cryogenian eukaryotes? *Geobiol.* 14 (6): 531–42. doi: 10.1111/gbi.12191. PMID 27422766.

Holland, H. D. (2002). Volcanic gases, black smokers, and the great oxidation event. *Geochim. Cosmochim. Acta* 66: 3811–26.

Johnson, J. E., et al. (9 July, 2013). Manganese-oxidizing photosynthesis before the rise of cyanobacteria. *PNAS USA* 110 (28): 11238–43. https://doi.org/10.1073/pnas.1305530110.

Kasting, J. F. (12 Feb., 1993). Earth's early atmosphere. *Science* 259 (5097): 920–6 doi: 10.1126/science.11536547

Kaufman, A. J., Johnston, D. T., Farquhar, J., Masterson, A. L., Lyons, T. W., Bates, S., & Buick, R. (2007). Late Archean biospheric oxygenation and atmospheric evolution. *Science* 317 (5846): 1900–3.

Kennedy, M., et al. (29 May, 2008). Snowball Earth termination by destabilization of equatorial permafrost methane clathrate. *Nature* 453 (7195): 642–5. doi: 10.1038/nature06961. PMID 18509441.

Kopp, R. E., et al. (2005). The Paleoproterozoic snowball Earth: A climate disaster triggered by the evolution of oxygenic photosynthesis. *PNAS USA* 102 (32): 11131–6. doi: 10.1073/pnas.0504878102. https://doi.org/10.1073/pnas.0504878102

Kramer, S. J., Bisson, K. M., & Fischer, A. D. (20 Nov., 2020). Observations of phytoplankton community composition in the Santa Barbara Channel during the Thomas Fire. *Journ. of Geophysical Research: Oceans.* https://doi.org/10.1029/2020JC016851.

Kump, L. R. & Barley, M. E. (30 Aug., 2007). Increased subaerial volcanism and the rise of atmospheric oxygen 2.5 billion years ago. *Nature* 448: 1033–6.

Lenton, T. M. (2012). Fires and the rise and regulation of atmospheric oxygen. EGU General Assembly 2012, held 22–27 April, in Vienna, Austria, p. 12930.

Lenton, T. M., Boyle, R. A., Poulton, S. W., Shields-Zhou, G. A., & Butterfield, N. J. (2014). Co-evolution of eukaryotes and ocean oxygenation in the Neoproterozoic era. *Nat. Geosci.* 7: 257–65.

Lenton, T. M. & Watson, A. J. (1 March, 2000). Redfield revisited: 2. What regulates the oxygen content of the atmosphere? *Global Biogeochem. Cycles* 14 (1): 249–68. doi: 10.1029/1999GB900076

Lenton, T. M. & Watson, A. J. (10 March, 2004). Biotic enhancement of weathering, atmospheric O_2 and carbon dioxide in the Neoproterozoic. *Geophys. Research Letters* 31 (5): L05202. Bibcode:2004GeoRL..31.5202L. doi 10.1029/2003GL018802

Lenton, T. M., et al. (30 Aug., 2016). Earliest land plants created modern levels of atmospheric oxygen. PNAS USA 113 (35) 9704–9. https://doi.org/10.1073/pnas.1604787113.

Liang, M.-C., et al. (Dec. 12, 2006). Production of hydrogen peroxide in the atmosphere of a Snowball Earth and the origin of oxygenic photosynthesis. *PNAS USA* 103 (50): 18896–9. https://doi.org/10.1073/pnas.0608839103. doi: 10.1073/pnas.0608839103.

Logan, G. A., et al. (4 July, 1995). Terminal Proterozoic reorganization of biogeochemical cycles. *Nature* 376 (6535): 53–6. doi: 10.1038/376053a0

Lyons, T. W., Reinhard, C. T., & Planavsky, N. J. (2014). The rise of oxygen in Earth's early ocean and atmosphere. *Nature* 506: 307–15.

Melezhik, V. A. (2006). Multiple causes of Earth's earliest global glaciation. *Terra Nova* 18 (2): 130–7. doi: 10.1111/j.1365-3121.2006.00672.x.

Mikami, O.K., Katsuno, Y., Yamashita, D.M., Noske, R., & Eguchi, K. (2010). Bowers of the Great Bowerbird (*Chlamydera nuchalis*) remained unburned after fire: is this an adaptation to fire? *Journ. Ethology* 28: 15– 20.

Mills, D. B. & Canfield, D. E. (Dec., 2014). Oxygen and animal evolution: Did a rise of atmospheric oxygen "trigger" the origin of animals? *Bioessays* 36 (12): 1–11.

Mills, D. B., et al. (18 March, 2014)). Oxygen requirements of the earliest animals. *PNAS USA* 111 (11): 4168–72. https://doi.org/10.1073/pnas.1400547111.

Nelson, M. (2001). A dynamical systems model of the limiting oxygen index test: II. Retardancy due to char formation and addition of inert fillers. *Combustion and Theory Modelling* 5 (1): 59–83.

Nisbet, E. G. & Fowler, C.M.R. (7 Dec., 1999). Archaean metabolic evolution of microbial mats. *Proc. of Royal Soc. B—Biol. Sciences* 266 (1436). doi: https://doi.org/10.1098/rspb.1999.0934.

Nugent, D. T., Leonard, S. W., & Clarke, M. F. (2014). Interactions between the superb lyrebird (*Menura novaehollandiae*) and fire in south-eastern Australia. *Wildl. Res.* 41: 203– 11.

Och, L. M. & Shields-Zhou, G. A. (Jan., 2012). The Neoproterozoic oxygenation event: Environmental perturbations and biogeochemical cycling. *Earth Science Reviews* 110 (1): 26–57. doi: 10.1016/j.earscirev.2011.09.004.

Peltier, W. R., et al. (6 Dec., 2007). Snowball Earth prevention by dissolved organic carbon remineralization. *Nature* 450: 813–8. doi: 10.1038/nature06354.

Pierrehumbert, R.T. (2004). High levels of atmospheric carbon dioxide necessary for the termination of global glaciation. *Nature* 429 (6992): 646–9. doi: 10.1038/nature02640. PMID 15190348.

Planavsky, N. J., Asael, D., Hofmann, A., Reinhard, C. T., Lalonde, S. V., Knudsen, A., & Rouxel, O. J. (2014). Evidence for oxygenic photosynthesis half a billion years before the Great Oxidation Event. *Nature Geoscience* 7: 283–6.

Pope, E. C., Bird, D. K., & Rosing, M. T. (2012). Isotope composition and volume of Earth's early oceans. *PNAS USA* 109: 4371–6.

Porada, P., et al. (Feb., 2014). Estimating impacts of lichens and bryophytes on global biogeochemical cycles. *Global Biogeochem. Cycles* 28 (2): 71–85. https://doi.org/10.1002/2013GB004705.

Poulton, S. W., Bekker, A., Cumming, V. N., et al. (29 March, 2021). A 200-million-year delay in permanent atmospheric oxygenation. *Nature* 592: 232–6.

Probert, J. R. et al. (October 2019). Anthropogenic modifications to fire regimes in the wider Serengeti-Mara ecosystem. *Global Change Biol.* 25 (10): 3406–23. https://doi.org/10.1111/gcb.14711.

Robinson, J. M. (1989). Phanerozoic O_2 variation, fire, and terrestrial ecology. *Palaeogeogr. Palaeoclimatol. Palaeoecol.* 75: 223–40.

Robinson, J. M. (1991). Phanerozoic atmospheric reconstructions: a terrestrial perspective. *Global and Planetary Change* 5: 51–62.

Roche, J. W., et al. (Oct., 2019)). Estimating evapotranspiration change due to forest treatment and fire at the basin scale in the Sierra Nevada, California. *Ecohydrology* 11 (7): e1978. doi: 10.1002/eco.1978.

Schwartzman, D. W. (1999). *Life, Temperature, and the Earth.* New York, Columbia Univ. Press, New York, NY, p. 241.

Sessions, A. L., Doughty, D. M., Welander, P. V., Summons, R. E., Newman, D. K. (28 July, 2009). The continuing puzzle of the Great Oxidation Event. *Curr. Biol.* 19 (14): R567–74. doi: 10.1016/j.cub.2009.05.054

Shen, B., Dong, L., Xiao, S., & Kowalewski, M. (2008). The Avalon explosion: Evolution of ediacara morphospace.

Science 319 (5859): 81–84. doi:10.1126/science.1150279. Bibcode:2008Sci...319...81S. PMID 18174439. S2CID 206509488.

Summons, R. E., et al. (2006). Steroids, triterpenoids and molecular oxygen. *Philosoph. Trans. of the Royal Soc. of London. Series B—Biol. Sciences* 361: 951–68.

Sverjensky, D., et al. (31 Oct.-3 Nov., 2010). The great oxidation event, mineral diversification, and mineral correlations with atmospheric composition through geologic time. GSA Denver Annual Meeting Paper No. 80–5.

Tang, W., Llort, J., Weis, J., et al. (15 Sept., 2021). Widespread phytoplankton blooms triggered by 2019–2020 Australian wildfires. *Nature* 597: 370–5. https://doi.org/10.1038/s41586-021-03805-8.

Waldram, M. S., Bond, W. J., & Stock, W. D. (2008). Ecological engineering by a mega-grazer: white rhino impacts on a South African savanna. *Ecosystems* (11): 101–12. https://doi.org/10.1007/s10021-007-9109-9.

Watson, A., Lovelock, J. E., & Margulis, L. (Dec., 1978). Methanogenesis, fires and the regulation of atmospheric oxygen. *Biosystems* 10 (4): 293–8. https://doi.org/10.1016/0303-2647(78)90012-6.

Wine, M. L., et al. (18 Aug., 2016). Hydrologic effects of large southwestern USA wildfires significantly increase regional water supply: fact or fiction? *Environ. Research Letters* 11 085006. doi: 10.1088/1748-9326/11/8/085006.

4 Species Profoundly Affect the Evolution of Other Species: Coevolution Is Fundamentally Important and Was Involved in Many Major Evolutionary Breakthroughs and Innovations

A great part of the evolution of every species is strongly influenced by other species, and every species strongly affects the evolution of other species. No species evolves in isolation. Evolution often involves coevolution, which is evolution in which there is interaction between two species, and each species affects the other. There is feedback between two coevolving species. *Many of the major breakthroughs and innovations in the evolution of species were influenced by other species, some of them via coevolution. Coevolution is of fundamental importance in biology. All evolution between species whose relationship is competition, plant-herbivore, predator-prey, parasite-host, or symbiosis is coevolution.* Organisms influence the evolution of other members of their species. And coevolution between life and the nonbiological environment also has a profound effect on evolution.

Proteins, DNA, RNA, and lipids (organic compounds whose functions include storing energy, signaling, and acting as structural components of cell membranes) had to coevolve with each other during the chemical evolution of life. None could evolve without interacting and coevolving with all three of the other molecules. And there was apparently constant exchange of genetic material between protocells (groups of molecules enclosed in membranes that later became cells) at this time (Damer, 2018). Protocells coevolved with each other and exchanged DNA as they evolved to become the first cells during the chemical evolution of life.

There is symbiotic coevolution and antagonistic coevolution. *Symbiotic coevolution* occurs when two coevolving species aid each other. It can result in big evolutionary breakthroughs, followed by large diversifications into many new species. An example of symbiotic coevolution is the evolution of pollination. Animal-pollinated flowers and their animal pollinators arose as a result of symbiotic coevolution with each other. Before there were flowers pollinated by animals, beetles ate the pollen of wind-pollinated flowers of conifers, the group that includes pine trees. The beetles inadvertently pollinated some of the conifers, although inefficiently at first. Some leaves evolved into colorful petals, and flowers evolved nutritious nectar. Both the colorful petals and sweet-smelling nectar attracted the beetle pollinators. Nectar guides evolved on the flowers, guiding the beetles to the nectar. The nectar rewarded the beetle with a tasty, nutritious meal. All of these adaptations by the flower benefited it, because it resulted in more beetle pollinators, and thus more efficient pollination. Pollination by animals is more efficient than pollination by the wind. In this way, flowers evolved. Flowers have no function other than to attract pollinators. In time, flowers adapted to be excellent at attracting pollinators, and pollinators evolved improvements in sight and smell to excel at locating flowers. Adaptations in beetles for finding and eating pollen and nectar arose as a result of this coevolution. Once there were animal-pollinated flowers and beetles adapted to finding and obtaining nectar and pollinating the flowers, these flowers and beetles each were in a new adaptive zone. An adaptive zone is a new evolutionary and ecological space of many empty available niches that provides an opportunity for diversification into many new species. Adaptive zones come about as a result of invasion of a new territory, such as fish colonizing land, or from new innovative evolutionary adaptations that occur with large changes, such as the evolution of flowers, flight, or the shell of a turtle. The new adaptive zone of animal-pollinated flowering plants resulted in the evolution of many new species of flowers, from daisies to orchids. The new adaptive zone of pollinating beetles led to tremendous diversification of pollinating beetles. Both these happened by coevolution of the flowers and beetle pollinators, as flowers and beetles formed new species by evolving together. Once there was a diversity of plants with flowers adapted to animal pollination, other insects, such as butterflies, bees, and flies coevolved with them and became new species that pollinated them. New flower species adapted to being pollinated by these insects evolved in this coevolution. And bats, monkeys, birds, and other animals coevolved with the flowers, creating both new species of vertebrate pollinators and new species of flowers that they pollinated. There was tremendous diversification of flowering plants and their animal pollinators, creating countless new species of both. This was followed by the evolution of many new animal species that benefitted from the flowering plants and their pollinators. This included animals that ate the new plant species or used them as shelter, and animals that ate or

DOI: 10.1201/9781003246640-4

formed symbiotic relationships with the new species of animal pollinators.

Some plants, including some trees, have symbiotic fungi in their roots that effectively increase the root length tenfold and greatly help the tree roots obtain soil nutrients and water. The plants give the fungi a home, and carbohydrates that the plants produce by photosynthesis. The fungi cannot carry out photosynthesis, so benefit from this. Rodents such as squirrels bury the fungi when they bury nuts that they eat that the trees produce. Burial of the nuts allows the fungi to join up with the tree roots, because some fungal spores settle on the nuts and get buried with them. So the rodents help the trees and fungi, and benefit from the trees by receiving food. There are also helper bacteria in the tree roots that help the fungus and tree root form their alliance, and protect trees from disease. Fungi help these bacteria because chemicals they exude serve as nutrients for the bacteria. The trees help the bacteria by making nutrients from soil minerals and organic matter available to them and providing them with a home in their roots. So all four members of this system of trees, fungi, rodents, and helper bacteria help and are helped by all of the other members of this four-way symbiosis, either directly or indirectly. Since there are many species of tree, fungus, rodent, and bacterium involved, it is really a system of symbiosis of many species. All four of these groups formed this system by symbiotic coevolution, and all species in this system affect the evolution of and are affected in their evolution by all of the other species in the system, either directly or indirectly. Many species evolved that benefit from the four members of this system. There are an especially high number of species that benefit from the plants, especially the trees. Examples include tree-dwelling insects, deer, and rodents that benefit from trees.

Almost all of the nitrogen in the atmosphere is molecular nitrogen, N_2, which cannot be used by plants. Some plants have bacteria, called nitrogen-fixing bacteria, in their roots that convert the nitrogen in the air to a form useable by plants. This is crucial, because nitrogen is a key, essential nutrient to all life, used in DNA, RNA, proteins, and carbohydrates. Plants provide a habitat to the bacteria, so each species involved here helps the other; thus, this is symbiosis. This relationship evolved by symbiotic coevolution, with both the bacteria and plants having profound effects on the evolution of the other. Many species evolved that benefit from the trees in this symbiosis. Various species of squirrels and birds are examples.

Seed-dispersing animals underwent symbiotic coevolution with plants, including trees, which evolved to produce fruits and nuts to attract the seed dispersers. The fruits selected for better vision in some seed dispersers, including some monkeys and some birds, because this allowed them to better locate and recognize the fruit. I hypothesize that in some cases, plants evolved fruits and nuts to better attract their seed predators to use them to disperse their seeds. In this way, seed predators were converted from plant enemies to plant allies that disperse their fruit. Seed dispersal by animals is highly beneficial to plants because a tree or other plant that grows right next to its parent is much more subject to disease and being eaten by herbivores than one dispersed to a distant area by an animal. Without this coevolution, there would not be any fruits or nuts.

Antagonistic coevolution occurs when the two coevolving species are in conflict with each other, as in two competing species, a predator and its prey, or a parasite and its host. For example, some venomous snakes have prey that evolved resistance to their venom. The snake species responded by evolving stronger venom. The prey evolved stronger resistance, and the snake evolved yet stronger venom, in an arms race of positive feedback. After this proceeded for a while, snakes as venomous as the cobra appeared. The need for such potent venom is impossible to explain without antagonistic coevolution, and almost certainly would not have evolved without it. Venom this powerful is not needed to kill the prey of any species of snake.

Viruses cannot exist without cells, so appeared after cells, and the first viruses were only able to evolve because of prokaryotes. Viruses of bacteria and archaea are called bacteriophage, or simply phage. All prokaryotes coevolved both antagonistically and symbiotically with phages. When pages and bacteria underwent antagonistic coevolution with each other, bacteria evolved defenses against phage, and phage evolved countermeasures to the defenses. Bacteria evolved cell walls partly to make it difficult for phage to enter and infect them. Phage evolved the ability to attach to specific molecules on the surface of the bacteria's cell wall in order to enter the cell. Some bacteria evolved changes in the site where the phage attaches to their cell wall, giving them resistance to phage infection. Phage also evolved enzymes that degrade the capsular outer layer of bacteria in order to enter and infect them. A genome is the complete set of genetic material of an organism. CRISPR is a family of DNA sequences found in the genomes of prokaryotes, both bacteria and archaea, derived from DNA fragments of phages that had previously infected the prokaryote. The CRISPR DNA sequences detect and destroy DNA from similar phages during subsequent infections, being part of a prokaryotic anti-viral immune system. The system is the result of the coevolution of prokaryotes with phage.

Bacteria and their viruses underwent symbiotic coevolution as well. This coevolution was of a special type. Here I will now coin a new term. *Symbiotic genetic coevolution* is symbiotic coevolution in which at least one species gives genetic material to the other, enhancing its genetic variability and helping its evolution. The coevolution of bacteria and phage was of this type. Segments of DNA jump between bacteria and their phage, in both directions. The phage took and utilized DNA from their bacteria hosts, leading to many evolutionary innovations. Today, phages of cyanobacteria called cyanophages have photosynthesis genes that they obtained from their bacterial hosts (Lindell et al., 2004; Millard et al., 2004; Sullivan et al., 2006). And phage gave many useful genes to their bacteria hosts, some of which led to innovations. The coevolution between prokaryotes and their phages shaped much of the evolution of both of them, and led to key innovations and large adaptive evolutionary changes in both. The innovations in prokaryotes would have made many new niches available to them, and would have been followed by great diversifications that filled these niches, greatly increasing their biodiversity. This diversification would have allowed phage to greatly diversify, evolving new species to utilize the many new species of prokaryotes.

Buckling and Rainey (2002) pointed out that antagonistic coevolution between hosts and parasites is thought to play an important role in host and parasite population dynamics, the evolutionary maintenance of sexual reproduction, the evolution of parasite virulence, and the rapid evolution in different directions between geographically isolated populations of both hosts and parasites. These authors showed a long-term arms race between the infectivity of a phage and the resistance of its bacterial host. The hosts evolved resistance to a wider range of viruses, and the phages evolved the ability to infect to a wider range of host genotypes. They ran experiments starting with many genetically similar populations of both bacteria and phage, and the different populations of both evolved on divergent trajectories between replicate communities. The bacteria adapted to their own phage populations, but not other ones. This shows coevolution can result in increased biodiversity in the form of greater variability between populations of the same species. If the diverging evolution continued, it could have potentially increased diversity by producing new species, although the experiment did not run long enough to show this.

Antagonistic coevolution of bacteria and fungi produced antibiotics that kill bacteria, such as penicillin, from fungi, and antifungal toxins from bacteria. Bacteria also evolved antibiotics that kill species of bacteria that compete with them. The bacterium *Rhodococcus fascians* does not produce the antibiotic rhodostreptomycin in the lab unless grown with *Streptomyces* bacteria, which themselves are prolific antibiotic producers. In eons of coevolution in which bacteria competed against each other, bacteria evolved many antibiotics to fight off their bacterial competitors, and these competitors responded with resistance to these antibiotics and antibiotics of their own. Both competing bacterial species responded to the other with resistance to their competitor's toxins and stronger toxins of their own, in a repeating process of positive feedback, an arms race of antagonistic coevolution. Some bacteria have evolved tiny syringes filled with toxins that kill their competitors or incapacitate their predators. Of course, they evolved these in response to the competitors and predators. The syringes are slightly different from one species to another, and display a good deal of diversity.

Alegado et al. (2012) found that the bacterial species, *Algoriphagus machipongonensis*, and its relatives make a molecule, RIF-1, which causes a species of choanoflagellate, a one-celled species of eukaryotic microorganism, to aggregate into colonies. The choanoflagellate preys on *Algoriphagus machipongonensis*, so this could be an adaptation of this predator to better capture and consume the bacteria. It normally forms colonies in nature. Choanoflagellates are among the closest living relatives of animals. *This suggests bacteria may have been instrumental in the evolution of multicellular organisms by influencing choanoflagellate evolution.*

Protozoa is an informal term for a group of unicellular eukaryotes; they are not a true taxonomic group. Examples of them are the *Amoeba*, *Paramecium*, and *Euglena*. Some of them feed on bacteria. Predation by protozoa that live in both freshwater and saltwater has profoundly affected the evolution of their aquatic bacteria prey, which have evolved anti-predator strategies, including the production of protective proteins, the formation of protective filaments, the ability to escape at high speeds, becoming extremely tiny, and the production of defensive poisons. *It is therefore conceivable that predation on bacteria by protozoa has shaped the evolution of these bacteria as profoundly as the evolution of oxygenic photosynthesis has* (Pernthaller, 2005). Protozoa have responded by evolving resistance to the toxins, becoming faster, becoming more efficient at ingesting bacteria, and in other ways. One adaptation for more efficient ingestion of prey is found in the ciliated protozoan genus *Paramecium,* which uses its hairlike cilia to sweep bacteria and other prey, along with some water, through its oral groove ("mouth") and into the cell. Protozoa and their bacterial prey have undergone antagonistic coevolution that has led to a number of key innovations in each of these groups.

Very small one-celled parasites called parasitoids in the sea attack phytoplankton, which are larger than they are. Protective armor of phytoplankton evolved at least partly as a protection against parasitoids, and parasitoids evolved to get around this. So there was antagonistic coevolution between parasitoids and phytoplankton, resulting in strong armor for the phytoplankton. The small number of unprotected forms such as amoebae in the open ocean is testimony to the importance, abundance, and ubiquity of parasitoids there. The armor is effective, since parasitoids do not succeed in penetrating it, but squeeze into areas where the armor is lacking. Parasitoids attack diatoms, which have silica armor. They also coevolved with phytoplankton that have armor made of carbon, such as dinoflagellates, coccolithophorids, and foraminiferans, so were important selective forces in the evolution of armor made of both silica and carbon compounds.

Similarly, there are small animals that swim in the sea, occur in large numbers, and eat phytoplankton. These include copepods, isopods, and krill. Krill are well known as important food for seals, penguins, and large whales. All of these groups of small animals underwent antagonistic coevolution with phytoplankton, selecting for stronger armor in phytoplankton to resist their grazing. Copepods are the dominant grazers in the ocean in terms of their weight, so they exert a strong selective pressure. Diatoms, the phytoplankton with silicon armor, affected the evolution of their predators. Many copepods have teeth reinforced with silica, indicating they coevolved with diatoms, as the teeth are likely used mainly to crush diatom armor. And krill are important grazers, which can crush phytoplankton with their gizzards. In addition to resisting parasitoids, phytoplankton armor evolved to avoid being crushed by mouths and gizzards, and to survive passage through the gut, since a substantial portion of phytoplankton pass through the guts of their grazers unharmed. It is clear that diatoms and dinoflagellates, and probably coccolithophores and foraminiferans, evolved armor at least partly to defend against their grazers. As the strength of phytoplankton armor increased, crushing and piercing tools of their grazers increased in strength in response, and armor strength increased again, and so on, in a positive feedback loop, until a stable state was reached whereby it was too costly to evolve yet stronger armor or jaws to crush it. Specific species of phytoplankton prey coevolved with specific grazer species, each with unique defensive and offensive mechanisms. Parasitoids and herbivores of phytoplankton with protective shells made mainly of carbon indirectly helped keep Earth favorably cool and at a favorable pH, by selecting for this carbonaceous armor on these phytoplankton, great quantities of which sank in the geologic past and sink today, sequestering carbon. This process regulated and still regulates carbon, temperature, and acidity of the sea today. Since selection pressure from the parasitoids and grazers helped cause the evolution of silicon armor in diatoms, parasitoids indirectly aided the sequestration of silicon when diatoms died and sank. Of course, this sequestration of carbon and silicon improved conditions for life and hence aided diversity (see Chapter 2).

Fungi that live on land made the colonization, evolution, and diversification of plants on land possible; these phenomena could not have happened without them. Land plants could not obtain the large quantity of nutrients and water from the soil that they require to thrive without their fungal allies. One of the biggest challenges plants have faced since their appearance on land has always been extracting water and nutrients from the soil. Symbiotic fungi that live in plant roots help plants acquire these nutrients and water from the soil. Plant roots accomplish this to a certain extent, but their symbiotic fungi allow them to obtain greater quantities of nutrients and water. Plants help the fungi by providing them with a home in their roots and giving them some of the carbohydrates they manufacture from photosynthesis.

Pirozynski and Malloch (1975) hypothesized that land plants are the product of an ancient and continuing coevolution and symbiosis between an aquatic fungus and a semi-aquatic ancestral green alga. This symbiotic algae-fungus association started in water, but it eventually colonized and lived on land. It may have been somewhat similar to a lichen before it developed into a land plant and its symbiotic root fungus. The fungus benefitted from the carbohydrates made by the alga by photosynthesis. The fungus at first acted as a primitive plant root. Many of the earliest land plants lacked true roots, having instead root-like structures that anchored them to the soil called rhizoids. So they would have greatly benefitted from the fungi in their roots, since these fungi expanded their root capacity. And fossilized plants from 400 mya show well-preserved structures that look identical to modern fungal structures called hyphae that penetrate root cells and form highly branched structures today (Remy et al., 1994). These fungal hyphae help plant roots obtain nutrients and water. Fungi also helped build the soil for plants.

Invertebrates could colonize land only because plants and fungi were there as food sources and habitats for them. Vertebrates were only able to colonize land because fungi, plants, and invertebrates were there, providing nutrition and, in the case of plants, habitat. All four cases of colonization of land—fungi, plants, invertebrates, and vertebrates—occurred with a huge evolutionary change and with the colonization of an entirely new habitat, as the four groups went from the sea to land. So each of the four groups was in a new habitat—land—that had no competitors, and no predators or herbivores to eat the colonizers of land, at least when they first became established there. As a result, the fungi, plants, invertebrates, and vertebrates, each at their time of early colonization of land, speciated and diversified into a tremendous number of new species, greatly increasing their diversity. This colonization and diversification were done first by fungi and plants, then later by invertebrates, and finally later by vertebrates. Every one of the four groups needed another group to colonize land, except perhaps fungi, which were aided by the green alga; it is not clear whether or not fungi needed the alga for colonization.

Lignin, a key innovation of trees, resulted in a great deal of CO_2 sequestration and oxygen increase in the air, because large amounts of reduced carbon were buried when large numbers of trees with lignin died over long time periods (see Chapters 2 and 3). Lignin was partly selected for by fire. But it was also favored by natural selection as a defense against animals that ate trees and to aid trees in increasing their height in the competition for light that occurred both within and between species of trees. So antagonistic coevolution between trees and their herbivores and between different tree species and trees of the same species aided the evolution of lignin, and hence the regulation of atmospheric CO_2, temperature, and oxygen.

Coevolution of herbivores, especially plant-eating insects, and plants, has produced numerous plant toxins and insect defenses against them. First, plants evolve toxins to defend against insects that eat them. When insects evolve resistance to the toxins, plants evolve stronger toxins. Eventually, some plants acquire extremely poisonous compounds as a result of this antagonistic coevolutionary arms race. Insects have often co-opted the poisons and used them as defense against their predators. They store the toxins in their bodies in areas where they cannot harm the insects. Then, insects tend to evolve bright, contrasting colors that warn predators that they are poisonous and dangerous to eat. An example of these colors is the bright orange and black of the monarch butterfly. This is often followed by the evolution of these warning color patterns in insect species that are not poisonous. The nonpoisonous insect evolves to resemble the poisonous one. This often fools the predator into avoiding it, since it resembles a poisonous species. This is called Batesian mimicry. Sometimes two or more poisonous species coevolve similar patterns and adapt to resemble each other. This protects all of the species that evolve the warning coloration, because the predator learns to avoid all of these species after trying to eat one and finding that it is poisonous and distasteful. This called Mullerian mimicry. In Batesian mimicry, the plant and herbivore eating it coevolve, the predator and poisonous prey coevolve, and the poisonous species has a major influence on the species mimicking it. In Mullerian mimicry, the plant and herbivore coevolve, the predator and prey coevolve, and the two or more prey species that mimic each other coevolve. Thus, there is a great deal of coevolution and of species influencing the evolution of other species.

Competition between plant species has resulted in coevolution, with one result being the production of toxic chemicals by plants to interfere with the growth of plant species that compete with them. Rainforests have a stunning diversity of toxic chemicals in myriad plant and tree species to resist herbivores and fight competitors. Hundreds of medicines come from plants as a result of their coevolution with their herbivores and competitors, especially in rainforests. The toxic chemicals can be used to treat various maladies when given in the right doses. This includes antifungal medicines, antibacterial medicines, anticancer medicines, heart medicines, and many others.

Grasses and the animals that grazed on them had an effect on diatoms in the sea, in an ecosystem far away from them. When grasslands evolved, grazing animals, including horses, deer, rhinoceroses, and antelopes, arose and diversified tremendously in tandem with them, taking advantage of this newly available food source. This is shown clearly in the fossil record in several elegant studies. For reviews, see Stebbins (1981), Janis (2000), MacFadden (2000), and Stromberg (2006). Grasses and their mammal herbivores coevolved, each influencing and adapting to the other. Grazing animals evolved teeth with high crowns and enamel edges to grind the tough, gritty grass. The grazers also coevolved with their predators, with both of these groups influenced immensely in this coevolution by their grassland habitat. Predator and prey could see, hear, and smell each other better on open grassland than in closed, dark forests. So it was advantageous to both groups to become better at seeing, hearing, and smelling, and both groups did. The open nature of the habitat allowed relatively unobstructed running, selecting for faster running predators and prey, each a selective agent on the other, as they coevolved. Then an amazing thing happened. The chemical breakdown of silicate rocks releases silica very slowly. So before grasses evolved, silica was in short supply. But after grasses appeared, more silica was available to the biosphere. The grazers' teeth could grind the silicate-rich grass. Silica in the grasses is twice as soluble in water as that in mineral forms bound up in rocks. The grazers freed the silica in the grass even more when they chewed it, and then deposited the silica on the ground when they defecated. Then the rain washed the silica into rivers, which carried it to the sea. So grazers and grasses together greatly increased the input of silica to the ocean. Diatoms require silica to build their famously gorgeous, protective shells. Thus, *the increased supply of silica to the oceans that resulted from grasses and their grazers is responsible for the great success and diversification of diatoms* (Falkowski et al., 2006)! The evolution of grasses caused the diversification of grazers. The grazers sent silica to the sea, causing the evolutionary advancement and diversification of diatoms. The increase and diversification of diatoms increased their ability to regulate temperature and increase oxygen levels by sequestering carbon, some of which was reduced carbon. The diversification of the diatoms also meant there were more species of diatoms at the base of the major oceanic food web for their herbivores to eat. Thus, more species of their herbivores, such as krill and copepods, evolved. This led to the evolution of more species that eat krill, copepods, and other species at their level of the food web. This increase in diversity continued up the food web. Thus, the number of species of animals in the sea increased at all levels of the food web higher up than diatoms.

The digestive systems of animals, including saliva, teeth, tongue, esophagus, stomach, intestines, spleen, and so on, would not have evolved without the plants and animals that they consume. The immune system of vertebrates evolved in coevolution with viruses, bacteria, fungi, parasitic worms, and other pathogens. Without these pathogens, there would be no immune system.

Predators and prey coevolve, resulting in novel innovations in both. The great speed of gazelles and cheetahs is a result of this. Shells and spikes on shells of clams, scallops, and mussels evolved as defense. These were countered by, for two examples, the strength and endurance of starfish and the venom and scaping structure full of tiny teeth called a radula of triton snails, which can drill through the shells of its prey (such as clams).

Echolocation is the detection of objects by sending sound waves out and detecting the sound waves reflected back off objects. It is a very important skill used by whales, dolphins, and bats. Lindberg and Pyenson (2007) gave good evidence that echolocation was selected for in toothed whales, which include sperm whales, dolphins, and porpoises, at first, about 32 mya, to locate nautiloid prey at night in waters not far from the surface. Nautiloids are animals related to squids and octopuses that have tentacles that protrude from the shells they live in. Nautiloids dominated the waters when whale sonar first appeared, but they then decreased in number, possibly due to toothed whale predation, and were only in protected reefs by about 10 mya. At that time, their scarcity and limited range meant toothed whales needed other prey to obtain enough nutrition to survive. This led to the evolution in these whales of a more sophisticated sonar that could detect not just the hard shells that nautiloids possessed, but the soft tissues of squids, in dark waters. The whales also evolved the ability to dive deep to seek squid, some of which live deeper than nautiloids. Some squid can be as deep as 1,000 meters (about 3,280 feet) during the day. Squids and nautiloids are the most abundant and high-energy resource for these whales in the ocean, eaten by 90% of all toothed whales (although only two genera of nautiloids exist today). Baleen whales, the large whales such as blue whales, which eat small organisms and not nautiloids or squid, and so do not need echolocation, do not have it. This suggests echolocation evolved in toothed whales as an adaptation to detect their nautiloid and squid prey. The authors extend their scenario of the evolution of echolocation for finding prey to other squid-feeding marine vertebrates, such as seals, and to marine reptiles of the Mesozoic era (252–266 mya), like Ichthyosaurs, the predatory "fish lizards." Renouf and Davis (1982) stated that harbor seals evolved enhanced visual acuity to better see their prey, and provided evidence that they use echolocation to locate prey in low light, darkness, and turbid water.

Insectivorous bats evolved sonar (echolocation) as a result of selection to detect flying insects at night. Fruit bats eat only fruit and do not have sonar. Insects evolved defenses against bat sonar. Many tiger moths produce ultrasonic clicks when they detect bat echolocation calls (Barber, 2006). For most tiger moth species, these are warning clicks that signal to bats that the moths are poisonous (Hristov, 2005). But some jam the bats sonar. A moth called Grote's bertholdia (*Bertholdia trigona*) produces clicks at the extraordinarily high rate of up to 4,500 per second that jam the bat's echolocation system (Corcoran et al., 2009). Jamming is the most effective defense against bats that has been documented; it causes a 10-fold decrease in bat capture success rate in the field (Corcoran & Conner, 2012). Moths also attempt to escape bats by escape maneuvers, which vary by species. They fly using zig-zags, loops, tight turns, passive dives, and abrupt power dives. The escape-tactic diversity hypothesis posits that species-specific differences in evasive behavior will increase the overall unpredictability experienced by predators with many prey species within a predator-prey community, enhancing protection for all prey species. Unpredictable flight reactions of the prey increase the uncertainty for the bats, making it difficult for them to adapt to them or to adapt to any one escape tactic. This means *the predator acts as a selective force for diversity in escape tactics of prey species*. An alternative interpretation is that escape-tactic diversity is a consequence of differences in the bodies and wings of the prey that correlate with evasive capabilities. Hügel and Goerlitz (2019) tested this hypothesis in eared moths, and showed differences between moth species in both overall flight strength and change of flight strength over time, supporting the escape-tactic diversity hypothesis. They even found strong differences in evasive flight to the same stimulus between individuals *within* some species. So bats and moths have undergone coevolution that led to innovations, and probably diversification, in both groups. Clearly, *echolocation evolved in toothed whales, seals, and bats as a result of selection provided by their prey species*.

Also, the evolution of organisms is influenced by other members of their own population, sometimes profoundly. For example, competition between males for females has produced a fantastic array of beautiful and diverse colors, songs, and dances in some species. It has also produced weapons such as antlers in deer. These serve to both intimidate other males and attract females. Selection for these colors, structures, and behaviors is called sexual selection. Birds of paradise are a striking example. The males of many species are very colorful and beautiful, and have elaborate songs and dances. One good example is the Raggiana bird of paradise (*Paradisiac raggiana*), shown in Figure 4.1. Several males congregate in leks, which in this species are areas of 30 to 100 meters (about 98.4 to 328.1 feet), where they display their beautiful plumage to females. The males compete for a group of tall, slender trees for prominent perches, and defend them from rivals. They display to the females to win them as mates on these perches. The display involves clapping the wings and shaking the head. Their gorgeous plumage and elaborate displays evolved as a result of competition between males and preference of the displays and bright coloration by the females. This is but one example of many spectacular displays of color, song, and dance in birds-of-paradise, which demonstrate the tremendous influence of other males and of females on the evolution of males of this group of birds by sexual selection.

FIGURE 4.1 The male Raggiana bird of paradise (*Paradisaea raggiana*) evolved its beautiful bright colors, song, and dance to intimidate other males and attract females.

Shutterstock Item ID: 1643261989

Title: Raggiana bird of paradise (*Paradisiac raggiana*).
Photo Contributor: winphong

Sexual selection can lead to greatly increased diversification and rapid speciation, and has done so in many groups. Males in different populations of the same species can be selected to have different colors, songs, and dances, with the preference of the female being the agent of natural selection. Birds of paradise have a closely related group, called a sister group, that do not undergo sexual selection. These birds are called manucodes. There are 33 known species of birds of paradise and only 5 species of manucodes (Mitra et al., 1996). Hummingbirds undergo intense sexual selection and have 319 known species, and their sister group that does not undergo sexual selection, swifts, have only 103 (ibid.). These two comparisons are strong evidence that sexual selection causes diversification into many species.

Sexual selection and other mechanisms by which organisms in the same population affect each other's evolution are discussed in more detail in Chapter 3 of the of the companion book to this one (Seaborg, 2022).

Interaction between individuals in a population may have been important in the evolution of intelligence. Sociality is correlated with intelligence. Honeybees are among the most social and most intelligent insects. Among mammals, whales and dolphins, primates, and elephants are among both the most social and intelligent. The social brain hypothesis of British anthropologist Robin Dunbar proposes that human intelligence evolved not primarily to solve ecological problems, but as a way of surviving and reproducing in large and complex social groups (Dunbar, 1998). Social behaviors that select for intelligence include reciprocal altruism (here, two individuals do each other favors, remembering favors and paying them back at a later time), friendships, coalitions, deception, cheating, and recognizing cheaters. The ability to understand the thoughts and emotions of others, called theory of mind, is an important adaptive trait of those living in groups, and it requires intelligence. Ruminants, which are animals such as antelope and deer, show that there are other factors than living in groups that select for intelligence, since they are social, but not highly intelligent. And those other factors might involve other species, not solely other individuals in the group. For example, the need of predators to outwit their prey could select for intelligence. But this too could involve social interaction, such as teamwork. Dolphins swim in groups in circles around schools of fish, blowing bubbles, entrapping the fish. This makes the fish easy to catch. Lions hunt as teams, with some chasing the prey toward others hidden and waiting in ambush. Of course, there can be challenges that select for intelligence that do not involve any social interaction. But it is clear that individuals in groups affect each other's evolution, and sociality under the right conditions can select for high intelligence.

Different species evolve together in groups as evolutionary suites. Any species, its microbiome, pathogens, predators, prey, competitors, species that it is symbiotic and commensal with, etc., all coevolve together. I will now introduce a new concept that I have originated, and I will coin a new term, the *holobiome*. It is an evolutionary unit of interacting, coevolving species all connected to one central species. Of course, it also includes the central species. All of the species in a holobiome coevolve with each other and exchange DNA with each other. My idea is that there are systems of interacting species that coevolve and exchange DNA, affecting and promoting each other's evolution, and, for the most part helping each other survive and diversify. There is one central species. The other species include its microbiome, predators, prey, competitors, commensals, symbiotic species, and so on—any species it interacts with. The central species exchanges DNA, in both directions, with each of these species. The other species also exchange DNA with each other. For example, the holobiome with the California kingsnake (*Lampropeltis californiae*) as its central species, which I call the California kingsnake holobiome, includes this snake, its microbiome, its prey (rodents, snakes, lizards), predators (birds of prey, coyotes, and so on), parasites, competitors (such as some other snake species, birds of prey, and other predators that eat the same prey as the California kingsnake eats), commensals (it sometimes uses burrows dug by desert tortoises, *Gopherus agassizii*), species it is symbiotic with, and so on. This holobiome includes all species that are directly interacting with the California kingsnake. The coyote (*Canis latrans*) is a predator of the kingsnake. The coyote holobiome, which has the coyote as its central species, includes gray wolves (*Canis lupus*), which compete with and sometimes kill coyotes, but do not normally eat or interact directly with kingsnakes. In this case, the wolf interacts indirectly with the kingsnake through interacting holobiomes. And the microbes in the microbiome of the kingsnake interact indirectly with the microbes in the microbiome of the wolf.

The species in holobiomes have three major ways they affect each other and each other's evolution. They provide a selective force that affects the evolution of the other species they interact directly with. This may be antagonistic, symbiotic, some of both, or something else, such as commensal. They affect the population size of the species they interact with, often regulating it to its benefit. Finally, they exchange genes with species they interact with, in both directions, providing variability for natural selection to act on and promoting the evolution of the interacting species.

Coevolution of life and the physical-chemical-geological environment was important in the past and still is important today. Some of this coevolution is covered in other parts of this book, principally in Chapters 2 and 3. Life coevolved with the atmosphere. This includes sequestration of carbon by life, thereby regulating temperature and pH in the seas and freshwater ecosystems. The resultant cooler temperatures and lower acidity in aquatic ecosystems allowed life to thrive and diversify. Life caused oxygen to greatly increase in the atmosphere through photosynthesis and burial of reduced carbon. Cyanobacteria, phytoplankton, and higher land plants played a large role in this. The high oxygen built the ozone layer, protecting life from ultraviolet light. All of these favorable conditions created by life in turn allowed life to increase, evolve, become more complex, and greatly diversify. This increase in complexity and diversification included the evolution of mammals, including humans. The increased abundance and diversity of life allowed it to bury more carbon, including reduced carbon, further cooling the planet, regulating acidity in aquatic ecosystems, and increasing oxygen levels. Minerals were important in the chemical evolution of life. The high oxygen increased mineral diversity, creating myriad niches for prokaryotes, allowing them to diversify. Life coevolved with the geological environment, including helping make the soil. Life broke down rocks, released minerals, and added organic matter, which helped build the soil, which was necessary for terrestrial life to thrive. The soil in turn helped life increase and diversify, which allowed life to aerate and improve the soil further for life, allowing life to increase and diversify further. Life's effects on the soil will be discussed in more detail in Chapter 6. The point of this paragraph is that life and the physical-chemical-geological environment coevolved in a positive feedback loop, and this resulted in higher biodiversity.

REFERENCES

Alegado, R. A., et al. (15 Oct., 2012). A bacterial sulfonolipid triggers multicellular development in the closest living relatives of animals. *eLife* 1: e00013. doi: 10.7554/eLife.00013.

Barber, J. R. (2006). Tiger moth responses to a simulated bat attack: timing and duty cycle. *Journ. of Experimental Biology* 209 (14): 2637–50. doi: 10.1242/jeb.02295. PMID 16809455. S2CID 11687445.

Buckling, A. & Rainey, P. B. (7 May, 2002). Antagonistic coevolution between a bacterium and a bacteriophage. *Proc. Biol. Sci.* 269 (1494): 931–6. doi: 10.1098/rspb.2001.1945 PMCID: PMC1690980. PMID: 12028776.

Corcoran, A. J., Barber, J. R., & Conner, W. E. (2009). Tiger moth jams bat sonar. *Science* 325 (5938): 325–7. Bibcode: 2009Sci...325..325C. doi: 10.1126/science.1174096. PMID 19608920. S2CID 206520028.

Corcoran, A. J. & Conner, W. E. (2012). Sonar jamming in the field: effectiveness and behavior of a unique prey defense. *Journ. of Experimental Biology* 215 (24): 4278–87.

Damer, B. (2018). Personal communication.

Dunbar, R. I. (1998). The social brain hypothesis. *Evol. Anthropol.: Issues, News, and Reviews* 6 (5): 178–90. doi: 10.1002/(SICI)1520-6505(1998)6:5<178::AID-EVAN5>3.0.CO;2-8.

Falkowski, P. G., et al. (2004). The evolution of modern eukaryotic phytoplankton. *Science* 305: 354–60.

Hristov, N. I. & Conner, W. E. (2005). Sound strategy: aposematism in the bat-tiger moth arms race. *Naturwissenschaften* 92 (4): 164–9. Bibcode:2005NW.....92..164H. doi: 10.1007/s00114-005-0611-7. PMID 15772807. S2CID 18306198.

Hügel, T. & Goerlitz, H. R. (2 July, 2019). Species-specific strategies increase unpredictability of escape flight in eared moths. *Functional Ecology* 33 (9): 1674–83. https://doi.org/10.1111/1365-2435.13383.

Janis, C. M., et al. (2000). Miocene ungulates and terrestrial primary productivity: where have all the browsers gone? *PNAS USA* 97: 7899–904.

Lindberg, D. R. & Pyenson, N. D. (Dec., 2007). Things that go bump in the night: evolutionary interactions between cephalopods and cetaceans in the tertiary. *Lethaia* 40 (4): 355–43. https://doi.org/10.1111/j.1502-3931.2007.00032.x.

Lindell, D., et al. (2004). Transfer of photosynthesis genes to and from *Prochlorococcus* viruses. *PNAS USA* 101 (30): 11013–8.

MacFadden, B. J. (2000). Cenozoic mammalian herbivores from the Americas: reconstructing ancient diets and terrestrial communities. *Annual Review of Ecology and Syst.* 31: 31–59.

Millard, A., et al. (2004). Genetic organization of the psbAD region in phages infecting marine *Synechococcus* strains. *PNAS USA* 101.30 (2004): 11007–12.

Mitra, S. H., et al. (1996). Species richness covaries with mating systems in birds. *Auk* 113: 544–51.

Pernthaller, J. (2005). Predation on prokaryotes in the water column and its ecological implications. *Nature Rev. Microbiol.* 3: 537–46.

Pirozynski, K. A. & Malloch, D. W. (Mar., 1975). The origin of land plants: a matter of mycotrophism. *Biosystems* 6 (3): 153–64.

Remy, W., et al. (1994). Four hundred-million-year-old vesicular arbuscular mycorrhizae. *PNAS USA* 91: 11841–3.

Renouf, D. & Davis, M. (16 Dec., 1982). Evidence that seals may use echolocation. *Nature* 300: 635–7. https://doi.org/10.1038/300635a0.

Seaborg, D. (2022). *How Life Increases Biodiversity: An Autocatalytic Hypothesis*. CRC Press/Taylor & Francis. Boca Raton, FL; London, UK; New York, NY.

Stebbins, G. L. (1981). Coevolution of grasses and herbivores. *Annals of the Missouri Bot. Garden* 68: 75–86.

Stromberg, C. A. E. (2006). Evolution of hypsodonty in equids: testing the hypothesis of adaptation. *Paleobiol.* 32: 236–58.

Sullivan, M. B., et al. (2006). Prevalence and evolution of core photosystem II genes in marine cyanobacterial viruses and their hosts. *PLoS Biology* 4 (8): e234. https://doi.org/10.1371/journal.pbio.0040234.

5 In Ecological Succession, Communities Create Favorable Environmental Conditions for Succeeding, Usually More Diverse and Complex, Communities

Ecological succession is a largely predictable change in the species of a community over time following a disturbance or the initial colonization of a new area. It is a sequence of discrete stages, each replacing the previous one. It is very common in nature. Succession involving higher organisms generally involves plants as the main drivers of the process. The general rule by which it proceeds is that simple, pioneer species that are good at colonizing and that tolerate a broad range of environments, but are poor competitors, colonize a barren area. They change the environment, making it better for the next stage, which then outcompetes and replaces them. The pattern continues, with the species of each stage acting as ecosystem engineers, improving the environment for the next stage, and sometimes making the environment worse for themselves. These changes may be done by lichens breaking down rock and building soil, changes in the acidity of soil by plants, or early-stage trees dropping leaves and enriching the soil. Each subsequent stage outcompetes the previous one. One way this can be accomplished is by later-stage trees that are tolerant to shade growing through the shade of trees of the stage before them and shading them out. The trees of the earlier stage are better colonizers and faster-growing than the subsequent stage, but the trees of the subsequent stage are more shade-tolerant than those of the earlier stage. The trees of the earlier stage make the soil better for the following stage, as their fallen leaves and trunks fertilize the soil, and their root bacteria often add useable nitrogen to the soil. As the process proceeds, the system generally becomes more biodiverse and complex, and has higher numbers of organisms and biomass. Progressing from early to later stages, the complexity of the most complex species in its stage increases, the ecosystem becomes more complex and information-rich and diverse, and each stage tends to have better competitors with a narrower range of environmental conditions that they can tolerate. This is all accomplished by the effects of the species of each stage on the environment.

Ecological succession ensures that there will be tremendous numbers of species both in a given area and through time. Disturbances cause most or all stages of succession to occur at any given time if the area is large, resulting in high biodiversity. The number of species increases with each stage except the last one.

An example of an ecological succession is: bare rock –> lichens –> mosses –> grasses and herbaceous flowering plants –> brush –> deciduous hardwood forest –> mixed deciduous-coniferous forest –> coniferous forest –> old growth coniferous forest. A deciduous tree is one that drops its leaves each fall. The community begins with relatively few fast-growing pioneer plants and animals, and goes through more or less discrete stages that increase in diversity, with each stage having more species than the one preceding it, except that the climax community, which is the last stage, is often less diverse than the stage immediately before it. It is noteworthy that the earlier stages alter the environment to the benefit of the next successional stage. Primary succession begins in unoccupied habitats, uninfluenced by pre-existing communities, while secondary succession follows disruption of a pre-existing community. A specific example of primary succession is the succession following the retreat of glaciers in Glacier Bay, Alaska (Cooper, 1939; Crocker and Major, 1955; Lawrence, 1958). Receding glaciers left an exceedingly nitrogen-poor soil, and initial colonizing plants were stunted with yellow leaves as a result of severe nitrogen deficiency. Only a few species could reproduce, and willows and cottonwoods were stunted and as small as shrubs. Among the first plants to thrive were alder trees and *Dryas*, a genus of evergreen dwarf shrubs in the rose family that grows best with little or no shade. Both alders and *Dryas* have bacteria in their roots that convert nitrogen in the air that is useless to plants to a form of nitrogen plants can use. Alders and *Dryas* formed dense stands on the low nitrogen soil (Crocker and Major, 1955; Lawrence, 1958), and their bacteria added a great deal of nitrogen that plants can use. This was the main reason for an enormous increase in soil nitrogen during the first 100 years of succession. The increase in nitrogen allowed the willows and cottonwoods to become erect and tall, and shade out the alders and *Dryas*. So willows and cottonwoods replaced alders and *Dryas* as the dominant trees, and this was the second stage of succession. Then the increased useable soil nitrogen provided by the alders and *Dryas* allowed shade-tolerant spruce to invade and

DOI: 10.1201/9781003246640-5

form a closed canopy and further shade out the alders. Two species of hemlock, also shade-tolerant, then appeared, and coexisted with the spruce, and a higher-diversity community was formed. The community changed from a habitat low in nitrogen but with a high availability of light to one higher in nitrogen, but with a closed canopy and low light availability. The alders and *Dryas* added nitrogen and caused its levels to greatly increase, aiding their competitors and causing their own destruction; their actions also caused a great increase in diversity.

Succession may be classified as allogenic or autogenic. Allogenic succession is not caused by vegetation, but external factors, such as soil changes due to the deposition of silt and clays, which can alter nutrient content and water relationships. Animals also influence allogenic succession; they greatly increase plant diversity, since they pollinate, disperse seeds and fruits, and keep plant populations from becoming too crowded or overtaxing their environment, by eating them. They increase soil nutrients when they defecate and die. Some animals, such as termites, help decompose plants, including trees. Some help plants by aerating, tilling, and moving soil; this is done by moles, earthworms, ants, and termites. Some animals, such as termites and wood-eating beetles that decompose dead trees, help create regeneration sites that favor certain plant species.

Autogenic succession is brought about by changes in the soil caused by the plants, as described above and exemplified in the Glacier Bay succession. An example of some of the stages of this type of succession is lichens growing on and breaking down rock, which forms soil for plants of later stages. Mosses can appear in the stage immediately following lichens, and some of these can grow on rocks. But plants of stages after lichens and mosses are unable to grow on rocks, but outcompete lichens after the lichens make soil, which these plants need and can exploit more effectively than lichens can. Thus, the lichens improve the environment for their competitors, leading to their decline. There is generally enough rock available that lichens do not completely disappear. But the communities that succeed the lichens and mosses are all more diverse and composed of more complex species than the lichen and moss communities. Another example is the accumulation of organic matter from leaves falling from trees, favoring the subsequent stage (also trees), which then displaces the earlier stage. Here too, the earlier stage alters the environment to the benefit of the subsequent stage, leading to its own demise, and the succeeding community is more diverse and has generally more complex species than the earlier community. Plants can also change the acidity of the soil, resulting in the subsequent, more diverse, and complex successional stage replacing the earlier stage, which is the one that changes the soil acidity.

Later stages can alter the environment to favor themselves over the stage they replace as well, as when the later stage consists of taller, shade-tolerant trees that grow and shade out the shorter, less shade-tolerant ones. In this case, the trees of the later stage change the environment by making it darker and with more shade, making the environment worse for the earlier stage, which is less tolerant to darkness and shade, but better for themselves because they decrease the population size of or eliminate tree species that compete with them. The later, shade-tolerant stage is more diverse than the preceding stage.

Generally, the pioneering species that are good dispersers and colonizers and fast-growing forms precede and are replaced by the slower-dispersing and slower-growing, better competitors. Earlier stages alter the environment to favor a more diverse and complex subsequent stage, with the later stage outcompeting and displacing the preceding stage. The species of most stages help the stage that follows their stage, and each stage is more diverse than the previous one, except for the climax stage. And the most complex species of each stage are more complex than the most complex species of the previous stage, except perhaps the climax stage. Life increases diversity by this mechanism. Spatially there can be higher diversity because disturbances can cause several or even all of the stages to occur simultaneously if the ecosystem occupies a large area.

> *A key point in autogenic succession is that each stage alters the environment with the species acting as ecosystem engineers that bring about and favor a subsequent stage that is more diverse and complex than it (with the exception of the subclimax compared to the climax community). This is a mechanism by which life increases biodiversity, and by which the ABH occurs.*

Animal diversity also increases with each autogenic successional stage, as new animals immigrate and take advantage of the increased plant diversity. In the lichen stage (in communities where they are the first stage), animal diversity is low, with mites, ants, and spiders living in the limited habitat available. The fauna increases in diversity, numbers, and biomass during the grass and forb stage. The animals found during this stage include ants and other insects, spiders, mites, nematodes, frogs, lizards, snakes, birds, rodents, and others. The animals increase and diversify tremendously as the first forest stage becomes established, with a full invertebrate community and all classes of land vertebrates, from amphibians to mammals. This trend of increasing animal diversity and complexity continues with each forest stage, except perhaps the last one. In ecological succession, from the first stage of simple species such as lichens through the forests and to the last stage, each stage has higher diversity and complexity than the stage just prior to it, with one exception.

This exception is that the final stage, called the climax stage or community, tends to have lower diversity than the stage just previous to it, called the subclimax stage or community. This is because superior tree competitors eliminate inferior ones through competition during the climax stage. Since competition has not fully gone to completion in the subclimax community, it is the most diverse successional stage.

The climax community usually is somewhat unstable and does not persist indefinitely. In fact, the idea of a climax community has been largely abandoned by many who think stable equilibria are never maintained in ecosystems. At any rate, the climax community often does not persist too long, and this is because of disturbances. There are many causes of disturbance, which can return a climax community to an earlier successional stage. The disturbance can return the climax community to the first stage. Succession then proceeds again. But disturbances often put the community in the subclimax stage, the one with the most species. The result of disturbances can be a combination of all successional stages if the ecosystem is in a sufficiently large area. Of course, this maximizes diversity.

One agent of disturbance is fire. Life tends to keep fire at a favorable level for life through negative feedback (see Chapter 3). Thus, the frequency and intensity of fires will tend to favor a mosaic, or combination, of all successional stages, and therefore maximum diversity. Herbivores are also agents of disturbance. They tend to disturb at an intermediate level, and this maximizes diversity. This is discussed in detail in Chapter 8 of the companion book to this one (Seaborg, 2022). If herbivores grazed plants too much, there would be less plants and less plant species. But if they grazed too little, dominant plants would outcompete and drive poorer competitors to extinction, reducing the number of plant species. Herbivores tend to maintain the community in the subclimax stage. Thus, herbivores tend to maximize plant diversity. Avalanches disturb forests on mountains, creating mosaics of different successional stages.

Hurricanes disturb coral reefs, increasing their diversity by creating subclimax communities and mosaics of communities of all successional stages, and their intensity may be optimized for diversity by life's regulation of CO_2 and methane, and hence temperature. Life's regulation of atmospheric carbon and hence temperature may keep the temperature high enough to have hurricanes, but not so high that they are too large.

Coral are not plants, but a symbiotic association of a small animal and a phytoplankton, but coral reefs do undergo ecological succession. Corals are the key group of species that most species in the reef benefit from. The species that coral aids are great in number. Coral reefs are the Earth's most biodiverse marine ecosystem. They contain about a quarter of all marine species, even though they cover less than1% of the Earth's seas, about half the area of France (Spalding & Grenfell, 1997). Corals provide a three-dimensional habitat for most of the reef species and food for many. The higher the number of species and biomass of coral in the reef, the more species of other animals there. This discussion refers to times before human-induced climate change increased the intensity of hurricanes. Increased temperatures do not increase the number of hurricanes, but their intensity. It appears that hurricanes tend to disturb coral reefs an intermediate amount, being intermediate in intensity, often putting coral reefs at a subclimax stage of succession. Hurricanes that are too forceful would reduce the number of coral species and the amount of coral because hurricanes destroy coral. This would decrease the diversity of the entire reef. But hurricanes that are not intense enough would allow coral to complete their competition between species, allowing the dominant species to drive the species that are not as effective at competition locally extinct. This would decrease both the number of coral species and number of total species on the reef. Thus, an intermediate intensity of hurricanes results in the highest biomass, and number of species, of coral, and the highest number of total species, on the reef. In natural systems, hurricanes apparently tend to occur at an intermediate intensity. However, this is not fully verified and needs further study to prove.

Under natural circumstances, predators of corals, such as the crown-of-thorns sea star, generally maintain coral reefs at subclimax, highest-diversity states by intermediate disturbance. The predators are controlled by their own predators, keeping them at optimal levels to consume coral at an intermediate level that maximizes coral diversity. This in turn increases the diversity of the many species that benefit from coral and hence the diversity of the entire coral reef ecosystem.

Succession occurs throughout microbial communities. It has been shown in cheese communities (Wolfe et al., 2014). Rind communities in Vermont cheese aged over 63 days start out with a pioneering community of bacteria of the phylum Pseudomonadota (formerly Proteobacteria), bacteria of the genus *Leuconostoc*, and candida yeast. *Staphylococcus* bacteria replace the Pseudomonadota bacteria within a week. As the cheese ripens, bacteria genera *Brevibacterium* and *Brachybacterium*, and fungi genera *Pernicillium* and *Scopulariopsis* become dominant. This pattern repeats itself whenever these organisms congregate on cheese in the laboratory. Archaea and protists also take part in microbial succession on cheese. Protists are a large group of eukaryotes that are mainly unicellular and include amoebas, diatoms, and slime molds. They are not a natural group; rather, the biological category protist is used for convenience. Viruses prey on all species in each of these microbial communities, adding to the diversity of the system because this adds more species in the form of viruses.

There is ecological succession of microorganisms in the soil in recycling a dead organism, as well as feces. Fungal, protist, bacterial, archaeal, and viral communities are replaced by different such communities. Bacteria secrete products that change the environment, often aiding their successors. The acidity may change, sometimes due to the activities of one stage, allowing the next stage to replace it. So earlier stages sometimes alter the environment to their detriment and the benefit of the next stage, as in plant succession. Both fungi and bacteria break down components at each stage. Of the fungi, the first to appear are those that use sugar and starches, followed by those that obtain nutrients by breaking down cellulose and lignins if the organism is a plant. Lignins are also broken down by white-rot fungi. Bacterial succession of species in recycling dead organisms starts with the first bacteria to appear on a dead animal, which tear down the muscle tissue into individual protein strands. Then bacteria that use fermentation break down proteins into their building blocks, called amino acids. Then bacteria called acetogens become dominant, and break the amino acids down into the following simpler molecules: acetic acid, hydrogen gas, nitrogen gas, and CO_2 gas. Next, the acetogens are replaced by methane-producing microbes that utilize and break down the acetic acid, producing methane and water. There is a sequential, successional division of labor in a multistep process of recycling nutrients, with each stage producing nutrients used by the stage that follows it. The succession also makes nutrients available to plants, and creates more soil and humus, the dark, organic component of soil necessary to its healthy function. However, biodiversity does not necessarily increase, and seems to even at times decrease, from earlier to later stages of succession. Pechal et al. (2014) found a significant decrease in the overall number of phyla and families of organisms as decomposition progressed. This is but one study, and more work needs to be done, but it may represent a general rule because diversity might decrease when the environment is being consumed and reduced, as is the case in decomposition. In general, there is no climax community in microbial succession.

Payne (1965) found there is ecological succession of arthropods on dead baby pigs. He found great diversity, observing a total of 522 species representing 3 phyla, 9 classes, 31 orders, 151 families, and 359 genera.

Animal scavengers arrive at carcasses in an ecological succession. Diversity over time is maximized because the scavengers arrive in reverse order of their competitive abilities, the least competitive arriving first and the best competitors last. Scavengers displace other scavengers in a consistent, sequential order. This is a form of niche partitioning. The less effective head-to-head competitors are the best at finding the carcass, and are displaced by the species that arrive immediately after them. If the species that was the best competitor was also the best at finding the carcass and the first to arrive, it would be the only species of scavenger in the area. In North America, the turkey vulture arrives at carcasses before the black vulture because it has the better sense of smell of the two. Black vultures observe turkey vultures to find carcasses, arriving after them They displace turkey vultures, being more aggressive. In some parts of the desert, caracaras arrive last, and displace both vulture species. In Africa, what is left of predatory cat kills are eaten after the cat eats its fill of the carcass. Species of jackal, vulture, hyena, and others arrive in sequential order according to the same principle, maximizing diversity over time.

Whale carcasses are especially noteworthy as ecosystems supporting diverse successional communities of decomposers. Many sharks are scavengers as well as predators; they and other scavenger fish are the first to consume and recycle some of the whale. It is believed that some sharks and fish are specialized to feed exclusively on dead whales. Sometimes the whale is washed up on the beach, where a succession of bacteria, archaea, protists, fungi, crustaceans that have many species that swim and that are typically two inches long called isopods, shrimp, other invertebrates, birds, raccoons, and bears consume and recycle it, usually with some of the nutrient going to land, some to sea. This process takes up to 10 years. On land or sea, the whale can support entire ecosystems because of its size. On land, much of the nutrient ends up in the soil, where it is utilized by plants, entering into the terrestrial food web.

Whale carcasses in the sea tend to sink because the bodies of most great whales are slightly denser than the surrounding seawater, and only float when the lungs are filled with air or when bacteria in the whale's digestive tract produce enough gas to cause them to float. If the carcass does float, this is ended and the whale sinks to the seafloor after the sharks and fish deflate its gassy interior. There are few scavengers in the water column, so the whale usually sinks to the seafloor mostly intact. Here it is called whale fall, and its huge body provides tremendous nutrient to the ecosystem of decomposers. The cold, deep water keeps the carcass cold and slows decomposition significantly. On the seafloor, the carcass can support entire complex ecosystems for over 50 years. Dead whales are a key method by which nutrients are transferred from shallow or intermediate-depth waters to the seafloor, an otherwise largely static, nutrient-poor part of the ocean. They produce islands of habitat rich in nutrients on the seafloor. This is one of many ways that whales are ecosystem engineers. There is a complex succession of microbes, with bacteria, archaea, fungi, and protists of many species succeeding one another, much like in plant succession. Deep-sea communities of animals pass through a series of overlapping successional stages that vary with carcass size, water depth, and environmental conditions. Evolutionary novelties include a diversity of grazers on sulfur bacteria that are on the whale carcass, and bone-eating worms and snails. Viruses attack each of these microbe and animal species, adding yet more diversity.

Whale falls have species of deep-sea bristle worms, giant isopods (three times or more as long as a typical isopod; Figure 5.1), lobsters, flat lobsters called squat lobsters, crabs, prawns, shrimp, sea cucumbers, hagfish (eel-shaped scavengers), and sleeper sharks. The name "sleeper shark" comes from their slow swimming, low activity level, and perceived nonaggressive nature. Whale fall has up to three successional stages, each differing from the others in species composition, mobility, and overall feeding structure (Smith and Baco, 2003). Whale carcasses reaching the deep ocean floor support high diversity by providing immense quantities of nutrient in the form of organic matter.

The first stage, called the mobile-scavenger stage, starts a month to a month and a half after the whale settles on the seafloor, and takes from four months to five years, depending on the size of the whale. Each stage helps prepare the carcass for the following stage by removing meat and leaving underlying meat that the following stage can then access. Sleeper sharks are the first to arrive and remove more tissue than any other species. Next come the most populous, impressive organisms, hundreds of hagfish, coming from areas of one to two square kilometers (about 0.39 to 0.77 square mile). As the tissue fragments decrease in size, they are consumed by successively smaller scavengers. Rattail fish (bizarre deep-sea fish with big heads and long, rat-like tails), then amphipods (small, swimming, segmented crustaceans), and finally copepods (very common and numerous small, swimming crustaceans) remove the remaining meat. The mobile-scavenger stage is finished when all of the soft tissue

FIGURE 5.1 Giant isopods live in the sea, but look like woodlice, which are also called sow bugs.

Shutterstock Item ID: 1776168140

Title: A shelled sea urhen, Giant isopod.
Photo Contributor: Darren Hugh Lynch

has been removed. This is almost 90% of the whale's wet weight. The larger scavengers of this stage disperse a great deal of nutrients when they defecate; these are used by other species on the seafloor.

The second stage is referred to as the enrichment opportunity stage, and commences when all the meat and blubber of the whale has been consumed, and what remains is bone and scraps, surrounded by rich sediments. The first stage removed the meat and blubber, allowing species of the second stage to gain access to the nutrients in the second stage. Colonization in and around the whale bones is accomplished by large and dense populations of bottom dwellers: sea anemones specific to whale carcasses, bristle worms, snails and their relatives, juvenile clams and mussels and similar groups, some crustaceans, and other groups. Bristle worms called zombie worms grow their roots into the bones and feed on lipids (fats and other compounds that cannot dissolve in water). Snails and shrimp feed on both the sediment and bone. This stage typically lasts about two years or less, but this varies with carcass size.

The third stage, the sulfophilic stage, is the most diverse stage, and the most diverse ecosystem of all hard surfaces in the deep sea. A huge community that is complex in its food web thrives on the skeleton. Sulfophilic bacteria break down lipids in the bones in the absence of oxygen, reducing dissolved sulfate and producing hydrogen sulfide. Only bacteria tolerant to hydrogen sulfide, which is toxic to most organisms, grow with the sulfonic bacteria. Bacterial mats form, which provide nutrient to clams, mussels, limpets, sea snails, tube worms, and other species, all of which use the carcass as habitat. Predators of the invertebrates are present too. The bacteria and invertebrates break down lipids embedded in the bone. Whale bones have a high lipid content, 4 to 6% of the whale's weight, and the lipids are broken down slowly, so this stage can last 50 to 100 years.

Some postulate a fourth stage, the reef stage, which occurs after the organic compounds have been exhausted and only minerals remain in the bones, providing a hard surface for suspension and filter feeders, which use the skeleton as an aid in water flow. A suspension feeder feeds on material, such as plankton, suspended in the water, and filter feeders are a subgroup of suspension feeders that feed by straining suspended matter and food particles from water, typically by passing the water over a specialized filtering structure. On small skeletons, the bones may decompose before this fourth stage occurs.

Scientific studies suggest that whale falls are hot spots of diversification into many species for a specialized set of animals adapted to them (Smith et al., 2015). They are evolutionary stepping stones for mussels that live in vents and seeps, and could have facilitated speciation in other animals that live in vents and seeps (ibid.). This is because whale carcasses provide isolated areas that are very high in nutrient, so animals that feed on them have thriving populations that are isolated from each other—ideal conditions for speciation.

Whale skeleton species richness is higher than that in any other deep-sea hard surface habitat. The following numbers do not count very small animal species. There is an average of 185 animal species on single whale skeletons. This approaches levels of the *total* number of animal species that live in cold seeps, which is 229 species, and exceeds the number of species of the most species-rich cold-seep vent field known (121 species) (Baco and Smith, 2003). Twenty-one species in five phyla are known to live exclusively on whale fall. Researchers think animals that live off whale fall have been exploiting this unique habitat for tens of millions of years, and have adapted to it, with special adaptations for the sulfophilic stage.

Succession of a similar type to that on dead whales, though with somewhat less diverse communities, occurs on the carcasses of large squids and sharks. Danise et al. (2014) found succession on an ichthyosaur on a shelf from the ocean of the late Jurassic period (about 201.3 mya to about 145 Mya). Ichthyosaurs are large, extinct, predatory marine reptiles (Figure 5.2). Fossils showed that there were microbial mats that attracted sea urchins and other mat-grazing invertebrates, and abundant cemented suspension feeders. So ichthyosaur falls apparently fulfilled similar ecological roles to whale falls. From what is presently known, microbial succession and succession on carcasses of whales and some other sea animals differ from plant community succession in that the former do not regularly experience disturbance and increased diversity due to it.

I will now coin a new term, *evolutionary succession*, which is succession on an evolutionary time scale, and can take hundreds of thousands of years or more. Like ecological succession, this involves discrete stages; earlier stages altering the environment to the benefit of later stages; replacement of earlier stages by later ones; and, as succession proceeds, an increase in biodiversity, complexity of the system, and the complexity of the most complex species of a given stage. Cyanobacteria were an early stage in an evolutionary succession. They are unusual in that they can both photosynthesize and fix nitrogen. Nitrogen fixation is the process by which molecular nitrogen, chemical symbol N_2, which life cannot use, is converted to a form of nitrogen that life can use, such as ammonia (NH_3). Earth's atmosphere is about 80% molecular nitrogen. Nitrogen-fixing bacteria

FIGURE 5.2 Ichthyosaurs are large, predatory marine reptiles that lived from about 250 to about 90 mya. Decomposers on their carcasses underwent ecological succession.

Shutterstock Item ID: 2030512373

Title: *Ichthyosaur stenopterygius*, large extinct marine reptiles from Early Triassic to Late Cretaceous, 3D science rendering. Illustration Contributor: Dotted Yeti

are bacteria that carry out nitrogen fixation, converting molecular nitrogen to a form useable by life. Some of these bacteria live in plant roots in a symbiotic relationship with plants, in which plants give them a home and carbohydrates, and they give plants a form of nitrogen that plants can use. Some live in the soil without any association with plant roots. So cyanobacteria were doubly important in the evolution and diversification of life, including higher life, because they carried out both photosynthesis, producing oxygen, and nitrogen fixation, producing a form a nitrogen that life can use. They converted molecular nitrogen to ammonia, a form of nitrogen that life can utilize, and fertilized the seas with it. Cyanobacteria thrive in low ammonia, and do not do well in high ammonia, so they made the sea environment worse for themselves. But their nitrogen fixation helped their competitors, since many eukaryotic phytoplankton thrive in high ammonia environments, since they use it as nutrition. The less complex prokaryotic cyanobacteria thus set the stage for more complex eukaryotic phytoplankton to thrive. The result was an evolutionary succession in which cyanobacteria decreased, and eukaryotic phytoplankton, such as green algae and diatoms, outcompeted cyanobacteria in much of the sea, and increased. Phytoplankton thrived and diversified as a result of cyanobacteria's adding useable nitrogen. Phytoplankton are more complex than cyanobacteria, so complexity increased. The high ammonia content of the water caused cyanobacteria to decrease in number, but they did not disappear. And they did not fully eliminate their low-ammonia niche. The result was a big increase in biodiversity, with cyanobacteria coexisting with many phytoplankton species, allowing the evolution of complex food webs with high diversity. For example, one food web that still exists today consisted of diverse zooplankton that ate the phytoplankton; many species of small animals like krill and copepods that ate the zooplankton; shrimp, crabs, and small fish that ate the smaller animals; and up to great numbers of species of larger fish, sharks, and seabirds at the top of the food web. And the phytoplankton evolved into multicellular forms, such as seaweeds, increasing the ocean biosphere's complexity and diversity; it would have been difficult for them to have done so wIthout the useable nitrogen provided by cyanobacteria.

In Coral Bluffs, Colorado, Lyson et al. (2019) documented the first million years of evolutionary succession after the Cretaceous-Paleogene (K-Pg) mass extinction of about 66 mya. This extinction killed off a great percentage of life, creating empty niches for evolutionary succession to occur from an early stage. It killed the dinosaurs, creating many empty niches available to mammals. Ferns appeared quickly, being pioneers that rapidly colonize disturbed habitats. Small, rat-like mammals lived among the ferns in the first 1,000 years after the extinction. The number of species of mammals doubled, and maximum body weight of the mammals increased to almost pre-extinction levels, which was the size of an average modern raccoon, within the first 100,000 years. Ferns built soil, allowing the next stage, palm trees, to appear and partially displace them. By about 200,000 years after the extinction, mammals were benefiting from the palm trees and other new vegetation, and increased again in diversity, and in maximum size, with rat-like mammals reaching the size of beavers, and eating nuts similar to walnuts. Palm trees built the soil further, helping the next stage, a more diverse community of trees, displace them. This community of diverse forests of large trees appeared about 300,000 years after the extinction, allowing mammals to increase their maximum body weight threefold and develop specialized diets. More large mammals appeared by about 700,000 years after the event, the same time as the first appearance of plants in the pea and bean family, which have bacteria in their roots that convert molecular nitrogen to a form useable by plants. Mammals thus had access to seed pods much like beans and pea pods, a great protein source that allowed them to increase in size. The largest mammals now were the size of a large dog. Mammal populations were large. It is reasonable to assume that the mammals aided the plants they ate at each stage by dispersing their seeds. Plants, including trees, that are planted far from their parent plants have a much better chance of survival, so seed dispersal by mammals is a big help to plants (see Chapter 4 of the companion book to this one, Seaborg, 2022). There were now mature, diverse forests. This example of evolutionary succession occurred very rapidly, in just 700,000 years. It involved earlier successional stages making the environment better for the next stages, just as in ecological succession. It involved symbiotic coevolution between plants and animals, as both diversified and the system recovered.

REFERENCES

Baco, A. R. & Smith, C. R. (2003). High species richness in deep-sea chemoautotrophic whale skeleton communities. *Marine Ecol. Progress Series* 260: 109–14. doi: 10.3354/meps260109.

Cooper, W. S. (1939). A fourth expedition to Glacier Bay, Alaska. *Ecology* 20: 130–55.

Crocker, R. L. & Major, J. (1955). Soil development in relation to vegetation and surface age at Glacier Bay, Alaska. *Journ. Ecol.* 43: 427–48.

Danise, S., et al. (10 Sept., 2014). Ecological succession of a Jurassic shallow-water ichthyosaur fall. *Nature Communications* 5, Article number 4789. https://doi.org/10.1038/ncomms5789.

Lawrence, D. B. (1958). Glaciers and vegetation in Southeastern Alaska. *Amer. Sci.* 46: 89–122.

Lyson, T. R., et al. (22 Nov., 2019). Exceptional continental record of biotic recovery after the Cretaceous-Paleogene mass extinction. *Science* 366 (6468): 977–83. doi: 10.1126/science.aay2268.

Payne, J. A. (1 Sept., 1965). A summer carrion study of the baby pig *Sus scrofa* Linnaeus. *Ecology* 46 (5): 592–602. https://doi.org/10.2307/1934999.

Pechal, J. L., et al. (Jan., 2014). The potential use of bacterial community succession in forensics as described by high throughput metagenomic sequencing. *International Journ. of Legal Medicine* 128 (1): 193–205.

Seaborg, D. (2022). *How Life Increases Biodiversity: An Autocatalytic Hypothesis*. CRC Press/Taylor & Francis. Boca Raton, FL; London, UK; New York, NY.

Smith, C. R. & Baco, A. R. (2003). Ecology of whale falls at the deep-sea floor. *Oceanography and Marine Biol.: An Annual Review* 41: 311–54.

Smith, C. R., et al. (Jan., 2015). Whale-fall ecosystems: Recent insights into ecology, paleoecology, and evolution. *Annual Review of Marine Science* 7: 571–96. https://doi.org/10.1146/annurev-marine-010213-135144.

Spalding, M. D. & Grenfell, A. M. (1997). New estimates of global and regional coral reef areas. *Coral Reefs* 16 (4): 225–30. doi: 10.1007/s003380050078.

Wolfe, B. E., Button, J. E., Santarelli, M., & Dutton, R. J. (July, 2014). Cheese rind communities provide tractable systems for in situ and in vitro studies of microbial diversity. *Cell* 158 (2): 422–33. doi: 10.1016/j.cell.2014.05.041.

6 Life Is the Main Creator of Soil, a Diverse Ecosystem That Benefits Life in the Soil and above It

Soil is amazingly biologically diverse and contains spectacular numbers of organisms. A single gram of typical soil has a million species, and 10 billion cells, of bacteria (Gans et al, 2005); 1,000 to 500,000 algae; up to 500,000 protozoa; 10 to 5,000 roundworms; and about a million fungi and 50 kilometers (about 31 miles) of fungal mycelia[1] in over 500 species. The life in soil includes myriad species of viruses, archaea, insects and other arthropods, and worms. It has over 425 million animals per acre, of which 90 to 209 million are mites, 28 to 119 million are springtails (small animals related to insects), 3 to 25 million are larvae of one insect family (the click beetles), 11 million are other arthropods, and a million are earthworms. There are more organisms in a gram of soil than people on Earth. About 90% of soil species are not yet named by science. *The life in soil causes tremendous biodiversity above and below the ground.*

Soil is not simply fertile dirt that acts as a medium for the growth of plants. *It is a highly diverse, complex, interacting ecosystem in which the species are highly connected and interacting, with many tight symbioses and commensalisms. It behaves in a manner similar to a superorganism, and is analogous to an organism. It is an intricately connected system.*

Soil is not formed exclusively by life. Volcanoes can send ash clouds rich in nutrients needed by life such as sulfur, nitrogen, and phosphorus compounds tens of kilometers into the atmosphere. These can fertilize soils halfway around the Earth. Volcanic dust, ash, and rocks decompose into soils with an exceptional ability to hold nutrients and water, making them very fertile. Volcanoes are sporadic, but do add significant nutrients to soil. Also, erosion of rocks by wind and water adds minerals to soil. Volcanoes and rock erosion can help build new soil, and have helped to build soil in the past.

However, biological activity is the primary factor in the formation and maintenance of soils and their life-enhancing physical and chemical properties. The first step of biological soil formation is the breaking down of rocks by lichens, and plant roots and their symbiotic fungi. Lichens are a symbiotic association between a photosynthesizing organism, usually a green alga or cyanobacterium, and a fungus. The physical action of the fungal mycelia of lichens penetrating the spaces within rocks combines with changes in temperature and moisture to break down rock into soil. Many lichens chemically degrade rock with acids, adding new minerals to soil while obtaining nutrient. Mycelia of fungi in lichens make oxalic acids and other acids, which break down rocks. Over time, this activity creates new fertile soil from lifeless stone. Plant roots and their symbiotic fungi break rock down physically and biochemically in helping build soil. They physically break rock into smaller pieces, causing weathering rates of surface rocks such as granite, basalt, and limestone to go 10 times as fast as they would without roots and their fungi. Roots and their fungi break up soil particles, aerating the soil, and making passageways for water flow. Iron, calcium, potassium, phosphorus, and various other minerals must be weathered from rocks before organisms can use them. Plants and their symbiotic fungi, as well as lichens, are crucial in the weathering that releases these nutrients, making them available to life, building soil. This weathering process is also done nonbiologically, although not as quickly and this does not build nearly as much soil in the same amount of time.

Soil is then enhanced by animal feces, fallen leaves, and dead organisms, and decomposers, including bacteria, fungi, worms, insects, and others, which break down the leaves, dead organisms, and feces. Fungal mycelia spread the feces, pieces of leaves, and nutrients around in the soil. Soil is made primarily of stone broken up and weathered mainly by organisms, and nutrients from the decomposition of numerous kinds of organisms and their waste products. The decomposition is accomplished by decomposers. And nitrogen-fixing bacteria, many of them symbiotic with plants and living in their roots, convert atmospheric nitrogen to a form useable by plants.

It is noteworthy that a significant amount of the soil's nitrogen, a key element for life, is provided by symbiosis (between nitrogen-fixing bacteria and plants).

The maintenance of fertile soil is one of the most important ecological services the soil organisms perform. Soil organisms make the soil useable and beneficial for plant growth, and plants need them to survive. This means the entire terrestrial biosphere depends on soil organisms because plants are the primary producers on land, directly or indirectly supporting all the species in the terrestrial food webs.

Soil community composition is a key factor in regulating the functioning of the soil and aboveground ecosystems. Wagg et al. (2014) showed that *the loss of soil biodiversity and the simplification of the soil community impair multiple ecosystem functions, including plant diversity, decomposition, the retention of nutrients, and nutrient cycling. Hence, soil*

DOI: 10.1201/9781003246640-6

biodiversity is crucial to soil ecosystem function and plant diversity. Since soil biodiversity is dependent on plants and these soil ecosystem functions, *soil organisms cause the diversity of the soil ecosystem to increase by aiding plants and the rest of the aboveground ecosystem, and the soil itself. There is a positive feedback loop between the belowground and aboveground ecosystems that increases the diversity of each until an equilibrium is reached. It is particularly noteworthy that soil biodiversity is needed for nutrient retention and recycling.*

Soil is a biological structure, built by life that secures its survival. Plants, bacteria, fungi, soil algae, and soil animals work together to build an immeasurably complex architecture, a structure can resist floods, droughts, and wind. If it were not for the structuring of soil by life, soil would be eroded away by water and wind.

The trillions of soil microbes are essential for soil health. They are necessary for plant growth and health, and supply plants with macronutrients and micronutrients. They are needed for water retention, for flood and drought control, water purification, soil aeration, decomposition of plant fiber, recycling nutrients and chemicals, defense against plant diseases, and preventing soil carbon from exiting into the atmosphere. Soils with their full complement of microorganisms draw down and sequester CO_2 and nitrous oxide. This is important because when in the atmosphere nitrous oxide breaks down the protective ozone layer in the stratosphere, and is 298 times as powerful a greenhouse gas as CO_2, molecule per molecule, over a 100-year period. It is also a pollutant and causes acid rain.

Bacteria and fungi break down organic matter on and in the soil, including leaf litter, feces, dead organisms, and other debris, releasing essential nutrients and CO_2 into the soil. Of course, CO_2 is a nutrient and source of carbon, the most important element to life. Bacteria and fungi produce organic compounds that bind soil particles together, and fungal hyphae entangle particles into aggregates. Bacteria and fungi thus keep nutrients from being washed and leached away, helping prevent soil erosion. Bacteria produce a slime that is beneficial to soil by preventing dissolved nutrients from leaching away. They make a cement, gluing mineral particles together, and create pores and passages through which water, oxygen, and nutrients can pass. The small clumps they build are used by animals in the soil to construct bigger passages. Bacteria and fungi make some nutrients more available to and useable by organisms. Bacteria draw nutrients from clay and humus particles in soil, and symbiotic root fungi transport these towards the plant roots. Humus is the organic component of soil, formed by the decomposition of leaves and other plant material by soil microorganisms. It is rich in nutrients, and very beneficial to soil life. Both bacteria and fungi recycle nutrients. The soil is infertile without bacteria and fungi. Bacteria play key roles in the nitrogen and phosphorous cycles. Bacteria also play a key role in purifying water in the soil. My companion book to this one discusses in more detail symbiosis between plant roots, their fungi, helper bacteria, and rodents; symbiosis between nitrogen-fixing bacteria and plants; and how lichens benefit topsoil (Seaborg, 2022).

The physical characteristics of soil are crucial in determining how useable it is to organisms, from bacteria to plants, and hence to the biodiversity and population levels it can support. Friability refers to the soil's texture and other properties. Friable soil is best for life. It is between grainy sand and hard or mushy clay. Friable soil drains water at a life-favorable intermediate level, draining enough that the soil does not become flooded with too much water, but also retaining a good amount of moisture, often able to hold up to 15 times its weight of this substance so essential to life. Friable soil is rich in minerals, carbon, and organic matter. Soil with good porosity allows oxygen and water movement, and friable soil has enough porosity for ample oxygen and water flow, aiding organisms. Friable soil contains large quantities of microbes that replenish its fertility. Plants that grow in friable soil tend to be healthy, to have healthy leaves and extensive root systems, and to be resistant to diseases and pests. Such soil is characterized by ease of fragmentation of undesirable large clumps, but difficulty of fragmentation of small aggregates into undesirable small grains. So its clumps tend to be of intermediate size, which is the best size for life. Numerous examples in this chapter will show that soil life, from microbes to plants, alters soil's physical properties, making it more friable and porous, and hence highly favorable to life.

In some deserts, lichens help bind and stabilize soil sand in dunes, so it is not easily blown or washed away; act as living biological soil crusts, essential for maintaining the soil structure; and contribute nitrogen to soils through being eaten by snails, which then defecate the nitrogen. And some lichens carry out nitrogen fixation.

Soil algae and cyanobacteria are key to soil health and ecology. They are important in making soil, stabilization of soil and soil structure, soil friability, and building nutrient content. And they photosynthesize, creating carbohydrates, a major chemical source of energy for life. Cyanobacteria occur in the upper layers of soils ranging in soil type from the desert to the forest floor. Some form compounds conducive to growth of plants; and some are helpful because of their involvement in nutrient cycling. Many in the genus *Anabaena* help release soil-bound phosphorous, making it available to organisms. Soil algae, some cyanobacteria, and some other soil bacteria are unique in their ability to fix nitrogen in the presence of oxygen at the soil and air interface, and excrete substances which glue soil particles into clumps, providing more aerated soil area for further growth of themselves and beneficial organisms associated with them in the soil. Cyanobacteria and some soil algae bind soil particles together, due to the production of chemicals with a number of sugar molecules bonded together on their outer cell surface called exopolysaccharides. Exopolysaccharides, widely distributed in soils, are the most stabilizing factor for small clumps of soil, which are helpful to soil function and soil life. Exopolysacchharides hold soil together and in place. They facilitate the attachment of nitrogen-fixing bacteria to soil particles and plant roots, aiding colonization of roots by these bacteria. They also aid symbiosis between plants and their fungal allies attached to their roots. They help make dissolved minerals available to life by transforming them into insoluble solids.

Mucilage is a thick, sticky substance produced by nearly all plants and some microorganisms. In plants, mucilage helps with food and water storage, seed germination, and thickening of membranes. One-celled soil eukaryotes employ it for locomotion. During the production of humus, bacteria and fungi secrete sticky gums and mucilages. Mucilage made by a fungus that decomposes lignin plays a role in soil aggregation and stabilization, holding particles together (Caesar-Tonthat, 2002). Mucilage incapacitates toxins such as heavy metals and excess nutrients, keeping them from harming the soil ecosystem (Huang et al., 2008).

Too little or too much nitrogen in the soil would reduce the amount of soil life and species diversity. Intermediate nitrogen levels are necessary for healthy, diverse ecosystems, both in the soil and sea. Too little nitrogen is associated with lower diversity and productivity because it is a limiting nutrient necessary for life. Too much nitrogen reduces biodiversity for several reasons. In aquatic ecosystems, it causes eutrophication, which occurs by the following mechanism. Nitrogen is a nutrient. Very high amounts of it cause blooms of great quantities of algae. The great numbers of algae cause the algae to be crowded, and they also deplete their nutrients. So the algae die in great numbers. Bacteria decompose the dead algae, obtaining nutrient and growing large populations. In decomposing algae, bacteria use great quantities of oxygen, depleting it. Oxygen levels become exceedingly low. So fish and invertebrates die in great numbers. This happens in both saltwater and freshwater ecosystems. There are now seven major dead zones in the Earth's seas caused by excess nitrogen and phosphorous from human activity. One is in the Gulf of Mexico and approaches the size of New Jersey. Algal blooms also make the water turbid, limiting the sunlight available to bottom-dwelling organisms. Conditions of extremely low oxygen in the sea promote the growth of bacteria that can only grow in the absence of oxygen, which include *Clostridium botulinum*, which produces the toxin causing botulism, killing birds and marine mammals. Extremely low oxygen in the ocean also allows sulfur-reducing bacteria, which also can only grow in the absence of oxygen, to thrive. These bacteria produce hydrogen sulfide, which is toxic to most life. If these bacteria reach large populations, the hydrogen sulfide they produce kills off a good portion of sea and land life. This can result in mass deaths of fish and invertebrates.

Simkin et al. (2016) measured nitrogen levels in over 15,000 forest, woodland, brush, and grassland sites, and found human-induced high nitrogen levels have caused a loss of plant diversity. The number of species increased at low nitrogen deposistion levels, but decreased with increased nitrogen input. Grasslands, shrublands, and woodlands were susceptible to species losses at lower nitrogen loads than forests, which are more diverse than these ecosystems. This supports the thesis that diversity helps maintain stability and resilience in ecosystems, even when they are challenged by factors such as excess nitrogen, which is compatible with the ABH/Pachamama Hypothesis. In some meadows, forests, and bogs that are characterized by low nitrogen content and species-rich, slowly growing vegetation adapted to these low levels, excess nitrogen input causes them to become overgrown by faster-growing and more competitive, species-poor vegetation, like tall grasses, that can take advantage of unnaturally elevated nitrogen levels. In uncommon soils called serpentine soils with specialized plants adapted to low nitrogen, adding nitrogen allows nearby plants to replace the serpentine plants, eliminating this unique community. This reduces diversity by eliminating a type of plant community. With too much nitrogen, species-rich fens are overtaken by species-poor reed or reedgrass species, and in forests, undergrowth is turned into an assemblage of thick nettle and bramble shrub, which has fewer species. Increased soil nitrogen can greatly decrease the number of species by adversely affecting the growth of plants' symbiotic fungi connected to their roots. Excessive nitrogen also leads to higher atmospheric levels of nitrous oxide, whose deleterious effects have already been discussed in this chapter.

The nitrogen cycle in soil is a symbiotic loop of species of bacteria, archaea, and protists in which wastes of organisms in one step in the loop are nutrients for the next step. Species in each step in the circle benefit from the species in the step preceding them, and benefit the species in the step following them. The cycle keeps nitrogen in the soil at intermediate levels beneficial to life. It proceeds as follows. When animals and plants die or expel waste, or trees drop their leaves, bound nitrogen is made available. Free-living soil bacteria, such as *Azotobacter*, convert this nitrogen to ammonia (NH_3), or ammonium ion (NH_4+), in a process called ammonification. This is part of the process of decomposition, which is responsible for the recycling of the dead organisms, feces, and leaf litter, preventing these resources from accumulating in large, unusable quantities, and making them available to plants and thus the ecosystem. Another source of ammonia is nitrogen-fixing prokaryotes, both symbiotic bacteria in plant roots and free-living species of bacteria and archaea in the soil, which perform nitrogen fixation. They convert molecular nitrogen in the atmosphere to ammonia (some of which gets converted to ammonium). Lightning fixes some nitrogen, but most fixation is accomplished by soil bacteria. Without nitrogen-fixing prokaryotes, there would not be enough usable nitrogen to support very much life. Next, some of the ammonium is taken up and used by plants and their symbiotic fungi. The rest is consumed by nitrifying bacteria in the soil in a two-step process called nitrification, which is the oxidation of ammonium to nitrates. In the first step, *Nitrosomonas* uses the ammonium as an energy source, forming nitrite as a by-product. In the second step of nitrification, nitrifying bacteria such as *Nitrobacter* use the nitrite made by *Nitrosomonas*, producing nitrate. This step is crucial, because if nitrites accumulated in the soil, they would be toxic to plant life. It is also of critical importance because nitrates can be used by plants as nutrient. Nitrifying bacteria are crucial to the entire ecosystem, since plants are dependent on them, and plants are essential to the ecosystem. The nitrogen cycle is "completed" by the "final" step, denitrification, when such bacteria as *Pseudomonas* and *Clostridium* use nitrate to make molecular nitrogen gas (N_2), which is not useable by plants, and return it to the air. Since the cycle started with nitrogen gas, this step completes the cycle. Many bacteria species accomplish denitrification from nitrate to produce molecular nitrogen gas while also making nitrous oxide. As previously pointed out, nitrous oxide has deleterious effects on the ecosystem, but under natural conditions, its concentration in the atmosphere is less than 1/1,000 that of CO_2, so it is not a problem. The species in each step benefit those in the next step. Denitrification is crucial because it prevents the accumulation of too much nitrogen in the soil, where it would cause severe problems, as discussed, and would also get washed to bodies of water and cause eutrophication. Denitrification is not really the final step, since the nitrogen cycle is a repeating loop. Since each species benefits the next species in the chain, and it is a cycle or loop, there are many indirect symbiotic relationships between species removed from each other in the cycle by two or more steps. In addition to the main species involved in the nitrogen cycle mentioned here, there are other species involved in the various steps in the pathway.

Intermediate soil nitrogen levels that are beneficial to life are maintained by soil organisms, mainly bacteria, which seem to keep the levels of bioavailable nitrogen compounds close to optimal for soil life, and thus keep the number of species of soil organisms, including plants, at a high level. The nitrogen cycle is generally a stable system featuring negative feedback, regulated by bacteria and other soil organisms, including plants. Each nitrogen compound in the nitrogen cycle in the soil is maintained at a beneficial level to life. When any of the nitrogen compounds in the cycle becomes too high or too low, the system's living components tend to respond and correct this, balancing the amount of nitrogen toward optimality for life, through negative feedback. For example, if there is an excess of ammonia in the soil, the nitrifying bacteria, which use it as a food source, will increase in response to their increased nutrient supply, and consume it. They will thus consume more of the ammonia than if it were at a low to average level, causing the amount of it to drop. Similarly, if there is too little nitrate in the soil, the denitrifying bacteria that eat it will have less to eat, and thus decrease in number. With less bacteria to consume the nitrate, it will increase to a higher level until it reaches equilibrium as bacteria such as *Nitrobacter* produce it. Thus, the levels of the various nitrogen compounds in the soil are regulated by the organisms within intermediate ranges beneficial for life and the maximization of the number of species in the soil. Life in the form of the species of microorganisms in the nitrogen cycle keeps these nitrogen compounds high enough to support life and yet not so high that they are toxic, and it does this by negative feedback and natural controls in the nitrogen cycle. The evidence for this is that when not influenced by human nutrient input, the levels of the various nitrogen compounds in the nitrogen cycle in the soil tend to remain at life-favorable intermediate levels in a dynamic equilibrium.

The nitrogen cycle in the sea is similar to that in the soil, though the species are different (Moulton et al., 2016). Cyanobacteria are the major nitrogen fixers of the sea, making it available to sea organisms. Phytoplankton excrete ammonia and a nitrogen compound called urea. Recall that the photic zone is the zone where the sea is shallow enough for light to penetrate, and hence where phytoplankton can carry out photosynthesis and survive. Nitrogen is delivered below the photic zone by the sinking of phytoplankton and other dead organisms, the sinking of feces of animals that migrate up and down the water column in the sea, and the mixing of water at different depths in the sea. This delivers ammonia and organic nitrogen to below the photic zone. Some bacteria add more ammonia by carrying out ammonification, converting the organic nitrogen to ammonia. The ammonia is then converted to nitrite and nitrate by bacteria below the photic zone. Upwelling and vertical mixing return nitrate to the photic zone, where phytoplankton use it. Denitrifying bacteria return nitrogen gas to the atmosphere. Each step in the process is regulated by the other steps in the same way that this occurs in the soil, in a stabilizing negative feedback system, regulated by life. This tends to keep each nitrogen compound at life-friendly intermediate levels. As in the soil, this is supported by the fact that if there is no human interference, the levels of the various compounds in the nitrogen cycle in the sea tend to remain at these intermediate levels that are beneficial to most life and high biodiversity, in a dynamic equilibrium.

Nitrogen enters the sea through precipitation, runoff, or as nitrogen gas from the air. Nitrogen levels are not evenly distributed in the ocean. But nitrogen input to the sea appears to be generally kept from being too low or being excessive, mainly by life. By the negative feedback mechanism discussed earlier in this chapter, microbes in the nitrogen cycle in the soil might keep the levels of nitrogen sent to the sea at intermediate levels favorable to most life under natural conditions. Thus, there is a subhypothesis of the ABH that needs testing: I propose that under natural conditions, life's regulation of temperature through carbon burial causes a beneficial, intermediate level of precipitation and runoff, rendering it highly probable that an intermediate nitrogen input favorable to most life would occur to both freshwater ecosystems and the sea. Also, upwelling supplies nitrogen to the photic zone, and that is aided greatly by life's regulation of the carbon cycle and hence temperature (see Chapter 2).

All of the species in soil act as ecosystem engineers, improving the soil for other species. Bacteria of several species make urine available to plants in a three-step process. First, bacteria break down animal urine into ammonia, which is useable by most organisms. Second, other bacteria species use and oxidize this to nitrite. Third, other bacteria use and oxidize the nitrite to nitrate, which is useable by plants. The bacteria species in each step benefit those in the next step in the chain, and the species of each step require the species before them. There is negative feedback, keeping the nitrogen compounds at favorable levels to the species in this cycle. The bacteria produce the nitrate gradually. And pores created by soil organisms and slime created by soil bacteria soak up the urine and swell, capturing it and aiding the slow release of nitrate. If nitrate were released all at once or too quickly, plants would not be able to benefit maximally, and would even be harmed.

Feces and dead organisms on the soil are decomposed by a very diverse community of many species of bacteria, fungi, arthropods, and other soil organisms. These organisms also release nutrients gradually, and it would be deleterious for plants if the nutrients were released in a short, large burst. Also, pores in the soil made by life are filled by feces, and this aids its gradual release. So microbes and other soil life act as ecosystem engineers, benefiting soil life and plants by ensuring the gradual release into the soil of nitrate and feces.

Pores and channels in the soil are necessary to hold and transport water, oxygen, CO_2, air, and nutrients. As organisms move through the soil, they create channels that improve the movement of air, water, and nutrients through the soil; that is, aeration and drainage of the soil. This makes soil more beneficial to life. Small pores are created by plant roots, fungal hyphae, bacteria, and protozoa. Hyphae of the symbiotic fungi that grow as extensions of plant roots also transport nutrients and chemical signals between plant roots, even between roots of different plant species. Larger soil pores are created by soil mesofauna, which are discussed in the third paragraph after this one. Earthworms, ants, termites, and large insect larvae greatly increase soil porosity and create larger channels that allow transport of greater amounts of water, air, nitrates, phosphorus, dissolved organic carbon compounds, and other nutrients, which can leach from the top down. The transport of air aerates the soil, and biologically generated pores and channels aid this. Soil must be kept moist to keep the organisms in it healthy, and its ability to hold water for this reason is crucial. Formation of pores by organisms increases soil's water-holding capacity tremendously.

Larger animals, such as frogs, toads, salamanders, desert and gopher tortoises, lizards including monitors, tuataras (strange, lizard-like reptiles), moles, gophers, kangaroo rats, prairie dogs, badgers, foxes, and other species dig burrows, in some species with long tunnels, for homes or part-time residence in the soil. These burrows and tunnels create yet larger channels that allow transport of even greater quantities of water, air, minerals, and organic nutrients. Small animals that do not not dig the burrows use them as shelter and homes. Aeration of the soil provides oxygen to organisms, and also suppresses methanogens, which produce methane and can only grow in the absence of oxygen. This controls the production of this powerful greenhouse gas, keeping atmospheric temperature low enough to be favorable to life, while conserving carbon.

It has been shown that primitive plants called bryophytes, which include modern mosses, enriched soil by increasing its mud content from about 458 to about 359 mya, and rooted plants further increased soil mud when they arose and began spreading about 430 mya (McMahan and Davies, 2018), eventually forming great forests approximately 382 mya. Mud has minerals, nutrients, and a consistency beneficial to soil and its living organisms.

Soil microfauna range from 1–100 micrometers (a micrometer is one millionth of a meter), and include protozoa. They control bacterial and fungal populations by consuming them. Without this population regulation, bacteria and fungi would grow exponentially and become so numerous that they would make soil into very a poor environment that would support only a very low number of species of soil organisms and plants. Soil microfauna positively affect soil structure and aggregation through interactions with fungi, yeast, and bacteria. This makes the soil better for organisms in general. They are a food source for soil mesofauna.

Soil mesofauna are invertebrates between 0.1 and 2 millimeters in size in the soil and in the leaf litter layer on its surface. They include mites, small spiders, harvestmen, small isopods (crustaceans that include woodlice, which are also called sow bugs, which should be honored [Seaborg, 2008]), springtails, insect larvae, small insect-like arthropods called proturans, pauropods (small, pale, millipede-like arthropods), myriapods (millipedes, centipedes, and their relatives), tardigrades (bizarre little eight-legged animals), pseudoscorpions (little arthropods with pincers, but no stinger; see Figure 6.1), rotifers (small animals that can contract and stretch out), roundworms, and segmented worms such as potworms. They are food for the soil macrofauna. They feed on and control populations of soil microfauna, microorganisms, fungi, algae, lichens, and spores. Without regulation of these organisms by soil mesofauna, their populations would explode and the soil system would lose a great amount of its diversity. Some species of soil mesofauna feed on and control dead animal matter, live or decaying plant material, and plant roots. By removing these, they open drainage and aeration channels in the soil. Notable soil benefactors among mesofauna are proturans, which consist of about 800 species that eat decaying organic matter, and pauropods, which consume mold, fungal hyphae, and the root hairs of plants. Fecal matter of soil

FIGURE 6.1 Pseudoscorpions are typical soil mesofauna. This is enlarged on the page of a book to show it well.

Shutterstock Item ID: 370146785

Title: Super macro of Pseudoscorpion on the page of a book.
Photo Contributor: Gallinago_media

mesofauna remains in channels and is a food source for small animals that break it down, and it becomes a fertilizer. Soil mesofauna create pores that account for a small portion of soil porosity. They help moisten the soil. They are important in the carbon cycle. Some soil mesofauna, including roundworms, eat and regulate soil bacteria. This is crucial to soil health, since an overpopulation of bacteria would cause the death of a great portion of soil life. This consumption of bacteria also considerably increases available soil nitrogen, since the nitrogen is released when the bacteria are eaten, broken down, and defecated.

Springtails (Figure 6.2) are notable mesofauna. They have six legs, but are not insects. They are very small. They are incredibly abundant and important in the soil. In sheer numbers, they are reputed to be one of the most abundant of all animals large enough to be visible to the naked eye; only roundworms, crustaceans, and mites are likely to have global populations of similar size. Most live in the soil and in leaf litter, although some species routinely climb trees. Many springtails are found in leaf litter and other decaying material, where they eat mainly dead organisms, decaying leaves, and feces, and very small prey, and are one of the main biological agents responsible for the control and dissemination of soil microorganisms (Ponge, 1991). Their controlling of microorganisms and spreading them around in the soil are crucial to the health of the soil ecosystem. Supporting numerous predator species, mainly arthropods, but also some vertebrates, they are a key food source for soil macrofauna.

FIGURE 6.2 Slender springtail (*Orchesella flavescens*) on a leaf. Springtails are notable soil mesofauna, amazingly numerous and important in the soil. They are reputed to be one of the most abundant of all animals large enough to be visible to the naked eye. They have six legs, but are not insects. Most, but not all, live in the soil and in leaf litter.

Shutterstock Item ID: 1659008374

Title: *Orchesella flavescens* is a species of slender springtail in the family Entomobryidae. Slender springtail, *Orchesella flavescens* on green leaf.
Photo Contributor: Ihor Hvozdetsky

Soil macrofauna control soil mesofauna populations, which is crucial to soil health. They are a food source for animals, such as anteaters, moles, rodents, birds, lizards, frogs, toads, salamanders, insects, and other animals. They include earthworms, ants, termites, and large insect larvae. They are ecosystem engineers, creating pore spaces to the point of greatly increasing soil porosity, mixing organic and mineral particles, redistributing organic matter and microorganisms, fertilizing with fecal pellets, increasing soil moisture, and stimulating microbial activity. All these functions are crucial to making soil fertile and functional, with the result that it supports more life and biodiversity.

Earthworms are soil macrofauna that are ecosystem engineers that profoundly transform soil into habitat beneficial to plants and other soil life, benefiting the soil biologically, chemically, and physically. They make tunnels, aerating the soil for other organisms and making it capable of absorbing heavy rain. They pull fallen leaves and manure lying on the ground down below the surface, and rotate the soil in the upper layer, such that leaf litter, feces, and other materials become buried, eventually migrating slowly to the base of the upper layer. Leaf litter and other material are more quickly and efficiently recycled if pulled below the soil surface. They shred the organic matter and partially digest it, saturating it with their intestinal secretions, mixing it with the soil in the process. This greatly improves soil fertility. By consuming and digesting it, they convert organic matter such as dead organisms and feces into rich humus. Earthworm castings, the end products of worm digestion, can potentially contain up to 40% more humus than the top nine inches of soil the animal lives in. Earthworms transform nutrients to make them available to plants. They even aid plants in nutrient uptake.

Ants, especially leafcutter ants, and termites are ecosystem engineers that make the soil far better for life in many ways, including aerating and fertilizing it. The benefits to the soil and ecosystem of leafcutter ants and termites are discussed in more detail in Chapter 4 and of earthworms in Chapter 5 of my companion book to this one (Seaborg, 2022).

This system of interaction between plants and the soil ecosystem sequesters large amounts of carbon through weathering and burial of dead organisms, mainly trees, regulating the temperature, and this carbon is stored in the soil, conserving it for the biosphere (see Chapter 2).

Nitrogen fixation is done by nitrogen-fixing bacteria. Some species live on their own in the soil; others live in plant roots and are symbiotic with plants, receiving a place to live and carbohydrates from the plants they live with. These bacteria, called nitrogen-fixing bacteria, take molecular nitrogen gas, which life cannot use, and convert it to a form life can use, such as ammonia, while deriving energy from the molecular nitrogen. This is crucial, since nitrogen is so important to life. Once plants obtain this useable nitrogen, the entire food web benefits. Still, about 30% of nitrogen mineralization is contributed by soil animals. Mineralization is the process by which chemicals present in organic matter are decomposed or oxidized into easily available forms for plants. This is done by decomposers, including bacteria, soil mesofauna, soil macrofauna, and other groups. Soil mesofauna and macrofauna break down plant residues, releasing nitrogen (Swift, 1979). Bacteria make the most nitrogen available to plants and the ecosystem, but soil mesofauna and macrofauna make a great deal of it available as well.

Soil macrofauna, so essential to soil health, would disrupt the soil ecosystem if not regulated. Moles are crucial in this function, feeding on earthworms and insect larvae. Each day, a mole eats 50 to 100% of its body weight; each year, it consumes about 40 pounds of worms. Shrews and some birds, such as robins, eat large numbers of earthworms and insect larvae. Giant anteaters (*Myrmecophaga tridactyla*) are crucial in regulating termite populations in the soil of South American grasslands.

Soil arthropods deserve special discussion, and include springtails, beetles, ants, termites, wood lice (sow bugs), spiders, scorpions, pseudoscorpions, mites, centipedes, and millipedes. Several thousand different species of arthropod live in a square mile of forest soil. One square yard of soil has 500 to 200,000 arthropods. Some arthropods, such as millipedes, woodlice, termites, mites, and cockroaches, increase the surface area of nutrient accessible to microbes by shredding dead plant material and making burrows in coarse woody debris, which increases the surface area of the woody debris available to microbes. Bacteria can then eat the dead plant material, including wood, and decompose it, adding the decomposition products to the soil, fertilizing it much faster as a result. Bacteria have limited mobility in soil. Arthropods aid them by distributing nutrients, bringing them closer to bacteria, and carrying bacteria on their exoskeletons and through their digestive systems. This mixing of microbes with their nutrients enhances decomposition of dead organisms, leaves, feces, and other organic matter. It is symbiosis because the soil arthropods transport bacteria to their nutrient sources, and bacteria improve the soil for them. A large percentage of the nutrients available to plants is a result of nutrient release by arthropods and other animals, as well as fungal and bacterial decomposers. Most soil arthropods eat bacteria, fungi, worms, or other arthropods, regulating their populations. Arthropods mix and aerate the soil, mineralize soil nutrients, and excrete nutrients in forms available to plants. They ingest bacteria and fungi that compete with and interfere with the growth of symbiotic fungi in plant roots and other fungi and bacteria that benefit the soil. Some predatory arthropods eat the grazers of bacteria and fungi, keeping the grazers from causing bacterial and fungal populations to decline too much, which would cause the soil ecosystem to crash due to a lack of bacterial and fungal food at the base of the food web and the loss of the many services these two groups provide to the soil. In grassland and forest soils, every particle in the upper several inches of soil has been through the digestive tract of numerous animals, mainly arthropods and earthworms, mixing it with organic matter and mucus and depositing it as fecal pellets. This enhances soil aggregation and fertility. The pellets are a highly concentrated source of nutrients needed by bacteria and fungi.

Some arthropods burrow, creating a tremendous enhancement of soil habitats. Burrowing increases soil porosity, the ability of the soil to hold water, and the ability of water to move through the soil. It aerates the soil. It creates tunnels that are habitat for many small soil species.

Plants, especially trees, benefit the soil tremendously. As discussed earlier, they create, build, and enrich soil. They make complex compounds that feed, aid, and signal soil bacteria that the plant is symbiotic with. The bacteria make nutrients upon which the plant depends available to it, breaking down complex compounds into simpler ones that the plant can absorb. They produce growth hormones and other chemicals that help plants grow. Plants release chemicals that aid the bacteria that help them in specific conditions. If they need nutrients or water, or the soil has an excess of salt, plants send chemicals that help bacteria with these needs. Symbiotic bacteria create a ring around the root system that defends the plant from disease-causing microbes and fungi. The plant's immune system is trained and primed by bacteria around its root, just as the immune system of animals depends on their microbiome to develop.

The leaves of trees provide shade, shielding soil from the drying effects of the sun. They also protect soil from rain, which can wash soil away, causing erosion. They keep soil cool when it is hot and conserve heat in cold weather. Their roots hold soil in place and prevent erosion from wind and rain. The roots stabilize river banks, affecting the course of rivers, benefiting river life. Just as their trunks and stems provide habitat for aboveground organisms, their roots, leaf litter, and dead plants, especially dead trees, provide habitat for organisms on and in the soil. Plants are primary producers that take nutrients from the soil and transfer them to aboveground ecosystems. This important function is crucial for all land-based, aboveground ecosystems and their organisms. But it takes tremendous amounts of nutrients out of the soil. Plants replenish these nutrients in conjunction with soil organisms. They feed the soil. Fifty percent of a plant's primary production from photosynthesis is sent to its root system, and some of this feeds soil organisms. Typically, another 10% of the aboveground portion replenishes the soil by the falling of leaves. This is a spectacular amount of nutrient given to the soil, but it is after the leaves have carried out their function of making carbohydrate for the plant. The entire plant is returned to the soil after its death. It is remarkable that plants sacrifice so much of their primary production to soil and soil organisms. But plants could not survive for long if they did not replenish the soil in these ways. The nutrients are used by bacteria, fungi, and all the soil organisms that aid plants by: nitrogen fixation; making the soil porous; increasing its friability; recycling nutrients, making them available to plants and other organisms; extending the root system (this is done by symbiotic fungi that connect to plant roots); and much more. Soil organisms decompose and recycle aboveground and belowground dead organisms, leaf litter, feces, urine, and organic matter. This allows both the aboveground and soil ecosystems to survive. This is a multispecies system of obligatory symbiosis, in which plants and their many allies in and on the soil could not persist without each other's aid. Plants would actually starve if they did not supply nutrient to the soil. Without recycling of leaves, dead trees, and other organic matter from the aboveground system, returning it to the soil, the soil would have insufficient nutrients for plants and soil organisms to survive. And because soil organisms would not survive to recycle organic matter, it would accumulate and litter the soil surface, making gas exchange, including oxygen entry into the soil, difficult at best. *If plants did not feed the soil OR soil organisms did not recycle soil nutrients and provide other services, the soil ecosystem would collapse. This would lead to the collapse of the aboveground ecosystem as well. The importance of plants, soil organisms, and their symbioses to the soil and all terrestrial life cannot be overstated.* Thus, plants are ecosystem engineers that maintain themselves and maintain the soil ecosystem and the aboveground ecosystem. Soil organisms are ecosystem engineers that maintain themselves, plants, and the aboveground ecosystem. The aboveground ecosystem and soil ecosystem maintain themselves and each other by ecosystem engineering and symbiotic positive feedback loops that continue until an equilibrium is reached. This maintains and increases the biodiversity of both the aboveground and soil ecosystems.

Overgrazing causes a destabilizing positive feedback loop, as follows. Overgrazing causes a decrease in plants, including trees. This decreases nutrients in the form of leaves, dead trees, and dead plants dropping to the soil. Thus, many soil organisms starve. The top soil decreases. Leached nutrients are not replenished. Many plants starve and die. Lack of plants results in soil erosion by water and wind, drying of the soil by the sun, and a paucity of soil nutrients. As a result, even more soil organisms die. Thus, there are less nutrients available to plants, and more plants die, in a positive feedback loop. This can lead to a tipping point beyond which the soil cannot recover. Therefore, predators of the grazers play a key role in protecting the soil ecosystem, including plants. Clear cutting forests results in the same positive feedback loop and catastrophic loss of biodiversity.

In desert and semidesert habitats, what is called cryptogamic soil consists of biological soil crusts, which are communities of organisms forming a hard, thin layer on the soil surface. They are found throughout the world with varying species compositions, depending on the climate, physical features of the area, soil characteristics, plant community, microhabitats, and type and amount of disturbance. Species that make up cryptogamic soil are mainly bacteria and algae. Their most important function is stabilization of the soil, preventing erosion. They also fix carbon and nitrogen, making them available to soil life; alter the reflectivity of the soil in a way that generally benefits life; aid the germination of seeds into young plants; make nutrients more available to plants; and aid soil water retention and flow. Physical soil crusts that lack organisms, on the other hand, often reduce the flow of water into soil, can inhibit plant establishment, and can be eroded rapidly if disturbed (Belnap et al., 2001). These negative impacts can be reduced in cases where physical crusts that lack organisms coexist with biological ones with life.

Ocean life also helps build the soil. Sea birds eat sea organisms and defecate nutrients, mainly nitrates and phosphates, where they roost, transferring nutrient from sea to land, creating dense guano (excrement) deposits. Dimethyl sulfide is transferred into the sea from plankton, evaporates into the atmosphere, and then is deposited into the soil by rain, adding sulfur, an element needed by life, to the soil. In the Pacific northwest, Grizzly bears eat salmon, leaving an estimated half of the salmon they capture uneaten on the forest floor, where they also defecate. These two actions add substantial nutrient to forest soil.

The sand of ocean bottoms and beaches can be considered a form of soil. Some of this is formed when rock is broken down and carried by rivers to the beach. Crushed lava from volcanoes forms black sand. But some sand is formed by life. Calcium carbonate is the primary component of the shells of clams, scallops, mussels, and sea snails, and of coral and the protective armor of some phytoplankton. When these shells, armor, and so on are broken down, they form sand, creating and maintaining beaches and seafloors, both important habitats. Overcollecting of sea shells or the destruction of coral reefs can decrease sand habitats, causing the number of species that live in them to decline. A few species of parrotfish, such as the green humphead parrotfish *(Bolbometopon muricatum)*, eat coral. The excrete white sand particles. This produces and distributes coral sands in the coral reef. They also eat the algae that grows on the coral, preventing it from overgrowing, smothering, and killing the coral. This keeps the species diversity of the corals and the many species that depend on them high.

The soil abounds with symbiotic relationships. Examples include plants and nitrogen-fixing bacteria; plants, their symbiotic root fungi, helper bacteria, and rodents that inoculate plants roots with the symbiotic root fungi (see companion book to this one, Seaborg, 2022); moles aerating soil, helping their earthworm and insect larva prey; and myriad species, notably earthworms, ants, and termites helping each other by aerating and increasing friability and porosity of the soil, and also helping plants, which help them by providing leaf litter, nutrients, and carbohydrates from photosynthesis, and preventing erosion with their roots. Soil bacteria live in huge associations of many diverse species, many providing what the others need and benefiting from them. For a detailed discussion of soil microbe symbioses, see Shrivastava et al. (2021). There are groups of microbial species that work together and have a division of labor, helping one another, in the soil and elsewhere (Hays, et al., 2015). Invertebrates have many symbioses with each other and soil microbes. The many indirect symbioses significantly increase this count of symbioses in the soil. There are also countless commensal relationships in the soil. There is a myriad of beneficiary commensal species that benefit from bacteria, from fungi, from plants, from various invertebrates, and from various vertebrates, in soil. Earthworms are keystone species that benefit many other species. Indirect commensalisms greatly increase the amount of commensalism in the soil.

Forests account for 70% of terrestrial biodiversity, and most of it is found underground. The ecosystem below the ground has roots, invertebrates, fungi, protists, bacteria, and archaea. Most of a forest's carbon storage and sequestration occurs underground. Forests account for 25% of the Earth's biomass, and the root systems of trees are responsible for 75% of forest biomass production. The purifying of air and water, provision of habitat for countless species, carbon sequestration, nutrient cycles (especially the carbon and nitrogen cycles), causing of rainfall and accumulation of moisture through evapotranspiration, holding soil in place, decomposition, and other ecosystem benefits cannot be accomplished without a great deal of help from the underground ecosystem. The underground is a habitat and source of nutrients. About half of animal biodiversity is underground. In fact, when all types of organisms are considered, the soil might be more diverse than the aboveground forest. When either the aboveground or belowground ecosystem is out of balance, the other one tends to be as well.

Aboveground and belowground ecosystems are intimately linked and mutually dependent. They greatly increase each other's diversity. This relationship is so complex and elaborate that an entire book is devoted to it (Wardle, 2002). Plants are a link between the two worlds via their underground roots and aboveground stems, trunks, and leaves. As primary producers, they are the base of food webs for both diverse aboveground and underground ecosystems. Trees and other plants transfer nutrients from the underground ecosystem to the aboveground one, and provide food, shelter, habitat, oxygen, and many other services to entire aboveground ecosystems, where they benefit many viral, prokaryote, protist, fungal, plant, and animal species. Trees and other plants are dependent on the aboveground biosphere; examples of animal groups they are obligately symbiotic with are pollinators and seed dispersers. They transfer nutrient from photosynthesis from their leaves above the ground into their roots, and via their leaf litter and decomposition when they die, feeding belowground ecosystems. They are dependent on the underground soil biosphere. Examples of groups of organisms they depend on are their symbiotic root fungi, nitrogen-fixing bacteria, and soil-builders such as earthworms. The aboveground biosphere is dependent on the belowground one, on soil. Soil is necessary for the growth of plants, which are the base of essentially all aboveground food webs. Prairie dogs and badgers live in burrows in soil. Snakes, lizards, tortoises, frogs, toads, and salamanders hibernate and/or estivate in soil. (Estivate means to spend a hot or dry period in a prolonged state of inactivity.) Snakes, lizards, and other animals bury their eggs in soil, which acts as an incubator. Various rodents, birds, amphibians, and lizards consume invertebrates on the soil surface or just below it. Aardvarks, anteaters, and egg-laying mammals called echidnas dig into the soil to eat soil organisms. Belowground ecosystems support trees, and, indirectly, all species that benefit from trees. This is not a comprehensive list of the ways the soil benefits the aboveground biosphere. The belowground biosphere is likewise dependent on the aboveground biosphere. Aboveground animals benefit soil by defecating on it and providing nutrient when they die. Aboveground decomposers decompose leaves, dead trees, and animals, adding nutrient to the soil and keeping it from being permanently covered over by feces, corpses, and leaves. Some animals burrow into the soil, aerating it and giving it water channels. This also is not a comprehensive list of how aboveground organisms help the soil. Many aboveground species that benefit and/or receive benefit from trees interact at least indirectly, in symbiosis or commensalism, with many soil species, most of which are underground, that help and/or are helped by trees. There is an amazing multispecies set of symbioses and commensalisms between the aboveground biosphere and the soil biosphere under the ground. The two ecosystems benefit each other, maintain each other, are mutually dependent, and work as interconnected parts of one system.

NOTE

1 A mycelium (plural mycelia) is the vegetative part of a fungus or fungus-like bacterial colony that does not make spores for reproduction. It consists of a mass of thin, thread-like, often white branches that grow in and on soil and many other substances, such as wood and bread. The thread-like branching structures are called hyphae (singular is hypha), and hyphae collectively make up the mycelium.

REFERENCES

Belnap, J., et al. (2001). *Biological Soil Crusts: Ecology and Management*. U.S. Dep. of Interior, BLM and U.S. Geol. Survey. Technical Ref. 1730-2, 118 pp.

Caesar-Tonthat, T. C. (Aug., 2002). Soil binding properties of mucilage produced by a basidiomycete fungus in a model system. *Mycol. Research* 106: 930–7. doi: 10.1017/S0953756202006330.

Gans, J., Wolinsky, M., & Dunbar, J. (26 Aug., 2005). Computational improvements reveal great bacterial diversity and high metal toxicity in soil. *Science* 309 (5739): 1387–90. doi: 10.1126/science.1112665.

Hays, S. G., et al. (Dec., 2015). Better together: Engineering and application of microbial symbioses. *Current Opinion in Biotechnology* 36: 40–9. https://doi.org/10.1016/j.copbio.2015.08.008.

Huang, D.L., et al. (2008). Degradation of lead-contaminated lignocellulosic waste by *Phanerochaete chrysosporium* and the reduction of lead toxicity. *Environ. Science and Technol.* 42: 4946–51. doi: 10.1021/es800072c.

McMahan, W. J. & Davies, N. S. (2 March, 2018). Evolution of alluvial mudrock forced by early land plants. *Science* 359 (6379): 1022–4. doi: 10.1126/science.aan4660.

Moulton, O. M., et al. (May, 2016). Microbial associations with macrobiota in coastal ecosystems: patterns and implications for nitrogen cycling. *Frontiers in Ecol. and the Environment* 14 (4): 200–8. doi: 10.1002/fee.1262. ISSN 1540-9295.

Ponge, J.F. (1991). Food resources and diets of soil animals in a small area of Scots pine litter. *Geoderma* 49: 33–62. doi: 10.1016/0016-7061(91)90090-G.

Seaborg, D. (2022). *How Life Increases Biodiversity: An Autocatalytic Hypothesis.* CRC Press/Taylor & Francis. Boca Raton, FL; London, UK; New York, NY.

Seaborg, D. M. (2008). *Honor Thy Sow Bug*. Beatitude Press, Berkeley, CA. ISBN 978-0-9795651-2-0. This is a humorous reference. This is in fact a book of poetry by the author of this book, available on Amazon and at davidseaborg888@gmail.com.

Shrivastava, N., Mahajan, S., & Varma, A. (Eds.) (2021). *Symbiotic Soil Microorganisms: Biology and Applications*. Springer: Berlin/Heidelberg, Germany. ISBN 978-3-030-51916-2.

Simkin, S. M., et al. (12 April, 2016). Conditional vulnerability of plant diversity to atmospheric nitrogen deposition across the United States. *PNAS USA* 113 (15) 4086–91. https://doi.org/10.1073/pnas.1515241113.

Swift, M. J. (1979). *Decomposition in Terrestrial Ecosystems.* Blackwell, Oxford, UK. ISBN 0-632-00378-2.

Wagg, C., et al. (8 April, 2014). Soil biodiversity and soil community composition determine ecosystem multifunctionality. *PNAS USA* 111 (14): 5266–70. https://doi.org/10.1073/pnas.1320054111.

Wardle, D. (2002). *Communities and Ecosystems. Linking the Aboveground and Belowground Components. Monogr. in Population Biol.* 304. Princeton Univ. Press: Princeton, NJ, and Oxford, UK.

7 Eukaryotes Are Complex Ecosystems with Diverse Microbiomes, Showing the Importance of Symbiosis and Commensalism, and Challenging the Concept of the Individual

7.1 ALL EUKARYOTES ARE ECOSYSTEMS WITH DIVERSE MICROBIOMES, WITH A GREAT DEAL OF SYMBIOSIS AND COMMENSALISM

All eukaryotes are diverse ecosystems providing habitats for tens to hundreds of thousands of species of bacteria, and many species of viruses, archaea, protists, fungi, and even some invertebrates. The eukaryote that has and provides a home for this diverse array of species is called their host. The complex ecosystem of small organisms of the host is called its microbiome, and it is highly diverse. Every eukaryote, from unicellular eukaryotes to humans, lives in association with a complex microbiome (unicellular eukaryotes have smaller microbiomes than multicellular ones). No eukaryotic species lives in isolation. Every species of eukaryote has coevolved and potentially exchanged DNA with every species in its microbiome, and is a superorganism and a highly diverse ecosystem. Every species in the microbiome has coevolved and potentially exchanged DNA with many other species in this shared diverse ecosystem. Some have suggested, and I agree, that symbiosis (I would add commensalism as well) between microbes and eukaryotes existed since the first appearance of eukaryotes. There are also a great number of symbiotic and commensal relationships between the species in the microbiome. All eukaryotic species need their microbiomes for normal health and function.

The number of papers on icrobiomee has risen exponentially. There is a book on it by a science writer (Yong, 2016). A DNA sequencing study found the human mouth alone contains about 500 species of bacteria (Paster et al., 2001). This is but one organ of one species; the same principle applies to all organs of all multicellular species.

The microbiome shows the pervasiveness of both symbiosis and commensalism in the living world, and is a robust mechanism for both maintaining and increasing biodiversity. As such, it strongly supports the Autocatalytic Biodiversity Hypothesis (ABH)/Pachamama Hypothesis.

Providing a place to live, protection, and nutrients are clear benefits provided by the host to the species of its microbiome living on and in it. Most of the species of the microbiome cannot exist without their host. The vast majority of the relationships between the host and its microbiome are symbiotic or commensal, with a very small minority being parasitic. Less than 1% of microbes cause disease, in spite of their reputation and the misplaced emphasis on this. Fewer than 50 bacterial species cause disease in humans, and there are thousands of species of bacteria in the human microbiome. We do not know at this writing the percentage of bacteria that are in symbiosis vs. commensalism with their host. Most appear to be commensal. However, research is continuing to show that many species help the host. The majority may be symbiotic, since there could easily be many as yet undiscovered benefits that many microbiome species provide their hosts. This discussion reviews some of the literature on benefits of the microbiome to its host.

7.2 UNICELLULAR EUKARYOTES HAVE MICROBIOMES

Unicellular eukaryotes have microbiomes. The host provides a habitat, nutrition, and perhaps other services yet to be discovered. The microbes provide nutrition, chemicals that promote growth of the host, and probably other services not yet discovered. We know that the microbiomes of one-celled microalgae provide vitamins and other nutrients that keep their hosts alive. Little is known about these microbiomes and the relationships they have with their hosts. It is known that the *Paramecium* microbiome differs from that of *Stentor*. These are both genera of unicellular eukaryotes called ciliated protozoa that have cilia that are like small whips used for movement and moving food, waste, water, and so on. Most species of both of these genera live in water. The phycosphere is a mucus region that is rich in organic matter surrounding a phytoplankton cell; it is a microenvironment. It is high in nutrients due to extracellular waste from the phytoplankton cell. It is part of the microbiome and consists of a diverse food web of microbes. It has been suggested that the bacterial assemblages within the phycosphere are species-specific and can vary depending on different environmental factors (Sapp et al., 2007). If they

DOI: 10.1201/9781003246640-7

vary according to species and environmental factors, there are a great many different phycospheres, each with their own set of species, and they represent tremendous biodiversity over broad geographical areas and through time. Phytoplankton rely on their microbiome for survival, receiving resources and growth cofactors from them. In this symbiotic relationship, microbial communities are known to benefit from chemicals that the phytoplankton excrete that accumulate within the phycosphere. These chemicals support the growth of the species of the microbiome. Shibl et al. (2020) showed that the ubiquitous unicellular phytoplankton *Asterionellopsis glacialis* secretes a suite of chemicals, including two unusual secondary metabolites, rosmarinic acid and azelaic acid, in response to bacteria. The main chemicals among these are used by both symbiotic bacteria and those that do not benefit the phytoplankton. However, rosmarinic acid promotes attachment of beneficial bacteria to the phytoplankton and simultaneously suppresses the attachment of opportunistic, unhelpful bacteria, and azelaic acid enhances growth of beneficial bacteria and inhibits the growth of opportunistic, unhelpful species. This shows that at least one unicellular eukaryote has powerful mechanisms for influencing the species of its microbiome. It also shows the importance of the microbiome to the host and how tightly the two are connected. The researchers also showed that the bacterial response to azelaic acid is globally distributed in the world's seas, but only occurs in a limited number of species of bacteria.

Unicellular eukaryotes are tremendously diverse. They comprise the majority of eukaryotic species on Earth. They help a tremendous number of other eukaryotic species, including unicellular ones. For example, Chapters 2 and 3 showed that phytoplankton are major primary producers in the sea, supporting large food webs, and are important in the carbon cycle, regulating temperature and acidity in the water by removing greenhouse gases from the atmosphere, and in oxygen production. So the microbiomes of all unicellular eukaryotic species indirectly benefit an astronomical number of both unicellular and multicellular species and are indirectly benefitted by the microbiomes of countless unicellular eukaryotic species. Thus, there is a tremendous amount of indirect symbiosis involving unicellular eukaryotic microbiomes. Of course, since all species of multicellular eukaryotes benefit other species, there is a great deal of indirect symbiosis involving all microbiomes. If species A and species C are both symbiotic with species B, but not directly symbiotic with each other, they are indirectly symbiotic with each other.

7.3 ABOUT THE MICROBIOME

Probiotics (probiotic is a noun and an adjective) are bacteria, yeast, and other microbes in the microbiome that are beneficial to humans. They are live microorganisms reputed to provide health benefits when consumed, generally by improving or at least partially restoring the digestive tract microbiome. In the human, about a third of all digestive tract bacterial species are in the genus *Bacteroides*. Some of these are commensal, some symbiotic. Other major genera of bacteria in the human digestive tract that may have probiotic species are *Lactobacillus*, which has a number of symbiotic species; *Escherichia* species, such as the famous and usually commensal *E. coli*; some *Clostridium* species; members of the *Bifidobacteria* genus; and lesser-known genera, such as *Peptostreptococcus*, *Peptococcus*, and *Ruminococcus*. Of the over 1,000 species of bacteria in the human digestive tract, there are only in a few phyla (Sommer and Bäckhed, 2013).

Symbiotic microbes are in such close association with their host that the host and microbes are one superorganism. These microbes have evolved to appear to the body's immune system as cells of the organism they are in. Species of bacteria in the genus *Bacteroides* remove sugar molecules from the surface of cells lining the large intestine, and coat themselves with it, thus disguising themselves as part of the organism. Hence, immune cells and antibodies do not attack them.

There are more microorganisms in the digestive tract than cells in the body of the host, including humans, and more microbes in and on the human body than stars in the Milky Way. There are estimated to be about 100 trillion bacterial cells in the human digestive system at any one time, making up over 1,000 species, at least 150 of which are in tight symbiosis with humans. The majority likely belong to approximately 50 species. The weight of bacteria in one human's digestive system is about 1 kilogram (about 2.2 pounds). The human genome has 20,000 to 25,000 genes, whereas the human microbiome has about 4.5 million genes. Five of the six kingdoms[1] of life are in the microbiomes of higher organisms.

Animals tend to begin life as sterile beings, because the womb does not normally have any bacteria in it. Baby organisms pick up bacteria by passing through the birth canal, by breastfeeding if they are mammals, by touching other members of their species, and by interacting with their environment. The latter includes human infants crawling on dirt and eating it, and putting objects in their mouths. By the time animals are infants, their bodies support ecosystems of amazing complexity.

Microbes inhabit several areas both on and in the body. They are not distributed equally between organs and tissues. Friendly bacteria in the digestive system occur mainly in the ileum, the part farthest away from the stomach of the three parts of the small intestine, and the colon, the longest part of the large intestine. The stomach and duodenum, the first section of the small intestine, the one closest to the stomach, are mostly bacteria-free because of the deterrent effect of the strong stomach acid and the high level of digestive enzymes. Though the intestines have the most species and numbers, microbes occur in many body areas, with each area having different species than the other areas. There are diverse microbes, although relatively few in number, in the esophagus. Different areas of the skin have different microbial species, each unique to its local area. The mouth has a great diversity of microbes. The genitals have many microbes. Bacteria also vary with time and conditions within the same individual. Digestive tract bacteria change in women during pregnancy.

They also change with changes in weight. Thus, microbiome diversity through time is high, since it changes a great deal through time. It is also high at any given time in the host. Said another way, the microbiome has high temporal *and* high spatial biodiversity.

This diversity does not end within individual humans. Microbiomes vary significantly between people in the same area, and even more so geographically. Rural people have very different microbiomes than urban ones. People of the United States generally have less microbial diversity than, for example, rural West Africans, because Americans use more antibiotics, eat more processed food (which has fewer bacteria), and are less exposed to bacteria. Rural West Africans have shorter lives and more disease than Americans, but they have, because of their richer microbiome, fewer chronic disorders like allergies, asthma, Type 2 diabetes, and cardiovascular disease.

7.4 THE MICROBIOME AIDS DIGESTION AND ABSORPTION, AND THE DIGESTIVE SYSTEM

Digestive tract bacteria are ecosystem engineers that are indispensable for digestion and absorption. These bacteria aid in the absorption of minerals like calcium, magnesium, and iron. They ensure that food is adequately broken down and that nutrients are absorbed by the body. Digestive tract bacteria have some enzymes that animals, including humans, lack, for degrading carbohydrates, so animals would not be able to utilize many carbohydrates without their digestive tract bacteria. There are symbiotic *Bacteroides* bacteria that help digest complex sugars in plant cell walls. Humans and many other animals would not be able to fully digest plants without these bacteria. Carbohydrates that humans and some other animals cannot digest without bacterial help include certain starches, sugars, fiber, and carbohydrates called oligosaccharides.

Remember from Chapter 2 that an enzyme is a protein that speeds up and/or helps with chemical reactions of organisms. This includes digestion. Digestive system microbes have carbohydrate active enzymes, or CAZymes, which supply humans with energy from polysaccharides (Hehemann, et al., 2010). A polysaccharide is a carbohydrate whose molecules consist of a number of sugar molecules bonded together. CAZymes break polysccaharides down into glucose and other small, simple, easily digested sugars. They act on polysaccharides from land plants, which were a major part of the diet throughout human evolution (ibid., and references therein). Humans lack these enzymes. The enzymes are very diverse in the microbiome. For example, *Bacteroides thetaiotaomicron*, a symbiotic bacterium in the digestive tract of humans, has 261 enzymes involved in digesting starch (ibid., and references therein). This bacterium allows many mammals to efficiently obtain nutrients from apples, potatoes, wheat germ, and other plant foods.

Bacteria ferment and break down carbohydrates, turning them into short-chain fatty acids (SCFAs), including acetic, propionic, and butyric acids. These products are a major energy and nutrient source for the cells of their hosts, including humans. SCFAs also help the body absorb essential dietary minerals, such as iron, calcium, and magnesium. Acetic acid is used by muscles; propionic acid helps the liver produce ATP, the molecule used by all organisms to store and produce energy; and butyric acid provides energy to intestinal cells and may prevent cancer. SCFAs stimulate the growth of intestinal cells, and help with the absorption, metabolism, and storage of lipids. Lipids are important biological molecules that include fats, oils, waxes, certain vitamins (such as A, D, E, and K), hormones, and most of the cell membrane that is not made up of protein. They function in structure, as energy-storage molecules, and as chemical messengers. Both SCFAs and the microbiome itself increase the digestive tract's absorption of water. The microbiome detoxifies substances that get into the body and are not normally there; it breaks down toxins and drugs, and hormones that are no longer needed; and it decreases the amount of toxins absorbed into the blood.

Microbes in the intestine aid its ability to allow substances to pass through it, regulate the acidity of the colon, and maintain optimal acidity in the digestive system, which is crucial for optimal digestion (Nicholson et al., 2012). Methanogens, which are archaea, remove the waste products of bacteria, such as hydrogen, controlling acidity. The methanogens benefit from the bacteria because they use the hydrogen as an energy source. And their removal of waste helps the bacteria, so this is one example of the many symbiotic relationships between microbes in the digestive tract. It is also an example of symbiosis between a microbe species and its host. The methanogen benefits from the habitat provided by the host, and it helps the host by regulating the acidity of its digestive tract.

Microbes in the digestive tract are in a balance much like species in a terrestrial ecosystem, and changing their numbers and species can reduce the host's ability to utilize carbohydrates and digestive acids, and can also cause diarrhea. Disrupting the balance and the normal healthy species community can allow disease-causing competing microbes to grow and sicken the host. The host can be harmed in other ways if bacterial numbers are thrown out of balance. The host produces enzymes that regulate the bacterial balance, which shows the importance of this balance and the coevolution of the host and microbiome.

7.5 THE MICROBIOME MAKES VITAMINS THAT ITS HOST NEEDS

Symbiotic microbes manufacture nutrients their hosts need. They regulate vitamin synthesis (Nicholson et al., 2012). All life needs vitamin B12 for cellular energy production, DNA synthesis, the manufacture of fatty acids, and many other functions. Only bacteria can make the enzymes needed to make this vitamin, meaning all eukaryotes need bacterial symbionts to make this vitamin for them.

Biotin, which is vitamin B7, is needed by virtually all life. Only bacteria, yeast, fungi, and molds can make it. Animals get biotin from ingestion of foods that have it and from symbiotic bacteria in their digestive tract that make it.

Lactobacillus acidophilus is a bacterium in the microbiome that helps synthesize niacin (vitamin B3), folic acid (vitamin B9), and vitamin K. It also aids in the digestion of lactose in the digestive tract. It helps protect against colonization of the digestive tract by deleterious bacteria. It aids in the recycling of amino acids from bile, an important digestive enzyme. It also grows in the vagina, discouraging pathogenic fungi from growing there by competing with them, explaining why women on antibiotics are susceptible to infections by yeast, which are fungi that grow as single cells. In addition to *L. acidolphilus*, other species of digestive tract bacteria synthesize and help the body absorb B vitamins and vitamin K. These vitamins cannot be made by humans and some are hard to get from food. Digestive tract microbes also make folate, a form of vitamin B9, needed to make red and white blood cells in the bone marrow, convert carbohydrates into energy, and produce DNA and RNA.

7.6 THE MICROBIOME HELPS THE INTESTINE OF ITS HOST

The digestive tract microbiome produces SCFAs that control the proliferation and differentiation of intestinal epithelial cells in a way that is beneficial to the host. Epithelial cells are cells that line the surfaces of the body.

The microbiome regulates the normal division of intestinal stem cells in zebrafish. Zebrafish that lack microbes have fewer intestinal epithelial cells, and lack certain kinds of cells and intestinal enzymes (Rawls et. al., 2004, 2006; Bates et al., 2006). Bacteria aid the health of the intestine by changing the expression of cell surface proteins, such as those that transport sodium and glucose in and out of cells (Sears, 2005). They alter cells in a way that decreases the probability of injury to the intestinal barrier of mucus, the body's first line of defense against pathogenic microbes (Keeley, 2004). They even increase the growth of digestive tract cells.

7.7 THE MICROBIOME IS NECESSARY FOR THE NERVOUS SYSTEM

A paradigm shift taking place proposes that the microbiome is needed for nervous system development (Cryan and O'Mahony, 2011). Heijtz et al. (2011) have concluded that "during evolution, the colonization of digestive tract microbiota has become integrated into the programming of brain development, affecting motor control, and anxiety-like behavior." Digestive tract bacteria help humans make neurotransmitters, which are signaling molecules secreted by neurons that affect other neurons, or gland or muscle cells. Neurons are nerve cells. Neurotransmitters carry messages between neurons and are the major way they communicate with each other. The microbiome is essential for brain function. This view also postulates that the microbiome is important in emotional and cognitive function and health. Microbes communicate with the brain via the vagus nerve, which connects the brain to the heart, lungs, and digestive tract. A strain of the microbiome bacterium *Lactobacillus rhamnosus* helps regulate emotional behavior through the vagus nerve, reducing behavior related to anxiety and depression (Bravo et al., 2011). It does so by regulating a form of gamma-aminobutyric acid, the chief inhibitory neurotransmitter in the developmentally mature mammalian central nervous system. This neurotransmitter's principal role is reducing neuronal excitability throughout the nervous system.

Some of the chemicals the microbiome helps make regulate stress levels and even temperament (Cryan and O'Mahony, 2011). Messaoudi et al. (2010) showed that a combination of the bacteria *Lactobacillus helveticus* and *Bifidobacterium longum* reduced anxiety-like behavior in rats and had beneficial psychological effects in humans, alleviating distress, depression, and anger. Humans with these two microbiome species increased problem-solving ability. They had a decrease in the body's main stress hormone, cortisol, in their blood. Besides helping make neurotransmitters, the microbiome also plays a role in making signaling molecules that carry messages in the immune and metabolic systems.

The positive effects of bacteria on the brain and for maintaining mental health are profound. Bacteria, even pathogenic ones in the digestive tract, can activate neural pathways and central nervous system signaling systems. Mice lacking bacteria because of treatment with antibiotics show exaggerated stress responses and reduced expression of brain-derived neurotrophic factor (BDNF) in the cortex and hippocampus. The hippocampus has a major role in learning and memory. BDNF is important for memory and for supporting neuron survival and aiding the growth and differentiation of new neurons and synapses. A synapse is a gap between neurons through which neurons pass chemical and electrical signals to communicate with each other. The mice lacking bacteria also had less of a protein needed to turn on genes that induce growth of neurons (Sudo et al., 2004). Recolonization with *Bifidobacteria* species reversed these effects (ibid.). The mice without bacteria also show impaired memory of previously seen objects. Levels of monocytes (immune cells) were lower, and injection of the monocytes increased new neurons, *indicating that digestive tract bacteria help neuron production via increasing monocytes*. Exercise and probiotic treatment with eight types of live bacteria increased neuron production and improved memory in antibiotic-treated mice. Mice separated from their mothers when young show signs of depression. They show reductions in the microbiome bacteria, *Lactobacillus* and *Bifidobacterium*; intestinal abnormalities; increased corticosterone (stress hormone) levels; and weight loss. Treating them with *Lactobacillus* lowered corticosterone levels and digestive tract abnormalities (Gareau et al., 2007). Giving mice separated from their mothers *Bifodobacterium infantis* reduces signs of depression, minimizes weight loss, and increases production of tryptophan (Desbonnet, et al., 2010), an amino acid that is a precursor to the neurotransmitter serotonin, which relieves depression. Stressed mice show a 10-fold increase in stress-related cytokines and MCP-1. These are proteins that can increase inflammation. The proportion of the symbiotic bacterium, *Bacteroides*, fell by 20 to 25%, while *Clostridium*, a genus of bacteria that often cause disease, increased a similar amount in the digestive tracts of stressed mice compared to controls (Bailey et al., 2011). So social stressors significantly affect digestive tract bacteria while increasing molecules that increase inflammation. Anxiety disorders are common in humans with lowered digestive tract bacteria, and increased digestive tract bacteria diversity reduces anxiety and enhances memory (Foster and McVey Neufeld, 2013; Cryan and Dinan, 2012).

One can conclude from the above paragraphs that the microbiome is involved in—is even essential for—the development of the brain and nervous system in animals. Stress, a significant factor in major depressive disorder, alters gastrointestinal tract microbes, lowering levels of the beneficial bacteria genera, *Lactobacillus* and *Bifidobacterium*. Logan and Katzman (2005) contend that probiotics may be used as an additional treatment to the standard care in major depressive disorder. So the microbiome is important in the regulation of emotional and cognitive behaviors. Furthermore, there is a positive feedback loop between stress and the microbiome. The way this works is that more stress leads to a less diverse and less healthy microbiome, which leads to more stress, which leads to a less diverse and less healthy microbiome, and so on.

Matthews and Jenks (2010) found some bacteria increase the intelligence of their host. Mice fed the soil bacterium *Mycobacterium vaccae* navigated a maze twice as fast, and with fewer anxiety behaviors, than control mice that were not fed this bacterium. The effects persisted for three weeks, but did not show statistical significance at that point. In nature, people and mice inhale and ingest this bacterium. Eating or breathing this bacterium also makes organisms more positive and alert. This is one reason people benefit emotionally from going out into nature. Also, ingesting the probiotic bacterium *Lactobacillus casei* improves mood in those feeling depressed.

The gut-brain-microbiome axis is now recognized as an important regulator of intestinal physiology. Smith et al. (2014) were able to normalize both local (intestinal) physiology and central (behavioral and brain) changes in mice by pretreatment with probiotics.

Alcock et al. (2004) reviewed mechanisms that digestive tract bacteria use to manipulate host eating behavior to their advantage, often at the host's expense. The microbes obtain benefit and increase their fitness by generating host cravings for foods that the microbes need for their survival or that suppress their competitors. The microbes induce discomfort until the host eats foods that enhance the microbes' fitness. Specific techniques include microbial influence on reward and satiety pathways, changes to taste receptors, production of toxins that alter mood, and hijacking of the vagus nerve. This shows the host and its digestive tract microbes may have conflicts at times, with the microbes reducing host fitness to increase their own, although in many, and probably most, cases, the host and its microbiome have common needs and work for mutual benefit. In addition, there is evidence for a positive feedback loop between the preferences of the host as influenced by the digestive tract microbiome and the composition of the digestive system microbiome that depends on the host's diet (Norris et al., 2013). That is, certain microbes cause the host to eat foods that favor them, so there are more of these microbes, so they influence the host to eat yet more foods that benefit them, and so on.

The digestive system has a circadian (about 24-hour) clock coordinated to both light and the timing of eating. Melatonin, the hormone produced at night that stimulates sleep, is made in the pineal gland in the brain, but is also in the digestive system, and many foods contain it. Paulose et al. (2016) found the digestive system bacterium *Enterobacter aerogenes* expresses circadian rhythm patterns because of its sensitivity to melatonin. When the bacteria were exposed to melatonin at levels similar to those found in the digestive tract, they communicated with each other and coordinated periods of swimming and dividing in a phenomenon known as swarming, every 24 hours. The researchers think this indicates that circadian organization in vertebrates might be due to multiple circadian pacemakers that are organized in a hierarchical system of clocks, with some or all of the microbiome as part of that hierarchy. Animals send signals such as melatonin to the bacteria in their digestive system, and perhaps mouths, skin, and so on, and, in a two-way communication feedback loop, the microbiome sends signals back that affect their animal hosts, from the molecular level to behavior.

Sampson et al. (2016) transferred digestive tract microbes making too much α-synuclein, the protein believed to cause Parkinson's when it clumps in the brain, from humans with Parkinson's disease into germ-free mice. The mice that received the microbes developed motor (movement) problems characteristic of Parkinson's disease when tested six or seven weeks after the transfer. Control mice receiving microbes from healthy humans had no such problems. Other work showed that mice with Parkinson's given a healthy microbiome had their symptoms improve markedly (Hou et al., 2021). There is now compelling evidence that an unhealthy digestive tract microbiome plays an important role in Parkinson's disease by facilitating the spread of α-synuclein aggregates from the intestinal nervous system to the brain. For a review that includes therapies such as probiotics and fecal implants, see Lubomski et al. (2020).

Parker et al. (2022) used fecal transplantation of microbiomes to exchange the intestinal microbiomes of young (3 months old), old (18 months), and aged (24 months) mice. The transfer of the microbiomes of the aged 24-month-old mice into young mice accelerated age-associated inflammation of the central nervous system, inflammation of the retina of the eye, and cytokine signaling, which is associated with age-induced degeneration, and promoted the loss of protein function in the eye. All of these effects are associated with a leakier intestinal barrier. These detrimental effects can be reversed by the transfer of the microbiomes of young mice to the affected recipient mice. These results show that aging of the digestive tract microbiome drives detrimental changes in the gut-to-brain and gut-to-retina interactions.

7.8 THE MICROBIOME IS ESSENTIAL TO THE FUNCTION OF THE IMMUNE SYSTEM, AND DIRECTLY COMPETES WITH PATHOGENS

The "immune self" model, the dominant paradigm until recently, portrays the immune system as a defensive network against a hostile exterior environment that fights against anything that is not "self." Recent studies support a fascinating modification of this paradigm, showing that an organism's immune system requires the microbiome for its development and is partly created by it (Lee and Mazmanian, 2010). Microbes have coevolved with the immune system, which favors friendly bacteria over pathogenic ones (ibid.). There is a symbiotic interplay between the immune system and its microbiome.

The immune system fails to function properly and has a greatly reduced repertoire when symbiotic microbes are absent in the digestive system (see Lee and Mazmanian, 2010; Round et al., 2010; Rhee et al., 2004; Lanning et al., 2005). As Lee and Mazmanian (2010) stated: "multiple populations of intestinal immune cells require the microbiota for their development and their function."

Digestive system bacteria help create the immune system. In vertebrates, they promote the early development of the digestive system's mucosal immune system and aid its operation later in life, and symbiotic bacteria stimulate and organize the lymphoid tissue associated with the digestive system's mucosa (Rhee et al., 2004; Lanning et al., 2005). The digestive system's mucosa can then produce antibodies to pathogenic bacteria. Lymphoid tissues are cells and organs that make up the lymphatic system, such as white blood cells, bone marrow, and the thymus gland, spleen, and lymph nodes. They are organized structures that support immune responses. The mucosa is the moist, inner lining of some organs and body cavities, such as the nose, mouth, lungs, and stomach. The microbiome also stimulates the entire immune system. These observations are amazing. *The traditional view in embryology is that an organism's cells direct differentiation of its other cells. Cells of a different species directing an organism's cell differentiation is an extraordinary demonstration of the tight symbiosis and connectedness of bacteria and animals, almost as if they are one species.*

Bacteria of the microbiome train the immune system to respond only to pathogens and parasites, including viruses, bacteria, fungi, and worms, and help direct its development. Thus, the immune system attacks only harmful bacteria, leaving the beneficial species alone. Some harmless bacteria can change their surface receptors to mimic those of host cells so the immune system does not attack them. Harmful bacteria also do this, but the host immune system adapts to recognize and attack them.

Normally, the inflammatory process helps restore health. When the body activates the immune system, it sends out inflammatory cells. These cells attack bacteria or heal damaged tissue. Inflammation is only unhealthy in chronic inflammation, where the body sends out inflammatory cells when the organism is not sick or injured. T helper 17 (Th17) cells are immune cells that play a key role in the inflammatory process. In a stunning discovery, it was found that *specific microbes direct the differentiation of Th17 cells in the mucosa of the small intestine* (Ivanov et al., 2008; Ivanov et al., 2009). Again, *this is bacterial cells directing the differentiation of the cells of the multicellular host.*

There is a symbiotic interaction that involves communication between the host's immune system and digestive tract, and the microbes in the latter. The immune system actually regulates the composition of digestive tract bacteria. Immunoglobulin A (IgA) is an antibody that plays a role in the immune function of mucous membranes. It attacks pathogens, but also regulates digestive tract bacteria and maintains the symbiotic balance between digestive tract microbes and the immune system, playing a key role in maintaining a sustainable host-microbial relationship (Peterson et al., 2007). It *helps* bacteria at times in so doing. Antibodies, of course, are blood proteins that normally attack such invaders as pathogenic bacteria. The existence of immunoglobulin A shows that beneficial bacteria are so valuable to the host that it evolved an antibody that sometimes *aids* bacteria. Communication between the immune system and bacteria cause the appropriate amount of IgA plasma cells. *Thus, the immune system and the microbiome influence each other, in a symbiosis that benefits both.*

Intestinal bacteria help toll-like receptors (TLRs) function in the intestines. *Bacteroides* bacteria use them to establish a symbiotic relationship with the host. These molecules help the host repair damage from injury and aid immunity. For instance, they help the immune system repair damage from radiation. TLRs help the digestive tract discriminate between pathogenic and harmless bacteria.

Clear evidence of the role of microbes in development of the immune system comes from the finding that laboratory mice, which have relatively few microbes and a microbiome that lacks diversity, have a less-developed immune system than wild mice. In addition, lab mice have less diverse microbiomes than pet store mice. Housing lab mice by pet store mice for a month changed the lab mice's immune systems to be more developed and like the pet store mice's (Beura et al., 2016).

Peyer's Patches are organized lymphoid nodules in the ileum, the lowest portion of the small intestine. Lymphoid nodules are small, localized collections of lymphoid tissue that contribute to the defense against microorganisms or foreign materials. Peyer's Patch antibodies fight opportunistic infections, and are additionally involved in the creation of an optimal environment for symbiotic and commensal microbes in their interior (Obata et al., 2010). The evolution of Peyer's Patches in the host shows the importance of its microbiome to it. Symbiotic and commensal microbes use the immunity pathways the host was born with and those that the host acquired to initiate their relationships with the host, leading Round et al. (2011) to conclude the host immune system can discriminate between pathogens and allies by recognizing specific molecules. By discriminating between allies and enemies, the immune system fights pathogens while helping symbiotic allies, while friendly microbes communicate with and utilize the immune system to their advantage (Agrawal, 2001; Hooper et al., 2001; Dale and Moran 2006; Tauber, 2017). Through evolution, the immune system has adapted to recognize which microbes to attack and which to allow entry and support. If accepted, the microbial ally can participate in the physiology, biochemistry, and development of the immune system of the host. Thus, the host is not an autonomous organism, or "self," nor are any of its microbes. The "self" is the *holobiont*, which is the composite ecosystem, to be discussed later in this chapter, and this too is dynamic and dependent on its context.

A spectacular interaction between the immune system and digestive tract bacteria shows symbiotic coevolution of host digestive and immune systems and bacteria in mice. If bacteria touched intestinal cells, it would cause the body to launch a full-blown immune response that would attack and kill good bacteria and strain the body. Cells in the small intestine sense when bacteria are getting too close to them, and produce a lectin called Regilly that kills them when they do, keeping the space around them bacteria-free (Vaishnava et al., 2011).

In squids (McFall-Ngai et al., 2010) and mammals (Hooper et al., 2012), parts of the host immune system have coevolved with the microbiome to support the colonization, persistence, and limitation of symbiotic bacteria within the host, and the immune system actively recruits symbiotic microbes.

In several insect species, bacteria of the genus *Wolbachia* seem to play an important role in protection of the host against viruses (Teixeira et al., 2008; Moreira et al., 2009; Hanson et al., 2011). Some fungi and bacteria that live in plants provide their host with increased pathogen immunity.

Beneficial bacteria also protect against and reduce the number of pathogenic bacteria and yeasts by competing with them (Guarner and Malagelada, 2003; Reid et al., 2001; Yoon et al., 2014). They compete for nutrients, and some compete for attachment sites on the lining of the colon. Beneficial species are better adapted to the digestive system than pathogens. They produce bacteriocins, which are toxins that inhibit growth of similar, competing, pathogenic bacterial strains. They can make substances that kill harmful microbes, and the levels of these can be regulated by enzymes produced by the host, in an amazing interaction between host and beneficial bacteria. This is a result of coevolution between the host, its microbiome, and its pathogens. Bacteria of the microbiome protect animals, including humans, from both harmful invading species and species normally present in low numbers that would cause disease if their growth were not kept in check by competition from the beneficial species. Symbiotic microbes are key in disease prevention in mammals (Lee and Mazmanian, 2010; Ballal et al., 2011). The highly pathogenic bacterium, *Clostridium difficile*, is one of many bacteria that can cause serious illness if antibiotics kill off the host's helpful bacteria, which normally keep it from establishing itself by out-competing it. It kills 14,000 Americans each year. Fecal transplants put a healthy person's microbes into a sick person's intestine. There are now techniques to add only the beneficial bacteria in pills without the rest of the fecal contents. These implants can effectively treat antibiotic-resistant *C. difficile*. *Lactobacillus brevis* showed significant anti-viral activity against the herpes virus (Mastromarino, et al. 2011). Beneficial bacteria also stimulate the growth of other helpful bacteria. These are cases of symbiosis and commensalism. A strain of the common intestinal bacterium, *Escherichia coli*, prevents poisoning from pathogenic *Salmonella* bacteria because it is more efficient than *Salmonella* at capturing the essential element, iron. And probiotic bacteria help to ensure that toxins are not absorbed into the blood.

The species of bacterium called *Bacteroides thetaiotaomicron* turns on a gene in human intestinal cells, causing synthesis of a protein that acts as an antibiotic against its major competitor, the bacterium *Listeria*, an intestinal pathogen. It is interesting that coevolution between the human host and *B. thetaiotaomicron* resulted in the symbiotic bacterium out-competing *Listeria*. Note that *the symbiotic bacterium turns on a human gene, rather than the human's system turning on its own gene. The bacterium is regulating a gene in the human host. This shows an intimate connection between the human host and its bacterial ally.* The protein produced by the gene that the bacterium turns on also stimulates the production of new blood vessels (Cash et al., 2006).

Amphibians of many species are dying off worldwide at an alarming rate from a virulent disease called chytridiomycosis, caused by two species of chytrid fungi, one of which is *Batrachochytrium dendrobatidis*. This involves great population declines and even extinctions of species. Habitat loss, exotic species, water pollution, climate change, and over-collecting are other reasons amphibians are dying off, so the fungus is not the only cause, but a major one. Some amphibians coexist with the fungus and are not harmed by it, partly because they produce antimicrobial skin chemicals. However, these amphibians are also protected from the fungus by symbiotic microbes on their skin, which resist colonization by the fungus or inhibit its growth, while being themselves resistant to the amphibians' antimicrobial skin chemicals (Woodhams et al., 2007). These microbes compete with the chytrid fungi.

7.9 THE MICROBIOME PROTECTS ITS HOST FROM ALLERGIES

Microbes also protect their hosts from allergies. The immune system evolved to fight disease organisms, such as viruses, bacteria, fungi, and parasitic worms. If a human or animal is never exposed to these, the immune system may attack the body, causing allergies. Some probiotic bacteria, such as *Lactobacillus casei*, *L. paracasei*, *L. acidophilus*, and *Bifidobacterium longum*, help in the treatment of nasal and sinus symptoms linked to allergies. Hill et al. (2012) have shown that symbiotic microbes provide developmental signals that limit the growth of cells that become a type of white blood cell called a basophil cell and thereby prevent allergies caused by too many of these cells. The many benign and few potentially pathogenic microbes associated with the natural world, including the dirt that was intimately associated with our preindustrial society, once guaranteed the establishment of normal background levels of microorganisms and kept our bodies from overreacting to foreign bodies. But now there is an alarming rise in allergies and autoimmune disorders, and other diseases due to low microbiome diversity. These include Type 1 diabetes, inflammatory bowel disease, multiple sclerosis, and others (see, for example, Ballal et al., 2011). Inflammatory bowel disease is a group of inflammatory conditions of the colon and small intestine; the principal types are Crohn's disease and ulcerative colitis. The National Institutes of Health estimates that autoimmune diseases collectively affect between 5% and 8% of the U.S. population. There is a good deal of evidence that the increase in these diseases is at least in part due to our lack of exposure to the high diversity of microorganisms that once covered our food and inhabited us. In today's more sterile world of hand sanitizers and wet wipes, it is hard to conceive of the preindustrial lifestyle that resulted in the daily intake of trillions of helpful organisms.

Babies and young children who do not later acquire allergies tend to have more beneficial digestive tract bacteria than those who do (Björkstén et al., 2001). This is probably because helpful bacteria stimulate and train the immune system to respond adaptively only to antigens that attack the body. However, it has not been proven that the difference in bacteria is a cause, rather than a result, of the absence of allergies (ibid.).

Some parents put their infant's pacifier in their own mouth after it drops on the floor, thinking they are cleaning it, although they are actually adding a good deal of bacteria to it. A study showed that the babies of such parents were less likely to have allergies like asthma and eczema than the babies of parents who did not put the pacifier in their mouths (Hesselmara, et al.). The interpretation of this is that the immune system is primed to attack, and needs to have the bacteria to attack; otherwise, it attacks the organism it is part of, causing allergies. Children with dogs or cats have less risk of gastroenteritis (a short-term illness triggered by the infection and inflammation of the digestive system), are 44% less likely to get ear infections, are 29% less likely to have used antibiotics during their first year, and acquire fewer allergies in childhood. One study found children with two dogs or cats got less allergies of the six that were tested (Ownby et al., 2002, and references therein). Babies raised with dogs are 31% more likely to be in good health than those raised without dogs, while babies with cats are 6% more likely to be healthy than those raised without them. This is thought to be due to endotoxins, the breakdown products of bacteria found in the animals' mouths, priming children to not have allergies by giving them the needed invaders to fight off. Again, without something from the outside to attack, the immune system sometimes attacks the body, causing allergies. The dirt and microbes brought into the house by pets increase the number of helpful bacteria, yeast, and other microbes in a baby's body. These pet-borne microbes help train the immune system.

Parasitic worms have successfully treated the autoimmune disease, Crohn's disease (Radford-Smith, 2005). Crohn's disease is caused by the immune system attacking the person with it, because it has no worms to attack in societies where worm infections are uncommon. Part of the immune system evolved to attack worms, and in some cases, it will attack people when there are no worms present. In addition, current research and available therapy with worms are targeted at the treatment of ulcerative colitis, multiple sclerosis, asthma, eczema, dermatitis, hay fever, and even some food allergies.

Bacteria can make the immune system less prone to overreact and cause allergies and autoimmune diseases. Probiotics can regulate immune responses and immunity, thus reducing the severity of allergies and susceptibilities of infants and children to them (Hsieh and Versalovic, 2008).

Probiotic bacteria can aid against the immune system disorder, celiac disease, an allergy or intolerance to gluten and/or gliaden, common components of many grains like wheat, and against rheumatoid arthritis (Yeoh et al., 2013). Celiac disease is associated with imbalances in digestive tract bacteria that can be fully explained by the known effects of glycophosphate, the active ingredient in the herbicide, Roundup, on digestive tract bacteria (Samsel and Seneff, 2013). Thus, the use of Roundup and other herbicides containing glycophosphate can cause celiac disease.

The amazing evolution of one species of bacterium to protect itself from our immune system is known to have resulted in its benefiting humans. T cells are cells of the immune system that recognize and attack invaders of the body, and play a major role in the generalized inflammatory response to infection—swelling, redness, and rising temperature. But T cells release toxic substances, and inflammation can be harmful if T cells react too strongly. Thus, soon after the body increases T cell production, it produces regulatory T cells, whose function is to counteract the normal T cells, controlling inflammation (Mazmanian et al., 2008). Regulatory T cells make a protein that dampens the effect of the pro-inflammatory T cells. This brings down inflammation and prevents the immune system from attacking the body's own cells. It is crucial to maintain a balance between the two types of T cells for the body to stay healthy. A common bacterium, *Bacteroides fragilis*, which lives in 70 to 80% of people, has many sugar molecules that protrude from its surface. One, called polysaccharide A (PSA), promotes maturation of the immune system. It is picked up by human immune cells called dendritic cells, which present it to cells that are destined to develop into T cells, but are not yet T cells. This stimulates them to develop into regulatory T cells. The bacteria and their PSA thus signal the immune system to make more regulatory T cells. This is accomplished by regulation of the genes of the host. *Thus, this system involves a bacterium regulating gene expression of the host.* Germ-free mice have diminished regulatory T cell function, but regain it and normal immune function when they are given this bacterium. This is an adaptive trait that the bacterium evolved, because it causes the pro-inflammatory T cells to stop attacking it. Strains of *B. fragilis* that lack polysaccharide A get attacked and do not survive.

7.10 PROBIOTIC BACTERIA HELP KEEP THE HOST FREE OF INFECTIONS, AND HELP IT IN OTHER WAYS

Symbiotic, probiotic bacteria species of the genera *Lactobacilli* and *Bifidobacteria* protect against disease and slow aging. Probiotic bacteria may counter urinary tract infections, respiratory infections, and other types of organ and tissue infections.

Probiotics may help balance intestinal bacteria, provide key nutrients, suppress the proliferation and virulence of infectious agents, and prevent disease. Supplementation of food with the probiotic bacteria, *Lactobacillus rhamnosus* and *Bifidobacterium lactis*, and canola oil, improved breast milk nutritional quality. Increasing plant food consumption, including the consumption of fruit, aids the growth of beneficial bacteria, which contributes to the health effects of these foods (Kieran et al., 2012).

Probiotic bacteria lower cytokines not only in the digestive tract, but throughout the bloodstream (O'Mahoney et al., 2005). Cytokines are chemicals important in carrying signals between cells. But too high a concentration of them is associated with anxiety, depression, and other maladies, and probiotic bacteria relieve these (O'Leary et al., 2009; Bergami et al., 2008). They also prevent cytokine-induced death of cells lining the intestine (Yan and Polk, 2002). Probiotic bacteria also decrease oxidative stress, and improve nutritional status (Logan and Katzman, 2005). Oxidative stress is chemical stress on cells from high levels of reactive oxygen compounds such as peroxides, and is harmful to health.

Adult carrion beetles of the genus *Nicrophorous* (Figure 7.1) put a dark microbial film in the body of the dead animal that their larvae eat. It has antimicrobial chemicals, but also has beneficial microbes from the parent's digestive tract that help the larvae grow as they eat the dead animal (Shukla et al., 2018).

FIGURE 7.1 Carrion beetles (*Nicrophorus orbicollis*) on a dead mouse at Rock Cut State Park in northern Illinois. Their larvae eat dead animals with tremendous quantities of beneficial bacteria from their parent.

Shutterstock Item ID: 1248195937

Title: Burying Beetles (*Nicrophorus orbicollis*) on a dead mouse at Rock Cut State Park in northern Illinois.
Photo Contributor: Jason Patrick Ross

7.11 THE SKIN IS A DIVERSE ECOSYSTEM OF MICROBES THAT ARE ESSENTIAL TO SKIN HEALTH

Trillions of viruses, bacteria, archaea, fungi, and small arthropods live on the skin surface and obtain nutrients from it. Called the skin microbiome, it is very diverse, with different species in each skin area. Organisms are on the back, forearm, scalp, neck, and other areas. The armpits and genitals each have a diverse microbe community that lives off nutrients in sweat and other substances. All these external microbes produce anti-inflammatories and antibiotics that protect themselves and their host from pathogenic microbes that could potentially colonize the skin. They also protect the skin by competing with pathogenic microbes that could attack the skin. They are generally better competitors than the pathogens. A healthy ecosystem of microbes is essential to skin health (Kong et al., 2009). Human navels have thousands of distinct species of bacteria, the vast majority of which are quite uncommon (Hulcr et al., 2012). New species of skin microbe are still being found. The commonest species are found in 70% of people, with no one species present in all humans. Bacteria in the nose likewise are diverse, and protect against nasal pathogens. The diverse microbes in different parts of the body signal, communicate, and interact with each other in one huge, interactive, symbiotic, interconnected ecosystem.

7.12 THE MICROBIOME IS INVOLVED IN GENE EXPRESSION

The microbiome is a normal and necessary part of the life cycle of all mammals. Incredibly, microbes regulate gene expression of newborn babies (Hooper et al. 2001). In the developing digestive tracts of mice and zebrafish, hundreds of genes are activated by symbiotic bacteria (ibid.; Rawls et al. 2004). *Development of higher animals is dependent on interspecies communication and on the microbiome.*

7.13 OTHER BENEFITS OF THE MICROBIOME

B. fragilis also protects animals in experiments from colitis, an inflammation of the colon induced by *Helicobacter hepaticus*, a commensal bacterium with pathogenic potential (Mazmanian et al., 2008). This is done by use of polysaccharide A, a molecule of the bacterium that I mentioned earlier in this chapter that protects its host from inflammatory disease (ibid.). The researchers who did the work on this state: "These results show that molecules of the bacterial microbiota can mediate the critical balance between health and disease." Imbalances in the composition of the microbiome are postulated to be a major factor in many human and animal diseases, including inflammatory bowel disease. In fact, evidence suggests probiotic digestive tract bacteria prevent inflammatory bowel disease, which is linked to the types and quantities of intestinal microbes present (Hugot et al., 2004). Some other forms of bacteria also prevent inflammation.

Symbiotic bacteria in the digestive system improve the bowels of those suffering from colon cancer, and may prevent it. *Lactobacillus* and *Bifidobacteria*, which are genera of bacteria in the microbiome, help prevent tumors. Digestive tract microbiota break down carcinogens, such as chemicals called heterocyclic amines (HCAs). Although produced by cooking high-protein foods like meat at high temperatures, HCAs are naturally occurring, and hence impossible to totally avoid. They can induce tumors in organs such as the breast, prostate, and colon. It is possible that protection from cancer by digestive tract bacteria is an adaptation to keep their host and hence their environment in existence longer, but this needs more investigation.

The microbiome benefits the liver, reducing low-grade inflammation and seemingly influencing fat metabolism in the liver (Gratz et al., 2010).

7.14 HELICOBACTER PYLORI CAUSES PEPTIC ULCERS, BUT ALSO HELPS ITS HOST

Helicobacter pylori, one of the few bacteria that thrive in the acidic environment of the stomach, has been shown to be a causative agent of peptic ulcers (sores on the lining of the stomach, small intestine, or esophagus), though not the only cause. It was once thought to be only a pathogen, with no benefits to the host. But it gives protection against irritable bowel syndrome, a common, chronic disorder that affects the large intestine and involves cramping, abdominal pain, bloating, gas, and diarrhea or constipation or both. Higgins et al. (2011) showed that *H. pylori* causes *Salmonella* infections, which can cause diarrhea, fever, and abdominal cramps, to be less severe. Cohen et al. (2002) found less incidence of *H. pylori* in patients with shigellosis than in controls without shigellosis. Patients with shigellosis have diarrhea (sometimes bloody), fever, and stomach cramps. Chen and Blaser (2007) presented evidence that childhood acquisition of *H. pylori* is associated with reduced risks of asthma and allergy. The hormone ghrelin signals the brain that the body needs food, inducing eating. Studies showed that *H. pylori* regulates ghrelin, and hence appetite. Ghrelin levels in the blood increased a great amount in subjects without symptoms treated with antibiotics to eliminate *H. pylori* (Nwokolo et al., 2003). This has the potential to increase the appetite and result in weight gain. Bravo et al. (2018) reviewed the effects of *H. pylori* on human health, showing that it coevolved with humans over a long time, and has both positive and negative effects on human health and both positive and negative correlations with the development of several diseases. It appears that overall, even ulcer-inducing *H. pylori* is primarily symbiotic with and helpful to humans. It is pathogenic only in unusual circumstances and only with people susceptible to peptic ulcers. Peptic ulcers are a side effect that occurs only in such people.

7.15 THE APPENDIX FUNCTIONS TO ENSURE THAT THE HOST DOES NOT LOSE ITS MICROBIOME

The appendix has long been regarded as an organ that had a function in the past, but no longer has one. However, there is now convincing evidence that it functions as a refuge for normal digestive tract bacteria that can recolonize the digestive tract if the vast majority of the microbiome is lost from severe diarrhea, cholera, or any other mechanism that depletes the microbiome (Smith et al., 2009). *That an organ evolved solely for the purpose of providing a refuge for the microbiome and ensuring the host does not lose its microbiome is strong evidence for the crucial importance of the microbiome to the health of the eukaryotic host.*

7.16 LOW MICROBIOME DIVERSITY IS CORRELATED WITH OBESITY AND OTHER PROBLEMS

Antibiotics are fed to cattle partly to increase their growth and weight. It appears this works by lowering their microbiome diversity, for obese mice have low digestive tract bacterial diversity, and fewer *Bacteroidetes* and more *Firmicutes* than normal mice (Ley et al., 2005, and references therein). When the digestive tract microbes of obese mice were transplanted into germ-free mice, the recipient mice gained weight despite a decrease in food consumption (Bäckhed et al., 2007). Germ-free mice gained weight when transplanted with intestinal microbes from an obese person, but not those from a thin person (Ridaura et al., 2013).

There is a correlation in humans between low digestive tract microbial diversity and obesity, heart disease, diabetes, and cancer, the last three regardless of weight. Indeed, thin people with low digestive tract diversity are prone to the latter three ailments. There is also a strong correlation between children's exposure to antibiotics and later obesity. Comparison of obese and lean twins showed a healthy, diverse microbiome can decrease the likelihood of obesity, since the obese members of the compared twins had very different bacteria and significantly reduced diversity of their microbiome bacteria (Turnbaugh et al., 2008). Eating an unhealthy diet or ingesting a good deal of antibiotics may be key elements in the epidemics of obesity, heart disease, diabetes, and cancer, due partly to decreased diversity of the digestive tract microbiome.

Two or three generations ago, 80% of Americans had *H. pylori* in them, but now only 6% of American children have it. Most U.S. children have had multiple treatments with antibiotics by age 15 for ear infections. The widespread use of antibiotics likely changed the microbiome of young people profoundly, and may at least partially explain the rise in childhood obesity. Cox et al. (2014) showed that low-dose penicillin treatment amplified diet-induced obesity. When the microbiomes altered by penicillin were transferred to a germ-free host, that host became obese, showing the effect was due to the microbes, not the penicillin. Stem cells are unspecialized cells that are able to develop into many different cell types, ranging from muscle cells to brain cells. One researcher in this study, Martin Blaser, believes that the various bacteria in our microbiome influence whether a certain class of the body's stem cells develop into fat, muscle, or bone. He thinks the intense use of antibiotics of today could eliminate certain bacterial species in the person taking the antibiotics, throwing off the signals that bacteria send to stem cells, causing more than a normal amount of them to become fat cells.

Suzuki and Worobey (2014) found people living in cold, northern latitudes have bacteria in their digestive tracts that predispose them to obesity. The proportion of bacteria in the obsese-associated Firmicutes phylum increased, while the proportion of the lean associated Bacteriodetes phylum decreased, with latitude, regardless of sex, age, or detection methods. This is adaptive to the host (and thus to the microbes) by providing better insulation and getting more fat and energy from food in cold areas, where these are needed. The optimal microbiome, then, varies geographically. Worobey speculates that the changing microbiome allows all types of animals to adapt to different environmental conditions.

7.17 OTHER BENEFITS OF A DIVERSE MICROBIOME

There is a high diversity of species of microbes in microbiomes because hosts provide a great number habitat types and niches (a niche is the role of a species in its ecosystem, including what it eats), from armpits to genitals to intestines, and even different habitats within organs, such as the intestines. The microbes diversify by specializing on specific habitats and niches provided by the host. And they partition niches in the host, which means they divide up the niches so that each species uses its own, unique niche; this avoids competition. It keeps the number of species high, because species do not get eliminated due to competition.

A microbiome with higher diversity makes its host healthier than one with lower diversity. This is demonstrated by a problem created by modern society: human microbiomes lack their former microbial diversity. More than 30% of U.S. births and almost two-thirds of those in urban China are now by Caesarian section, preventing babies from picking up their mother's microbes by going through the birth canal. There has been a great increase in the use of antibiotics over the last few decades. Smaller family sizes throughout the world have meant fewer siblings to expose young children to germs. Cleaner water has added to reduced inoculation with harmless bacteria. Dirt floors of ages past have been replaced with wood and tile, so children do not pick up as many microbes from dirt as in the past. We wash our clothes and bodies much more often than people of the past. Our food is inspected and treated for germs. Farm animals are isolated in barns. All these changes, laudable as some are, mean the human microbiome is more impoverished than it has been historically, with severely deleterious consequences. These include more of the following: allergies, autoimmune diseases, cardiovascular disease, reduced immunity, and other known and possibly other unknown effects.

Preliminary results of studies of hunter-gatherers in the remote Amazon who have had little contact with technological society and no exposure to antibiotics, processed food, and modern birth methods indicate that their pristine microbiome has much higher biodiversity than ours, with a number of species that have never before been sequenced (Clemente et al., 2015). This may at least partially explain their markedly lower rates of allergies and chronic conditions such as Type 2 diabetes and cardiovascular disease.

Wibowo et al. (2021) looked at the genomes of microbes in fossil feces of ancient humans from the southwestern United States and Mexico that were 1,000 to 2,000 years old, and had well-preserved DNA. They compared these to 789 present-day human digestive tract microbiome samples from eight countries. The ancient microbial populations had more jumping genes and fewer genes related to antibiotic resistance than the ones of industrialized people. They also featured fewer genes that produce proteins that degrade the intestinal mucus layer than modern microbiomes; this degradation can produce inflammation that is linked with various diseases. The ancient feces showed preindustrial people had a much more diverse microbiome than modern human microbiomes, and the differences were striking. A stunning 39% of the species of microbes in the ancient feces are not present in modern industrialized microbiomes. For example, the bacterium *Treponema succinifaciens* was not in a single microbiome of industrialized people that the team analyzed, but was in every one of the eight ancient microbiomes. The ancient microbiomes were more similar to the microbiomes of nonindustrialized than industrialized people.

Cavities were nonexistent in hunter-gatherers, just as they are in wild animals. Cooking and cleaning food greatly reduced our microbiome's diversity. Hunter-gatherers from 7,000 years ago had far more microbial diversity in their mouths than did Stone Age agriculturalists. Our mouths have less diversity and fewer friendly bacteria than in the past. The change in our oral microbiome and high-sugar diet, which causes bacteria that produce chemicals that attack teeth to thrive, resulted in cavities.

Children with type I diabetes have unstable digestive tract microbiomes with reduced species diversity and much more Firmicutes bacteria than Bacteroidetes bacteria (Giongo et al., 2011), which is the opposite of healthy children.

Malaria mosquitoes (several species of genus *Anopheles* transmit malaria to humans; here I am referring to *Anopheles gambiae sensu stricto*) are more attracted to and tend to bite people with less diverse microbiomes on their skin; diverse skin microbiomes are a deterrent to these mosquitoes, because their sweat has different chemicals and hence smells less attractive to them (Verhulst et al., 2011). Without bacteria, sweat would have very little smell. *This could be selection at the ecosystem level, favoring more diverse ecosystems of skin bacteria*, since malaria is sometimes lethal.

One hypothesis concerning why the French kiss evolved is that it is a means of transferring bacteria; that is, the microbiome had a profound effect on human sexual behavior. As many as 80 million bacteria are transferred during a ten-second oral kiss (Kort, 2014). This adds diversity to the microbiomes, mainly in the oral cavities, of the people kissing, aiding their immune systems and perhaps other microbiome functions.

7.18 PLANT MICROBIOMES

Plant microbiomes live all over the plant, including within flowers, on leaves, and in roots. It is estimated these microbiomes have a thousand to a million times as much genetic diversity as their plant hosts, and that 30,000 microbial species live in a typical plant's rhizosphere, the narrow zone of soil surrounding its roots (Smith, 2014). The microbiome helps the plant access nutrients and suppress disease. Some microbes assist with plant growth and photosynthesis, some help plants weather droughts and other stressors, and some protect the plant from herbivores. Some strains of *Bacillus subtilis* bacteria improve size, chlorophyll content, and photosynthesis of the plant called Dang Shen (*Codonopsis pilosula*). Chlorophyll is the pigment plants use to carry out photosynthesis. In some plants, a great amount of *Streptomeces* bacteria, which are often used by humans to produce antibiotics and may protect plants against infection, can be found living in a single plant. The atmosphere is 80% molecular nitrogen gas (N_2). Nitrogen-fixing bacteria in the roots of plants convert molecular nitrogen to a form of nitrogen that plants can use. Symbiotic fungi and helper bacteria in plant roots help plants obtain nutrients and water and aid plants in many other ways, as discussed in Chapter 4 of the companion book to this one (Seaborg, 2022).

Soil bacteria help determine the flowering time of *Boechera stricta*, a wild relative of mustard (Wagner, 2014). Flowering time is sensitive to both microbes and the nonbiological properties of different soils. Varying soil microbes also alters patterns of natural selection on flowering time. Soil microbes potentially contribute to the ability of plants to vary their flowering time according to environmental conditions, and to the effects of natural selection on plants in different habitats.

To thwart aboveground pathogens, a plant called rockcress (genus *Arabis*) signals to microbes using the plant hormone salicylic acid (Lebeis et al., 2015). Interestingly, salicylic acid is a precursor to and a metabolite of aspirin. Bacterial communities change in response to salicylic acid signaling in the root zone as well (ibid.), no doubt to the benefit of the plant. Plant roots grow within extremely diverse soil microbial communities, but assemble a limited root-associated microbiome. Lebeis et al. also found that the abundance of some root-colonizing bacterial families increased at the expense of others, partly as a function of whether salicylic acid was used as an immune signal or as a carbon source for microbial growth.

Methylobacterium strains are common in soils, and on the surfaces of leaves of a wide variety of plants. There is evidence that they utilize methanol (an alcohol, CH_3OH) emitted by plant stomata. Stomata are minute pores on the surface of leaves, stems, and other plant parts. Stomata allow gases such as carbon dioxide, water vapor, and oxygen to pass into and out of the internal tissues of plants. Some species of *Methylobacterium* stimulate seed germination and plant development, perhaps by producing phytohormones, chemicals usually produced by plants that regulate their growth, development, reproductive processes, longevity, and death. One strain of this bacterium makes the cytokine zeatin, a chemical that promotes growth of side buds and stimulates cell division to produce bushier plants when experimentally sprayed on plant tissues called meristems. Some of these bacteria produce indole acetic acid, which is a signaling molecule necessary for development of plant organs and coordination of growth. Indole acetic acid has many other effects, such as inducing cells to get longer and to divide. Some *Methylbacterium* protect against disease-causing fungi. It has been suggested that some contribute to the flavor of strawberries, which would mean they aid in seed dispersal for their hosts, because this would cause animals to eat the strawberries. The animals would then move away and defecate the seeds far from the mother plant, dispersing the seeds. One strain of this bacterium forms a nitrogen-fixing symbiosis with a plant in the pea family (Lidstrom and Christoserdova, 2002, and references therein, for all of this paragraph). The bacterial strain fixes nitrogen for the plant, and the plant provides it with a home in its roots.

Shadea et al. (2013) showed a plant microbiome displayed ecological succession, with six successional stages with coherent dynamics whose abundances peaked at different times before and after flower bud opening. They looked at thousands of microbial groups, and the groups had significant associations with other community members. So, since the microbiome changes over time, there is a great amount of temporal microbial diversity over the life of the plant, in addition to the spatial diversity.

7.19 THE MICROBIOMES OF INVERTEBRATES

Symbiotic microbes that have been transmitted from parent to offspring have persisted for millions of years in invertebrates and have played an important role in their evolution. Arthropods often acquire their microbiome from the mother as well as from the environment. Insects, the most diverse animal class, show various types of associations with their microbiomes, ranging from obligate symbiosis, in which the host and symbiotic microbe need each other to survive, to facultative parasitism, where the parasite does not depend on its host and can survive without it (Kikuchi, 2009). Microbes are key symbiotic organisms in insect physiology (Weiss, et al., 2012). Many microbes inherited through insect eggs cause morphological[2] changes with no deleterious effects on their host (Buchner, 1965). In some insect groups, symbiotic bacteria and fungi called endosymbionts live in specialized insect cells called bacteriocytes (also called mycetocytes), and are transmitted only by the mother. Some of these endosymbionts are symbiotic organisms that provide nutrients such as amino acids, the building blocks from which proteins are made, that the host cannot otherwise obtain; others are symbiotic with the host because they metabolize insect waste into safer forms.

In the pea aphid, three symbiotic bacteria are notable. *Rickettsiella* bacteria provide color change, turning genetically red aphids green through the synthesis of chemicals called quinones (Tsuchida et al., 2010). Variants of bacteria called *Buchnera* provide heat tolerance at the expense of the number of offspring that the host can produce at normal temperatures, in a trade-off favoring one color of aphid or the other, depending on the temperature (Dunbar et al., 2007). The population as a whole does better with both variants, because this allows it to survive a range of temperatures. Greater genetic variability is maintained in the aphid by this mechanism. Genetic variability is one measure of biodiversity. So this mechanism maintains higher biodiversity. *Buchnera* are also thought to synthesize essential amino acids that aphids cannot acquire from their diet of plant sap (Douglas, 1998). Finally, some variants of the bacteria species *Hamiltonella defensa* provide immunity against very small wasps that are parasites of the aphids (Oliver et al., 2009). So microbial genomes undergo natural selection in concert with their hosts. Symbiotic bacteria actually provide phenotypic[3] variation in the aphid that is subject to natural selection, and this includes color, tolerance to temperature, and resistance to tiny wasp parasites; and these traits allow some aphids to out compete others in some environments.

Similarly, human microbiomes cause variability and natural selection in humans. The human genome contains only about 20,000 to 25,000 protein-coding genes, which is not much more than the fruit fly genome, while the microbiome of the human digestive tract alone has about 3.3 million genes (Qin et al., 2010)! The entire human microbiome has about 4.5 million genes. The genomes of the microbes in the human microbiome provide certain traits to humans, making it unnecessary for humans to evolve them on their own (Gill et al., 2006). It is likely that natural selection can act on these in nontrivial ways. The human microbiome and the factors that influence the distribution and evolution of its microorganisms must be characterized in order to understand the range of human genetic and physiological diversity (Turnbaugh et al., 2007). The Human Microbiome Project is striving to achieve this as one of its main goals (ibid.). Although more research is needed to show it, it is a reasonable assumption that other multicellular organisms have a similarly higher number of genes in their microbiomes than in their own genomes, and that their microbiomes affect the variability and evolution of each of them. It appears but needs further proof that the microbiome helps its multicellular host evolve much more quickly than it would without it.

A surprising finding is that *Burkholderia* bacteria dwelling in the digestive tracts of bean bugs (*Riptortus pedestris*) give this insect resistance to the common organophosphorus insecticide fenitrothion (Sato et al., 2021). The bacteria degrade this insecticide with an enzyme into a noninsecticidal compound. This compound is toxic to bacteria, but does not harm them because the insect excretes it before it can affect them. This is the first example of symbiosis providing insecticide resistance. This is very important because it suggests the possibility that bacteria may provide other advantages to their hosts through their coevolution with them. These benefits help the bacteria by benefiting the organism that gives them a habitat. This is natural selection on the entire group, not merely the individual, as the standard theory of natural selection states, and is called group selection, which is a controversial idea, not universally accepted by biologists. Group selection might be a general phenomenon that occurs in all microbiomes because the microbes in all microbiomes might undergo selection to help their host, which benefits them as a group because their host is their habitat. If shown to be the case, this would reveal that group selection is more widespread than thought, and that its importance needs re-evaluating. Hence, study of the microbiome has the potential to cause a spectacular change in evolutionary theory. *Burkholderia* bacteria also provide nutritional benefits and resistance against insecticides to stinkbugs, defend the eggs of beetles of the genus *Lagria* in the darkling beetle family (Tenebrionidae) against pathogenic fungi, and are possibly involved in nitrogen metabolism in ants (Kaltenpoth and Flórez, 2020).

Wigglesworthia glossinidia brevipalpis, a symbiotic bacterium in the tsetse fly (genus *Glossina*), has lost a large part of its genome because much of its needs are provided by the fly. It has one of the smallest genomes of any cellular organism (a cellular organism is any organism that is not a virus). It synthesizes important B vitamins the tsetse fly does not obtain from the blood it eats (Akman et al., 2002). Without the vitamins produced by *Wigglesworthia*, the fly has greatly reduced growth and reproduction, and can even be sterile (Nogge et al., 1976).

The immune system of insects and their bacterial allies coevolved in an interacting symbiosis, much like occurred in vertebrate immune systems. Many insects rely on symbiotic bacteria for proper immune system function, and this has been particularly well-demonstrated for tsetse flies (Weiss et al., 2012).

Cockroaches are dependent on certain microbes to survive. Bacterial genus *Blattabacterium* can recycle nitrogen from urea and ammonia, the breakdown products of uric acid, which is the chemical cockroaches produce as their nitrogen waste product, just as humans produce urea in their urine. The bacterium uses the recycled nitrogen to synthesize proteins, repair cell walls and membranes, and perform other metabolic functions (Sabree et al., 2009). This recycling allows the roach to live on low-nitrogen diets, which is critical to the global distribution and success of roaches, which consist of nearly 5,000 species, on every continent, including Antarctica. Cockroaches can no longer make their own amino acids, the building blocks of proteins, as other animals do. The cockroaches' bacteria make their amino acids for them, saving them energy.

The microbiome can play a critical role in mate selection, which could potentially lead to speciation. Groups of fruit flies (*Drosophila melanogaster*) were raised on different diets. When they were adults, flies in the different groups mated only with other fruit flies raised on the same diet. Antibiotics abolished these preferences. Hence, altering the microbiome, not the diet alone, caused the changes (Sharon et al., 2010).

Wolbachia is a genus of bacteria that live within cells and range from parasites to symbiotic organisms in their hosts. They are sexually transmitted through the eggs of many insect and roundworm species. They can cause incompatibility in the eggs. They can convert sexual organisms to ones where females can lay eggs and produce offspring without males, and can change male offspring to females, producing all-female populations. This enhances their own transmission and reproduction (Werren, 2005) because an all-female population can increase faster than a sexual one with males, and of course this causes more *Wolbachia* because they have more hosts, which means more habitat to live in. These microbes frequently pass some of their DNA to their insect hosts (Dunning Hotopp et al., 2007). They can give resistance to some viral infections in fruit flies (Teixeira et al., 2008). They can aid rapid speciation and high biodiversity of insects and nematodes (see, for example, Brelsfoard and Dobson, 2009). Thus, coevolution with this symbiotic bacterium can cause great diversification in insects and roundworms, and significantly increased the number of species in each of these groups. In a spectacular case of symbiosis and transfer of DNA between species, Husnik et al. (2013) showed that a mealybug (*Planococcus citri*) has an endosymbiont bacterium, *Tremblaya princeps*, which itself has its own bacterial endosymbiont, *Moranella endobia*, in a three-way symbiosis. *T. princeps* has the smallest reported bacterial genome, and this is likely a result of its having acquired its endosymbiont. The mealybug has at least 22 genes that it acquired from various microbes, which likely complement missing genes of the two endosymbiont bacteria in it. The 22 genes are from multiple diverse bacteria; none are from *Tremblaya*, showing that genome reduction in this symbiont is not from it transferring its genes to its host. This three-way symbiosis is dependent on genes from at least six lineages of organisms, and reveals a path to intimate endosymbiosis distinct from that followed by mitochondria and chloroplasts, discussed in Chapter 4 the companion book to this one (Seaborg, 2022).

One particular roundworm has a toxic bacterium living inside it. The roundworm infects and kills insects, using a toxin produced by the bacterium. The insect must be dead for the roundworm to be able to consume it. This is an example of obligate symbiosis, because the roundworm and bacterium need each other to survive. This roundworm-bacterium combination also controls the insect population, aiding the ecosystem.

A 6-foot shipworm (*Kuphus polythalamia*), a mollusc (remember, clams and snails are molluscs), harbors a symbiotic bacterium in its gills that feeds on hydrogen sulfide gas, which is emitted from deep lagoons with rotting wood where the shipworm lives (O'Connor et al., 2014). Hydrogen sulfide gas is toxic to most organisms. The bacterium produces enzymes that allow the mollusc to digest the cellulose in the wood. The shipworm and the bacterium are also commensal beneficiaries of the trees that supply the wood. Many of the shipworm's organs have atrophied. This is a somewhat unusual find in that the bacterial symbionts live outside the host's digestive tract. The authors think there may be many more symbiotic relationships of this nature. There are many symbiotic relationships between invertebrates and their bacteria at deep-sea vents and cold seeps on the seafloor, in which the bacteria often provide food for their hosts via chemosynthesis and get a habitat from their host. These are discussed in Chapter 4 of the companion book to this one (Seaborg, 2022). Chemosynthesis produces organic food molecules using the oxidation of chemicals as a source of energy, rather than sunlight, as is done in photosynthesis.

Nearly 40% of the volume of some sponges are bacteria, which contribute significantly to their metabolism (Taylor et al., 2007). Symbiotic systems between the microbiome and host in corals, pea aphids, and cactuses provide tolerance to heat for the composite organism of the host and its many microbiome species (Gilbert et al., 2010).

7.20 RUMINANTS HAVE COMPLEX MICROBIOME ECOSYSTEMS

Ruminants are large, hoofed, plant-eating mammals that are able to acquire nutrients from plant-based food by fermenting it in a specialized stomach prior to digestion, principally through microbial actions. The approximately 200 species of ruminants include both domestic and wild species, including cattle, goats, sheep, giraffes, deer, gazelles, bison, water buffalo, camels, and antelope. Ruminants, like all animals, cannot digest cellulose on their own. They digest the plants they eat and cellulose with the help of a complex, diverse ecosystem of archaea, bacteria, fungi, and protozoa in their digestive systems. Animals that lack the proper microbiome cannot digest cellulose and hence cannot eat such plants as grass. This includes humans. Ruminants have coevolved with their microbiomes. Plant cell walls need to be broken down in order for microbes to access the cellulose. Ruminants evolved teeth adapted for crushing and grinding that start this process. They have very long digestive tracts, and the plants take a long time to pass through them, helping the microorganisms digest them, benefiting both herbivores and microbes, and providing an ideal habitat for the microbiome. They have a special compartment in the enlarged stomach called the rumen, which contains dense populations of several species of bacteria, protozoa, and sometimes yeasts and fungi. It is estimated that one millimeter of rumen contains 10–50 billion bacteria, one million protozoa, and several yeasts and fungi. These organisms digest the cellulose, so that the ruminants can extract nutrients and energy from plants.

Many of the microbes of ruminants are key symbiotic allies with them, receiving a habitat from the ruminant and helping the ruminant with digestion. A study of the microbiome of in cows revealed that they collectively have at least 100 times as many base sequences as the human genome (Hess et al., 2011).

The microbes of cows interact with each other in numerous symbiotic and commensal relationships in a complex, diverse ecosystem. For example, some fermenting microbes that do not use oxygen in the rumen and other parts of the rumen's gastrointestinal tracts are capable of degrading organic matter to short chain fatty acids and hydrogen. The hydrogen would accumulate and inhibit the microbes' ability to continue degrading organic matter, but other microbes of many types consume the hydrogen, obtaining energy, and allowing continued growth of their fermenting symbiotic allies. A study found the stomachs (rumens and abomasums) of lambs that had protozoa in them had more fungi that grow in the absence of oxygen, more methanogens, a major bacterium that breaks down cellulose (*Fibrobacter succinogenes*), and greater bacterial diversity, than lamb stomachs deprived of protozoa (Belanche et al., 2012). This shows protozoa help fungi, methanogens, and bacterial species in the digestive tracts of ruminants, in a relationship that is either commensalism or symbiosis, depending on whether protozoa receive help from the other groups. It is likely, but not certain, that there is some direct help received by the protozoa from some of these microbes. At any rate, many species in the other groups benefit the lamb, and thus help the protozoa in an indirect symbiosis. There are many other such symbiotic relationships and commensalisms in the digestive tracts of rumens, as well as all other higher organisms.

7.21 THE MICROBIOME CAN AID SPECIATION

The microbiome can facilitate speciation in multicellular organisms. When two closely related species of jewel wasp, *Nasonia giraulti* and *N. vitripennis*, interbreed, most of their male larvae die, maintaining the two as separate species. Yet when the hybrid larvae are treated with antibiotics that kill off their digestive tract bacteria, they survive as well as pure-bred larvae (Brucker and Bordenstein, 2013). This indicates the bacteria are helping keep the species separate, acting as barriers to reproduction between the two species. Moreover, a family tree of several species of these wasps based on differences in their DNA coincides with such a tree constructed from differences in their digestive tract bacteria. Thus, the microbiomes of these wasps coevolved with their hosts, and helped speciation and evolutionary change in the wasps. It is not unreasonable to hypothesize that this is generally applicable, holding true for many to all eukaryotes and their microbiomes. Further research is needed to test this hypothesis.

7.22 MORE FACTS ABOUT THE MICROBIOME

Human breast milk has some oligosaccharides, which are complex carbohydrates that infants do not have the enzymes to digest. This seems to make no sense, since evolution would select such unnecessary chemicals out of the gene pool as wasteful of energy. However, the dilemma of why milk has them was solved with the revelation that these nutrients evolved to nourish a digestive tract bacterium, *Bifidobacterium infantis*, which is well adapted to use them (Zivkovic et al., 2011). The oligosaccharides allow the *Bifidobacterium* to proliferate and dominate, and out compete and suppress pathogenic microbes. The *Bifidobacterium* also keeps the lining of the digestive tract healthy. This indicates that with respect to chemicals in milk, selection favored not only nutritional qualities, but the microbes in the digestive system. Milk is a food for infants, but also a food for microbes and a probiotic full of microbes beneficial to the infant. Bottle-fed babies have microbial populations that are less optimal for them than those of breastfed babies.

The epithelium is our "internal skin" that mediates our relationship with the outside world. All food passes through it. It is a protective barrier that prevents bacteria from passing into the blood, where even symbiotic and commensal ones could harm the host. Microbes play a crucial role in maintaining its health. Both the bacterium *Lactobacillus plantarum* and the bacterial genus *Bifidobacterium* enhance its function. Digestive tract bacteria produce SCFAs when they ferment plant fiber in the large intestine. These fatty acids feed the colon's epithelial cells. If the epithelium is not properly nourished, bacteria, toxins, and proteins can slip through it into the blood stream, causing an immune response from the body, resulting in low-grade inflammation, which affects the whole body. This can possibly lead in time to metabolic syndrome and other chronic diseases linked to it. Metabolic syndrome is a cluster of several unhealthy conditions that occur together, including increased blood pressure, high blood sugar, excess body fat around the waist, and abnormal cholesterol or triglyceride levels. Fermentation by bacteria increases colon acidity, making an environment conducive to helpful species, and bad for harmful ones. Greater acidity also may help the elimination of cancer-causing substances. Even more, digestive tract microbes actually increase the growth of digestive tract cells. Byproducts from symbiotic microbes regulate the function of the gastrointestinal barrier, the barrier that allows beneficial nutrients into, while excluding toxins and pathogens from, the interior of the digestive tract (Venkatesh et al., 2014).

Moeller et al. (2017) found that mammalian digestive tract bacteria differed between hosts of the same species that live in different areas in the Americas, indicating spatial limits on bacterial dispersal generate and maintain geographic mammalian digestive tract bacterial diversity across the western hemisphere. But within communities of mammals, there is widespread sharing of these bacteria between pairs of species of predators and their prey, indicating transfer of the bacteria through mammalian food chains. The bacteria in these predator-prey species pairs are similar as a result of the sharing of these microbes. This indicates a potential that the microbiome could play a large part in causing populations of the same species that live in different areas to evolve into separate species. It also suggests that DNA could be passed between mammalian species in close proximity to each other, such as predators and their prey, via bacteria.

Three viruses—polio, a retrovirus, and a mouse mammary tumor virus—use intestinal bacteria to infect mice. The viruses were severely impeded from infecting mice treated with antibiotics and lacking their normal microbiome. Polio viruses used molecules on the bacteria's surface to get past the body's defenses. Thus, although digestive tract bacteria are by far more beneficial than harmful, viruses can exploit them and cause a harmful effect on the host. Even these bacteria, however, may maintain and even increase the biodiversity of multicellular species to the extent that they enable some viruses to control host populations.

The extremely deadly chemical called tetrodotoxin of the moon snail, blue-ringed octopus, pufferfish, porcupinefish, rough-skinned newt, and other animals defends these animals from predators, and is produced by symbiotic bacteria of the genera *Actinomyces*, *Aeromonas*, *Alteromonas*, *Bacillus*, *Pseudomonas*, and *Vibrio* (Lago et al., 2015). These animals evolved warning colorations and behaviors that warn predators not to attack them, so the bacteria affected the evolution of the color and behavior of these animals, and, indirectly, the behavior of their predators.

Even human testicles share in the microbiome, having small numbers of bacteria of only a few species. Preliminary studies showed men with fewer but more diverse bacteria are more fertile. Fertile men have four bacteria species. Those with essentially no sperm have more bacteria, but less diversity, generally only two species (Massimo et al., 2018). Since, all else being equal, more fertile men have more offspring, *this could be selection at the ecosystem level for a more diverse ecosystem*, in this case a more diverse ecosystem of bacteria in the human male testicles.

Microbial genes can be transferred to their hosts. This usually is harmful to the host, but sometimes helps the host, and occasionally it results in a novel function in the host genome. Microbes even helped build their hosts' genomes. This is discussed in Chapter 9. So the microbiome affects host evolution by providing a selective force on it and coevolving with it, altering the host and providing it with variability, and transferring genes to the host, including building the host genome.

Hosts can control pathogens and keep the microbiome beneficial, and this can affect natural selection for the choice of mates and characteristics of the host! Rowe et al. (2011) showed that male mallard ducks (*Anas platyrhynchos*) that had semen with stronger antibacterial activity had more colorful bills, which females prefer. There was selection for this preference in females because it allows them to avoid sexually transmitted pathogens and acquire partners whose sperm suffer less bacteria-induced damage.

The microbiome likely played a key role in human evolution. Dunn et al. (2020) found it is likely that the social behavior of ancestors of humans was influenced by microbes that survived better when there were more interactions among individual human ancestors. Their findings indicate that the digestive tract microbiome of human ancestors adapted rapidly to new environmental conditions, and could have been critical for human dispersal, allowing human ancestors to survive in a range of different environments, including new geographic areas. This allowed human ancestors to spread around the world. The social sharing of microbes might have led to adaptations of the microbes in local areas. They posit that digestion in the human line began externally by fermentation, allowing early humans to store food and stay in one place for a longer time, facilitating the persistence of larger groups living together. Thus, the microbiome allowed human ancestors to settle and develop large groups, and yet disperse and diversify.

Viruses are part of the microbiome. In viviparous mammals (mammals that give live birth, as opposed to the few mammals that lay eggs), endogenous retroviruses, which were formerly infectious viruses, suppress the mother's immune system, protecting the embryo from attack by it. They were key in the evolution of the placenta and placental mammals. Viviparous mammals evolved from an ancestor that was infected with a retrovirus that gave the ancestor these key adaptations (Villareal, 2008, 2016). (See Chapter 8.)

7.23 IMPLICATIONS OF THE MICROBIOME

Four major implications result from the microbiome. First, biological definitions of individuality are being challenged. Second, the view that I advocate and argued extensively for in the companion book to this one (Seaborg, 2022), and that is really part of the ABH, that symbiosis and commensalism are of primary importance, is supported by the existence of the microbiome. Third, the microbiome exemplifies and supports group selection. Fourth, there is division of labor in multispecies symbiotic relationships, not just within individual organisms.

The view that organisms are isolated individuals that survive, function, and evolve as autonomous units is obsolete. So is the concept that one organism corresponds to one genome. Gilbert et al. (2012) state, "The discovery of symbiosis throughout the animal kingdom is fundamentally transforming the classical conception of an insular individuality into one in which interactive relationships among species blurs the boundaries of the organism and obscures the notion of essential identity." If all multicellular organisms are in fact ecosystems composed of many organisms and even species, and individuals with distinct genomes do not even exist, then multicellular organisms are in some ways not individuals. Since individuality is a basic philosophical concept *and* a fundamental concept in biology, this insight is profound.

A new paradigm is emerging called the hologenome hypothesis, which proposes that the unit of natural selection is neither the gene or individual organism, but the holobiont, which is the multicellular organism together with its associated microbiome communities, or, said another way, an ecosystem of organisms that functions as an integrated whole (Zilber-Rosenberg and Rosenberg, 2008; Gilbert, 2014; Rosenberg and Zilber-Rosenberg, 2013; Rosenberg and Zilber-Rosenberg, 2016; Bordenstein and Theis, 2015). The balance of these communities affects the survival of the host. In this view, the holobiont exists, and the individual organism does not exist, on the physiological, anatomical, immunological, developmental, genetic, biochemical, and evolutionary levels, so individual selection does not exist, and in fact, has no meaning. All the genomes of the holobiont collectively form a hologenome. The hologenome can evolve by natural selection and random genetic drift, just as happens in standard Darwinian natural selection. Genetic variation and genetic changes in the hologenome can be brought about by both changes in the host genome and the microbiome, including acquisitions of new microbes, changes in abundance of different microbiome species within hosts, and the exchange of genes between organisms in the microbiome and the host. The hologenome concept includes the vast symbiotic complexity of the genomes of the host and its microbiome. The host and its microbes alter one another's gene expression and exchange DNA. The microbiome is a second mode of genetic inheritance, providing genetic variation for natural selection. The system evolves as a unit. The hypothesis is controversial and not universally accepted. But there is a growing consensus among evolutionary biologists that one cannot separate an organism's genes from the context of its resident microbes.

While each human shares 99.9% of his or her DNA with other humans, most people only share about 10% of their microbes with other people, so this is a source of population variation. It is possible that microbes in the microbiome regularly regulate host genes, and that changes in this regulation can cause adaptive, beneficial changes in the host, even large ones. Further research is required to determine how common and important this is. It is common and important enough to state that the concept of the individual is limited in the study of evolution; the holobiont and hologenome are better considered the units of selection than the individual host organism and the host genome.

Developmentally, the standard view has been that multicellular organisms go through stages from fertilized egg to adult and to the next fertilized egg as individuals. Again, this has to be revised to take into account that we now know the so-called organism develops as a consortium or ecosystem of cells of the multicellular organism and microbes in intimate association (Fraune and Bosch, 2010; Pradeu, 2011). The ovaries of a minute parasitic wasp undergo cell death if their symbiotic bacteria *Wolbachia* are not present (Pannebakker et al., 2007). Organ development in many organisms is dependent on chemical signals from symbiotic microbes in the microbiome. The Hawaiian bobtail squid cannot develop its light organ without its bioluminescent bacterium, *Vibrio fisheri* (Visick et al., 2000). The light organ's function is to provide a place to hold the bioluminescent bacteria, which give off light that camouflages the squid by hiding its silhouette when viewed from below by matching the amount of light hitting the top of the squid. I discussed the microbiome's role in the development of the immune and nervous systems earlier. So, developmentally, the idea of the autonomous individual organism is not fully valid, for a consortium of many organisms take part in the host's development.

Now I will discuss the second implication, that symbiosis and commensalism are of primary importance. Symbiosis is the rule, not the exception. Commensalism is almost for sure the most common interspecific relationship. The microbiome, and the commonness of symbiosis and commensalism between eukaryotes, between eukaryotes and their microbiomes, and between microbes in the microbiomes of eukaryotes, demonstrate that interspecific relationships are at least as important as individual species. The importance of symbiosis and commensalism in structuring ecosystems; maintaining healthy individuals, species, and ecosystems; and promoting high biodiversity, is demonstrated in the microbiome, just as it is in terrestrial and aquatic ecosystems. Cooperation between species is a fundamental feature of evolution. This raises the question: How would population biology be different and what different questions would we ask if it were accepted that symbiosis is the rule, not the exception? And how would it differ if commensalism were seen as the most common interspecific interaction?

We are only beginning to understand the many relationships of the various microbial species with each other and with their hosts. We do not know what percentage are symbiotic as opposed to commensal. We do know that the few microbiomes that have been looked at reasonably thoroughly show a large percentage of symbiosis between microbial species and the host, and between different microbial species in the microbiome. The available evidence suggests that all eukaryotes would not function close to optimally without their microbiomes. And we know that only a very small percentage of microbial species are pathogens.

The third implication is that these symbiotic and commensal relationships between the host and its microbiome appear to fulfill the criteria for, exemplify, and give support to group selection, which posits that traits can spread throughout a population because of the benefits they bestow on groups, even if they have a deleterious effect on individuals. Group selection is discussed in more detail in Chapter 10. It is a controversial theory not universally accepted. In the case of the holobiont, the common habitat of the microbes *is* the host. Thus, the microbes that increase host fitness benefit themselves, others of their species in their population, and other microbe species they share a host with, by benefiting their common habitat. In fact, the group in group selection needs to be redefined here. Single species models can still apply if the researcher is interested in group selection of a single species in the host. But a more realistic view might be to look at the entire community as a group, in which case group selection would be viewed in a different light than most discussions of it presented thus far (see, for example, Williams, 1966, © renewed 1992; Lewontin, 1970; Hull, 1980). Certainly, the microbiome provides a model for the study of group selection as well as an argument for its common occurrence.

Also, the microbiome throws new light on the problem of cheaters, those individuals that multiply at the expense of others and the group. The problem may be at least partially resolved if the individuals and species are under the social control of the whole, the holobiont. This control that unifies, and enforces socialization may be the immune system, which may be viewed as consisting of two parts. One protects from foreign pathogens, the other from microbes within the host that may pose a threat (Tauber, 2017). The immune system keeps host and microbiome cells functioning to each other's mutual benefit. Obedience to it is being a "citizen" of the holobiont, while escaping its control is becoming autonomous, as cancers or pathogens do. Cancer is autonomous, uncontrolled, proliferating cells at the individual level of selection, and pathogens are microbes at the individual selection level that have similarly escaped immune-enforced social cooperation. For example, most species of the bacterium called *Neisseria* are symbiotic or commensal with their host, but the pathogens *Neisseria gonorrhoeae* (which causes gonorrhea and some other diseases) and *Neisseria meningitides* (which can cause meningitis and other forms of meningococcal disease, such as meningococcemia, a life-threatening sepsis) evade immune control and social control by the microbiome (Mulks and Plaut, 1978; Welsch and Ram, 2008). Symbiotic and commensal microbes, on the other hand, are under the control of the host's immune system and the microbiome. The species of the microbiome help control its various microbial species. Thus, the immune system and microbiome together control individual species in the microbiome, fighting them if they become pathogenic to the host. Pathogens are microbes that escaped policing by both the host's immune system and its microbiome.

The fourth implication of the microbiome concerns physiology and genetics. Here, the classical view has held since the 1800s that individuals exist and function through a division of labor between their cells, and each organism is derived from one fertilized egg with one genome. In fact, it has been long known that *division of labor is commonly accomplished through multispecies symbiotic relationships*, as seen in lichens; coral; some bioluminescent species; nitrogen-fixing bacteria and plants; plants, their symbiotic root fungi, helper bacteria, and rodents; and other symbiotic relationships between species. Such division of labor is common in microbiomes. There are several examples of it in this chapter. It is even true that microbe genomes are sometimes integrated into the genomes of their hosts, where they can exhibit division of labor. For example, let us revisit the example discussed earlier of the mealybug and the two endosymbionts. The mealybug actually has a composite genome with DNA of the betaproteobacterium *princeps* in it. *T. princeps* has DNA of the gammaproteobacterium, *Moranella endobia*, in its genome. The genomes of all three species have been altered by this symbiotic arrangement. Amino acid synthesis is coordinated between the genomes of the insect and the two microbes. Three enzymes required to make the amino acid phenylalanine are coded for by *Moranella*, five by *Tremblya*, and one by the insect. None of the three species in this complex has a complete genome on its own, each needing the other two for a complete genome (Husnik et al., 2013, and references therein). None of the three is an individual, autonomous organism. The holobiont has the complete genome, and is the unit of selection.

The effect of the microbiome on host evolution has just begun to be explored, and all indications are that it is profound. Since the microbiome can transfer genes to the host, it provides variability for selection and innovation. It affects adaptation, evolution, and speciation. This applies to humans as well.

Our sense of who we are, our definition of health, and our attitude toward bacteria must be re-evaluated in this major paradigm shift. Health of humans—indeed, the health of all eukaryotes—must now be considered a collective property of the host and its associated microbiome. It is a function of the community, not the individual.

Eukaryotes serve as habitats for diverse ecosystems of viruses, archaea, bacteria, protists, fungi, and invertebrates, many of which aid them and each other. This system of pervasive symbiosis and commensalism is a mechanism of maintaining and generating high biodiversity through ecosystem engineering, supporting the ABH.

NOTES

1 A kingdom is a very high level of classification. Animals make up a kingdom; so do plants.

2 Morphology (adjective is morphological) is the form of an organism, including its structure, size, shape, color, etc.

3 Phenotypic refers to the phenotype, which is the characteristics and traits of an organism that can be seen. The term covers the organism's physical form and structure, as well as developmental processes, biochemical and physiological properties, and the organism's behavior. The phenotype includes the outward appearance (shape, structure, coloration, pattern, size, weight), and the form and structure of the internal parts like bones and organs. It is distinguished from the genotype, which is the complete set of genetic material, the entire genome, of an organism.

REFERENCES

Agrawal, A. A. (2001). Phenotypic plasticity in the interactions and evolution of species. *Science* 294: 321–6.

Akman, L., et al. (2002). Genome sequence of the endocellular obligate symbiont of tsetse flies, *Wigglesworthia glossinidia*. *Nature Genetics* 32 (3): 402. doi:10.1038/ng986. PMID 12219091.

Alcock, J., Maley, C. C., & Aktipis, C. A. (2014). Is eating behavior manipulated by the gastrointestinal microbiota? Evolutionary pressures and potential mechanisms. *BioEssays* 36 (10): 940–9. doi: 10.1002/bies.201400071.

Bäckhed, F., Manchester, J. K., Semenkovich, C. F., & Gordon., J. I. (Jan., 2007). Mechanisms underlying the resistance to diet-induced obesity in germ-free mice. *PNAS USA* 104 (3): 979–84. doi: 10.1073/pnas.0605374104. PMC 1764762. PMID 17210919. http://www.pnas.org/cgi/pmidlookup?view=long&pmid=17210919.

Bailey, M. T., et al. (2011). Exposure to a social stressor alters the structure of the intestinal microbiota: implications for stressor-induced immunomodulation. *Brain, Behavior, and Immunity* 25 (3): 397–407.

Ballal, S. A., et al. (2011). Host and gut microbiota symbiotic factors: lessons from inflammatory bowel disease and successful symbionts. *Cellular Microbiol.* 13: 508–17.

Bates, J. M., Mittge, E., Kuhlman J., Baden, K. N., Cheesman, S. E., & Guilemin, K. (2006). Distinct signals from the microbiota promote different aspects of zebra-fish gut differentiation. *Developmental Biol.* 297: 374–86.

Belanche, A., de la Fuente, G., Pinloche, E., Newbold, C. J., & Balcells, J. (Nov., 2012). Effect of diet and absence of protozoa on the rumen microbial community and on the representativeness of bacterial fractions used in the determination of microbial protein synthesis. *Journ. Animal Sci.* 90 (11): 3924–36. doi: 10.2527/jas.2011-4802. Epub 4 June 2012.

Bergami, M., Rimondini, R., Santi, S., Blum, R., Gotz, M., & Canossa, M. (2008). Deletion of TrkB in adult progenitors alters newborn neuron integration into hippocampal circuits and increases anxiety-like behavior. *PNAS USA* 105: 15570–5.

Beura, L. K., Hamilton, S. E., Bi, K., et al. (28 Apr., 2016). Normalizing the environment recapitulates adult human immune traits in laboratory mice. *Nature* 532: 512–6. doi: 10.1038/nature17655.

Björkstén, B., Sepp, E., Julge, K., Voor, T., & Mikelsaar, M. (Oct., 2001). Allergy development and the intestinal microflora during the first year of life. *Journ. Allergy Clin. Immunol.* 108 (4): 516–20. doi: 10.1067/mai.2001.118130. PMID 11590374.

Bordenstein, S. R. & Theis, K. R. (18 Aug., 2015). Host biology in light of the microbiome: ten principles of holobionts and hologenomes. *PLOS Biol.* 13 (8): e1002226. doi: 1 0.1371/journal.pbio.1002226. PMC 4540581. PMID 26284777.

Bravo, J. A., et al. (2011). Ingestion of *Lactobacillus* strain regulates emotional behavior and central GABA receptor expression in a mouse via the vagus nerve. *PNAS USA* 108: 16050–5.

Bravo, D., et al. (2018). *Helicobacter pylori* in human health and disease: Mechanisms for local gastric and systemic effects. *World Journ. Gastroenterol.* 24 (28): 3071–89. PMID: 30065554. doi: 10.3748/wjg.v24.i28.3071. PMCID: PMC6064966. PMID: 30065554.

Brelsfoard, C. L. & Dobson, S. L. (2009). *Wolbachia*-based strategies to control insect pests and disease vectors. *Asia Pacific Journ. of Molecular Biol. and Biotechnol.* 17: 55–63.

Brucker, R. M. & Bordenstein, S. R. (18 Jul, 2013). The hologenomic basis of speciation: gut bacteria cause hybrid lethality in the genus *Nasonia*. *Science* 341 (6146): 667–9. doi: 10.1126/Science.1240659.Epub 2013 Jul 18.

Buchner, P. (1965). *Endosymbiosis of Animals with Plant Microorganisms*. Interscience Publishers, New York, NY.

Cash, H. L., Whitman, C. V., Behrendt, C. L., & Hooper, L. V. (2006). Symbiotic bacteria direct expression of an intestinal bactericidal lectin. *Science* 313: 1126–30.

Chen, Y. & Blaser, M. J. (23 April, 2007). Inverse associations of *Helicobacter pylori* with asthma and allergy. *Archives of Internal Med.* 167 (8): 821–7. doi:10.1001/archinte.167.8.821.

Clemente, J. C., et al. (2015). The microbiome of uncontacted Amerindians. *Science Advances* 1 (3): e1500183. doi: 10.1126/sciadv.1500183.

Cohen, D., et al. (15 Feb., 2002). An inverse and independent association between *Helicobacter pylori* infection and the incidence of Shigellosis and other diarrheal diseases. *Clinical Infectious Diseases* 54 (4): e35–e42. https://doi.org/10.1093/cid/cir916.

Cox, L. M., et al. (14 Aug., 2014). Altering the intestinal microbiota during a critical developmental window has lasting metabolic consequences. *Cell* 158 (4): 705–21. https://doi.org/10.1016/j.cell.2014.05.052.

Cryan, J. F. & Dinan, T. G. (Oct., 2012). Mind-altering microorganisms: the impact of the gut microbiota on brain and behavior. *Nature Revs. Neuroscience* 13: 701–12. doi: 10.1038/nrn3346.

Cryan, J. F. & O'Mahony, S. M. (Mar., 2011). The microbiome-gut-brain axis: from bowel to behavior. *Neurogastroenterology & Motility* 23 (3): 187–92. doi: 10.1111/j.1365-2982.2010.01664.x.

Dale, C. & Moran, N. A. (2006). Molecular interactions between bacterial symbionts and their hosts. *Cell* 126: 453–65.

Desbonnet, L., Garrett, L., Clarke, G., Kiely, B., Cryan, J. F., & Dinan, T. G. (2010). Effects of the probiotic *Bifidobacterium infantis* in the maternal separation model of depression. *Neuroscience* 170 (4): 1179–88. doi: 10.1016/j.neuroscience.2010.08.005. PMID 20696216.

Douglas, A. E. (1998). Nutritional interactions in insect-microbial symbioses: Aphids and their symbiotic bacteria *Buchnera*. *Ann. Rev. of Entomol.* 43: 17–38. doi: 10.1146/annurev.ento.43.1.17. ISSN 0066-4170. PMID 15012383.

Dunbar, H. E., Wilson, A. C. C., Ferguson, N. R., & Moran, N. A. (10 April, 2007). Aphid thermal tolerance is governed by a point mutation in bacterial symbionts. *PLOS Biology*. http://dx.doi.org/10.1371/journal.pbio.0050096.

Dunn, R. R., et al. (19 Feb., 2020). The internal, external and extended microbiomes of Hominins. *Frontiers in Ecol. and Evol.* https://doi.org/10.3389/fevo.2020.00025.

Dunning Hotopp, J. C., et al. (2007). Widespread lateral gene transfer from intracellular bacteria to multicellular eukaryotes. *Science* 317: 1753–6.

Foster, J. A. & McVey Neufeld, K.-A. (May, 2013). Gut-brain axis: how the microbiome influences anxiety and depression. *Trends in Neurosciences* 36 (5): 305–12. doi: 10.1016/j.tins.2013.01.005.

Fraune, S. & Bosch, T. C. G. (2010). Why bacteria matter in animal development and evolution. *Bioessays* 32: 571–580.

Gareau, M. G., Jury, J., MacQueen, G., Sherman, P. M., & Perdue, M. H. (2007). Probiotic treatment of rat pups normalises corticosterone release and ameliorates colonic dysfunction induced by maternal separation. *BMJ Journals; Gut* 56 (11): 1522–8. doi: 10.1136/gut.2006.117176. PMC 2095679. PMID 17339238.

Gilbert, S. F. (April, 2014). Symbiosis as the way of eukaryotic life: The dependent co-origination of the body. *Journ. of Biosciences* 39 (2): 201–9.

Gilbert, S. F., McDonald, E., Boyle, N., Buttino, N., Gyi, L., Mai, M., Prakash, N., & Robinson, J. (2010). Symbiosis as a source of selectable epigenetic variation: taking the heat for the big guy. *Philosoph. Transactions of the Royal Soc. B: Biol. Sciences.* 365: 671–8.

Gilbert, S. F., Sapp, J., & Tauber, A. I. (Dec., 2012). A symbiotic view of life: we have never been individuals. *Quarterly Rev. of Biol.* 87 (4): 325–41. http://www.jstor.org/stable/10.1086/668166.

Gill, S. R., et al. (2006). Metagenomic analysis of the human distal gut microbiome. *Science* 312: 1355–9.

Giongo, A., et al. (2011). Toward defining the autoimmune microbiome for type 1 diabetes. *ISME Journ.* 5 (1): 82–91. doi: 10.1038/ismej.2010.92. PMC 3105672. PMID 20613793.

Gratz, S. W., Mykkanen, H., & El-Nezami, H. S. (2010). Probiotics and gut health: A special focus on liver diseases. *World Journ. Of Gastroenterology* 16 (4): 403–10. doi: http://dx.doi.org/10.3748/wjg.v16.i4.403.

Guarner, F. & Malagelada, J. R. (Feb., 2003). Gut flora in health and disease. *Lancet* 361 (9356): 512–9. doi: 10.1016/S0140-6736(03)12489-0. PMID 12583961.

Hanson, R. D. E., et al. (2011). A worm's best friend: recruitment of neutrophils by *Wolbachia* confounds eosinophil degranulation against the filarial nematode *Onchocerca ochengi*. *Proc. of the Royal Soc. B: Biol. Sciences* 278: 2293–302.

Hehemann, J.-H., et al. (8 April, 2010). Transfer of carbohydrate-active enzymes from marine bacteria to Japanese gut microbiota. *Nature* 464: 908–12.

Heijtz, R. D., et al. (2011). Normal gut microbiota modulates brain development and behavior. *PNAS USA* 108: 3047–52.

Hess, M, et al. (28 Jan., 2011). Metagenomic discovery of biomass-degrading genes and genomes from cow rumen. *Science* 331 (6016): 463–7. doi: 10.1126/science.1200387.

Hesselmara, B., Sjöberg, F., Saalmanb, R., Åberga, N., Adlerberthc, I., & Wold, A. E. (2013). Pacifier cleaning practices and risk of allergy development. *Pediatrics* 131 (6): e1829–e1837. doi: 10.1542/peds.2012-3345.

Higgins, P. D. R., et al. (1 June, 2011). Prior *Helicobacter pylori* infection ameliorates *Salmonella typhimurium*-induced colitis: Mucosal crosstalk between stomach and distal intestine. *Inflammatory Bowel Diseases* 17 (6): 1398–408. https://doi.org/10.1002/ibd.21489.

Hill, D. A., et al. (2012). Commensal bacteria-derived signals regulate basophil hematopoiesis and allergic inflammation. *Nature Medicine* 18: 538–46.

Hooper, L. V., Littman, D. R., & Macpherson, A. J. (2012). Interactions between the microbiota and the immune system. *Science* 336 (6086): 1268–73.

Hooper, L. V., Wong, M. H., Thelin, A., Hansson, L., Falk, P. G., & Gordon, J. I. (2001). Molecular analysis of commensal host-microbial relationships in the intestine. *Science* 291: 881–4.

Hou, Y.-f., Shan, C., Zhuang, Sy. et al. (2021). Gut microbiota-derived propionate mediates the neuroprotective effect of osteocalcin in a mouse model of Parkinson's disease. *Microbiome* 9 (34). https://doi.org/10.1186/s40168-020-00988-6.

Hsieh, M. H. & Versalovic, J. (Nov.-Dec., 2008). The human microbiome and probiotics: implications for pediatrics. *Curr. Probl. Pediatr. Adolesc. Health Care* 38 (10): 309–27. doi: 10.1016/j.cppeds.2008.09.001.

Hugot, J. P. (June, 2004). Inflammatory bowel disease: a complex group of genetic disorders. *Best Pract. Res. Clin. Gastroenterol.* 18 (3): 451–62. doi: 10.1016/j.bpg.2004.01.001. PMID 1515 7820.

Hulcr, et al. (2012). A jungle in there: bacteria in belly buttons are highly diverse, but predictable. *PLOS One* 7 (11): e47712.

Hull, D. L. (1980). Individuality and selection. *Annual Review of Ecol. and Systematics* 11: 311–32. https://www.jstor.org/stable/2096911.

Husnik, F., et al. (20 June, 2013). Horizontal gene transfer from diverse bacteria to an insect genome enables a tripartite nested mealybug symbiosis. *Cell* 157 (7): 1567–78. https://doi.org/10.1016/j.cell.2013.05.040.

Ivanov, I. I., de Llanos Frutos, R., Manel, N., Yoshinaga, K., Rifkin, D. B., Sartor, R. B., Finlay, B. B., & Littman, D. R. (6 Oct., 2008). Specific microbiota direct the differentiation of IL-17-producing T-helper cells in the mucosa of the small intestine. *Cell Host & Microbe* 4 (4): 337–49. doi: 10.1016/j.chom.2008.09.009. PMC 2597589. PMID 18854238.

Ivanov, I. I., et al. (30 Oct., 2009). Induction of intestinal Th17 cells by segmented filamentous bacteria. *Cell* 139 (3): 485–98. doi: http://dx.doi.org/10.1016/j.cell.2009.09.033.

Jiang, C., et al. (20 Sept., 2018). Dynamic human environmental exposome revealed by longitudinal personal monitoring. *Cell* 175 (1): 277–91.e31. https://doi.org/10.1016/j.cell.2018.08.060.

Kaltenpoth, M. & Flórez, L. V. (Jan., 2020). Versatile and dynamic symbioses between insects and *Burkholderia* bacteria. *Annual Review of Entomology* 65: 145–70. https://doi.org/10.1146/annurev-ento-011019-025025.

Keeley, J. (2004). Good bacteria trigger proteins to protect the gut. *Anaerobe* 11(5): 247–51.

Kieran, M., et al. (2012). Up-regulating the human intestinal microbiome using whole plant foods, polyphenols, and/or fiber. *Journ. Agric. Food Chem.* 60 (36): 8776–82. doi: 10.1021/jf2053959.

Kikuchi, Y. (2009). Endosymbiotic bacteria in insects: their diversity and culturability. *Microbes and Environments* 24 (3): 195–204.

Kong, H. H. & Segre, J. A. (March, 2012). Skin microbiome: looking back to move forward. *Journ. of Investigative Dermatology* 132 (3), Part 2: 933–9. https://doi.org/10.1038/jid.2011.417.

Kort, R., et al. (2014). Shaping the oral microbiota through intimate kissing. *Microbiome* 2 (41).

Lago, J., et al. (2015). Tetrodotoxin, an extremely potent marine neurotoxin: Distribution, toxicity, origin and therapeutical uses. *Marine Drugs.* 13 (10): 6384–406. doi: 10.3390/md13106384. PMC 4626696. PMID 26492253.

Lanning D. K., et al. (2005). Intestinal bacteria and development of the B-lymphocyte repertoire. *Trends in Immunol.* 26: 419–25.

Lebeis, S. L., et al. (2015). Salicylic acid modulates colonization of the root microbiome by specific bacterial taxa. *Science* 349 (6250): 860–4.

Lee, Y. K. & Mazmanian, S. K. (2010). Has the microbiota played a critical role in the evolution of the adaptive immune system? *Science* 330: 1768–73.

Lewontin, R. C. (1970). The units of selection. *Annual Review of Ecol. and Systematics* 1: 1–18. doi:10.1146/annurev.es.01.110170.000245.

Ley, R. E., Turnbaugh, P. J., Klein, S, & Gordon., J. I. (Dec., 2006). Microbial ecology: human gut microbes associated with obesity. *Nature* 444 (7122): 1022–3. doi: 10.1038/4441022a. PMID 17183309.

Lidstrom, M. E. & Chistoserdova, L. (2002). Plants in the pink: cytokinin production by *Methylobacterium*. *Journ. of Bacteriol.* 184 (7). Amer. Soc. Of Microbiol. doi: 10.1128/JB.184.7.1818.2002.

Logan, A. C. & Katzman, M. (2005). Major depressive disorder: probiotics may be an adjuvant therapy. *Med. Hypotheses* 64: 533–8. https://doi.org/10.1016/j.mehy.2004.08.019.

Login, F. H., Balmand, S., Vallier, A., Vincent-Mone´gat, C., Vingneron, A., Weiss-Gayet, M., Rochat, D., & Heddi, A. (2011). Antimicrobial peptides keep insect endosymbionts under control. *Science* 334: 362–5.

Lubomski, M., et al. (2020). Parkinson's disease and the gastrointestinal microbiome. *Journ. of Neurology* 267: 2507–23.

Massimo, A., et al. (July, 2018). Testicular microbiome in azoospermic men—first evidence of the impact of an altered microenvironment. *Human Reproduction* 33 (7): 1212–7. https://doi.org/10.1093/humrep/dey116.

Mastromarino, P., et al. (2011). Antiviral activity of *Lactobacillus brevis* towards herpes simplex virus type 2: role of cell wall associated components. *Anaerobe* 17 (6): 334–6.

Matthews, D. M., & Jenks, S. M. (2010). Effect of *Mycobacterium vaccae* on learning in mice. *Proc. of the Ameri. Soc. for Microbiol.* 110th Gen. Meeting. San Diego, CA: Poster. Vol. 641.

Mazmanian, S. K., Round, J. L., & Kasper, D. L. (2008). A microbial symbiosis factor prevents intestinal inflammatory disease. *Nature* 453: 620–5.

McFall-Ngai, M., Nyholm, S. V., & Castillo, M. G. (2010). The role of the immune system in the initiation and persistence of the *Euprymna scolopes-Vibrio fischeri* symbiosis. *Seminars in Immunol.* 22: 48–53.

Messaoudi, M., Lalonde, R., Violle, N., et al. (2010). Assessment of psychotropic-like properties of a probiotic formulation (*Lactobacillus helveticus* R0052 and *Bifidobacterium longum* R0175) in rats and human subjects. *British Journ. Nutr.* 26: 1–9.

Moeller, A. H. (26 Dec., 2017). Dispersal limitation promotes the diversification of the mammalian gut microbiota. *PNAS USA* 114 (52) 13768–73. https://doi.org/10.1073/pnas.1700122114.

Moreira, L. A., et al. (2009). A *Wolbachia* symbiont in *Aedes aegypti* limits infection with dengue, chikungunya, and *Plasmodium*. *Cell* 139: 1268–78.

Mulks, M. H. & Plaut, A. G. (2 Nov., 1978). IgA protease production as a characteristic distinguishing pathogenic from harmless Neisseriaceae. *New England Journ. of Medicine* 299: 973–6. doi: 10.1056/NEJM197811022299180.

Nicholson, J. K., Holmes E., Kinross J., Burcelin R., Gibson G., Jia, W., & Pettersson, S. (2012). Host-gut microbiota metabolic interactions. *Science* 336: 1262–7.

Nogge, G. (1976). Sterility in tsetse flies (*Glossina morsitans* Westwood) caused by loss of symbionts. *Experientia* 32: 995–6.

Norris, V., Molina, F., & Gewirtzc, A. T. (Feb., 2013). Hypothesis: bacteria control host appetites. *Journ. Bacteriol.* 195 (3): 411–6. doi: 10.1128/JB.01384-12.

Nwokolo, C. U., et al. (2003). Plasma ghrelin following cure of *Helicobacter pylori*. *BMJ Journals; Gut* 52 (5): 637–40. http://dx.doi.org/10.1136/gut.52.5.637.

Obata, T., et al. (2010). Indigenous opportunistic bacteria inhabit mammalian gut-associated lymphoid tissues and share a mucosal antibody-mediated symbiosis. *PNAS USA* 107: 7419–24. doi: 10.1073/pnas.1001061107.

O'Connor, R. M., et al. (2014). Gill bacteria enable a novel digestive strategy in a wood-feeding mollusk. *PNAS USA* 111 (47): E5096–E5104.

O'Leary, O. F., Wu, X., & Castren, E. (2009). Chronic fluoxetine treatment increases expression of synaptic proteins in the hippocampus of the ovariectomized rat: role of BDNF signaling. *Psychoneuroendocrinol.* 34: 367–81.

Oliver, K. M., et al. (6 Sept., 2005). Variation in resistance to parasitism in aphids is due to symbionts not host genotype. *PNAS USA* 102 (36): 12795–800. https://doi.org/10.1073/pnas.050613110.2.

O'Mahony, L., et al. (2005). *Lactobacillus* and *Bifidobacterium* in irritable bowel syndrome: symptom responses and relationship to cytokine profiles. *Gastroenterol.* 128: 541–51.

Ownby, D. R., Johnson, C. C., & Peterson, E. L. (2002). Exposure to dogs and cats in the first year of life and risk of allergic sensitization at 6 to 7 years of age. *Journ. of Amer. Med. Assoc.* 288 (8): 963–72.

Pannebakker, B. A., et al. (2007). Parasitic inhibition of cell death facilitates symbiosis. *PNAS USA* 104 (1): 213–5.

Parker, A., Romano, S., Ansorge, R., et al. (29 April, 2022). Fecal microbiota transfer between young and aged mice reverses hallmarks of the aging gut, eye, and brain. *Microbiome* 10 (68). https://doi.org/10.1186/s40168-022-01243-w.

Paster, B. J., Boches, S. K., Galvin, J. L., Ericson, R. E., Lau, C. N., Levanos, V. A., Sahasrabudhe, A., & Dewhirst, F. E. (June, 2001). Bacterial diversity in human subgingival plaque. *Journ. Bacteriol.* 183 (12): 3770–83. doi: 10.1128/JB.183.12.3770-3783.2001.

Paulose, J. K., Wright, J. M., Patel, A. G., & Cassone, V. M. (11 Jan., 2016). Human gut bacteria are sensitive to melatonin and express endogenous circadian rhythmicity. *PLOS ONE*. http://dx.doi.org/10.1371/journal.pone.0146643.

Peterson, D. A., McNulty, N. P., Guruge, J. L., & Gordon, J. I. (2007). IgA response to symbiotic bacteria as a mediator of gut homeostasis. *Cell Host and Microbe* 2: 328–39. doi: http://dx.doi.org/10.1016/j.chom.2007.09.013.

Pradeu, T. (2011). A mixed self: the role of symbiosis in development. *Biol. Theory* 6: 80–8.

Qin, J., et al. (2010). A human gut microbial gene catalogue established by metagenomic sequencing. *Nature* 464: 59–65.

Radford-Smith, G. L. (2005). Will worms really cure Crohn's disease? *Gut* 54: 6–8. doi: 10.1136/gut.2004.044917. doi: 10.1001/jama.288.8.963.

Rawls, J. F., Mahowald, M. A., Ley, R. E., & Gordon, J. I. (2006). Reciprocal gut microbiota transplants from zebrafish and mice to germ-free recipients reveal host habitat selection. *Cell* 127: 423–33.

Rawls, J. F., Samuel, B. S., & Gordon, J. I. (2004). Gnotobiotic zebrafish reveal evolutionarily conserved responses to the gut microbiota. *PNAS USA* 101: 4596–601.

Reid, G., Howard, J., & Siang Gan, B. (Sept., 2001). Can bacterial interference prevent infection? *Trends in Microbiol.* 9 (9): 424–8.

Rhee, K.-J., et al. (2004). Roles of commensal bacteria in development of gut-associated lymphoid tissues and preimmune antibody repertoire. *Journ. of Immunol.* 172: 1118–24.

Ridaura, V. K., et al. (6 Sept., 2013). Gut microbiota from twins discordant for obesity modulate metabolism in mice. *Science* 341 (6150): 1241214. doi: 10.1126/science.1241214.

Rosenberg, E. & Zilber-Rosenberg, I. (2013). *The Hologenome Concept: Human, Animal and Plant Microbiota.* https://doi.org/10.1007/978-3-319-04241-1. Springer International Publishing, Switzerland. Print ISBN: 978-3-319-04240-4. Online ISBN: 978-3-319-04241-1.

Rosenberg, E. & Zilber-Rosenberg, I. (4 May, 2016). Microbes drive evolution of animals and plants: the hologenome concept. *mBio* 7 (2): e01395–15. doi: 10.1128/mBio.01395-15. ISSN 2150-7511. PMC 4817260. PMID 27034283.

Round, J. L., Lee, S. M., Li, J., Tran, G., Jabri, B., Chatila, T. A., & Mazmanian, S. K. (2011). The toll-like receptor 2 pathway establishes colonization by a commensal of the human microbiota. *Science* 332: 974–7.

Round, J. L., et al. (May, 2010). Coordination of tolerogenic immune responses by the commensal microbiota. *Journ. of Autoimmunity* 34 (3): J220-J225. https://doi.org/10.1016/j.jaut.2009.11.007.

Rowe, M., et al. (23 March, 2011). Sexual ornamentation reflects antibacterial activity of ejaculates in mallards. *Biol. Letters* 7 (5). https://doi.org/10.1098/rsbl.2011.0276. PubMed: 21490006.

Sabree, Z. L., Kambhampati, S. & Moran, N. A. (2009). Nitrogen recycling and nutritional provisioning by *Blattabacterium*, the cockroach endosymbiont. *PNAS USA* 106 (46): 19521–6.

Sampson, T. R., et al. (1 Dec., 2016). Gut microbiota regulate motor deficits and neuroinflammation in a model of Parkinson's disease. *Cell* 167 (6): 1469–80.e12. http://dx.doi.org/10.1016/j.cell.2016.11.018.

Samsel, A. & Seneff, S. (2013). Glyphosate, pathways to modern diseases II: Celiac sprue and gluten intolerance. *Interdiscip. Toxicol.* 6 (4): 159–84. doi: 10.2478/intox-2013-0026. Published online in: www.intertox.sav.sk & www.versita.com/it.

Sapp, M., et al. (31 Jan., 2007). Species-specific bacterial communities in the phycosphere of microalgae? *Microbial Ecology* 53 (4): 683–99. doi:10.1007/s00248-006-9162-5. PMID 17264999.

Sato, Y., Jang, S., Takeshita, K., et al. (2021). Insecticide resistance by a host-symbiont reciprocal detoxification. *Nature Commun.* 12 (6432). https://doi.org/10.1038/s41467-021-26649-2.

Seaborg, D. (2022). *How Life Increases Biodiversity: An Autocatalytic Hypothesis.* CRC Press/Taylor & Francis. Boca Raton, FL; London, UK; New York, NY.

Sears, C. L. (2005). A dynamic partnership: celebrating our gut flora. *Anaerobe* 11 (5): 247–51. doi: 10.1016/j.anaerobe.2005.05.001. PMID 16701579.

Shadea, A., McManusb, P. S., & Handelsmana, J. (2013). Unexpected diversity during community succession in the apple flower microbiome. *mBio* 4 (2): e00602–12.

Sharon, G., Segal, D., Ringo, J. M., Hefetz, A., Zilber- RosenbergI, & Rosenberg, E. (2010). Commensal bacteria play a role in mating preference of *Drosophila melanogaster*. *PNAS USA* 107: 20051–6.

Shibl, A. A., et al. (3 Nov., 2020). Diatom modulation of select bacteria through use of two unique secondary metabolites. *PNAS USA* 117 (44): 27445–55. doi: 10.1073/pnas.2012088117.

Shukla, S. P., et al. (April, 2018). Burying beetles regulate the microbiome of carcasses and use it to transmit a core microbiota to their offspring. *Molecular Ecol.* 27 (8): 1980–91. https://doi.org/10.1111/mec.14269.

Smith, C. J., et al. (15 Oct., 2014). Probiotics normalize the gut-brain-microbiota axis in immunodeficient mice. *Amer. Journ. of Physiol.* 307 (8): G793-G802. https://doi.org/10.1152/ajpgi.00238.2014. PubMed25190473.

Smith, H. F., Fisher, R. E., Everett, M. L., Thomas, A. D., Bollinger, R. R., & Parker, W. (2009). Comparative anatomy and phylogenetic distribution of the mammalian cecal appendix. *Journ. of Evol. Biol.* 22: 1984–99.

Smith, P. A. (Aug., 2014). Foliage friendships. *Scientific Amer.* 311 (2): 24–5 (secondary literature).

Sommer, F. & Bäckhed, F. (2013). The gut microbiota — masters of host development and physiology. *Nat. Review Microbiol.* 11, 227–38. https://doi.org/10.1038/nrmicro2974.

Sudo, N., Chida, Y., Aiba, Y., Sonoda, J., Oyama, N., Yu, X. N., Kubo, C., & Koga, Y. (2004). Postnatal microbial colonization programs the hypothalamic-pituitary-adrenal system for stress response in mice. *Journ. of Physiol.* 558 (Pt. 1): 263–75. doi: 10.1113/jphysiol.2004.063388. PMC 1664925. PMID 15133062.

Suzuki, T. A. & Worobey, M. (2014). Geographical variation of human gut microbial composition. *Biol. Letters* 10 (2): 20131037.

Tauber, A. I. (2017). *Immunity: The Evolution of an Idea.* Oxford Univ. Press, Oxford, UK.

Taylor, M. W., Radax, R., Steger, D., & Wagner, M. (2007). Sponge-associated microorganisms: evolution, ecology, and biotechnological potential. *Microbiol. and Mol. Biol. Rev.* 71: 295–347.

Teixeira, L., Ferreira A., & Ashburner, M. (2008). The bacterial symbiont *Wolbachia* induces resistance to RNA viral infections in *Drosophila melanogaster*. *PLOS Biology* 6:e1000002.

Tsuchida, T., Koga, R., Horikawa, M., Tsunoda, T., Maoka, T., Matsumoto, S., Simon, J.-C., & Fukatsu, T. (2010). Symbiotic bacterium modifies aphid body color. *Science* 330: 1102–4.

Turnbaugh, P. J., Hamady, M., Yatsunenko, T., Cantarel, B. L., Duncan, A., Ley, R. E., Sogin, M. L., Jones, W. J., et al. (2008). A core gut microbiome in obese and lean twins. *Nature* 457 (7228): 480–4. doi: 10.1038/nature07540. PMC 2677729. PMID 19043404.

Turnbaugh, P. J., Ley, R. E., Hamady, M., Fraser-Liggett, C. M., Knight, R., & Gordon, J. I. (2007). The human microbiome project: exploring the microbial part of ourselves in a changing world. *Nature* 449.7164: 804–10.

Vaishnava, S., et al. (14 Oct., 2011). The antibacterial lectin RegIIIgamma promotes the spatial segregation of microbiota and host in the intestine. *Science* 334 (6053): 255–8. doi: 10.1126/science.1209791.

Venkatesh, M., et al (2014). Symbiotic bacterial metabolites regulate gastrointestinal barrier function via the xenobiotic sensor PXR and Toll-like receptor 4. *Immunity* 41 (2): 296–310.

Verhulst, N. O., et al. (2011). Composition of human skin microbiota affects attractiveness to malaria mosquitoes. *PLOS One* 6 (12): e28991. doi: 10.1371/journal.pone.0028991. Epub 2011 Dec. 28. PMID: 22216154 PMCID: PMC3247224. doi: 10.1371/journal.pone.0028991.

Villareal, L. P. (2008). The viruses that make us: A role for endogenous retrovirus in the evolution of placental species. Lecture, UC Irvine.

Villareal, L. P. (2016). Viruses and the placenta: the essential virus first view. *Apmis*, 124 (1–2): 20–30.

Visick, K. L., et al. (2000). *Vibrio fischeri* lux genes play an important role in colonization and development of the host light organ. *Journ. of Bacteriol.* 182 (16): 4578–86.

Wagner, M. R., et al. (2014). Natural soil microbes alter flowering phenology and the intensity of selection on flowering time in a wild *Arabidopsis* relative. *Ecol. Letters* 17 (6): 717–26.

Weiss, B. L., Maltz, M., & Aksoy, S. (2012). Obligate symbionts activate immune system development in the tsetse fly. *Journ. of Immunol.* 188: 3395–403.

Welsch, A. & Ramb, S. (30 Dec., 2008). Factor H and Neisserial pathogenesis. *Vaccine* 26: I40–5. https://doi.org/10.1016/j.vaccine.2008.11.060.

Werren, J. H. (2005). Heritable microorganisms and reproductive parasitism. In *Microbial Evolution: Concepts and Controversies*, pp. 290–315. Ed. Sapp, J. Oxford Univ. Press, Oxford, UK.

Wibowo, M.C., Yang, Z., Borry, M., et al. (12 May, 2021). Reconstruction of ancient microbial genomes from the human gut. *Nature* 594: 234–9 (2021). https://doi.org/10.1038/s41586-021-03532-0.

Williams, G. C. (1966, copyright renewed 1992). *Adaptation and Natural Selection. A Critique of Some Current Evolutionary Thought.* Princeton Univ. Press, Princeton, NJ and Oxford, UK.

Woodhams, D. C., Rollins-Smith, L. A., Alford, R. A., Simon, M. A., & Harris, R. N. (2007). Innate immune defenses of amphibian skin: antimicrobial peptides and more. *Animal Conservation* 10 (4): 425–8. doi: 10.1111/j.1469-1795.2007.00150.x.

Yan, F. & Polk, D. B. (27 Dec., 2002). Probiotic bacterium prevents cytokine-induced apoptosis in intestinal epithelial cells. *Journ. Biol. Chem.* 277 (52): 50959–65. doi: 10.1074/jbc.M207050200.

Yeoh, N., Burton, J. P., Suppiah, P., Reid, G., & Stebbings, S. (Mar., 2013). The role of the microbiome in rheumatic diseases. *Current Rheumatology Reports* 15: 314.

Yong, E. (2016). *I Contain Multitudes: The Microbes Within Us and a Grander View of Life.* Harper Collins, New York, NY.

Yoon, M. Y, Lee, K., & Yoon, S. S. (2014). Protective role of gut commensal microbes against intestinal infections. *Journ. Microbiol.* 52 (12): 983–9. doi: 10.1007/s12275-014-4655-2. PMID 25467115.

Zilber-Rosenberg, I. & Rosenberg, E. (2008). Role of microorganisms in the evolution of animals and plants: The hologenome theory of evolution. *FEMS Microbiol. Revs.* 32: 723–35.

Zivkovic, A. M., German, J. B., Lebrilla, C. B., & Mills, D. A. (2011). Human milk glycobiome and its impact on the infant gastrointestinal microbiota. *PNAS USA* 108: 4653–8.

8 Viruses Are by Far the Most Genetically Variable and Biodiverse Group of Organisms, Generate High Diversity in Cellular Organisms, and Are Key Drivers of Major Adaptive Macroevolutionary Breakthroughs in Cellular Organisms

Because virus classification has differences from that of cellular organisms, many scholars refer to viral "species" as "types," and I will do likewise throughout this chapter. Viruses differ from other organisms in that they are the only organisms not made of cells. Organisms that are not viruses are called cellular organisms. Viruses cannot exist without cellular organisms, which they use to survive and reproduce.

Viruses are the major form of life and by far the most biodiverse group on Earth by any measure. This fantastic diversity can only exist because of the many species of cellular organisms that host them, demonstrating the importance of life to other life and the validity of the Autocatalytic Biodiversity Hypothesis (ABH). Some viruses infect more than one species, but every species of cellular organism is host to several viral species. Thus, there are many more types of viruses than species of all cellular organisms combined. This is spectacular considering the number of species of prokaryotes alone. Even a conservative estimate of 10 virus types infecting each species of cellular organism would mean there are 10 times as many virus types as all cellular species of organism on Earth, from bacteria to humans. Viruses are also by far the largest group in terms of numbers of individual organisms. Viruses transfer genes between organisms, even organisms of different species, and influence nutrient cycling, the carbon cycle, climate regulation, algal and prokaryote species numbers and distributions, control of phytoplankton blooms, and formation of the climate-regulating chemical dimethyl sulfide (Fuhrman, 1999). Although they are not the only mechanism of population regulation, they regulate the populations of all cellular organisms. They also affect evolution of the species they infect. In fact, they have caused some major evolutionary breakthroughs in multicellular organisms.

8.1 VIRUSES ARE THE MOST DIVERSE AND ABUNDANT LIFE FORM

Viruses are one of the largest reservoirs of unexplored genetic diversity on Earth (Suttle, 2007). An analysis of the genomes of two marine viral communities found that over 65% of their DNA was not significantly similar to the DNA of any known viruses, suggesting a majority of the viruses in these marine communities were previously unidentified types (Breitbart, 2002). The total population in both communities that were analyzed was estimated to contain between 374 and 7,114 viral types, amazingly high diversity, given that the two areas sampled were only 200 liters (about 211 quarts) of surface seawater.

The genomes of 10 soil-dwelling phages were studied, identifying 1,600 genes (Pedulla et al., 2003). About half of those genes are unique, not matching any previously described genes in any other organism. This means they are new virus types, previously unknown to science. The genome of bacteriophage G, which has 684 genes, many more than some bacteria, was analyzed. The proteins coded for by nearly 500 of those genes do not match any known proteins (Hendrix, 2009). Studies of viral DNA have found a plethora of new viral types. Typically, between 60 and 80% of marine phage DNA is different from known DNA, meaning between 60 and 80% of marine viruses studied are typically different from any known virus types, so are new virus types to science. Angly et al. (2006) did an analysis of viral communities that included between 41 and 85 individual samples from the Arctic Ocean, the coastal waters of British Columbia and the Gulf of Mexico, and one sample from the Sargasso Sea, obtaining about 1.8 million genetic sequences. On average, over 90% had no recognizable similarity to previously reported genetic sequences. Studies show coastal waters contain perhaps several thousand viral genotypes in 200 liters (about 211 quarts) of seawater (Breitbart et al.,

DOI: 10.1201/9781003246640-8

2002), and marine sediments harbor a million viral genotypes per kilogram (Breitbart et al., 2004). And 60 to 80% of the genetic sequences of marine viruses looked at were not similar (E-value 0.001) to those in databases, contrasting with 90% similarity of putative genes from data for prokaryotic communities to database sequences (Edwards and Rohwer, 2005), showing that marine viral communities contain great genetic richness, much greater, although less sampled, than their prokaryotic counterparts.

Some marine viruses infecting one-celled eukaryotes have almost no recognizable similarity to known DNA sequences (Nagasaki et al., 2005). Of marine viral DNA gathered directly from environmental samples, 65–95% of it is not similar to previously described DNA sequences (Angly et al., 2006; Breitbart et al., 2002; Breitbart et al., 2004), as opposed to only about a 10% difference from known cellular DNA gathered from the environment (Venter et al., 2004). Labonté and Suttle (2013) found 84% of single-stranded DNA viruses they sampled in the sea had no evident similarity to sequenced viruses. All of this implies that we have only begun to scratch the surface of marine viral diversity. The oceans are also a reservoir of tremendous RNA virus diversity, unknown until recently (Culley et al., 2006). In the relatively few cases of genome data available for tailed bacteriophages with very large genomes, called jumbo phages, the genomes are diverse, and the majority have no matches in current sequence databases (Hendrix, 2009). Pope et al. (2015) sequenced and compared 627 phages that infect one bacterial species, *Mycobacterium smegmatis*, finding great genetic diversity, with 28 distinct genomic types (clusters) that had related DNA sequences. All this means new virus types are discovered at great rates every time people look, implying incredible numbers of virus types in the world. Most are not yet discovered.

Analysis of proteins showed phage types or strains exchange genetic material frequently. Quantification of relatedness within and between genomic clusters showed a continuum of genetic diversity. They found the mycobacteriophage population is not closed, having a constant influx of genes from other sources.

Viruses have a greater range of variation than all cellular groups combined. All prokaryotes and eukaryotes have double-stranded DNA as their genetic material. But viruses differ from each other in the basic nature of their genetic material. There are some viruses with double-stranded DNA, some with single-stranded DNA, some with double-stranded RNA, and some with single-stranded RNA. Eukaryotic DNA is linear; bacterial DNA is circular. Viral DNA may be linear or circular. This is extraordinary variability in something as fundamental as the genetic material. Because of this variation, viruses have more variability in their genetic material than all other organisms combined. The way genetic material, such as DNA, makes new copies of itself is called replication. The mechanisms by which viruses replicate their genetic material are more varied than those of all other organisms. Each of the 83 virus families has a unique replication method that uses proteins and enzymes specific to it. In addition, viruses vary greatly in how specialized they are. They may specialize on one host, or infect a huge host range. The variola virus, the causative agent of human smallpox, infects only the human species. By contrast, the rabies virus seems capable of infecting any mammal species. Viruses are extraordinarily diverse in shape and structure. A shell of protein called a capsid surrounds and contains their genetic material. The capsid can be enveloped with one to several lipid bilayers (a lipid bilayer is two layers of lipids forming a membrane), or nothing. Their genome can be protected and enclosed by one, two, or even three capsids, which vary in shape, and are made of proteins. Viruses have great variation in size. At the small end, there are viruses that attack other viruses, including capsid-containing virophage that parasitize and kill the viruses they replicate on (Fischer and Suttle, 2011), and viroids, which consist of nucleic acid (DNA or RNA) without a protein capsule enclosing it. Related to these are prions, which are not viruses, but of relevance because they are protein and no more, and act as infectious particles (an example of a prion is the one that causes mad cow disease). At the large extreme, there are giant viruses about a hundred times bigger than typical viruses, with 1,000 or more genes. The largest known one, *Megavirus chilensis*, has 1,259,197 base pairs, 6.5% longer than any other known giant virus, and 1,120 genes, hundreds more genes than many bacteria have.

Viruses are the numerically most abundant form of life and have a correspondingly large effect on ecosystems. They are the most abundant and diverse form of DNA (or RNA) replicating agent in the world, and the biosphere's major life form. They are the world's largest genetic reservoir overall (Kimura, 2008). They are also the largest genetic reservoir in aquatic environments (ibid.). They are the most abundant biological entities in the oceans by at least a factor of 10 (Breitbart, 2002), estimated to be 10 to 15 times as abundant as bacteria and archaea combined (Wigington et al., 2016), typically numbering 10 billion particles per liter (Fuhrman, 1999), although their numbers vary tremendously from one location to another, and can reach 100 billion viruses per liter at the surface. They are 94% of all entities in the ocean that have DNA or RNA. Suttle (2005) made a calculation of the number of viruses in the sea, arriving at 4×10^{30}. If all of the viruses in the ocean were placed end to end, they would span about 10 million light years. This is about 100 times the distance across our own galaxy, or farther than the nearest 60 galaxies. They are the second largest component of biomass after prokaryotes in the sea, where 95 to 98% of biomass is prokaryotes. The vast majority have not been identified or characterized. Microbiologists found spectacular viral numbers in soil. The soil environment is a more diverse habitat for viruses than aquatic ones (Kimura, 2008). There are estimated to be 10^{30} bacteria worldwide (Whitman et al., 1998), and an estimated ten times as many phage as bacteria, for a total of 10^{31} phage worldwide (Hendrix et al., 1999; Hendrix, 2003). There are 10^{23} stars in the Universe. So there are estimated to be 100 million times more bacteriophages on Earth than there are stars in the Universe! If a phage were as big as an insect, they'd cover the earth several miles deep. The total number of virus particles is 10 times the total number of cells on earth. Thus, in terms of both number of types (species) and number of individuals, viruses are nature's most successful organisms.

8.2 THERE IS A CONTINUUM OF REPLICATING SEGMENTS OF GENETIC MATERIAL FROM SELFISH PARASITE TO FULL COOPERATOR WITH RESPECT TO THE HOST

Lysogeny (the adjective is lysogenic) is the integration of viral DNA into host DNA; the DNA of the virus inserts itself into the host DNA, and the DNA of both virus and host become one double strand of DNA. Lysogenic viruses do not harm the host at all. Lysogeny is widely distributed and very common (Williamson et al., 2007).

Koonin and Starokadomskyy (2016) pointed out that within each organism, there is a continuum of biological replicators from completely selfish to fully cooperative forms. Viruses that kill their hosts represent the selfish extreme, while lysogenic viruses and various DNA segments that jump within genomes and between organisms occupy positions closer to the middle of the range. Selfish replicators are intrinsic to any evolving system of replicating segments of genetic material. Koonin and Starokadomskyy assert that no living system can evolve without the emergence of parasites, including jumping segments of DNA and viruses. (In fact, mobile DNA segments and viruses seem to be the main parasitic entities they were referring to when they made this statement. Also, viruses are not always parasites, since they are sometimes lysogenic, and bacterial viruses often act more like predators, although, like parasites, they use the machinery of their hosts to reproduce.) Parasites are important factors, although not the only ones, that drive the evolution of biological complexity. Life's entire history consists of incessant parasite-host coevolution, with both an antagonistic arms race and various forms of cooperation between host and replicator, both happening constantly. All organisms are communities of interacting, coevolving replicating segments of genetic material ranging the full spectrum from fully parasitic to fully cooperative. Supporting this idea, Mizuuchi et al. (2022) performed long-term evolution experiments on RNA that replicates using a self-encoded RNA replicase, an enzyme that duplicates RNA molecules. The RNA diversified into multiple coexisting host and parasite lineages, whose frequencies in the population fluctuated at first, but then gradually stabilized. The final population, which consisted of five RNA lineages, formed a replicator network with diverse interactions, including cooperation to help the replication of all other members. This was an empirical demonstration that natural selection and evolution naturally lead to hosts coexisting with both replicators that are symbiotic and those that are parasitic. This is compatible with Koonin and Starokadomskyy's suggestion. It is likely that emergence of the entire range of replication strategies, from selfish to cooperative, is intrinsic to biological evolution. Thus, a coevolutionary interplay of selfish conflict between parasite and host and cooperation between cooperative replicating segments of genetic material played a major role in the evolution of complex forms and evolutionary breakthroughs. The breakthroughs gave access to new empty niches and were thus followed by diversification. Viruses were major players in this, both in lysogenic forms and forms that killed their hosts.

8.3 VIRUSES INCREASE THE DIVERSITY OF CELLULAR ORGANISMS

The effect of viruses on increasing diversity of cellular organisms is spectacular. All viral types are ecosystem engineers that positively affect biodiversity, both of prokaryotes and eukaryotes. Marine phages kill 20 to 40% of ocean bacteria every day (Fuhrman, 1999; Weinbauer, 2004; Suttle, 2005; Proctor and Fuhrman, 1990)! A liter of seawater collected in marine surface waters typically contains at least 1 billion microbes and at least 10 billion viruses. Every second, approximately 10^{23} viral infections occur in the ocean. This releases vast amounts of nutrients into the sea, as will be discussed starting with the next paragraph. This cause of nutrient recycling by viruses attacking and killing prokaryotes and phytoplankton is called the viral shunt. It is at least as important as unicellular predators of microbes at regulating microbial populations.

The viral shunt profoundly promotes the productivity and diversity of marine prokaryotes and phytoplankton in other ways than regulating their populations. It releases large quantities of nitrogen, phosphorous, iron, and micronutrients into the sea (Suttle, 2007; Gobler et al., 1997), keeping these nutrients in the photic zone. This means viruses are important in the phosphorous, nitrogen, carbon, and other nutrient cycles. The abundance of these nutrients determines the number of unicellular organisms, including phytoplankton, and these nutrients act as fertilizer for the growth of prokaryotes and phytoplankton, just as nutrients from upwelling do. The viral shunt is analogous to a forest fire fertilizing a forest, releasing nutrients, making them available, and allowing new life to flourish. The nutrients stimulate the growth of marine prokaryotes, phytoplankton, and algae, thus greatly increasing the amount of photosynthesis in the ocean. They increase photosynthesis even more because phage-infected cyanobacteria exhibit a higher rate of photosynthesis. This helps the virus obtain more host nutrient for reproduction. And it increases the amount of carbon incorporated into the living portion of the ecosystem. The viral shunt aids blooms of unicellular organisms. Enough iron is released by the viral shunt to supply the needs of phytoplankton (Poorvin et al., 2004).

Marine viruses are likely profoundly affecting the biogeochemical cycles that run the planet (Suttle, 2005), and are agents for global nutrient cycles through the viral shunt (Suttle, 2007), making nutrients available to marine food webs. Marine virus populations could make an important contribution to the reservoir and cycling of oceanic phosphorus (Jover et al., 2014), making it available to microbes. The viral shunt also enhances nitrogen cycling and regeneration, which makes this valuable nutrient available to life in the sea, by the killing by viruses of the abundant species of the cyanobacteria genus *Synechococcus* (An-Yi et al., 2014). Additionally, the presence of viruses has a positive effect on the daytime frequency of cell division of *Synechococcus* species in the coastal waters of the western subtropical Pacific Ocean. Removal of viruses slowed cell growth and proliferation of *Synechococcus*, presumably because of less nutrients released by viruses bursting open the cells of *Synechococcus* (Weinbauer et al., 2011). As much as 30% of *Synechococcus* cells observed divided in natural virus-containing samples compared to 5% in virus-diluted treatment groups. This tremendously enhances nitrogen recycling and its availability to the ecosystem. *Synechococcus* is important in supporting food webs as one of the species at their base.

The fact that the viral shunt enhances the growth of unicellular organisms has been documented in experimental model systems using microbes that predominantly occur near the ocean surface. Poorvin et al. (2004) showed that the killing of bacteria by viruses in the lab released organic iron complexes that were rapidly taken up by other bacteria and diatoms that normally live in the ocean, enhancing their growth rates. Shelford et al. (2012) did experiments in the low-nutrient Indian Ocean and nutrient-rich False Creek, Vancouver, B.C., comparing bacteria with low virus levels to those with high. The high-virus treatments had higher ammonia concentrations, lower bacteria, and higher phytoplankton, consistent with the high ammonia concentrations (because phytoplankton can use ammonia as a nutrient). The killing of bacteria by the viruses caused higher ammonia production, which caused higher phytoplankton and primary productivity. Ammonia is a form of nitrogen that life can use. This shows viruses play a key role in the sea's nitrogen cycle, and that the killing of bacteria by viruses likely provides a significant portion of global nitrogen requirements for phytoplankton. Bacteria of the rod-shaped genus *Vibrio* were infected with a bacreiophage, PWH3a-P1, and the dissolved organic nitrogen from the killed bacteria was added to cultures containing the aforementioned photosynthetic cyanobacterium genus *Synechococcus* (Shelford and Suttle, 2017). It was inferred that 74% of the dissolved organic nitrogen was ingested by the *Synechococcus* based on how much this bacterium increased. Water that the *Vibrio* killed by the phage had lived in was also added to natural microbial communities in seawater, which took it up. This showed that dissolved organic nitrogen from bacteria killed by phage can be efficiently other microbes, and provides yet more evidence that the viral shunt is important in nitrogen recycling in aquatic systems, making nitrogen available to the biosphere.

Middelboe et al. (1996) grew marine *Vibrio* bacteria infected with viruses with uninfected bacteria that live at the sea surface in the lab. They found that the nutrients released from the bursting of the bacteria by the phage caused an increase in metabolic activity and cell production by reproduction of the noninfected bacteria. In cultures limited by phosphorous, the presence of viruses increased the dissolved organic carbon uptake by the noninfected bacteria by 72% and the activity of their alkaline phosphatase, an enzyme that removes phosphorous in the form of phosphate from chemical compounds, by 89%, compared with control cultures lacking viruses. Organic carbon is carbon in nature that comes from or is in living organisms. The results indicate an increase in available phosphorous from the killing of *Vibrio* by the phage resulted in an increase in bacterial nutrient uptake.

The increase in prokaryotes and phytoplankton, the primary producers and base of marine food webs, as a result of the viral shunt, enhances the stability and maintains the diversity of these ecosystems. With more organisms at the base of a food web, it is a healthier, more stable ecosystem with larger, healthier populations of essentially all species at all levels of the food web up to the top predators. Of course, this is up to a point, since an overpopulation of prokaryotes or phytoplankton would cause a large decrease in diversity by eutrophication and other mechanisms.

Viruses are essential for carbon and nutrient cycling in marine environments. This makes carbon and nutrients available to organisms. They are among the most important mechanisms for recycling carbon in the sea. An amazing 5 to 25% of the carbon used in photosynthesis is estimated to enter into the microbial loop via virus-induced attacks in aquatic environments (Kimura et al., 2008). As much as 25% of the organic carbon in the sea flows through the viral shunt (Wilhelm and Suttle, 1999; Jiao et al., 2010). Evans et al. (2009) found the killing of bacteria by viruses equated to the release of 26.5 micrograms of carbon per liter per day in the eastern subarctic zone. Mojica and Brussaard (2015) showed that 33% in the north and 80% in the south of the carbon used in photosynthesis moved through the viral shunt into the dead particulate and dissolved matter pool in the Northeast Atlantic Ocean. Therefore, viruses are essential for keeping ecosystems running smoothly by helping to recycle carbon and other essential nutrients.

Marine viruses directly and indirectly have a significant influence on biogeochemical cycles and carbon sequestration in the sea (Danovaro et al., 2011). We have seen that sequestration of carbon removes greenhouse gases and keeps the planet at cooler, more favorable temperatures for life. Viruses have helped cool the Earth to a large degree while conserving a great quantity of carbon by storing it in their proteins and genetic material throughout life's history. Viruses also cool the Earth while conserving carbon because when they kill unicellular organisms, dimethyl sulfide is released (Fuhrman, 1999). This gas is involved in a negative feedback loop that regulates Earth's temperature. See Chapter 2 for more on the points made in this paragraph.

The viral shunt tends to cause larger populations of prokaryotes, phytoplankton, and animals. It causes more nutrients to be available to prokaryotes and phytoplankton, allowing them to have higher populations. This allows higher populations of their predators, and higher populations of the next level up the food web, and so on. Small populations go extinct with higher probability than large ones. Large populations undergo more mutations than small ones, so have greater genetic variability. Such greater variability is one measure of biodiversity. And the greater variability of large populations allows them to adapt to environmental challenges better, and more readily change and speciate. It increases the probability of adaptive mutations that can lead to evolutionary innovations. Thus, the viral shunt contributes to higher diversity of prokaryotes, phytoplankton, and animals up the food webs by increasing their population sizes in marine ecosystems.

The systems of bacteria and their phage have been experimentally shown to be remarkably stable in terms of the phage types (species) and bacterial species involved, while being dynamic at the more fine-grained level of phage and bacterial strains (Rodriguez-Brito, 2010). Strains are a lower level than species; they are similar to varieties. The dynamic changes in strains are results that are consistent with previous laboratory studies observing limited numbers of viral and microbial pairs (Bull et al., 2006; Lennon and Martiny, 2008, not to be confused with Lennon and McCartney, but let it be). At the finer-grain level, the process of strains of bacteria and phage continually changing during a period lasting over a billion years gives a high probability of macroevolutionary breakthroughs via the accumulation of many new, adaptive mutations, most small and a few large, until the new adaptive form of bacterium and/or virus arises. Natural selection often favored change in both the bacteria and their phages, and there was no doubt often positive coevolutionary feedback between the phages and bacteria, often resulting in one or each undergoing rapid evolutionary breakthroughs. Each likely helped cause evolutionary innovations in the other by providing selective pressure and exchanging genes.

One might expect the viral shunt to promote toxic algal blooms that are lethal or at least deleterious to many animal species, since the death of the unicellular organisms due to viruses adds a great deal of nutrient to the water, and would allow the explosion of algal populations. Some of these could be of the toxic variety. However, the viral shunt generally does not cause toxic algal blooms, and algal viruses are in fact the main agents responsible for the rapid destruction of harmful algal blooms (Suttle, 2005). This is yet another way marine viruses maintain the diversity of marine systems.

Luna et al. (2013) found direct empirical evidence that viruses enhance microbial biodiversity in nature. They studied microbes in marine surface and sub-surface sediments down to one meter (about 3.3 feet) depth in vegetated sediments (seagrass meadow) and nonvegetated sediments. They found that viruses promote differences in microbial species between different subsurface layers, and play a role in increasing microbial diversity within and between vegetated and nonvegetated sediments.

Marine viruses also maintain and increase the diversity of their bacteria, archaea, and phytoplankton prey by the mechanism called kill the winner (Weinbauer and Rassoulzadegan, 2004; Kirchman, 2013). This is the same mechanism by which animal predators promote diversity of their prey species by selectively attacking the most abundant and successful competing prey species, as illustrated in Paine's work on the sea star system, covered in Chapter 9 of the companion book to this one (Seaborg, 2022). Marine viruses selectively attack the most successful, abundant bacterial species among their competing prey species, causing its population to decline, and preventing it from driving its competitors locally extinct. The most abundant bacterial species is selectively attacked the most by viruses and its population therefore declines before it can eliminate its competitors. The least abundant species are attacked less by viruses, and increase in numbers in response to the decrease in the numbers of the most abundant, most competitive bacterial species. This way, viruses maintain the number of species of bacteria in the ecosystem, keeping this number high. The same is true for viruses that attack plankton. This keeps species diversity of marine prokaryotes and phytoplankton high. Kill the winner is central in models of the population dynamics of phage-bacteria systems. It is the current working paradigm for microbial-phage community dynamics (Rodriguez-Brito, 2010). Several studies reported dramatic changes in the relative concentrations of viruses and their prey, supporting the kill the winner model (Wommack et al., 1999; Fuhrman and Schwalbach, 2003), and some of these studies support kill the winner in natural environments (Chen et al., 2009; Short and Short, 2009; Winget and Wommack, 2009).

Kill the winner is supported by convincing observational evidence in phytoplankton systems (Bratbak et al., 1993; Brussard, 2004; Tomaru et al., 2004), where blooms of a single species seem to result in very high infection rates (Bratbak et al., 1993; Brussard, 2004; Tomaru et al., 2004). This can lead to collapse of the bloom (Bratbak et al., 1993), resulting in the maintenance of the number of phytoplankton species because the less abundant species are not driven extinct by the most abundant phytoplankton species. Viruses range from specialists to generalists. They range from viruses that infect just one strain of a species to viruses that attack many species. In systems with many specialist viruses each limited to one host species, kill the winner occurs because the virus of the host species that is most abundant increases in response to the high host population, and kills it proportionately more than its competitors. The "winner's" population plummets. The former dominant host's competitors increase in response to the low population of their viruses and the decline of their formerly successful competitor. Then the viruses of new "winners" increase and selectively kill them. In the process, new viral types arise frequently and rapidly, and previously rare viral types can quickly increase in abundance. Where generalist viruses are predominant, kill the winner tends to happen because generalist viruses attack cells that they come into contact with, and they bump into host cells that are more abundant more often. They kill the microbe species that is most abundant, switching to the microbe species that replaces it when the latter becomes the predominant microbe species. Populations of viral types may remain relatively constant, but they switch their hosts according to host species abundance. Of course, systems often have a combination of specialist and generalist virus types.

Viruses also maintain diversity of both prokaryotes and eukaryotes by preventing population explosions, as they clearly regulate their prey's numbers. They are especially important in controlling marine and freshwater bacteria and archaea populations. If prokaryotic and phytoplankton populations were not regulated, their populations would explode, and they would deplete their resources and pollute their environment with their wastes. They would die in great numbers. Bacteria that decompose them would deplete oxygen—eutrophication would occur. So there would be massive deaths of fish and invertebrates of many species due to the low level of oxygen in the water. And when the numbers of prokaryotes or phytoplankton increase greatly, the species that are better competitors drive the species that are inferior competitors locally extinct. This loss of diversity would profoundly decrease the diversity of all levels up to the top predators in the food webs, since prokaryotes and phytoplankton are important organisms at the base of of freshwater and marine food webs. These effects would occur in both freshwater ecosystems and the sea, so viruses maintain stable prokaryote and phytoplankton populations, stable ecosystems, and high diversity in both of these ecosystems. Viruses also regulate numbers of all species of organisms, including phytoplankton, fungi, plants, and animals. They regulate populations of all of these groups through negative feedback, the same way that predators regulate to their prey, as discussed in Chapter 9 of the companion book to this one (Seaborg, 2022). The negative feedback works as follows. When the host population becomes dense, be it prokaryote, phytoplankton, or multicellular plant or animal, it is subject to more viral infections because of increased stress from crowding, and because of increased contact each organism has with other individuals in the population, increasing the odds it will come into contact with an infected individual. Hence, more hosts become infected with the virus, bringing the high host population down. When the host population is less dense, there is less stress and less contact between healthy and diseased individuals, so the host population ends up with fewer infected organisms, and can more easily increase. This negative feedback decreases the probability of local extinctions of both host and virus, just as it does in predator-prey systems. So viruses tend to prevent both prokaryote and eukaryote populations from exploding and then overexploiting their environments, and then crashing as a result of the lack of food and space the population explosion creates. They also exert less pressure when their host population is low, allowing its recovery.

Viruses also decrease the random fluctuations of their host populations, which decreases the chance that they will go locally extinct. If a population increases and decreases randomly, it will eventually go extinct if there is an infinite amount of time. This is a statistical certainty. Negative feedback removes random fluctuations in a population. This removes the inevitability of eventual extinction. Negative feedback reduces the probability that a perturbation will lead to an extinction.

Viruses also help many eukaryotes from phytoplankton to plants to lizards to humans by controlling pathogens, such as disease-causing bacteria and fungi, and parasites, such as tapeworms and various species of roundworms.

8.4 VIRUSES EXCHANGE GENETIC MATERIAL, INCLUDING GENES, WITH THEIR HOSTS

Viruses also greatly promote their hosts' evolution and diversification by providing DNA, including genes, to their hosts. This has profoundly affected and accelerated the evolution of many different groups, causing the appearance of evolutionary innovations and novelties. Innovations tend to put the species that has acquired them into a new adaptive zone, with many new niches available to it. So innovations tend to be followed by diversification into the many new niches, creating many new species. This is also adaptive for the virus because a greater number of host species provides it with more habitats and niches. This leads to diversification of the virus into many new types. Transfer of DNA by viruses is particularly common in bacteria (Canchaya et al., 2003), but also occurs in eukaryotes. It promotes prokaryotic diversity (Kirchman, 2013).

The DNA of lysogenic viruses sometimes leaves the host DNA and becomes free again. In this case, it starts making new viruses and is pathogenic to the host. When lysogenic viruses leave their hosts, they can take some of the host's DNA with them. They can then integrate into a new host, becoming lysogenic again. This gives the new host the DNA the virus took from its previous host. This is called transduction. This is a way hosts can increase their DNA. Also, the DNA given to the new host could code for an adaptive new function, giving the host an adaptive advantage. For example, it might give the host a gene for an enzyme that allows the host to digest a new kind of food. Transduction usually occurs between hosts of the same species, but viruses can carry DNA from one species to another. They are a key natural mechanism for transferring genes between species (Liu et al., 2010), which increases genetic diversity, potentially drives evolution, and allows the possibility of evolutionary innovation and rapid evolutionary change. *It is thought that viruses played a central role in very early evolution, before the diversification of bacteria, archaea, and eukaryotes, and at the time of the last universal common ancestor of all cellular life on Earth!* Transduction would have played a key role in this.

A key process in the evolution of photosynthesis, the uniting into the same species of photosystems I and II, was likely accomplished by gene transfer via a phage. This will be explained in more detail in Chapter 9. There is evidence from genome analyses of tailed bacteriophages that viruses sometimes evolve by acquiring DNA from their hosts. The most realistic view appears to be that viruses did not simply evolve from cells, but are partners in a mutual coevolution with them (Hendrix et al., 2000). Viruses have likely been important in the transfer of DNA through the entire history of life. Viruses carry small pieces of DNA called transposons (also called jumping genes, and they will be discussed in detail in Chapter 9) between hosts, protecting them in their protein coats; without being encased in protective protein coats, transposons cannot move to new hosts.

Filée et al. (2003) showed numerous transfers of genes between viruses and cells, in both directions, in the genes for four proteins involved in the replication of DNA, and the manufacture of molecules used to make DNA. They suggested that the transfer of genes from viruses to the cells of organisms, and the replacements of the genes of cells by the genes of viruses are important sources of new genetic material in the evolution of organisms, and that viruses are major players in the evolution of the genomes of organisms.

Evidence now indicates that viruses routinely use genes taken from their hosts to control host metabolism to better produce virus particles (Thompson, 2010). Metabolism is the set of life-sustaining chemical reactions in organisms, including obtaining energy from food, building molecules like proteins, and elimination of wastes. Genes for photosynthesis are examples of genes viruses take from their hosts to control them (Sullivan et al., 2006; Zeidner et al., 2005; Lindell et al., 2004). Phages of photosynthetic cyanobacteria commonly have photosynthesis genes (ibid.; Sullivan et al., 2006), even acting as genetic storage places of these genes for their hosts. The phage normally shuts down most of its host's activity, and phage photosynthesis genes allow the phage to generate the energy necessary for its reproduction in its host. Thus, the possession of photosynthesis genes is beneficial for the phage. This also means selection can favor viral incorporation of host genes, the first step in transduction. The alteration of host metabolism by the virus for its own benefit can be significant, even speeding up the biological pump. This is because some phage that infect cyanobacteria increase the photosynthetic rate of the cyanobacteria. This increases carbon sequestration, since more carbon sinks when the viruses kill the cyanobacteria.

I will now coin a new term: *symbiotic genetic coevolution*, which is coevolution between two symbiotic species in which at least one of the species has its variability enhanced by the transfer of genes from the other species. This could involve the transfer of genes between both species, in both directions. It can result in adaptive evolution. Viruses and their hosts underwent (and still undergo) symbiotic genetic coevolution, whereby viruses benefit by getting a place to live and the use of host genes to control host metabolism, while hosts benefit because viruses increase host variability, progressive evolution, adaptation, and diversification. Both host and virus benefit by obtaining genetic material from the other. The host benefits at the level of evolution, unlike the case in most symbiosis, in which both species benefit by receiving an immediate, ecological advantage. The virus benefits at both the ecological and evolutionary levels. See Chapter 9 for a more detailed discussion of gene transfer. Of course, the virus also hurts its host.

It is now clear that viruses promote the diversity of prokaryotes by the mechanism of kill the winner and the transfer of genetic material (Kirchman 2013), as well as by regulating their populations and providing nutrients through the viral shunt. They do the same for phytoplankton by the same four mechanisms.

8.5 BACTERIOPHAGE RELEASE NUTRIENTS IN DEEP-SEA SEDIMENTS

Scientists found that bacteriophage are important in deep-sea sediments (Danovaro et al., 2005). They sampled sediments from scores of ocean sites, at depths from 183 meters (about 600.4 feet) to 4,603 meters (about 15,101.7 feet). This is the dark, nutrient-poor, inhospitable deep sea, the last great unexplored ecosystem on Earth. Below 1,000 meters (about 3,280.8 feet), prokaryotes make up 90% of the total biomass. They found an astonishingly high viral count. There are between 5×10^{12} and 10^{13} phage per square meter in deep-sea ecosystems. Viruses infect and kill huge numbers of prokaryotic cells in the ocean depths, liberating and recycling nutrients, and making them available to other prokaryotes, keeping the system functioning, and maintaining prokaryotic diversity. Phage make the system healthier, much like the viral killing of prokaryotes in the viral shunt at the sea's surface, and like forest fires liberating forest nutrients. Thus, viral infections in the ocean depths are responsible for transferring a tremendous amount of nutrients from living forms into organic matter dissolved in seawater and detritus, which then is used by other living prokaryotes. This explains why nutrients are recycled so efficiently and at such high rates in deep-sea sediments. Although the deep sea receives a great quantity of nutrients raining down on it from the shallower water above it, nutrients are often limiting there, and the nutrients liberated by viruses are a necessary, substantial addition to the nutrients of the deep-sea ecosystem. Surprisingly, the nutrients released from the death of prokaryotes by viruses are significantly more important to deep-sea ecosystems than the nutrient input from the sea's shallow waters above (ibid.). This system is self-sustaining, allowing the ocean depths to overcome severe nutrient limitations. Since carbon in the deep sea can be locked up and not available to life for ages, the liberating of carbon and making it available to organisms by viral attacks is crucial to the health and diversity of life in the deep sea (ibid.). Small deep-sea eukaryotes can utilize the nutrients that result from viral attacks on prokaryotes at the sea's bottom. The nutrients then flow up the food webs, as larger organisms eat smaller ones. The researchers of this study also found a great portion of viral-liberated deep-sea nutrients are carried by animals to the shallower sea, and contribute substantially to the ecosystems in the photic zone and which have phytoplankton at the base of their food webs. This is a surprising discovery, that deep-sea viruses help feed the ecosystems of the upper ocean, sustaining large fish and invertebrates, including seafood people eat. And it means the photosynthetic zone near the sea surface and the deep sea feed nutrients to each other, in a recycling system in which each ecosystem keeps the other healthy and diverse. Amazingly, the two ecosystems, far apart, are interconnected and aid each other.

Viral attack of prokaryotes in the deep ocean closes the loop in the carbon cycle, recycling carbon back into the photic zone. This is of crucial importance, because it keeps a substantial amount of carbon from being buried under the seafloor, keeping it available to the biosphere, without sending a significant amount into the air. It conserves carbon in the biological system without significantly heating the Earth, helping solve the atmospheric temperature versus carbon conservation dilemma (see Chapter 2). And deep-sea ecosystems are very important in the biogeochemical cycles of the Earth, including the carbon, nitrogen, sulfur, and phosphorus cycles. Viruses do this in a way that aids life, making more nutrients available to organisms.

8.6 VIRAL INFECTION REDUCES PHYTOPLANKTON NUTRIENT QUALITY, CONTRADICTING THE ABH

Phytoplankton infected with a virus tend to contain a smaller proportion of polyunsaturated fatty acids, which are needed by all organisms. This makes them a poorer quality food, which is passed up entire food webs. Yet, for reasons that are not clear, tiny organisms tend to prefer to graze on infected phytoplankton cells. This results in poorer nutrition to marine food webs. Though viruses greatly increase nutrients and biodiversity overall, this aspect of them decreases availability of some nutrients, the fatty acids, to entire food webs. This apparent contradiction to the ABH is not well understood, and needs to be clarified with further research.

8.7 COEVOLUTION OF PHAGE WITH THE BACTERIA GROUP CALLED SAR11 IS IMPORTANT IN NUTRIENT CYCLES

The SAR11 group is also called the Pelagibacteraceae, and consists of very small bacteria. They have the smallest known genetic structure of any independent cell on earth. They are very abundant (Morris et al., 2002; Zhao et al., 2013), comprising about a third of all cells on the sea's surface. They can live in water too low in nutrients for most other organisms to survive. They consume carbon, and produce the bulk of the CO_2 that phytoplankton need for photosynthesis. They also recycle organic matter, making it available to phytoplankton, indirectly feeding them. They therefore help phytoplankton tremendously, and are of crucial importance to the health and diversity of the entire biosphere, since phytoplankton are the base of the sea's food webs, the planet's major oxygen producers, and key players in carbon sequestration. Pelagiphages are strange-looking phages that attack SAR11, killing millions of them every second. They are estimated to be among the most abundant viral groups in the sea (Zhao et al., 2013). These two groups have coevolved for hundreds of millions of years, both evolving rapidly. Because of their huge numbers, these bacteria and phages are assumed to affect the carbon cycle and climate immensely. The carbon cycle affects all life on the planet. So the effects on life of these two groups is tremendous, with great implications for the ABH. However, the nature and strength of the effects are not known at this writing. SAR11 profoundly aids life on Earth, and greatly increases biodiversity. It is thought to have a large effect on atmospheric CO_2, but whether it sequesters it and how much are not known. Whether its phages enhance or hurt biodiversity, and by what amount, is unknown. One would assume that Pelagiphages have the same effect as the viral shunt, releasing nutrients and making them available to other prokaryotes and phytoplankton by killing SAR11 bacteria, increasing productivity and at least maintaining diversity, if not increasing it. It is highly likely that regulation of SAR11 populations by Pelagiphages is crucial to ocean ecosystem health, for if these bacteria had population explosions, the consequences for them and their ecosystems could be catastrophic. But these hypotheses have yet to be tested.

8.8 VIRUSES CAUSED MACROEVOLUTIONARY INNOVATIONS AND TREMENDOUS DIVERSIFICATION IN PROKARYOTES AND EUKARYOTIC PHYTOPLANKTON

Viruses seem to have been instrumental in many major macroevolutionary breakthroughs in both prokaryotes and phytoplankton. There is no fossil record of viruses and a limited one for prokaryotes and phytoplankton, but much can be learned from sequencing and other studies of all three groups. When viruses carry out kill the winner, they cause there to be a higher number of species of prokaryotes; that is, of bacteria and archaea. If there are more species, there are more opportunities for macroevolutionary breakthroughs. The more species an ecosystem has, the more likely it will have one or more species that has or have an adaptive evolutionary breakthrough. This principle is especially true with viruses and prokaryotes, since the number of types of viruses and the number of species of their prokaryotic prey, are generally much higher than those of animals and plants. If we add to this the fact that viruses also cause a great deal of adaptive transfer of genes to prokaryotes, we can see that viruses potentially caused a great number of macroevolutionary breakthroughs in procaryotes, and could cause more in the future. One can infer that this occurred throughout life's history on Earth, starting from the beginning. Cellular life started as prokaryotes at least 3.5 bya. It is not known exactly when viruses first appeared, but it was likely approximately when the first cells appeared. Unicellular eukaryotes first appeared between 1.6 and 2.1 bya, and the first multicellular organisms between 1.6 and 1 bya. So, from 3.5 bya or earlier to between 1.6 and 2.1 bya, marine phages were attacking prokaryotes exclusively, and from between 1.6 and 2.1 bya to between 1.6 and 1 bya, viruses were attacking only prokaryotes and unicellular eukaryotes. So, the period that the only cellular organisms that existed were unicellular, and hence only unicellular forms were killed by viruses, probably at very high rates, was from at least 3.5 bya to between 1.6 and 1 bya, which is much more than half of the time life has been on Earth. Such a long time yields a high probability of some adaptive genetic changes, resulting in key innovations and evolutionary breakthroughs. It is reasonable to assume the dynamics of viruses and their hosts were generally the same then as now. Thus, kill the winner, and the killing of numbers close to 20 to 40% of prokaryotes per day with the resulting fertilization of the sea, likely applied in the sea ever since the appearance of the first viruses.

This would have maintained if not increased the diversity of unicellular organisms. Nutrient released by the viral shunt allowed bacterial and phytoplankton populations to increase to very high levels at times. High populations increased the probability of genetic changes leading to key innovations. Major adaptive genetic changes in any given unicellular organism would be rare, but the large populations and long time spans maximized the odds that they would occur in some of the unicellular organisms at some times. Viruses aided the evolution of prokaryotes and phytoplankton by fertilizing the sea, maintaining diversity of unicellular species via kill the winner, transferring genes between individual organisms and between species, and acting as agents of natural selection. Unicellular organisms, including those of different species, exchanged DNA segments without viral aid as well. Marine viruses acquired genes from prokaryotes and other viruses, including those of other types, through gene transfer as well. They also had high populations, so high chances of adaptive mutations. The long time span also increased the chances of adaptive innovative mutations for viruses. So viruses also had evolutionary innovations and diversified in this symbiotic genetic coevolution. The great fluctuations in population size between very small and very large in both viruses and their hosts caused by kill the winner and standard virus-host population fluctuations could have led to evolutionary innovations if one accepts certain evolutionary models (Ohta, 1972, for example). Small populations can result in evolutionary changes by random factors, rather than natural selection (this is called genetic drift). Large populations allow for more mutations, more gene transfer, and more novel forms from more different genetic recombination events. Genetic recombination is the exchange of genetic material between organisms of the same species, generally as a result of sex. Today viruses promote microbial diversity via kill the winner and gene transfer (Weinbauer and Rassoulzadegan, 2004), and there is empirical evidence that viruses promote microbial diversity in marine sediments. There is no reason to think this was not the case throughout early life's history. Selection by the viruses would have made them active agents of bacterial macroevolutionary breakthroughs for another reason in addition to all the others in this paragraph. There are bacteria called mutators, which have enhanced mutation rates, are frequently found in natural and laboratory populations, and are often associated with clinical infections. They have increased adaptability to environmental challenges, but are subject to the accumulation of deleterious mutations. Rapidly changing environments, among the most common being those created by antagonistic coevolution with phages, probably drive the long-term maintenance of high bacterial mutation rates. Pal et al. (2007) showed 25% of laboratory populations of the bacterium *Pseudomonas fluorescens* coevolving with phages evolved 10- to 100-fold increases in mutation rates owing to mutations in repair genes, in fewer than 200 bacterial generations. These are mutator populations. Control populations that had no phage had no significant changes in mutation rates. Mutator populations also had a higher likelihood of sending their phage populations extinct. Since phage are everywhere, the authors conclude they may play a key role in the evolution of bacterial mutation rates, increasing them. This would increase bacterial population variability, potentially increasing speciation, evolutionary innovation, and macroevolution. *Selection provided by phage likely increased bacterial mutation rates and variability, and hence probably increased evolutionary rates and evolutionary innovations.* It is possible that the increased mutation rates resulting from coevolution with viruses contributed to the evolution of such innovations as photosynthesis and cellular respiration in bacteria. This is merely a hypothesis of mine at this point. It needs to be investigated. Macroevolutionary breakthroughs in bacteria as a result of any of the mechanisms discussed here resulting from coevolution with phage would then be followed by adaptive radiation into many new bacterial species. All of the general ideas discussed in this paragraph apply to archaea and phytoplankton as well.

All this likely caused a positive feedback loop of coevolution and diversification of microorganisms and viruses, each evolving and diversifying and causing the other to evolve and diversify. The new microbes created new niches for new viruses to exploit, causing them to diversify, while new viruses may have promoted the evolution of microorganisms to new forms by changing selective pressures on them, causing fluctuations in their populations, and providing them with segments of DNA. So the evolution and diversification of unicellular organisms was promoted by viruses ever since the origin of viruses, and the evolution and diversification of viruses was promoted by unicellular organisms, also since viruses originated.

Many evolutionary breakthroughs and key innovations appeared in both prokaryotes and their phages as a result of both antagonistic and symbiotic genetic coevolution between the two. There is good evidence that separate parts of the oxygenic photosynthetic system evolved and existed in different species of bacteria when photosynthesis was evolving, and perhaps viruses brought these parts together into two species of bacteria by gene transfer to create two photosynthetic systems. Photosystem I was in one species of bacterium, and photosystem II was in the other. Neither could carry out the full oxygenic photosynthesis that creates oxygen. A phage may have transferred genes between these bacteria to bring photosystems I and photosystem II together into one species of cyanobacteria, and the efficient photosynthesis that creates oxygen and carbohydrate that green plants use today was born (see Chapter 9). This is incredible. *Viruses might have helped transfer genes to allow the evolution of oxygenic photosynthesis!*

The CRISPR-Cas system is a prokaryotic immune system that confers resistance to foreign genetic elements, such as not only phages, but other entities as well. It could not have evolved in prokaryotes except as a result of antagonistic coevolution with bacteriophage. Phages were also likely a factor in the evolution of bacterial cell walls because they likely evolved at least partly to keep phage out of bacterial cells. Viruses might have played a role in the increase in size of prokaryotic genomes (Koonin, 2016).

Innovations that viruses may have obtained as a result of antagonistic coevolution with their hosts include, but are not limited to, many types of capsids, many forms of replication, lysogeny, the ability to incorporate host genes, and other mechanisms of overcoming host defenses.

The current state of evidence supports the idea that viruses promoted evolutionary innovations and diversification in prokaryotes since viruses first appeared, and this includes perhaps the major evolutionary advances leading to the eukaryotic cell, and that viruses and cells underwent symbiotic coevolution ever since the two have existed. The same is likely true for viruses and phytoplankton and other eukaryotic plankton ever since these eukaryotes first appeared. Viruses likewise had evolutionary innovations and diversification as a result of this symbiotic coevolution. Some of the innovations on both sides were probably large, adaptive macroevolutionary breakthroughs.

8.9 VIRUSES WERE IMPORTANT IN PROMOTING MAJOR MACROEVOLUTIONARY BREAKTHROUGHS IN CELLULAR ORGANISMS

There are now several hypotheses that put viruses at center stage for major early evolutionary transitions and breakthroughs (Claverie, 2006, and references therein). Viral agents play critical roles within the roots and stem of the tree of life (Villarreal and Witzany, 2010). It has been suggested that viral proteins replaced cellular proteins during the early evolution of the nuclear DNA replication system in prokaryotes (ibid.), and that viruses played a key role in other major evolutionary transitions, such as the origin of DNA replication mechanisms (ibid.). Furthermore, viruses might have helped the evolution of cells, the origin of DNA as the genetic material, increase in prokaryote and eukaryote genome size and complexity, the evolution of the eukaryotic nucleus, and more generally, increase in the diversity of organisms of life (Koonin, 2016).

Szathmáry and Demeter (1987) and Szathmáry and Maynard Smith (1997) showed that parasites, including viruses, promote natural selection between entire groups of cooperating replicating segments of DNA, which can lead to the evolution of cells. Of course, other factors than viruses helped the evolution of cells as well. Takeuchi et al. (2011) showed that a population of replicating segments of DNA that were in a uniform area, before cells evolved, is prone to be doomed to collapse from parasites, and the only way for such a population to be stable is compartmentalization, which would occur by the appearance of cells. The replicating DNA segments would be separated into cells, and the cell membranes would separate the DNA segments from each other, making it more difficult for a parasite such as a virus to spread from one DNA segment to another. Thus, this slows the rate of transfer of parasites to new hosts by creating barriers to their spread, and creates conditions for cooperation between replicating segments of DNA of the host because they would be in the same cell and they would all benefit from its survival (Higgs and Lehman, 2015; Shay et al., 2015). Thus, viruses seem to have played a key role in the evolution of the first cells.

When only precells on their way to evolving into cells and selfish replicating genetic segments existed, these two entities would have coevolved in an arms race of defense and counter-defense, leading to an increase in complexity of both hosts and selfish elements (Koonin, 2016). This caused an increase in genome size of both precell hosts and selfish elements. The selfish elements caused selection for cellularity, since cells with impermeable membranes created a strong barrier to the spread of any form of selfish parasite. This put selective pressure on selfish genetic elements to radiate into a diversity of types varying in a spectrum in cost to the cell and how easily they are transmitted from cell to cell (Jalasvuori and Koonin, 2015; Koonin and Starocadomskyy, 2016). On one extreme of the continuum are types that are hardly at all deleterious to the host and do not transmit to new hosts efficiently, such as circular pieces of DNA called plasmids that contain genes and exist in bacterial cells. At the other extreme are viruses that kill the host and transmit to a new host very efficiently. The host cells created a selective pressure for the most selfish elements to evolve mechanisms to combat host defenses. Often these mechanisms would increase the complexity of the selfish genetic element; an example is the protein coat of viruses. Thus, the diversity and sometimes complexity of the selfish genetic elements, which included viruses, increased due to selection pressure from their host cells. Viruses and other selfish elements put selective pressure on hosts to increase in complexity to produce more enzymes and other defenses against the parasites, and provided them with material to increase their genome size by transferring genes to them. The diversity of parasites increased host diversity, especially the diversity of mechanisms of defending against them. Thus, coevolution of parasites, including viruses, and their hosts increased diversity and complexity of both in a positive feedback loop.

Takahashi and Marmur (1963) posited that viruses caused DNA to replace RNA as the genetic material. RNA acts as both genetic material and enzymes, and preceded DNA in the chemical evolution of life (Atkins et al., 2010; Bernhardt, 2012). It was almost certainly the first genetic material, and was replaced by DNA, which is now the genetic material in all organisms except some viruses. Some think RNA viruses infecting cells that had RNA as their genetic material in the early evolution of life evolved the ability to convert their RNA to DNA to avoid being degraded by enzymes of the cell that attacked the RNA of the virus to defend the cell from it. RNA viruses would have hypothetically evolved the two major pathways to convert RNA to DNA. In this hypothesis, archaea, bacteria, and eukaryotes each obtained their DNA from a different virus (Takahashi and Marmur, 1963). Mathematical modeling indicates that defense against parasites like viruses is most effective when the genetic replicator (such as DNA and RNA) and catalyst (protein that helps the genetic replicator duplicate itself) are separated into two different types of molecules with specialized systems, such as DNA and proteins (Takeuchi et al., 2011). RNA can act as both replicator and catalyst, causing the two to not be separated, but one molecule. But DNA cannot be the catalyst, so with DNA as the genetic material, there are two separate molecules: DNA as replicator, protein as catalyst. So defense against viruses is more effective. *This indicates that the coevolution of prokaryotes and phage aided the evolution of DNA as the genetic material.* However, this was also selected for because it conferred other advantages. For example, the replacement of RNA by DNA was favored by natural selection because DNA has greater stability, and double-strandedness, which allows the ability for repair (ibid.). (Some RNA is double-stranded, but some is single-stranded.) Thus, phage likely played an important role, but other selective factors were probably involved. The replacement of RNA with DNA allowed the evolution of larger, more complex genomes (ibid.) because of the increased stability of DNA compared to RNA. The stability and larger, more complex genomes allowed for diversification into the great many species that evolved and the great evolutionary breakthroughs and increases in complexity that occurred since the appearance of DNA as the genetic material. RNA genomes would likely not have been stable enough to allow these macroevolutionary advances, this tremendous complexity, or these larger genomes, to occur. Thus, RNA genomes would not have allowed the opportunity for the great diversifications that evolutionary breakthroughs and complexity cause to occur. Thus, the fact that the replacement of RNA by DNA occurred partly because of the coevolution of prokaryotes and phage supports the ABH.

The origin of the sophisticated type of sex done by eukaryotes at the cellular level called meiosis, in which the chromosomes line up at the center of the cell and move apart, is a key part of the major transition that led to the origin of eukaryotes. Koonin (2016) hypothesized that mobile genetic elements that could move from one part of the DNA to another were fundamental in the evolution of meiosis. This would allow the possibility that viruses, having similarities to mobile genetic elements, played a key role in the evolution of meiosis and eukaryotic sex. Significantly, this is further supported by the observation that genes that were transferred to animals by viruses play a role in the fusion of the sperm and egg in animals, an observation that in itself is an example of the importance of viruses in the evolution of animals. The idea that viruses were important in the evolution of meiosis and sex in eukaryotes is supported by evidence, but needs further research to prove it definitively.

Martin and Koonin (2006) proposed that viruses selected for the nucleus of the cell. The nucleus is the part of the cell that contains and encloses the DNA. It only exists in eukaryotes; prokaryotic cells do not have a nucleus. Koonin and Krupovic (2020) showed that large bacteriophages protect themselves from enzymes made by the cell to attack them by encapsulating their genome inside the cell in a shell that is like a "nucleus" and that is impenetrable by some of these enzymes. This suggests a possible viral origin of the cell's nucleus.

The viral eukaryogenesis (VE) hypothesis proposes that the nucleus of the cells of eukaryotes evolved from a complex DNA virus (Bell, 2001; Bell, 2006). It proposes that eukaryotes are a composite that evolved from three unrelated organisms: a virus that evolved into the nucleus, a cell from the archaea that evolved into the cytoplasm (the cytoplasm is a thick solution that fills the cell and contains all the parts of the cell, such as the nucleus and mitochondria), and a bacterium that evolved into the mitochondrion. (In cells that have chloroplasts and carry out photosynthesis, there would be a fourth organism unrelated to the other three: a cyanobacterium that became the chloroplast that performs photosynthesis for such photosynthetic organisms as plants.) It has been extended to propose that the eukaryotic cell cycle (the life cycle of the cell that includes the cell dividing to form two cells) arose as a consequence of the nucleus having come from a lysogenic DNA virus (ibid.). There are a number of observations supporting the VE hypothesis. The DNA of both eukaryotes and viruses has specialized areas at the end of it. All eukaryotic DNA is linear; viral DNA may be linear or circular; all bacteria DNA is circular. Thus, the DNA of eukaryotes is like the DNA of some viruses, but the DNA of bacteria is not similar to eukaryotic DNA. The apparatus that caps the messenger molecule occurs in both eukaryotes and viral genus *Mimivirus*, but not in the archaea group that is the nearest relative of eukaryotes (Bell, 2020). The giant phage 201Phi2-1 forms a structure similar to a nucleus when it infects a bacterial cell (Chaikeeratisak et al, 2017). This nucleus segregates proteins according to function (ibid.). Bell (2001) further proposed phagocytosis and other processes based on the fusion of membranes are derived from viral membrane fusion processes, and coevolved with the nucleus. Phagocytosis is the process by which certain cells in the body eat bacteria or other foreign substances that invade an animal and cause disease. The VE hypothesis is not universally accepted and is controversial at this time.

Viruses seemingly contributed to the evolution of mitochondria (Filée and Forterre, 2005). Early in eukaryote evolution, the typical molecule that makes RNA from DNA (RNA polymerase) in bacteria and had multiple subunits was apparently replaced by a phage molecule with this function that had only one subunit and that now produces RNA from DNA in mitochondria in most eukaryotes (Shutt and Gray, 2006). This replacement is apparently a key component of the streamlining of the mitochondrion's information processing system. And recent findings indicate that SAR11, a bacterium discussed earlier in this chapter, may be one of the bacteria that coevolved into the mitochondrion (Thrash et al., 2011). (More than one species of bacterium evolved into mitochondria in different eukaryotes.) SAR11 has coevolved with its pelagiphage parasite for eons, no doubt evolving and changing substantially and continually over the ages, as a result of an arms race with its phage parasite. The phage could have been the selective pressure for some (not all) of the SAR11 in the sea to take refuge in another microbe, which was one of the archaea, to escape viral infection. Then the symbiotic coevolution between SAR11 and its host cell that led to the mitochondrion could have proceeded. The great numbers of SAR11 and its short generation time provided a larger probability for the needed mutation(s) for an SAR11 to be able to survive in a bacterial host, allowing it to evolve into the mitochondrion. The hypothesis that phages were instrumental in the genesis of the mitochondrion from SAR11 is thus feasible, but requires more evidence.

Witzany (2008) posited a viral origin of telomeres and telomerase, key elements of chromosomal structure and cell replication in eukaryotes. Telomeres are tiny caps at the ends of the DNA molecules that make up chromosomes. They protect the ends of chromosomes and keep them from fraying or sticking to each other, functioning much like the plastic tips on the ends of shoelaces. Chromosomes are structures in the nucleus of the cells of eukaryotes that contain the DNA as well as proteins associated with the DNA. Telomerase is an enzyme that adds DNA to telomeres to maintain their length. Without telomerase, telomeres and the chromosomes that they are on would gradually get shorter and shorter. This would be very bad for health and eventually lethal for the organism that contains them.

Thus, *viruses were major contributors to the origin of eukaryotic cells!* This was a major evolutionary breakthrough. Eukaryotic cells are about 1,000 times bigger by volume than prokaryotic ones; contain an inner membrane system with organelles such as the mitochondrion and the nucleus surrounded by membranes; and are fully compartmentalized so that movement of molecules between compartments is greatly limited. Prokaryotic cells do not have these features. There would have been spectacular diversification into *very many* new species after the appearance of the eukaryotic cell. The many new species provided habitats and niches for other species that evolved to use these habitats and niches, further increasing the number of species on Earth.

Forterre (2006) claims ancient viruses were at the origin of the three domains of life. These are the three biggest groups of life other than viruses. The three domains of life are archaea, bacteria, and eukaryotes. This would be amazing if viruses gave rise to the three major groups of cellular life! His hypothesis is called the "3 RNA cells, 3 DNA viruses" hypothesis. It may explain why there are three discrete lineages (the domains of life) and types of modern cells, rather than a gradual gradation from, say, bacteria to eukaryotes; why there are three canonical kinds of protein-making factories (called ribosomes), one in each domain; and why there are key differences between the systems used to replicate DNA between eukaryotes and archaea. However, there may be other explanations for these observations. For example, ancestors that were intermediate between archaea and bacteria or between eukaryotes and bacteria or archaea could have gone extinct long ago without leaving a fossil record. So the exciting idea has evidence supporting it, but is not yet proven.

Viruses of the genus *Mimivirus* are the largest known DNA viruses. One *Mimivirus* has 1.2 million base pairs (base pairs are the units that make up a DNA molecule), which is huge for a virus, and 911 genes that code for proteins, also a tremendous number for a virus, and has many types of genes never before encountered in a virus, including genes for the manufacture of proteins that were previously thought to be exclusive to cellular organisms (Claverie et al., 2006). The authors propose such viruses are closer to small parasitic prokaryotes than to regular viruses and *represent a fourth domain of life. Fourth domain! That is the largest category for classifying a group of life that there is! Then the four domains of life would be the Mimivirus, bacteria, archaea, and eukaryotes*. I will add that in my opinion, in reality, viruses already make up another, separate domain of life themselves. Thus, if *Mimivirus* is its own separate domain, I would say that there are five domains of life. This means viruses were involved in the origin of four or five domains of life.

There is clear evidence that components of viruses themselves actually became essential elements of the immunity that organisms use against viruses, that viruses provided crucial and coordinated features of the immune system in all organisms, from simple prokaryotes to complex eukaryotes (reviewed in Villarreal, 2011).

Viruses were important in the evolution from one-celled forms of life to many-celled forms. Multicellularity evolved multiple times in eukaryotes, and four times in complex ones with specialized cells and tissues: in animals, plants, fungi, and brown algae. It evolved in some prokaryotes, including some cyanobacteria (Schirrmeister, 2011; Kaiser, 2013).

Multicellularity is a huge evolutionary breakthrough, transition, and key innovation. It allows an increase in size, complexity, lifespan, and the kinds of actions that can be done. It opened up a new adaptive zone, and hence a myriad of new niches. As such, as soon as it evolved, multicellular forms underwent tremendous diversification into many new species. These new multicellular species provided many new habitats and niches that other unicellular and multicellular species evolved to use, further increasing the number of species on Earth.

Programmed cell death (PCD) is part of and fundamentally linked to multicellularity. It is the death of a cell as a result of events inside of a cell, programmed by the cell. It is generally good for the organism. It controls pathogens and cell proliferation, and is part of normal development in multicellular eukaryotes. For example, fingers and toes cannot develop properly in a human embryo without PCD. Iranzo et al. (2014) have a mathematical model showing that for unicellular organisms, including bacteria, under high virus load and imperfect immunity, joint evolution of the aggregation of cells and PCD is the optimal evolutionary strategy. Infected bacterial cells evolved to altruistically cause their own deaths, preventing other cells near them from being infected. This benefits the group—the aggregate of cells—at the expense of the altruistic cells that commit suicide. A model based on natural selection between entire groups with the groups with the most altruists cooperating and so defeating selfish groups that do not have internal cooperation cannot work because cheating cells that do not commit suicide would be selected for over the altruists that undergo PCD. The Iranzo et al. model works because many bacterial colonies are genetically identical clones, so the cells around the suicidal cell are genetically identical to it. So the altruistic cells that commit suicide are passing on their genes because they are helping cells next to them that are genetically exactly the same as they are. This is very Darwinian. It is natural selection favoring the cell that commits suicide because it helps other cells that share its genes. So it passes on its genes to the next generation through the cells next to it that have exactly the same genes as it does, and it so is an evolutionary success. This phenomenon of an organism passing on its genes by helping relatives that share its genes is called kin selection and will be discussed in more detail in Chapter 10. The model of the linking of PCD and kin selection requires multicellularity. A single cell cannot help the development of a group of cells that are separate from it. On the other hand, if the cell is part of a multicellular organism and commits suicide, it helps, for example, fingers separate and develop into a hand. Because the organism is multicellular, the cellular suicide helps nearby cells and the organism it is part of. Therefore, the model shows that multicellularity and PCD are fundamentally linked, that one cannot evolve without the other. Multicellularity and PCD are also linked to kin selection, and the system only works if the cells are closely related genetically, as in cells of the same organism. The model also implies that viruses were important in the evolution of multicellularity because PCD is linked to it, and PCD is crucial in fighting viruses. In fact, Quistad et al. (2017) proposed a model whereby PCD mechanisms were utilized to remove cells infected by viruses in the early evolution of animals, with the viruses acting as agents of natural selection for programmed cell death. As in Iranzo et al.'s model, cells infected by viruses sacrifice themselves by PCD, killing the viruses that are infecting them, thus helping control the viral infection.

Importantly, altruistic suicidal PCD promotes the survival of the genes of dying cells by kin selection in the eukaryotic social amoeba, *Dictylostelium* (Gilbert et al., 2007; Kuzdzal-Fick, et al., 2011; Ho et al., 2013). This is significant because this amoeba is sometimes unicellular, but can become multicellular under certain conditions. Yet, the role of viruses has yet to be shown directly in the evolution of eukaryotic multicellularity. However, Segovia et al. (2003) showed PCD in a unicellular alga, suggesting in their view that PCD did not originally evolve in animals. From this work, they hypothesize that ancient viruses transferred critical DNA segments of the PCD pathway to the DNA of early unicellular eukaryotes in the Precambrian Ocean before the evolution of multicellular organisms, and they were then passed on to multicellular plant and animal lines as they evolved. The Precambrian supereon spans from about 4.6 bya to about 541 mya.

Thus, it appears that kin selection and viruses were both essential in the evolution of both prokaryotic and eukaryotic PCD, which is important in the evolution of multicellularity. Coevolution between viruses and their hosts was almost certainly one of the key factors behind the evolution of simple multicellularity, possibly on more than one occasion. This prepared the ground for the emergence of advanced multicellular organisms.

Viruses also played a major role in the evolution of advanced multicellularity in complex animals from unicellular eukaryotes. Viruses may have brought cells together to better spread to uninfected cells. (This would apply to their role in the evolution of multicellularity in prokaryotes as well.) Viruses transferred genes from simpler organisms that are critical in the development and specialization of multicellular tissues and organs in higher animals. Two types of viral proteins that were important in the evolution of multicellularity have been identified thus far. First are syncytins, proteins discussed later in this chapter in the discussion on the origin of live birth in mammals. The other is EFF1, which helps form the skin of a species of roundworm. Significantly for the evolution of multicellularity, EFF1 does the work of linking one cell to another in viral infections (Jamin et al., 2014). *All known molecules that fuse cells together came from viruses, suggesting viruses were crucial to the evolution of multicellularity and to the communication systems between cells that enabled multicellularity. Genes in animals that came from viruses apparently give cells the ability to grow into tissues and organs, and even undergo sexual reproduction. Without these genes, animals could not have evolved beyond simple blobs of cells* (Slezak, 2014). *Without the capacity for cellular fusion, colonies of cells could have formed, but nothing even as complex as simple sponges could have existed.* However, recall Alegado et al.'s work with the choanoflagellate indicating that bacteria were instrumental in the evolution of multicellularity in animals. It is likely viruses were important in the evolution of multicellularity in most lineages and bacteria were important in one or more lineage. Additionally, there are several other hypotheses and factors unrelated to viruses that might be important to the evolution of animals.

Quistad et al. (2017) argued that viruses of eukaryotes coevolved with their hosts. Those viruses that provided the last common ancestor of higher animals with immune protection were selected for by the host in a group selection coevolution model. Host cells with pathogenic viruses were removed by PCD. Group selection is selection on groups instead of individuals and is discussed in more detail in Chapter 9.

Viruses played key roles in major evolutionary breakthroughs in animals as well. Proteins that are important in the regulating of animal development called hedgehog proteins were obtained from a peculiar class of parasites called inteins that combine jumping segments of DNA with shuffling parts of proteins around (protein splicing), implying a viral or other parasitic, selfish DNA contribution to the origin of animals and their development (Koonin, 2016).

Endogenous retroviruses (ERVs) are viral elements and a type of jumping gene in the genome that resemble viruses and can be derived from them. ERVs sometimes get packaged and moved within the genome with the result that they end up serving a vital role in gene expression and regulation (Khodosevich et al., 2002; Kim et al., 2004). Apparently, some of them evolved from viruses, although some researchers think some viruses evolved from ERVs, since ERVs can mutate and become separate from their cell, or even pathogenic (Cotton, 2001). It is not known which genes and regions of the genome that regulate genes (turning them on and off) came from ERVs that were originally viruses as opposed to originating in their host genome. I am using my best judgement in listing those I think originated from viruses in the paragraphs below. I may have erred in categorizing the ones I list here as of viral origin. However, even those ERVs that originated from their host's genome and not from viruses demonstrate the tendency of the genome, and hence life, to promote evolutionary breakthroughs and diversification, and thus the validity of the ABH.

ERVs are prominent in the genomes of vertebrates that have jaws. They are estimated to be 7 to 8% of the human genome (Lander et al., 2001; Belshaw et al., 2004; Nelson et al., 2004; Jern and Coffin, 2008). ERVs may play a big role in many evolutionary events. The human genome project found several thousand ERVs, which are organized into 24 families (Villarreal, 2001). Since ERVs are in the great majority of cases bad for and disruptive to the host genome and the host, mutations in host genes that inactivate or delete ERVs tend to be favored by natural selection. And ERVs and their hosts have undergone antagonistic coevolution. This resulted in the duplication and increase in the number of host repressor genes that repress ERVs, stopping them from expressing themselves and from making proteins. In some cases, these repressor genes have adapted to repressing host genes. When repressor genes turn off the right genes in an organism, adaptive evolution and even large evolutionary changes can occur. This is because repressing host genes can cause new, adaptive functions. For example, the gene that is turned off might stop growth of a tissue, so turning it off would make the tissue grow larger, possibly leading to a new, adaptive function.

ERVs have been important in evolution, and are used a great deal in the regulation and expression of genes (Khodosevich et al., 2002; Kim et al., 2004). They are often repurposed by the host to act as regulators of host genes, turning them on or off, often producing variants that regulate the development of specific tissues. ERVs are sometimes repurposed by the host to serve new functions for the host, especially in reproduction and development. ERVs can recombine, exchanging DNA with each other, if they are of like kind, sometimes generating novel genetic variants and genetic variation. This has the potential to sometimes create new functions and even evolutionary breakthroughs in their host organism.

The gene AMY1C codes for an enzyme important in digestion in humans called amylase. Amylase is in the saliva of humans and some other mammals, where it begins the chemical process of digestion. It breaks down starch into sugars. The gene for it has a complete ERV sequence in its control region that turns it on and off, showing that this ERV was repurposed for regulation of this digestive enzyme. A long terminal repeat is a pair of long identical repeating segments of DNA in eukaryotic genomes on either end of a jumping gene, ERV, or similar DNA entity. It is usually a DNA segment that proteins bind with to turn genes on, with 64% of them expressed in reproductive organs (Romanish et al., 2007). A long terminal repeat associated with the EVR in the amylase gene regulates the expression of amylase in the salivary glands (Ting et al., 1992). The primary switch that turns on and off an enzyme that participates in the production of bile acid is from a long terminal repeat of an ERV (van de Lagemaat et al., 2003). Bile acids facilitate digestion of dietary fats and oils.

The insertion of a single long terminal repeat of an ERV may have resurrected a gene important in the human immune system (Jern and Coffin, 2008). The direct placement and integration of an EVR into a gene helped build a receptor in cells in humans. A receptor is a structure made up of protein on the surface of a cell that binds with a chemical that transmits a signal to the cell. This can cause functions to be carried out in the body to help the organism. The receptor and chemical that binds to it fit like a lock and key. The human leptin hormone receptor is a receptor for the hormone leptin. Leptin suppresses hunger and causes people to want to stop eating when they are full. An endogenous retrovirus inserted itself into human DNA and helps us control our appetites by regulating leptin production. An upstream long terminal repeat of an EVR helps turn on a protein that humans need to utilize fats (Oliver and Greene, 2011). Note that the EVRs discussed in this paragraph likely came from viruses that once infected humans and/or human ancestors, but which were repurposed by humans or human ancestors to take on the important functions listed here.

There is a long terminal repeat that has become a gene for an important enzyme used to produce estrogen that is expressed in the brain and reproductive organs of most mammals (van de Lagemaat et al., 2003). A variant of this in primates is responsible for controlling estrogen levels during pregnancy (ibid.).

Human endogenous retroviruses (HERVs) are remnants of past retroviral infections of the germline cells of human ancestors (Hughes and coffin, 2001, and references therein). They make up as much as 8% of the human genome and possibly outnumber genes (ibid.). In one type of these, the HERV-K (HML-2) group, at least 16% of its elements have seemingly undergone apparent rearrangements that may have resulted in large-scale deletions, duplications, and reshuffling of chromosomes during the evolution of the human genome (ibid.). They have therefore greatly increased variability and affected evolution in the human line.

There are many visible differences between humans and chimpanzees, our closest relatives. Obvious differences include greater amount of hair on the bodies of chimpanzees and the fact that humans walk on two legs, not four. Many of these differences are a result of differences in the regulation of genes shared by these species. These genes are basically the same in humans and chimpanzees, but they are regulated differently. Some are turned on in humans and turned off in chimpanzees, or vice versa. Human-specific endogenous retroviruses and their solitary long terminal repeats are probable candidates for the regulators of these genes. Thus, human-specific endogenous retroviruses are likely responsible for many of the differences between humans and chimpanzees that are visible to us.

In humans, the major histocompatibity complex (MHC) genes are very important to the immune system. They have a high number of ERVs (Gogvadze, et al., 2009). Sznarkowska t al. (2020) stated: "Viral-derived elements … are extremely lense in the MHC region, accounting for flexible expression of the … genes and adjusting the level of immune response o the environmental stimuli … . The complex regulatory network behind MHC expression is largely controlled by virus-derived elements." Thus, regulation of the MHC region s the result of the repurposing of viruses by their hosts. In other words, a major part of our immune system is derived from viruses. Also in humans, human-specific endogenous retroviruses provide protection against a wide range of antigens. An antigen is a toxin or other foreign substance which induces an immune response in the body. Viruses are antigen. Human-specific endogenous retroviruses may have given humans an immune system superior to other primates (Dawkins et al, 1999). Also, a gene in humans called MER41. AIM2 that regulates other genes is derived from an ERV. It is necessary for an inflammatory response to infection (Chuong et al., 2013). An inflammatory response is needed in the body to repair tissues that were injured by bacteria, trauma, toxins, heat, or any other cause. So viruses were very important in the evolution of the human immune system.

Viruses that are specific to each species make up a substantial proportion of all organisms' genomes, including the portion that regulates genes (Sznarkowska et al., 2020, and references therein). Endogenous retroviruses and defective versions of them that were derived from them have been shown to actively shape the structure and regulation of basically all species' genomes (ibid.). This means the structure and regulation of the genomes of virtually all species has been and still is actively shaped by viruses.

Most scientists believe at this time in the "host comes first" perspective, which hypothesizes that functional and defective ERV's are viral remnants that were co-opted (repurposed) by their hosts as both structural genes (regular, active genes that code for proteins) and regulatory genes that turn structural genes on and off. Careful studies of DNA have shown that elements in host DNA that do not code for any functional proteins and that are derived from viruses that have become defective and that no longer function as viruses can function as elements that regulate genes (ibid.). As a result, Villarrreal (2016) and later Sznarkowska et al. (2020) argued that immunity and identity in cellular organisms was probably largely created by viruses. They argued that the persistent relationship between the host and its virus is a driving force of the evolution of both host and virus, and leads to the formation of the new virus-host entity that acts and evolves as one unit, having its own novel genetic identity and an enhanced immunity. The virus or virus-derived element integrates into the host genome and stays there without harming the host, causing stable persistence. They call this an "addiction module," in which the lethal, cell-killing effect of the virus is stopped by an antikilling feature of the same virus. This is accompanied by the evolution of immune responses of the host to the virus. And persistent DNA viruses, ancient ERVs, and their derivatives are the source of RNA molecules that regulate genes. As the parasitic virus-derived elements interact with each other and readily disperse over the genome, they establish and extend gene regulatory networks. Regulatory RNAs encoded by these viral-derived elements often control cell proliferation and immunity as they are used in the establishment of persistence of the virus via the addiction module. Remember that if cell proliferation is not controlled, cancer can result, and if immunity is not controlled, allergies can occur. Thus, viral-derived elements are important to an organism's health. *Much of the regulation of the genes of an organism is accomplished by elements derived from viruses!*

Viruses were key in the evolution of reproduction and live birth in mammals that bear live young. Not all mammals bear live young. There is a group of mammals called monotremes, which lay eggs. The duck-billed platypus is a monotreme. It is believed that the ancestors of modern live-bearing mammals evolved from an infection with a virus, perhaps improving the ability of the fetus to survive the immune system of the mother (Li et al., 2012). During pregnancy in live-bearing mammals, ERVs are activated and produced in high quantities during the implantation of the embryo. They are hypothesized to carry out the function of suppressing the mother's immune system, protecting the embryo from the immune system of the mother. The ERVS keep the mother's immune system from attacking the embryo or rejecting it and causing a spontaneous abortion. The ERVs apparently can suppress the mother's immune system because they are derived from infectious viruses that evolved the ability to suppress their host's immune system to defend themselves from it.

ERVs play an important role in the development of the placenta of mammals. The placenta is crucial in keeping the fetus alive in live-bearing mammals. It provides oxygen and nutrients to the growing fetus and removes waste products from its blood. Most cells have just one nucleus. Remember the nucleus is the organelle of the cell that contains the DNA. In mammals, proteins called syncytins are responsible for the formation and function of cells with many nuclei called syncytiotrophoblasts (Black et al., 2010), which maintain nutrient exchange between the mother and the fetus. And they separate the fetus from the mother's immune system (ibid.). Retroviral fusogenic envelope proteins cause cells to fuse together to form the syncytiotrophoblasts. They were originally used by viruses to enter cells to parasitize and infect them. It is probable that their ability to fuse cells together helped viruses get from one cell to another to infect uninfected cells when they reached a high population in the cell they were already infecting. At any rate, viruses and these fusogenic proteins that they make were co-opted by the mammals that they once infected, and the fusogenic proteins are now used to fuse cells together and form the cells with many nuclei that maintain the exchange of nutrients between the mother and her fetus, and keep the fetus separate from the mother's immune system in placental mammals. Thus, the mother's immune system does not attack the fetus. Also, the placental syncytium is necessary in order to limit the exchange of migratory cells between the developing embryo and the body of the mother, for such cells would harm the fetus. It is thought that live-bearing mammals evolved from an ancestor that was infected with a virus, in time giving the ancestor these key adaptations (Villarreal, 2016). We can see that *viruses played a key role in the evolution of live-bearing mammals and their placentas*. Also see Chapter 9 on the role of jumping DNA segments in the evolution of the uterus.

The gene called Arc is essential for long-term memory and learning in mammals. This gene codes for a protein that forms capsids much like the capsids viruses have that surround their genetic material. The capsids made by the proteins coded for by the Arc gene carry Arc RNA molecules between nerve cells. The RNA molecule is needed for nerve cell function. Arc is a repurposed, now inactive, jumping gene. Originally the viral protein formed a capsid to protect and transport the genetic material of the virus. The viral protein capsid was converted to carrying Arc RNA, and became necessary for long-term memory and learning in the vertebrate brain about 375 mya (Day and Shepherd, 2015; Pastuzyn et al., 2018). Fish do not have Arc, but amphibians do. So Arc first appeared in vertebrates in amphibians. Amphibians first appeared about 370 mya. So here we have a virus that infected vertebrates and then was co-opted by vertebrates and put to use for learning and memory.

It is important to remember that *all of the ERVs discussed above were probably originally viruses that were repurposed by their host to provide an important adaptive function for the host.*

Viruses may have provided eukaryotes with histones. Histones are especially important proteins that play key roles in numerous functions of eukaryotic cells. They act as spools around which DNA winds to create compact structural units called nucleosomes. DNA must be compact in order to regulate its production of RNA and to separate chromosomes during cell division. Histones prevent DNA from becoming tangled and protect it from damage. Histones prevent DNA from clumping together and so keep it functional. Histones allow unpacking of the appropriate amount of DNA when needed. They help regulate gene expression and play an important role in DNA replication. In eukaryotes, they are the linchpins for complex life and countless evolutionary innovations.

Valencia-Sánchez et al. (2021) presented evidence that histones in eukaryotes came from viruses. Giant viruses of the Marseilleviridae family have histones that are related and surprisingly similar to the four main eukaryotic histones. The only difference is that in the viral versions, the histones that normally pair up in eukaryotes are already fused together into doublets. The fused viral histones form structures that are virtually identical to typical eukaryotic nucleosomes. But this evidence is not universally accepted. The idea requires that the fused viral histones became unfused in eukaryotes because there are no fused doublets of the same histones as those fused in viruses in living eukaryotes. And inferring evolutionary relationships of viruses from viral sequences has the problem that viruses mutate readily. And it is possible that histones in eukaryotes evolved from histones in archaea. Finally, it is possible that an early species that had histone doublets was an ancestor to both the giant viruses and eukaryotes, and could have passed the proteins along to both lines of organisms a very long time ago. However, Čuboňová et al. (2005) presented evidence that histones evolved in the common ancestor of archaea and eukaryotes. This leaves open the possibility that this common ancestor acquired its histones from a virus.

Brunk and Martin (2019) suggested that histones may have played a major role in the evolution of the mitochondrion into a cellular organelle by providing the basic functional unit with which eukaryotes organize DNA into nucleosomes, exert control of gene expression, transcribe genes with CCAAT-box promoters, and a cell cycle with condensed chromosomes. This would mean histones and the mitochondrion coevolved, and viruses played an indirect role in the origin of the mitochondrion and eukaryotic cell.

In addition to the viruses and former viruses discussed above, bornaviruses integrate into mammalian and invertebrate DNA, and are a source of genetic variability and innovation. These viruses infect only nerve cells, infecting their host's brain. *Two bornaviruses in humans code for proteins* (Feschotte, 2010; Horie et al., 2013). Whether they helped with the evolution of intelligence remains to be determined.

8.10 VIRUSES ARE OFTEN SYMBIOTIC WITH ANIMALS, AND GREATLY INFLUENCE THE EVOLUTION OF PROTEINS IN CELLULAR ORGANISMS: A LARGE PORTION OF EUKARYOTIC GENOMES ARE DERIVED FROM VIRUSES

Viruses are frequently symbiotic with animals. After the active infection is over, some viruses enter a dormant state known as latency, which persists for the life of the host. During latency, the individual is well and has no symptoms, but is at risk for subsequent viral reactivation and disease. Barton et al. (2007) showed that either mouse gamma herpes virus 68 or mouse cytomegalovirus, which are genetically very similar to the human pathogens Epstein-Barr virus and human cytomegalovirus, respectively, confer immunological protection from the disease-causing bacteria *Listeria monocytogenes* and *Yersinia pestis*, when these viruses are latent in mice. *Yersinia pestis* causes plague in humans and caused the Black Death, the deadliest pandemic in recorded history. Both bacteria are pathogenic to mice. This is adaptive for the virus because it protects its host, which is its habitat. This is symbiosis because the virus protects its host from pathogenic bacteria and the host gives the virus a habitat.

The bacteriophage T4 lives in the mucous of essentially all animal species. They are symbiotic with the animals that have the mucous that they live in: they receive a place to stay where they have easy access to their bacterial hosts, and they protect their animal hosts from the disease-causing bacteria that they attack and kill. The phage is also protecting its habitat. The surfaces of mucous in the animals and the phage coevolved to maintain the ability of the phage to adhere to these surfaces (Barr et al., 2013).

The influence of viruses on the evolution and structure of proteins that occur in cells is astounding. Enard et al. (2016) found an astonishing *30% of all protein adaptations since the divergence of humans from chimpanzees have been driven by viruses, and that not only immune system proteins, but essentially all proteins that contact viruses adapt to them. There is as much adaptation to viruses by proteins outside the immune system as in it. This analysis of 1,300 mammalian proteins showed that adaptive changes have occurred three times as frequently in proteins that interact with viruses as those that do not. The vast majority of cell machinery was altered through coevolution with viruses.* Since viruses take over and control most of the machinery of their host cells, this finding is not really surprising. It is reasonable that this adaptation of proteins to viruses applies to all cellular species from prokaryotes to humans. This may account for some of the evolution of different mechanisms to perform identical functions in related species, such as the making of membranes and DNA replication. These mechanisms may have changed and adapted to viruses in different ways in different species.

About 8% of the human genome originates from viruses, not from our vertebrate ancestors. These were viruses that attacked us from the outside. Another 40% is made up of repetitive segments of DNA that are also thought to have a viral origin. So it is thought that about 48% of our genome ultimately comes from viruses. For comparison, only about 2% of our genome codes for proteins. So it is thought that the human genome has up to 24 times as much DNA that it acquired from viruses as DNA that codes for proteins! Over 100,000 DNA sequences in the human genome are derived from ERVs. As many as 85 genes derived from viruses might be functioning in the brain and during pregnancy in humans. It is reasonable to assume that similar percentages hold for other eukaryotes, and that all higher organisms are chimeras and a large part of them is viruses.

Viruses are the most abundant, diverse, and variable form of life. They are key in biogeochemical cycles, fertilize the ocean, maintain the species diversity of their hosts by regulating their populations and the mechanism of kill the winner, regulate climate while conserving carbon, and caused many key macroevolutionary innovations and incredible diversification in all major groups of cellular organisms, while their hosts drove them to key innovations and great diversification in a symbiotic yet simultaneously antagonistic coevolutionary process. Viruses may be the life form most iconic of the ABH/Pachamama Hypothesis.

REFERENCES

Angly, F. E., et al. (7 Nov., 2006). The marine viriomes of four oceanic regions. *PLOS Biology* 4: e368.18. https://doi.org/10.1371/journal.pbio.0040368.

An-Yi, T., et al. (Dec., 2014). Importance of the viral shunt in nitrogen cycling in *Synechococcus* spp. growth in subtropical western Pacific coastal waters. *Terrestrial, Atmospheric & Oceanic Sciences* 25 (6): 839–46. doi: 10.3319/TAO.2014.06.11.01(Oc).

Atkins, J. F., Gesteland, R. F., & Cech, T. R. (eds). (2010). *RNA Worlds: From Life's Origins to Diversity in Gene Regulation*, 3rd ed. Cold Spring Harbor Lab. Press, Cold Spring Harbor, NY.

Barr, J. J., et al. (25 June, 2013). Bacteriophage adhering to mucus provide a non-host-derived immunity. *PNAS USA* 110 (26): 10771–6. https://doi.org/10.1073/pnas.1305923110.

Barton, E. S., et al. (17 May, 2007). Herpesvirus latency confers symbiotic protection from bacterial infection. *Nature* 447: 326–9.

Bell, P. J. L. (2001). Viral Eukaryogenesis: Was the ancestor of the nucleus a complex DNA virus? *Journ. of Mol. Evol.* 53: 251–6.

Bell, P. J. L. (7 Nov., 2006). Sex and the eukaryotic cell cycle is consistent with a viral ancestry for the eukaryotic nucleus. *Journ. of Theoretical Biol.* 243 (1): 54–63. https://doi.org/10.1016/j.jtbi.2006.05.015.

Bell, P. J. L. (Nov., 2020). Evidence supporting a viral origin of the eukaryotic nucleus. *Virus Research* 289: 198168. https://doi.org/10.1016/j.virusres.2020.198168.

Belshaw R., et al. (April, 2004). Long-term reinfection of the human genome by endogenous retroviruses. *PNAS USA* 101 (14): 4894–9. doi: 10.1073/pnas.0307800101.

Bernhardt, H. S. (2012). The RNA world hypothesis: the worst theory of the early evolution of life (except for all the others). *Biol. Direct.* 7: 23. doi: 10.1186/1745-6150-7-23.

Black, S. G., et al. (2010). Endogenous retroviruses in trophoblast differentiation and placental development. *Amer. Journ. of Reproductive Immunol.* 64 (4): 255–64. PMID 20528833. doi: 10.1111/j.1600-0897.2010.00860.x.

Bratbak, G., et al. (1993). Viral mortality of the marine alga *Emiliania huxleyi* (Haptophyceae) and termination of algal blooms. *Marine Ecol. Prog. Ser.* 93: 39–48.

Breitbart, M., et al. (2002). Genomic analysis of uncultured marine viral communities. *PNAS USA* 99:14250–5.

Breitbart, M., et al. (2004). Diversity and population structure of a near-shore marine-sediment viral community. *Proc. Royal Soc. Lond.* B 271: 565–74.

Brunk, C. F. & Martin, W. F. (Aug., 2019). Archaeal histone contributions to the origin of eukaryotes. *Royal Trends in Microbiology* 27 (8): 703–14. https://doi.org/10.1016/j.tim.2019.04.002.

Brussard, C. P. D. (2004). Viral control of phytoplankton populations--a review. *Journ. Eukaryot. Microbiol.* 51: 125–38.

Bull, J. J., Millstein, J., Orcutt, J., & Wichman, H. A. (2006). Evolutionary feedback mediated through population density, illustrated with viruses in chemostats. *Amer. Nat.* 167: E39–E51.

Canchaya, C., Proux, C., Fournous, G., Bruttin, A., & Brussow, H. (2003). Prophage genomics. *Microbiol. and Molecular Biol. Reviews* 67 (2): 238–76.

Chaikeeratisak, V., et al. (13 Jan., 2017). Assembly of a nucleus-like structure during viral replication in bacteria. *Science* 355 (6321): 194–7. doi: 10.1126/science.aal2130. PMC 6028185. PMID 28082593.

Chen, F., et al. (2009). Diverse and dynamic populations of cyanobacterial podoviruses in the Chesapeake Bay unveiled through DNA polymerase gene sequences. *Environ. Microbiol.* 11: 2884–92.

Chuong, E. B., et al. (3 Mar., 2016). Regulatory evolution of innate immunity through co-option of endogenous retroviruses. *Science* 351 (6277): 1083–7. doi: 10.1126/science.aad5497.

Claverie, J.-M. (2006). Viruses take center stage in cellular evolution. *Genome Biol.* 7 (6): 110. doi: 10.1186/gb-2006-7-6-110.

Claverie, J.-M., et al. (April, 2006). Mimivirus and the emerging concept of "giant" virus. *Virus Research* 117 (1): 133–44. https://doi.org/10.1016/j.virusres.2006.01.008.

Cotton, J. (2001). Retroviruses from retrotransposons. *Genome Biol.* 2 (2): 6.

Čuboňová, L., et al. (1 Aug., 2005). Histones in *Crenarchaea. Journal of Bacteriology* 187 (15). doi: https://doi.org/10.1128/JB.187.15.5482-5485.2005.

Culley, A. I., Lang, A. S., & Suttle, C. A. (2006). Metagenomic analysis of coastal RNA virus communities. *Science* 312: 1795–8.

Danovaro, R., Corinaldesi, C., Dell'anno, A., Fabiano, M. & Corselli, C. (2005). Viruses, prokaryotes and DNA in the sediments of a deep-hypersaline anoxic basin (DHAB) of the Mediterranean Sea. *Environ. Microbiol.* 7: 586–92.

Danovaro, R., Corinaldesi, C., Dell'anno, A., Fuhrman, J. A., Middelburg, J. J., Noble, R. T., & Suttle, C. A. (Nov., 2011). Marine viruses and global climate change. *FEMS Microbiol. Rev.* 35 (6): 993–1034.

Dawkins, R., et al. (1999). Genomics of the major histocompatibility complex: haplotypes, duplication, retroviruses and disease. *Immunol. Rev.* 167: 275–304. doi: 10.1111/j.1600-065X.1999.tb01399.x.

Day, C. & Shepherd, J. D. (2015). Arc: building a bridge from viruses to memory. *Biochem. Journ.* 469 (1): e1–e3. https://doi.org/10.1042/BJ20150487.

Edwards, R. A. & Rohwer, F. (2005). Viral metagenomics. *Nature Rev. Microbiol.* 3: 504–10.

Enard, D., et al. (2016). Viruses are a dominant driver of protein adaptation in mammals. *eLife* 5: e12469. doi: 10.7554/eLife.12469.

Evans, C., et al. (27 Oct., 2009). Viral-mediated lysis of microbes and carbon release in the sub-Antarctic and Polar Frontal zones of the Australian Southern Ocean. *Environ. Microbiol.* 11 (11): 2924–34. https://doi.org/10.1111/j.1462-2920.2009.02050.x.

Feschotte, C. (2010). Bornavirus enters the genome. *Nature* 463 (7277): 39. doi: 10.1038/463039a.

Filée, J. & Forterre, P. (2005). Viral proteins functioning in organelles: a cryptic origin? *Trends Microbiol.* 13: 510–3. doi: 10.1016/j.tim.2005.08.012.

Filée, J., et al. (May, 2003). The role played by viruses in the evolution of their hosts: a view based on informational protein phylogenies. *Research in Microbiol.* 154 (4): 237–43. https://doi.org/10.1016/S0923-2508(03)00066-4.

Fischer, M. G. & Suttle, C. A. (8 April, 2011). A virophage at the origin of large DNA transposons. *Science* 332 (6026): 231–4. doi: 10.1126/science.1199412. www.sciencemag.org/content/332/6026/231.short.

Forterre, P. (7 March, 2006). Three RNA cells for ribosomal lineages and three DNA viruses to replicate their genomes: A hypothesis for the origin of cellular domain. *PNAS USA* 103 (10): 3669–74. https://doi.org/10.1073/pnas.0510333103.

Fuhrman, J. A. (10 June, 1999). Marine viruses and their biogeochemical and ecological effects. *Nature* 399 (6736): 541–8.

Fuhrman, J. A. & Schwalbach, M. S. (2003). Viral influence on aquatic bacterial communities. *Biol. Bull.* 204: 192–5.

Gilbert, O. M., et al. (2007). High relatedness maintains multicellular cooperation in a social amoeba by controlling cheater mutants. *PNAS USA* 104: 8913–7. doi: 10.1073/pnas.0702723104.

Gobler, C. J., Hutchins, D. A., Fisher, N. S., Cosper, E. M., & Sanudo-Wilhelmy, S. (1997). Release and bioavailability of C, N, P, Se, and Fe following viral lysis of a marine Chrysophyte. *Limnol. Oceanogr.* 42: 1492–504.

Gogvadze, E., et al. (2009). Human-specific modulation of transcriptional activity provided by endogenous retroviral insertions. *Journ. Virol.* 83 (12): 6098–105. doi: 10.1128/JVI.00123-09.

Hendrix, R. W. (2009). Jumbo bacteriophages. *Curr. Topics Microbiol. Immunol.* 328: 229–40. PMID: 19216440.

Hendrix, R. W. (Oct., 2003). Bacteriophage genomics. *Curr. Opin. Microbiol.* 6 (5): 506–11.

Hendrix, R. W., Lawrence, J. F., Hatfull, G. F., & Casjens, S. (1 Nov., 2000). The origins and ongoing evolution of viruses. *Trends in Microbiol.* 8 (11): 504–8.

Hendrix, R. W., Smith, M. C., Burns, R. N., Ford, M. E., & Hatful, G. F. (2 Mar., 1999). Evolutionary relationships among diverse bacteriophages and prophages: All the world's a phage. *PNAS USA* 96 (5): 2192–7.

Higgs, P. G. & Lehman, N. (2015). The RNA world: molecular cooperation at the origins of life. *Nat. Rev. Genet.* 16: 7–17. doi: 10.1038/nrg3841.

Ho, H.-I., et al. (2013). Kin recognition protects cooperators against cheaters. *Curr. Biol.* 23: 1590–5. doi: 10.1016/j.cub.2013.06.049.

Horie, M. (19 Sept., 2013). Comprehensive analysis of endogenous bornavirus-like elements in eukaryote genomes. *Philosoph. Trans. of the Royal Soc. B. Biol. Sciences* 3681 (1626). https://doi.org/10.1098/rstb.2012.0499.

Hughes, J. & Coffin, J. M. (2001). Evidence for genomic rearrangements mediated by human endogenous retroviruses during primate evolution. *Nat. Genet.* 29 (4): 487–92. doi: 10.1038/ng775.

Iranzo J., et al. (1 Oct., 2014). Virus-host arms race at the joint origin of multicellularity and programmed cell death. *Cell Cycle* 13: 3083–8. doi: 10.4161/15384101.2014.949496.

Jalasvuori, M. & Koonin, E. V. (2015). Classification of prokaryotic genetic replicators: between selfishness and altruism. *Annu. N. Y. Acad. Sci.* 1341: 96–105. doi: 10.1111/nyas.12696.

Jamin, M., et al. (10 April, 2014). Structural basis of eukaryotic cell-cell fusion. *Cell* 157 (2): 407–19.

Jern, P. & Coffin, J. M. (2008). Effects of retroviruses on host genome function. *Annu. Rev. Genet.* 42: 709–32. doi: 10.1146/annurev.genet.42.110807.091501.

Jiao, N., et al. (2010). Microbial production of recalcitrant dissolved organic matter: long-term carbon storage in the global ocean. *Nat. Rev. Microbiol.* 8: 593–99.

Jover, L. F., et al. (2014). The elemental composition of virus particles: implications for marine biogeochemical cycles. *Nature Reviews Microbiol.* 12: 519–28. doi: 10.1038/nrmicro3289.

Kaiser, D. (2013). Are *Myxobacteria* intelligent? *Front. Microbiol.* 4: 335. doi: 10.3389/fmicb.2013.00335.

Khodosevich, K., et al. (Oct., 2002). Endogenous retroviruses and human evolution. *Comparative and Functional Genomics* 3 (6): 494–8. doi: 10.1002/cfg.216.

Kim, F. J., et al. (2004). Emergence of vertebrate retroviruses and envelope capture. *Virology* 318 (1): 183–91. doi: 10.1016/j.virol.2003.09.026.

Kimura, M., et al. (Feb., 2008). Ecology of viruses in soils: Past, present and future perspectives. *Soil Science & Plant Nutrition* 54 (1): 1–32. doi: 10.1111/j.1747-0765.2007.00197.x.

Kirchman, D. L. (2013). Microbial oceanography: Killers of the winners. *Nature* 494 (7437): 320–1. doi: 10.1038/nature11951.

Koonin, E. V. (19 Aug., 2016). Viruses and mobile elements as drivers of evolutionary transitions. *Philosoph. Trans. of the Royal Soc. B: Biol. Sciences* 371 (1701). This article is part of the themed issue, "The major synthetic evolutionary transitions." doi: 10.1098/rstb.2015.0442.

Koonin, E. V. & Krupovic, M. (2020). Phages build anti-defence barriers. *Nature Microbiol.* 5: 8–9.

Koonin, V. & Starokadomskyy, P. (Oct., 2016). Are viruses alive? The replicator paradigm sheds decisive light on an old but misguided question. *Studies in History and Philosophy of Science Part C: Studies in History and Philosophy of Biological and Biomedical Sciences.* Volume 59, pp. 125–34. Elsevier, Amsterdam, the Netherlands. https://doi.org/10.1016/j.shpsc.2016.02.016G.

Kuzdzal-Fick, J. J., et al. (2011). High relatedness is necessary and sufficient to maintain multicellularity in *Dictyostelium*. *Science* 334: 1548–51. doi: 10.1126/science.1213272.

Labonté, J. M. & Suttle, C. A. (2013). Previously unknown and highly divergent ssDNA viruses populate the oceans. *ISME Journ.* 7: 2169–77. doi: 10.1038/ismej.2013.110. www.nature.com/ismej/journal/v7/n11/full/ismej2013110a.html.

Lander, E. S., et al. (International Human Genome Sequencing Constortium). (15 Feb., 2001). Initial sequencing and analysis of the human genome. *Nature* 409 (6822): 860–921. doi: 10.1038/35057062.

Lennon, J. T. & Martiny, J. B. H. (2008). Rapid evolution buffers ecosystem impacts of viruses in a microbial food web. *Ecol. Lett.* 11: 1178–88.

Li, J., et al. (March, 2012). Mouse endogenous retroviruses can trigger premature transcriptional termination at a distance. *Genome Res.* 22 (5): 870–84. doi: 10.1101/gr.130740.111. PMC 3337433. PMID 22367191.

Lindell, D., et al. (2004). Transfer of photosynthesis genes to and from *Prochlorococcus* viruses. *PNAS USA* 101: 11013–8.

Liu, H., et al. (Nov., 2010). Widespread horizontal gene transfer from double-stranded RNA viruses to eukaryotic nuclear genomes. *Journ. Virol.* 84 (22): 11876–87. doi: 10.1128/JVI.00955-10.

Luna, G. M., Corinaldesi, C., Rastelli, E., & Danovaro, R. (5 Oct., 2013). Patterns and drivers of bacterial α- and β-diversity across vertical profiles from surface to subsurface sediments. *Environ. Microbiol. Rep.* 5 (5): 731–9. doi: 10.1111/1758-2229.12075. Epub 11 July, 2013.

Martin, W. & Koonin, E. V. (2006). Introns and the origin of nucleus-cytosol compartmentalization. *Nature* 440: 41–5.

Middelboe, M., et al. (June, 1996). Effects of viruses on nutrient turnover and growth efficiency of noninfected marine bacterioplankton. *Appl. Environ. Microbiol.* 62 (6): 1991–7.

Mizuuchi, R., Furubayashi, T., & Ichihashi, N. (2022). Evolutionary transition from a single RNA replicator to a multiple replicator network. *Nature Communications* 13 (1460). https://doi.org/10.1038/s41467-022-29113-x. www.nature.com/naturecommunications.

Mojica, K. D. A. & Brussaard, C. P. D. (2015). *The Viral Shunt in a Stratified Northeast Atlantic Ocean*, Ch. 7, pp. 207–22. Univ. of Amsterdam, Amsterdam, the Netherlands.

Morris, R. M., Rappé, M. S., Connon, S. A., et al. (2002). SAR11 clade dominates ocean surface bacterioplankton communities. *Nature* 420 (6917): 806–10. doi: 10.1038/nature01240.

Nagasaki, K., et al. (2005). Previously unknown virus infects marine diatom. *Appl. Environ. Microbiol.* 71: 3528–35.

Nelson, P. N., et al. (Oct., 2004). Human endogenous retroviruses: Transposable elements with potential? *Clinical and Experimental Immunol.* 138 (1): 1–9. doi: 10.1111/j.1365-2249. 2004.02592.x.

Ohta, T. (Dec., 1972). Population size and rate of evolution. *Journ. Mol. Evol.* 1 (4): 305–14.

Oliver, K. R. & Greene, W. K. (2011). Mobile DNA and the TE-Thrust hypothesis: supporting evidence from the primates. *Mob. DNA* 2 (1): 8. doi: 10.1186/1759-8753-2-8.

Pal, C., Maciá, M. D., Oliver, A., Schachar, I., & Buckling, A. (13 Dec., 2007). Coevolution with viruses drives the evolution of bacterial mutation rates. *Nature* 450: 1079–81. doi: 10.1038/nature06350.

Pastuzyn, E. D., et al. (11 Jan., 2018). The neuronal gene Arc encodes a repurposed retrotransposon Gag protein that mediates intercellular RNA transfer. *Cell* 172 (1–2): 275–88. e18. https://doi.org/10.1016/j.cell.2017.12.024.

Pedulla, M. L., et al. (18 April, 2003). Origins of highly mosaic Mycobacteriophage genomes. *Cell* 113: 171–2.

Poorvin, L., Rinta-Kanto, J. M., Hutchins, D. A., & Wilhelm, S. W. (2004). Viral release of iron and its bioavailability to marine plankton. *Limnol. Oceanogr.* 49: 1734–41.

Pope, W. H., et al. (April, 2015). Whole genome comparison of a large collection of mycobacteriophages reveals a continuum of phage genetic diversity. *eLife* 2015;4:e06416. doi: 10.7554/eLife.06416.

Proctor, L. M. & Fuhrman, J. A. (1990). Viral mortality of marine bacteria and cyanobacteria. *Nature* 343: 60–2.

Quistad, S. D., et al. (2017). Viruses and the origin of microbiome selection and immunity. *ISME Journ.* 11: 835–40. doi: 10.1038/ismej.2016.182. Pub. online 16 Dec., 2016.

Rodriguez-Brito, B. (2010). Viral and microbial community dynamics in four aquatic environments. *ISME Journ.* 4: 739–51. doi: 10.1038/ismej.2010.1.

Romanish, M. T., et al. (2007). Repeated recruitment of LTR retrotransposons as promoters by the anti-apoptotic locus NAIP during mammalian evolution. *PLOS Genet.* 3 (1): e10. doi: 10.1371/journal.pgen.0030010.

Salcher, M. M., et al. (2011). Seasonal bloom dynamics and ecophysiology of the freshwater sister clade of SAR11 bacteria 'that rule the waves' (LD12). *ISME Journ.* 5: 1242–52. doi: 10.1038/ismej.2011.8. Pub. online 17 March, 2011.

Schirrmeister, B. E. (2011). The origin of multicellularity in cyanobacteria. *BMC Evol. Biol.* 11: 45. doi: 10.1186/1471-2148-11-45.

Seaborg, D. (2022). *How Life Increases Biodiversity: An Autocatalytic Hypothesis.* CRC Press/Taylor & Francis. Boca Raton, FL; London, UK; New York, NY.

Segovia, M., et al. (May, 2003). Cell death in the unicellular chlorophyte *Dunaliella tertiolecta*. A hypothesis on the evolution of apoptosis in higher plants and metazoans. *Plant Physiol.* 132 (1): 99–105. doi: https://doi.org/10.1104/pp.102.017129.

Shay, J. A., et al. (2015). The origin and spread of a cooperative replicase in a prebiotic chemical system. *Journ. Theor. Biol.* 364: 249–59. doi: 10.1016/j.jtbi.2014.09.019.

Shelford, E. J. & Suttle, C. A. (2018). Virus mediated transfer of nitrogen from heterotrophic bacteria to phytoplankton. *Biogeosci.* 15: 809–19. https://doi.org/10.5194/bg-15-809-2018.

Shelford, E. J., et al. (2012). Virus driven nitrogen cycling enhances phytoplankton growth. *Aquat. Microb. Ecol.* 66: 41–6. doi: 10.3354/ame01553.

Short, S. M. & Short, C. M. (2009). Quantitative PCR reveals transient and persistent algal viruses in Lake Ontario, Canada. *Environ. Microbiol.* 11: 2639–48.

Shutt, T. E. & Gray, M. W. (Feb., 2006). Bacteriophage origins of mitochondrial replication and transcription proteins. *Trends Genet.* 22: 90–5. doi: 10.1016/j.tig.2005.11.007.

Slezak, M. (1 Mar., 2014). No viruses? No skin or bones either. *New Scientist* 2958: 16.

Sullivan, M.B., et al. (2006). Prevalence and evolution of core photosystem II genes in marine cyanobacterial viruses and their hosts. *PLOS Biol.* 4: e234.

Suttle, C. A. (15 Sept., 2005). Viruses in the sea. *Nature* 437: 356–61. doi: 10.1038/nature04160.

Suttle, C. A. (Oct., 2007). Marine viruses—Major players in the global ecosystem. *Nature Reviews Microbiol.* 5 (10): 801–12. doi: 10.1038/nrmicro1750. PMID 17853907.

Szathmáry, E. & Demeter, L. (21 Oct., 1987). Group selection of early replicators and the origin of life. *Journ. Theor. Biol.* 128: 463–86. doi: 10.1016/S0022-5193(87)80191-1.

Szathmáry, E. & Maynard Smith, J. (1997). From replicators to reproducers: the first major transitions leading to life. *Journ. Theor. Biol.* 187: 555–71. doi: 10.1006/jtbi.1996.0389).

Sznarkowska, A., et al. (4 May, 2020). MHC Class I regulation: The origin perspective. *Cancer* 12 (1155): 1–23. www.mdpi.com/journal/cancers.

Takahashi, I. & Marmur, J. (1963). Replacement of thymidylic acid by deoxyuridylic acid in the deoxyribonucleic acid of a transducing phage for *Bacillus subtilis. Nature* 197: 794–5.

Takeuchi, N., Hogeweg, P., & Koonin, E. V. (24 Mar., 2011). On the origin of DNA genomes: evolution of the division of labor between template and catalyst in model replicator systems. *PLOS Comput. Biol.* 7 (3): e1002024. doi: 10.1371/journal.pcbi.1002024.

Thompson, L. R. (2010). Auxiliary metabolic genes in viruses infecting marine cyanobacteria. Ph. D. thesis. Available in the Inst. Archives and Special Collections, Dept. of Biol., M. I. T. http://hdl.handle.net/1721.1/57562.

Thrash, J. C., et al. (June, 2011). Phylogenomic evidence for a common ancestor of mitochondria and the SAR11 clade. *Scientific Reports* 1, Article number: 13. doi: 10.1038/srep00013.

Ting, C. N., et al. (1992). Endogenous retroviral sequences are required for tissue-specific expression of a human salivary amylase gene. *Genes & Dev.* 6 (8): 1457–65. doi: 10.1101/gad.6.8.1457.

Tomaru, Y., et al. (2004). Quantitative and qualitative impacts of viral infection on a *Heterosigma akashiwo* (Raphidophyceae) bloom in Hiroshima Bay, Japan. *Aquat. Microb. Ecol.* 34: 227–38.

Valencia-Sánchez, M. I., Abini-Agbomson, S., Wang, M., et al. (29 April, 2021). The structure of a virus-encoded nucleosome. *Nature Structural & Molecular Biol.* 28: 413–7. https://doi.org/10.1038/s41594-021-00585-7.

van de Lagemaat, N., et al. (Oct., 2003). Transposable elements in mammals promote regulatory variation and diversification of genes with specialized functions. *Trends Genet.* 19 (10): 530–6. doi: 10.1016/j.tig.2003.08.004.

Venter, J. C., et al. (2004). Environmental genome shotgun sequencing of the Sargasso Sea. *Science* 304: 66–74.

Villarreal, L. P. (1997). On viruses, sex, and motherhood. *Journ. Virol.* 71 (2): 859–65. PMC 191132. PMID 8995601.

Villarreal, L. P. (Oct., 2001). Persisting viruses could play role in driving host evolution. *ASM News* (Amer. Soc. for Microbiol.) 67: 501–7.

Villarreal, L. P. (20 Oct., 2011). Viral ancestors of antiviral systems. *Viruses* 3 (10): 1933–58. https://doi.org/10.3390/v3101933.

Villarreal, L. P. (June, 2016). Persistent virus and addiction modules: An engine of symbiosis. *Curr. Opinion in Microbiol.* 31: 70–9. https://doi.org/10.1016/j.mib.2016.03.005.

Villarreal, L. P. & Witzany, G. (21 Feb., 2010). Viruses are essential agents within the roots and stem of the tree of life. *Journ. of Theoretical Biol.* 262 (4): 698–710. https://doi.org/10.1016/j.jtbi.2009.10.014.

Weinbauer, M. G. & Rassoulzadegan, F. (2004). Are viruses driving microbial diversification and diversity? *Environ. Microbiol.* 6: 1–11. doi: 10.1046/j.1462-2920.2003.00539.x.

Weinbauer, M. G. (2004). Ecology of prokaryotic viruses. *FEMS Microbiology Reviews* 28: 127–81.

Weinbauer, M. G., et al. (2011). *Synechococcus* growth in the ocean may depend on the lysis of heterotrophic bacteria. *Journ. Plankton Res.* 33: 1465–76. doi: 10.1111/j.1574-6976.2010.00258.x.

Whitman, W. B., Coleman, D. C. & Wiebe, W. J. (1998). Procaryotes: The unseen majority. *PNAS USA* 95 (12): 6578–83.

Wigington, C. H., et al. (2016). Re-examination of the relationship between marine virus and microbial cell abundances. *Nature Microbiol.* 1 (15024). doi: 10.1038/nmicrobiol.2015.24. PMID 27572161.

Wilhelm, S. W. & Suttle, C. A. (1 Oct., 1999). Viruses and nutrient cycles in the sea: Viruses play critical roles in the structure and function of aquatic food webs. *BioScience* 49 (10): 781–8. https://doi.org/10.2307/1313569.

Williamson, K. E., Radosevich, M., Smith, D. W., & Wommack, K. E. (Oct., 2007). Incidence of lysogeny within temperate and extreme soil environments. *Environ. Microbiol.* 9 (10): 2563–74.

Winget, D. M. & Wommack, K. E. (2009). Diel and daily fluctuations in virioplankton production in coastal ecosystems. *Environ. Microbiol.* 11: 2904–14.

Witzany, G. (2008). The viral origins of telomeres and telomerases and their important role in eukaryogenesis and genome maintenance. *Biosemiotics*. doi: 10.1007/s12304-008-9018-0.

Wommack, K. E., et al. (1999). Population dynamics of Chesapeake Bay virioplankton: Total-community analysis by pulsed-field gel electrophoresis. *App. Environ. Microb.* 65: 231–40.

Zeidner, G., et al. (Published online 23 June, 2005). Potential photosynthesis gene swapping between *Prochlorococcus* and *Synechococcus* via viral intermediates. *Environ. Microbiol.* doi: 10.1111/j.1462-2920.2005.00833.x).

Zhao, Y., et al. (2013). Abundant SAR11 viruses in the ocean. *Nature* 494: 357–60.

9 Genomes and Their Behaviors Promote Genetic Variability, Evolution, Large Adaptive Evolutionary Innovations, and Diversification: The Amount and Nature of Genetic Variability Is Subject to Natural Selection

9.1 THE BEHAVIORS OF GENOMES PROMOTE VARIABILITY, DIVERSITY, EVOLUTION, AND EVEN LARGE, ADAPTIVE MACROEVOLUTIONARY BREAKTHROUGHS

Genetic material performs genetic engineering and behaves in a way that tends to cause increased genetic variability within populations, rapid evolution, large evolutionary breakthroughs, diversification into many species, and, as a result of the latter, increased biodiversity. It sometimes causes increase in DNA content and complexity of the genome. Watson and Szathmáry (2016) even posit that evolution can learn from past experience to better evolve; this too involves behavior of the genome. Genomes of organisms are primed to evolve new forms and to diversify. Surprisingly large numbers of evolutionary breakthroughs and diversifications are in part driven internally, by the behavior of the genome itself. Rearrangement of the genome is thought to be a driving force in evolution, since it gives rise to novel gene combinations (Aguilera and Gomez-Gonzales, 2008). There are several mechanisms by which genomes function to increase variability and diversity, and to promote adaptive evolutionary change. Genomes of all sizes are a patchwork of genetic material coming from many different kinds of organisms and species with different histories.

9.2 NATURAL SELECTION OPTIMIZES THE AMOUNT AND NATURE OF GENETIC VARIABILITY

I will now discuss the evidence for my hypothesis that natural selection favors an optimal amount and type of variability, favoring changes that are compatible with the organism's phenotype and genotype.

Remember that DNA is in structures called chromosomes. The movement of entire sections of DNA from one part of a chromosome to another, between chromosomes, between individuals, and even between species is a major mechanism by which genetic variability is generated in populations of organisms. Sometimes macroevolution and key innovations occur as a result. It happens by a number of mechanisms. The key is that it causes an increase in variability that is not entirely random with respect to natural selection. Organisms are well-adapted to their environment as a result of long periods of natural selection. Even the simplest organisms are complex, highly integrated and coordinated systems, in which each protein and each trait affects and is affected by all the others. Likewise, the genome is highly integrated and coordinated, with genes interacting and affecting each other and the entire genome, which in turn affects each gene. Genes interact with each other, illustrating this inter-connectedness of the genes in the genome. Mutations such as point mutations in which only one DNA base is changed and deletions (loss of segments of DNA), both resulting from external influences such as chemicals and UV radiation, and mutations that occur when the DNA is replicated, have a very high probability of being incompatible with the genome they take place in and with the rest of the genes, and thus of being deleterious to the organism. It follows from this that the larger the effect of this type of mutation, the more likely it will be unfavorable to the organism. A complex, functioning car engine with its inter-related parts will almost surely function more poorly as a result of any random change to it. It is the same with organisms. A major, although not the sole, reason that most mutations are bad for the organism is that they are not compatible with an already well-adapted, complex, integrated, inter-related system. This is true of the genotype and phenotype, including behavior. Both the genotype and phenotype are integrated, inter-connected systems that tend to be disrupted and function more poorly if altered by random mutations.

On the other hand, phenotypic and genetic changes that are not fully random because they are integrated and compatible with the organism and its genome are more likely to be adaptive. Additionally, if a segment of DNA has undergone long periods of natural selection within genomes, it has a

DOI: 10.1201/9781003246640-9

better chance of causing an adaptive, beneficial change in the genome it moves into. The kind of moving DNA segments that this chapter will be discussing and that often occur in nature meet these criteria. The DNA bases in the segments that move within and between genomes also tend to be compatible with each other, because these segments have undergone long periods of selection favoring this self-compatibility. Thus, the moving DNA segments were selected to be both adaptive *and* to be compatible with the genome and chromosome that they integrate into. They still have a higher probably of being deleterious to the organism than favorable to it because they are causing a large change in an already well-adapted, well-integrated genetic system and organism. However, they are more likely to cause adaptive changes than are fully random mutations. They cause genetic changes that are not wholly random with respect to natural selection, in contrast to random mutations, such as chemically induced ones. They are likely to be compatible with the genome that they integrate into because they underwent selection for a long time in a complex, integrated genome. When they jump from one place to another in the same genome, they have already undergone natural selection to be compatible with the very same genome they are moving in and integrating into. Of course, this also means that the change they cause in the phenotype will have a higher likelihood of being adaptive, and that the phenotype will have a higher probability of remaining a well-adapted and well-integrated system with its parts remaining compatible with each other. Examples of these moving DNA segments that cause changes that are compatible with the rest of the genome are the parts of chromosomes that are exchanged in sexual reproduction, and segments of DNA that jump within and between genomes. An example of another genome-compatible genetic change is a change in a gene that regulates other genes. A change in such a gene could cause the entire organism to change in a coordinated, integrated, self-compatible way, such as having a uniform increase in size of the entire organism.

Another key mechanism by which natural selection optimizes the amount and nature of variability is that *mutations are not random with respect to natural selection, and adaptive mutations are more likely to occur than expected if this were the case*. It has long been believed that mutations are in fact random with respect to selection. However, Melamed et al. (2022) found evidence that this is not always the case. They showed a long-term directional mutational response to environmental pressure. Malaria has arguably exerted one of the strongest selection pressures acting on humans in the last 10,000 years, often killing over a million people per year in Africa in the recent past. The researchers looked at a region of six base pairs in the human hemoglobin subunit beta-globin gene and compared it to the identical, paralogous[1] hemoglobin subunit delta-globin region in sperm cells, with this latter region serving as a control. The human hemoglobin subunit region they looked at includes the site of the human hemoglobin S (HbS) mutation, which is a point mutation[2] that protects against malaria. They showed that the overall point mutation rate to the HbS mutation that gives resistance to malaria was much higher than expected if it were from random mutations. Furthermore, the rate of generation of the HbS mutation is higher in people from Africa, where malaria is endemic, than in people from Europe, where it is not. Additionally, the rate of generation of the HbS mutation is higher in the beta-globin gene, where it gives resistance to malaria, than in the control gene, the delta-globin gene, where it does not give resistance to malaria. Finally, the rate of generation of the HbS mutation when it appears in hemoglobin subunit beta gene is significantly higher than expected from the genome-wide average for mutations of this type. Nine instances of mutation to it were observed in the African hemoglobin subunit beta gene, where it is of adaptive significance, representing at least three independent originations; no instances were observed elsewhere. Hence, the mutation rate to HbS is higher in the population and in the gene where it is of adaptive significance in conferring resistance to malaria. Thus, the HbS mutation does not occur at random with respect to selection. These results are interesting. The authors hypothesized that evolution is influenced by two sources of information: Natural selection and internal information that is accumulated in the genome through the generations, and that this impacts the origination of mutations. They suggest that the genome can accumulate information over generations that influences the types and rates of mutations that occur. This ability is adaptive and favored by selection. Environmental pressures can influence this information and thus influence types and rates of mutations. This is a stunning demonstration of how the behavior of the genome helps organisms respond to environmental challenges and hence maintains and perhaps increases diversity.

A major thesis of this chapter that many of the ideas in it are dependent on is therefore an idea that is recognized occasionally in the literature, but given far too little importance. It is my hypothesis that *natural selection optimizes the amount and type of variability. Selection favors an optimum amount of genetic and phenotypic variability, as well as variability that is compatible with the rest of the genotype and phenotype, and with the organism.* There is ample evidence for this hypothesis. Some of this evidence has been and more will be discussed in this chapter.

9.3 SEXUAL REPRODUCTION INCREASES VARIABILITY WHILE MAINTAINING THE INTEGRATION AND SELF-COMPATIBILITY OF THE GENOTYPE AND PHENOTYPE, AND CAN PROMOTE EVOLUTIONARY INNOVATION AND DIVERSITY

This hypothesis, which I call the selection for optimal variability hypothesis is supported by the fact that sexual reproduction evolved, was selected for, and is the rule in higher organisms. The green alga *Spirogyra* reproduces without the use of sex in favorable conditions, and switches to sexual reproduction when environmental conditions change to a less favorable state, supporting the thesis that at least one function of sex is increasing variability to respond to unfavorable conditions. This is natural selection affecting the amount of variability in a population, one of the two theses in the hypothesis. When the environment is unfavorable, it is adaptive to vary more because this could lead to a change that adapts the organism's lineage to the unfavorable environment, whereas staying the same gives no possibility of changing to adapt to and better survive and reproduce in the unfavorable environment.

Sexual reproduction is the first mechanism by which the genome's behavior promotes diversity while maintaining the compatibility of the parts of the genome to each other that this chapter will discuss. It is among the most common ways that variability, adaptive macroevolutionary breakthroughs, and diversification result from the genome's behavior. It keeps populations from going extinct by keeping pace with environmental changes. It aids the speciation process. And the formation of new species by hybridization between species is a mechanism by which sexual reproduction can increase diversity that is not commonly recognized.

Sexual reproduction is costly to organisms that use it as a means of reproduction. It decreases the amount of genetic material an average organism passes on to its offspring by half. That is, on average, a sexually reproducing organism passes on half as much DNA, and half the number of genes, compared to how much it would if it reproduced without sex, to its offspring. This is because a female that produces an offspring without sex passes on all of her DNA, all of her genes, to her offspring. But if she reproduces by sex, she passes on only half of her genes to her offspring, since the male passes on the other half of the genes that the offspring inherits. There are other costs to sexual reproduction. For example, the organism must find, attract, and select a mate. This can be costly in time and energy. It can also increase the risk of being eaten by a predator. It can be costly in requiring the evolution of elaborate structures to intimidate and fight off other males and attract mates. For example, it can require the evolution of bright colors, songs, and dances in males to intimidate rival males and attract females. Yet most animals reproduce sexually. This apparent paradox is a major mystery in evolutionary biology. There are several hypotheses, many by prominent thinkers, to explain the predominance of sexual over asexual reproduction in multicellular organisms, given the loss of 50% of an organism's genetic material and other costs in this mode of reproduction. It is such a well-recognized question that it has a name: the problem of sex. I propose that a major solution to this problem is that *sexual reproduction generates genetic variability that natural selection acts on and that allows adaptation and evolution, while preserving the integration of and compatibility of the parts of both the phenotype and genotype of the organism and genetic system.*

When variability is generated, the genetic changes that generated it must not disrupt the integration of the organism, and must allow the different parts of the genetic system to remain compatible with each other and with the genome as a whole. The variability generated by the change must also allow the different parts of the phenotype to remain compatible with each other. An arm, hand, or wing must be compatible and function well with the rest of the body: the eyes, the heart, the brain, and so on. Sexual reproduction generally creates change and variability upon which natural selection can act while maintaining a well-adapted, integrated, self-compatible system. It maintains the genetic and phenotypic systems that have passed the test of eons of natural selection while generating needed, potentially adaptive variability. This contrasts with random mutations, which usually generate variability at the expense of the integration of the interrelated genetic system and compatibility of its parts, and at the expense of a well-integrated, self-compatible phenotype. Random mutations will disrupt a well-adapted system with a much higher probability than will changes generated by sexual reproduction. Sexual reproduction also sometimes increases DNA content.

How is sexual reproduction particularly effective at increasing variability and DNA content while preserving the integration of the genome and phenotype, and compatibility of their parts? The combining of two separate sets of chromosomes from two separate organisms, one set from each parent, creates variability. At the same time, it also preserves genetic integration and compatibility very well, since entire chromosomes are shuffled, shared, and combined into one fertile egg. These chromosomes have undergone eons of natural selection, so the offspring is inheriting well-adapted sets of chromosomes. Chromosomes exchange segments of DNA with chromosomes that are their analogues, that are similar to them. This exchange of DNA between chromosomes is called recombination (or sexual recombination). It increases the variability that natural selection can act on, allowing the population to evolve. Recombination preserves the integration of the organism while increasing variability because *large segments of chromosomes that are already compatible with the rest of the genome are exchanged between chromosomes, resulting in a new chromosome well integrated and compatible with the rest of the genotype. The chromosomal segments involved in recombination are as a rule relatively large and kept in their original form, with their DNA sequences basically unaltered. And the DNA segments that are exchanged move to essentially the same part of the receiving chromosome as they were in the donating chromosome; that is, there is generally no significant change in the location of the exchanged segments relative to the other genes and parts of the chromosome they end up on. The basic structure, fitness, integration, and self-compatibility of the genetic system are preserved.* Thus, sexual reproduction results in a relatively high probability of adaptive change. This contrasts with a random mutation, which provides variability for natural selection to act on, but that is much more likely to disrupt the integration and self-compatibility of the genetic system and be unfavorable to the organism. And if many small pieces of each chromosome were randomly exchanged and shuffled around, this would also be more likely to disrupt the integration and compatibility of the system. Sexual reproduction has evolved to exchange DNA segments that are close to the optimal size to generate variability while preserving self-compatibility of the genotype and phenotype.

Therefore, we can say that the generation of nonrandom variability that is compatible with the genotype and phenotype is one reason why natural selection favors sexual reproduction. Sexual recombination is actually favored by natural selection, supporting the thesis that there is selection for generating variability that maintains genetic integrity and self-compatibility. Evidence for this is seen from the fact that recombination is carried out by enzymes that evolved to cause it to occur. There are enzymes that break chromosomes for the exchange and enzymes that join exchanged segments of chromosomes together and fill in gaps of missing DNA created by the exchange. Recombinase is an enzyme involved in the exchange of segments of DNA between chromosomes. The fact that these enzymes evolved shows that the exchange of chromosomal segments was selected for. This suggests that selection favors a given amount of variability and variability that is compatible with the genome. Further evidence for selection for a favorable amount of variability comes from the fact that rates of recombination vary tremendously between different regions in the human genome, suggesting selection for more variability in some areas of the genome than others (Lander et al., 2001). Each chromosome has two segments called arms, one long arm and one short one. The rate of recombination generally tends to be a great deal higher toward the ends of chromosomes and on shorter chromosome arms (ibid.). This pattern promotes a minimum of one recombination per chromosome arm in each sexual event at the level of the cell. This is a high rate of recombination. There are recombination hotspots in the genome, where higher rates of recombination occur than is normally the case. Here, rates of recombination can be hundreds of times more than in the surrounding region (Jeffreys et al., 2001). These hotspots occur in both sperm and egg cells reproducing sexually and nonsexual cells that are dividing into two cells normally. There are more than 30,000 recombination hotspots within the human genome (Baudat et al., 2010), where the average number of recombination events per hotspot is one per 1,300 sexual cell divisions, which is far higher than average, and the most extreme hotspot has a frequency of one recombination per 110 sexual cell divisions (Myers et al., 2006). There are other sites that undergo recombination unusually frequently, which are associated with repeats of the four chemical bases that make up DNA. These bases are designated as A, C, G, and T, and N is used to indicate that any of the four bases can be present. These hotspots with high recombination rates are these DNA base sequences: CGG-CCG, GAG-CTG, GAA-TTC, and GCN-NGC (Aguilera and Gomez-Gonzales, 2008). These fragile sites are conserved in mammals and in yeast (ibid.).

An area associated with the PRDM9 gene is a recombination hotspot that codes for a protein. It contributes to at least two recombination hotspot locations (Baudat et al., 2010). The gene occurs in mice, chimpanzees, humans, and other species. Surprisingly, humans and chimpanzees have very few recombination hotspots in common, despite having 99% of their gene sequences in common (Auton et al., 2012). It is astonishing that *natural selection has acted on genomes, creating some areas on chromosomes where recombination is much more frequent than in other areas. Clearly, it is advantageous to recombine and thus generate variability more in some areas of the genome than others.* This is yet more evidence that natural selection optimizes the amount and type of genetic variability in organisms.

There are other advantages of sexual reproduction. Many are mechanisms by which the genome's behavior (the behavior in this case is the behavior of the chromosomes in sexual reproduction) promotes variability, evolutionary innovations, and diversity—the theme of this chapter. For example, when segments of DNA are exchanged by chromosomes, sometimes the segments they exchange are not the same size, and one chromosome receives more DNA than it gives. If the organism that ends up with more DNA by this mechanism is favored by natural selection and survives, it will have more DNA than its parents had. This increases DNA content. Unequal exchange of DNA could even cause an organism to have two copies of a gene (this is called gene duplication). This can allow significant adaptive changes in the organism, as will be discussed later in this chapter.

Sexual reproduction can also bring together multiple novel innovations without disruption, leading to macroevolutionary breakthroughs. In a hypothetical example that would not likely happen exactly as stated here, but is meant solely to illustrate the point, a fish that was an ancestor to the first amphibian and that had lungs that breathes air could mate with another fish of the same species that has legs for walking on land, resulting in an offspring that has both of these adaptations for life on land. Or an ancestor of a flowering plant with more developed reproductive flower structures could mate with one with more developed petals. After an evolutionary breakthrough such as those illustrated in these two examples, many new niches would be available to the new form, and it would diversify into many new species that would then fill these new niches. This is not a comprehensive list of the potential advantages of sexual reproduction. And there are many other hypotheses attempting to explain the predominance of sexual reproduction in multicellular organisms. Mine is not the only one, and the full explanation may be a combination of a number of the hypotheses. However, mine is consistent with the known facts and is reasonable.

To reiterate, the hypothesis is that *sex was selected for at least in part because it results in higher variability while maintaining the integration and self-compatibility of the genome and phenotype.*

9.4 HORIZONTAL GENE TRANSFER

The second mechanism by which the genome promotes variability, evolutionary innovation, and diversity that will be discussed in this chapter is horizontal gene transfer (HGT), also called lateral gene transfer. This is the lateral movement of genetic material between organisms. It is distinct from the transmission of DNA from parent to offspring, which is called vertical gene transfer. It is an important factor in evolution. HGT can occur between organisms of the same species, between species, and even between unrelated species in different groups. It can occur between groups regardless of how related they are. Thus, it can occur between different domains, which means DNA can be transferred from archaea or bacteria to eukaryotes, including humans. The major mechanisms of HGT are bacterial conjugation, transformation, gene transfer agents, transduction, and transposable elements (including integrons and gene cassettes in those that have them). All of these will be explained in this chapter. HGT frequently involves lysogenic bacteriophages and plasmids. Plasmids are small circular pieces of DNA in cells that are separate from the chromosome, which is larger than them and has most of the DNA. Plasmids replicate by themselves. They contain genes, often ones that benefit the organism. They are mostly in bacteria. They can transfer to other cells. They can integrate into the cell's chromosome (bacteria only have one chromosome). DNA can also jump from one part of the genome to another within one organism, although this is not referred to as horizontal gene transfer. This will be discussed later.

The code used by DNA to direct the production of, or code for, proteins, is called the genetic code. Amazingly, it is the same for all animals, plants, fungi, bacteria, and archaea, with a few minor exceptions. It has been proposed that natural selection favored one universal code that is the same in essentially all life because this allows DNA to jump between different species (Kubyshkin, 2018). This would be advantageous to all species because they could have their variability increased while maintaining the self-compatibility and integration of their genomes and phenotypes. Any species with a different genetic code would not be able to have DNA segments jump from another species into its DNA, because DNA with a different genetic code in its DNA would be disruptive and would probably kill the organism receiving the DNA. Thus, that species would be at a big evolutionary disadvantage compared to other species. Therefore, it is possible that the evolutionary advantage of the genetic code being largely the same in all species is that this allows different species to exchange genes and DNA segments, providing variability, new genes, or even sudden, adaptive evolutionary change. Additionally, such rapid exchange of genetic material would allow populations to keep up with rapidly changing environments, and rapidly adapt to exploit the varying local areas of nonuniform environments. Syvanen (1985) correctly said that HGT was involved in shaping evolutionary history from the beginning of life on Earth.

The genomes of essentially all species are chimeras, mosaic mixtures constructed with genes acquired from many different taxa.[3] HGT is so common that some scientists consider it to be the most serious hidden and underestimated hazard of genetic engineering (Ho, 2007).

Virulence is a pathogen's ability to cause damage to its host, to cause disease. Bacteria often acquire virulence via HGT, from their phage. There are many cases in which lysogenic phage have genes that increase the virulence of bacteria, mainly by helping the bacteria invade and survive in the host. These include genes for the bacterium to adhere to the host, genes for entering into the host, and genes for resistance to antibiotics. In fact, HGT is the primary reason for bacterial antibiotic resistance, often via plasmids and/or bacteriophages (Kay, 2002; Koonin et al., 2001). It plays an important role in the evolution of the ability of bacteria to break down and destroy chemical compounds the bacterium has never before encountered, such as human-created toxins and antibiotics. HGT also helped with the evolution of resistance in bacteria to natural antibiotics produced by fungi and species of bacteria that competed with the bacterial species that evolved the antibiotic resistance. (Bacteria often produce antibiotics that attack species of bacteria that compete with them.) Group A *Streptococcus,* a bacterium which causes a variety of diseases, acquired virulence factors encoded by phages or phage-like elements, likely helping it cause epidemics at specific times and places in the past (Banks et al., 2002). Viruses often contain genes helpful to bacteria. Bacteria incorporating these genes extend their ecological niches. This is adaptive to the virus, since it helps its bacterial host, which is its habitat.

Viruses of the genus *Bracovirus* are symbiotic with tens of thousands of species of tiny parasitic braconid wasps whose larvae develop in the bodies of butterfly and moth larvae. The many species of this wasp collectively parasitize caterpillars of virtually every butterfly and moth species. *Bracoviruses,* injected along with the wasp's eggs into the caterpillar host's body, integrate into its host cell's DNA. The viruses destroy host immune defenses against the wasp by disrupting the caterpillar's cytoskeletons (skeletal structures of the cell), and manipulate host development and physiology, and so enable wasp larvae to develop in an otherwise potentially harmful environment (Bézier et al., 2009). Interestingly, segments of *Bracovirus* DNA have been repeatedly incorporated into the genomes of the butterflies and moths, indicating this viral DNA can also enter reproductive cells, both sperm and egg cells (Gasmi, 2015). Amazingly, the genes transferred to the butterflies and moths protect them against the lethal baculovirus, which is a different virus than the similarly sounding *Bracovirus*. Some of the transferred *Bracovirus* genes originated from the wasp genome, demonstrating that there is exchange of genes between wasps and butterflies/moths, which diverged from each other about 300 mya. Also, the reason that infection with *Bracovirus* prevents infection by baculovirus is that *Bracovirus* disrupts the caterpillar's cytoskeleton, which baculovirus needs to use for it to achieve a successful infection. Thus, *Bracovirus* helps itself by both hurting one of its viral competitors and helping its primary host, the wasp.

Baculoviruses infect caterpillars (that are not protected with incorporated *Bracovirus* genes or infected with *Bracovirus*) that eat them because they are on the leaves the insects eat. Once in the caterpillar, baculoviruses reproduce in great numbers, sometimes multiplying to the point where the host swells from having so many viruses in it. The virus affects the caterpillar's brain and behavior, causing it to have an uncontrollable urge to crawl to the top of a plant, where it holds on tightly and hangs downward (Han et al., 2017; van Houte at al., 2015). Then the virus releases an enzyme that dissolves the insect. The many viruses in the dead caterpillar fall out onto the leaves below, where another caterpillar is likely to eat them. So the virus affects the caterpillar's behavior in a way that maximizes the chance that the next generation of the virus will be eaten by and infect another caterpillar host. One of the genes crucial for this adaptive manipulation of the caterpillar causes the insect to start wandering in response to light. There is good evidence that the virus stole the gene from the caterpillar, since the gene has a striking resemblance to a gene in insects that is involved in foraging for food in response to light. It seems an ancestral baculovirus picked up the caterpillar gene, which then evolved from a food-seeking function to one that causes the larva to crawl to an elevated position on the leaf, both functions being in response to light (Hoover et al., 2011).

It is exciting that the evidence supports the idea that once photosynthesis evolved, it spread by HGT (Blankenship, 1992; Raymond et al., 2002; Hohmann-Marriott and Blankenship, 2011; Blankenship, 2017, and references therein). Photosynthesis depends on a series of very large, complex molecules and a set of complex systems of molecules all working together. Raymond et al. (2002) looked at five unrelated taxa of bacteria that carry out photosynthesis from very different, unrelated evolutionary lines. They found a set of 188 genes inherited from common ancestors of the five groups. It appears the different species shared their genes by HGT. Photosynthesis did not evolve through a simple, straight path of increasing complexity, but through the merging of separate evolutionary lines that brought together independently evolving chemical systems, through the giving and taking of blocks of genetic material between different bacterial species. Evidence that photosynthesis evolved by HGT is seen in the fact that genes involved in photosynthesis in these bacterial species had different evolutionary histories and pathways. *Different pieces of the system evolved separately in different species. The genes were then brought together into one species, making a new combined system, by the transfer of DNA segments between species of bacteria.* Bringing together of genes from different species into one species and the evolution of photosynthetic systems likely occurred several times. Whether this involved viruses as agents of transfer of DNA between species of bacteria is unknown at present.

Changing the function of a structure or a gene to a new function is called exaptation. The same structure or gene takes on a new function. Feathers probably evolved in dinosaurs to keep the animals warm and to attract mates and intimidate rivals. They then changed function to be used in flight when birds evolved from dinosaurs. This is an example of exaptation. A leading researcher in the evolution of photosynthesis, Robert Blankenship, thinks that a good deal of the shared genes in photosynthesis were originally used by bacteria for purposes other than photosynthesis, and that these genes then changed functions to become part of photosynthesis systems (see Blankenship references in this chapter). This of course would be exaptation.

But most importantly, there are two major systems of photosynthesis that evolved separately in completely different evolutionary lines of bacteria, called Type I and Type II (abbreviated RCI and RCII). Organisms can photosynthesize with one photosystem or the other, but no organism performs *oxygenic* photosynthesis without both systems. Recall that oxygenic photosynthesis is the form of photosynthesis that produces oxygen, and the other kind of photosynthesis, anoxygenic, does not produce oxygen, and that oxygen provided the energy for the evolution of higher organisms. Cyanobacteria are the only prokaryotes with both photosystems and hence the only prokaryotes that carry out oxygenic photosynthesis. The evidence supports the idea that RCI and RCII evolved independently in two separate lines of cyanobacteria, and the two systems were then brought together by HGT, creating oxygenic photosynthesis (Hohmann-Marriott and Blankenship, 2011). We do not know whether this was facilitated by a virus (or viruses). Once the two photosystems came together, cyanobacteria that possessed them could very efficiently make carbohydrate for their food and energy. It was a huge key innovation and macroevolutionary leap that created a new adaptive zone with many new niches. Thus, it was followed by a tremendous diversification creating many new photosynthetic species of cyanobacteria. This was followed by the evolution of several additional new species that utilized or benefitted from the newly evolved photosynthetic species.

In fact, oxygenic photosynthesis in prokaryotes and then eukaryotes evolved by many biologically driven genetic processes in addition to HGT. These include the duplication of genes, deletion of DNA, movement and relocation of genes, and coevolution of a photosynthetic bacterium inside a prokaryote that resulted in a eukaryote that could carry out photosynthesis (Green, 2007). For a more detailed explanation and discussion of the last of these processes, see Chapter 4 of my companion book to this one (Seaborg, 2022). Oxygenic photosynthesis is one of the most important chemical processes life has ever developed. It is ultimately the source of energy (carbohydrates), directly or indirectly, for the vast majority of the organisms on Earth's surface and in the sea. It is the basis of a great number of Earth's food webs. As discussed in Chapter 3, the evolution of oxygenic photosynthesis in cyanobacteria is one of the two factors responsible for the GOE and the production of an oxygen-rich atmosphere that allowed the evolution of complex life. *In this dramatic demonstration of the ABH, HGT between species of bacteria, possibly mediated by viruses carrying the DNA, and other biologically driven mechanisms resulting from the behavior of the genome, led to oxygenic photosynthesis, which produced oxygen, an energy source that allowed the evolution of higher life!*

Jumping segments of DNA also played a key role in the evolution of nitrogen fixation. The filamentous cyanobacterium, *Microcoleus chthonoplastes*, a cosmopolitan prokaryote that often builds layered mats of microbes in a variety of different environments, acquired its cluster of nitrogen-fixing genes through HGT early in its evolution (Bolhuis, et al., 2010). This is tremendously significant in showing the importance of HGT in evolution and biodiversity, since nitrogen fixation is so important in helping support so many different ecosystems. HGT is thus a fundamental process in structuring ecosystems and greatly increasing biodiversity.

Genes moved from the chloroplasts of an unidentified plant to the mitochondria of the wild bean (genus *Phaseolus*) (Woloszynska et al., 2004). And HGT has occurred within eukaryotes, with genes moving from the chloroplast and mitochondrial genomes to the nuclear genome.

The largest gene transfer known thus far is the transfer of the entire mitochondrial genomes from three species of green algae and one species of moss into the mitochondrial genome of a small tree with no common name found only on the main island of New Caledonia. Its scientific name is *Amborella trichopoda* (Rice et al., 2013). New Caledonia consists of several islands in the South Pacific. The small tree is one of the two oldest lines of flowering plants. Also, the *Amborella trichopoda* mitochondrial genome has acquired entire mitochondrial genomes of varying sizes from a broad range of land plants and green algae, and not lost them for tens of millions of years. This small tree is more exposed to the foreign mitochondria of plants growing on it than most other species of plants and trees, which shows physical closeness of species to each other facilitates HGT between species. It was also found that the fusing together of mitochondria is the driving force for mitochondrial gene transfer.

HGT is common in prokaryotes, and a large percentage of the genes in their genomes were acquired by this transfer of DNA between species. Most of the increases in the number of protein families and acquisitions of new protein functions are due to HGT in prokaryotes (Grassi et al., 2012). Their great diversity, ability to adapt rapidly, and ability to occupy a great variety of niches and habitats are to a large extent due to HGT. This is significant because prokaryotes evolved the vast majority of life's biochemical functions. These include anoxygenic and oxygenic photosynthesis, respiration, nitrogen fixation, and the ability to break down and digest a good deal of plant nutrients that animals cannot digest. In some cases, the type, quantity, and source of the DNA acquired is associated with the lifestyle of the prokaryotic species. For example, bacteria adapted to extremely hot environments appear to have captured genes from archaea, which have many species adapted to very hot environments, to a greater extent than other bacteria have. And transfer of certain classes of genes from eukaryotes is most common in bacteria that are parasitic or symbiotic with eukaryotes, and so closely associated with them. The fact that HGT genes are so successful and persisted for long time periods in prokaryotes indicates that they confer advantages and are favored by natural selection. The advantage they confer is not known in many cases, but it is known in some. For example, *Chlamydia* bacteria captured genes by HGT for enzymes that break down proteins and that may be implicated in its ability to live in its hosts, where it causes disease.

Blanc et al. (2010) showed that *Chlorella variabilis* NC64A, a unicellular photosynthetic green alga that lives symbiotically in *Paramecium bursaria* (a one-celled organism that swims in water), likely acquired its ability to produce cell walls made of chitin from the capture of metabolic genes by HGT from prokaryotes, fungi, or viruses of algae. Chitin is a complex fibrous substance that forms the major constituent in the outer skeleton of arthropods like insects and crabs, and the cell walls of fungi. Their relationship is symbiosis because the alga produces food by photosynthesis for the *Paramecium*, which gives the alga a home.

Marine snow has microbial communities with many different species associated with it (see Chapter 2). Bacteria transported on marine snow may exchange genes with populations of bacteria that inhabit the seafloor. Turley (2001) calculated that a trillion live bacteria are transported from the shallow waters of the ocean to every square meter (about 10.8 square feet) of the deep seafloor each year. The seafloor covers about 60% of the Earth's surface. The bacteria transported in this way bring new genetic material to the otherwise genetically isolated bacteria on the seafloor. This gives the bacteria of the marine snow and those on the seafloor the opportunity to exchange DNA. This permits a potential increase in the genetic variability and biodiversity of both groups.

The soil-dwelling bacterium *Rhodococcus fascians* has a strain that produces an antibiotic that kills bacteria species that compete with it for space and nutrients, and its genome contains a large region of DNA from bacteria of the genus *Streptomyces*. Apparently, it acquired the antibiotic gene and/or a region of DNA that regulates this gene from the *Streptomyces* bacteria (Kurosawa, 2008). Brewer's yeast (*Saccharomyces cerevisiae*) and a fungus that grows in filaments with no common name (*Ashbya gossypii*) have acquired genes from bacteria (Hall et al., 2005). It appears that brewer's yeast acquired its gene that codes for an enzyme important in making nucleic acids, the building blocks of DNA and RNA, from lactic acid bacteria (bacteria that produce lactic acid, which has many uses). Brewer's yeast and a relative of it apparently acquired the gene BDS1 from bacteria called the alpha-proteobacteria. This gene allows organisms to use sulfate, $(SO_4)^{-2}$, from several organic sources, to obtain energy.

Ropars et al. (2005) sequenced the genomes of ten species of the mold *Penicillium*. Some members of this genus make the antibiotic, penicillin. Six of the species they sequenced grow on milk; some of those six are used to make cheese. Wild *Penicillium* typically grows on decaying plant matter, not milk. The researchers found up to 5% of the DNA of each mold they studied was made up of DNA from another species. This DNA had jumped from that other species to the species of mold the researchers found it on. These genes help the mold species that contain them grow better and faster on cheese. The jumps of these genes all happened in the last few centuries, after humans started making cheese. Humans selected for these adaptations, which occurred rapidly. Wallaby, the longest DNA segment acquired by the molds, contains 250 genes, and CheesyTer, the second longest, has about 60. These segments are not in wild strains, which are not used to make cheese. There is a gene in CheesyTer that appears to help break down lactose, the natural sugar in milk, and slows the mold's growth on simple sugar. The transfer of genes from other species of *Penicillium* allowed the six species that grow on milk to exploit a novel niche (milk).

Water bears, also called tardigrades, are microscopic animals that can tolerate extreme dryness. Hashimoto et al. (2016) found a few of the genes that allow one species of water bear to survive without water and keep from drying out were acquired by jumping from other species into its DNA. One of these genes codes for a protein that helps make a chemical called trehalose, which is thought to act as a protective gel preventing disruption and destruction of internal parts of cells in both plants and animals. However, the percent of the water bear genome acquired by DNA transfer is only 0.7%.

Bacteria produce toxins that attack competing bacterial species. Chou et al. (2015) found that genes for these toxins have jumped to ticks, mites, and other animals. The organisms carrying the bacterial toxins were incredibly diverse and included several species of mites and ticks. The genes have been transferred to eukaryotes at least six times, and have been preserved for hundreds of millions of years by natural selection, implying they have an adaptive function. In fact, the animals use the toxins to defend against pathogenic bacteria, including *Borrelia burgdorferi*, which causes Lyme disease. Evidence for this is that when the researchers reduced production of the toxin in ticks, levels of the Lyme disease pathogen rose significantly. The resistance to disease-causing bacteria from these genes allowed great diversifications into many species in these ticks and mites.

Remarkably, the soil bacteria genus, *Agrobacterium*, is adapted to transfer a part of its own DNA, called T-DNA, to plants. This plays an important role in plant evolution (Quispe-Huamanquispe et al., 2017). Sweet potatoes from all over the world have acquired this T-DNA from *Agrobacterium* (Kyndt et al., 2015). This DNA apparently was instrumental in adapting the sweet potato to human use. The gene transfer can also cause tumors in plants. The T-DNA also carries genes for enzymes making unusual amino acids that are not used in proteins, such as octopine and nopaline; for the production of important plant chemical hormones that control plant growth and cell division; and for the production of opines, which provide a carbon and nitrogen source for *Agrobacteria* that most other microbes cannot use, giving *Agrobacterium* a selective advantage. The bacterium aids its host plant, helping itself. Perhaps T-DNA started as selfish DNA adapted to jump to different species, including eukaryotic ones, then later aided its bacterium and plant hosts, helping itself. Three levels of selection (T-DNA, *Agrobacterium*, and plants) were involved in the evolution of this three-way symbiosis. It is also noteworthy that the extremely important plant hormones, auxins and cytokinins, came to plants from *Agrobacteria* via HGT. Auxins play a crucial role in coordination of many growth and behavioral processes in plant life cycles and are essential for plant body development. Cytokinins promote cell division in plant roots and shoots. They are primarily involved in cell growth and differentiation (specialization) into mature cells with specific functions. Additionally, cytokinins affect the dominance of the main, central stem over side stems; bud growth; and the aging of leaves.

High gene transfer rates also occur between soil bacteria and the plant root system and germinating seed (Sengelov, et al., 2001). *This could have greatly affected the evolution of both plants and soil bacteria.*

The bacterium *Wolbachia pipientis*, which lives in the cells of invertebrates without hurting them, transferred DNA into the genomes of four insect and four roundworm species in amounts ranging from nearly the entire *Wolbachia* genome (over a million DNA bases) to short (less than 500 bases) insertions (Dunning Hotopp et al., 2007), indicating *Wolbachia* is a major source of DNA for insects and roundworms. Potential transfers of DNA from *Wolbachia* were also detected in three other insect genomes that were sequenced. There are some inserted *Wollbachia* genes that code for proteins in insect cells that do not even have *Wolbachia* in them. A *Wolbachia* species has transferred all of its 1,206 genes into the genome of a fruit fly (Salzberg, 2005). Since *Wolbachia* occurs in 25 to 70% of all insect species, this suggests many insects may have *Wobachia* genes. These findings show that genes enter into eukaryotic hosts by HGT from prokaryotes that live in their cells. This potentially provides a mechanism for the acquisition of new genes and functions in eukaryotes.

Gene transfer between plants and fungi has been hypothesized for a number of events. Rotifers, commonly called wheel animals, make up a phylum of microscopic and near-microscopic animals, common in freshwater environments throughout the world, with a few saltwater species. Bdelloid rotifers are a group of rotifers that reproduce without sex and can survive in dry, harsh environments by entering a state of inactivity. About 8% of the genes of Bdelloid rotifers are from bacteria, fungi, and plants; they have the most genes from cellular animals imported from other taxa of any kind of animal (Watson, 2012). HhMAN1 is a gene that is thought to have been transferred from bacteria to the digestive tract of the coffee berry borer beetle (*Hypothenemus hampei*) (Acuña et al., 2012). This gene codes for an enzyme used to digest a complex carbohydrate that is found in coffee beans. Crisp et al. (2015) found the genomes of forty animals, including four roundworms, twelve fruit flies, and ten primates have genes originating from bacteria and fungi by gene transfer. DNA also travels from higher to lower organisms, for one of the pathogens that cause malaria, *Plasmodium vivax*, acquired genetic material from humans that might help facilitate its long stay in the body (Bar, 2011).

Ferns thrived until flowering plants, some of which are tall trees in forests where ferns live, evolved and shaded them out, about 125 to 100 mya, denying them access to sunlight, and causing many fern species to go extinct. But the ferns that survived acquired a gene for a light-sensing protein called neochrome, which allows ferns to detect dim levels of both red and blue light, enabling them to grow toward any light that reaches them through the forest (Li et al., 2014). This likely enabled them to prosper on shady forest floors. Ferns acquired this gene from hornworts, which are simple, rootless plants that descended from some of the first land plants (ibid.).

The reproductive cells of both ferns and hornworts congregate on moist areas on the forest floor, where they come into close contact, and gene transfer can readily occur. This gene transfer and adaptive response to low light led to a tremendous diversification of ferns about 100 mya. About 80% of living fern species are a result of this diversification.

Deuterostomes are higher animals that consist of three phyla, or major groups: hemichordates (acorn worms); chordates, which include vertebrates such as reptiles and mammals; and echinoderms, which include sea stars and sea urchins. They have 30 genes that are not found in other animals, but are in marine algae and bacteria, several of which are involved in modifying sugars on cell membranes (Simakov, 2015). The functions of these genes are thought to be associated with physiological, metabolic, and developmental adaptations of the filter-feeding ancestor. It is unlikely these genes were passed to animals vertically from algae and bacteria and then lost by nondeuterostomes. *They were very probably acquired by horizontal gene transfer from algae and bacteria to the deuterostome ancestor at least 570 mya. This raises the possibility that deuterostomes would not have been able to come into being without these HGTs from bacteria and algae. At minimum, deuterostomes would likely be substantially different without the gene transfers.*

A species of bacterium in the intestines of Japanese people, but not other races, has two genes that code for enzymes that break down porphyrans, which are unique components of sugars of marine algae (Heheman, 2010). It also has genes that code for enzymes that break down agar, a jelly-like substance consisting of sugars obtained from the cell walls of some species of red algae. It also has proteins associated with all of these enzymes. The bacterium obtained these genes through HGT from a member of the marine bacteria phylum, Bacteroidetes, that is active on a sugar from marine red algae. The benefit of this to Japanese people is unknown, but it is likely that the breakdown of porphyrans produces short-chain fatty acids useable by humans for calories that would not otherwise be available. The bacterium benefits by having a host and thus a stable habitat. Thus, the bacterium and human are likely in a symbiotic relationship. The Japanese eat a good deal of sushi wrapped in seaweed,, which has the bacterium that is the source of the genes. This is likely the only known case of HGT between ecosystems thus far, and indicates we will likely find more cases of organisms humans eat transferring genes to their intestinal bacteria to the benefit of humans and these bacteria. Digestive system microbes supply the human body with energy from complex dietary sugars through carbohydrate active enzymes, which are absent in the human genome. The findings suggest that contact with nonsterile food may be a general factor in the acquisition of carbohydrate-active enzymes. It may also increase the number of species in the human digestive tract microbiome. Hehemann suggested there might be a frequent update of our digestive tract bacterial ecosystem through gene transfer and that this increases its genetic diversity.

Plant cell walls are enclosures around each plant cell. They protect the plant cell. They are among Earth's largest repositories of organic carbon (Duchesne and Larson, 1989, and references therein). To utilize this resource, bacteria secrete enzymes that break down plant cell walls. These enzymes can break down pectin, a unique fiber found in fruits and vegetables, and cellulose, an important structural component of the cell walls of green plants and many forms of algae. Eukaryotes generally cannot produce such enzymes, so cannot digest pectin or cellulose, but rely on their bacterial symbionts in their microbiome to do it for them. Most animals do not have the enzymes or the appropriate microbes to digest pectin and plant cell walls, so cannot digest certain plants such as grasses. The microbes produce enzymes the host cannot. Shelomi et al. (2016) have found that genes that code for enzymes that digest pectin have moved horizontally to stick insects, resulting in the ability of these insects to produce their own enzymes that break down pectin. Stick insects, also called walking sticks, are long, thin insects that camouflage themselves by looking like sticks or leaves (Figure 9.1). Their genomes contain multiple genes for enzymes that break down and digest pectin, allowing them to break down plant cellulose cell walls and obtain nutrients from them. This frees them from dependence on microbes for this function. And it increases their digestive efficiency because it is more efficient for them to digest plant cell walls themselves, rather than rely on microbes to do it for them. These genes are found in gamma-proteobacteria, the most common bacteria type in the stick insect microbiome, and on leaves the insects eat. They thus obtained the genes either from their digestive tract bacteria or from such bacteria that deposited them on their leaf food. Tests show that, while some of the new genes

FIGURE 9.1 A stick insect. This one is the Northern Walking Stick (*Diapheromera femorata*), perched on a tree branch, in Ontario, Canada. Stick insects are masters of camouflage, making it difficult for their predators to find them. They acquired genes for enzymes to digest pectin from gamma-proteobacteria, allowing them to eat more kinds of plants and greatly diversify.

Shutterstock Item ID: 109854143

Title: Northern Walking Stick (*Diapheromera femorata*) Perched on a Tree Branch—Ontario, Canada.
Photo Contributor: Brian Lassenby

that digest pectin kept their original function, others actually evolved new, yet unknown, functions. Analysis showed the transfer of genes occurred between 110 and 60 mya, before stick insects rapidly diversified into the approximately 3,000 species living today. This indicates that the gene transfers that conferred the new ability to digest pectin was a key innovation causing or at least greatly aiding the great diversification of stick insects. Shelomi's Enzyme Expansion Hypothesis posits that the sudden appearance of new enzyme abilities, through either HGT or mutation, can change the diet of a species to specialize on a single new food source. This is not part of Shelomi's hypothesis, but I will add that the change in diet could lead to speciation and diversification. Also bear in mind that the microbiome provides a tremendous reservoir of DNA that can jump from the microbiome to hosts ranging from worms to mammals, and provide a spectacular number of new functions and abilities. It appears that HGT and the microbiome together are very important drivers of the evolution of multicellular organisms.

Kirsch et al. (2012, 2014) showed that a gene that encodes an enzyme that breaks down pectin was horizontally transferred from a sac fungus (phylum Ascomycota) to a common ancestor of the superfamily Chrysomeloidea, the leaf beetles and long-horn beetles, and the superfamily Curculionoidea, the weevils, about 200 mya. This was followed by independent gene duplications in these two lineages. These are all plant-eating insects. This HGT allowed these insects to digest pectin and obtain energy and nutrition from it. This gene diversified into a large family of genes that code for enzymes that break down pectin in these lineages of insects. There are now active and no-longer-active enzymes for breaking down pectin in the leaf beetles, long-horn beetles, and weevils. As a result of this key innovation, all of these groups ended up in new adaptive zones with many new available niches. This allowed these groups to greatly diversify, which they did. Diversification was especially immense in leaf beetles and weevils, more so than in long-horn beetles, after the HGT.

McKenna et al. (2016) found in a sequencing study of DNA that increasing the number of and evolutionary divergence of genes associated with specialized feeding on plants contributed to the digestive capabilities of the Asian long-horned beetle (*Anoplophora glabripennis*), certain other plant-eating beetles, and to a lesser degree, other plant-eating insects, and this was partly accomplished by horizontal gene transfer from bacteria and fungi. This aided the evolution of genes that code for enzymes involved in the digestion of woody plant tissues and the detoxification of toxic chemicals plants produce to defend against being eaten by insects. It also helped the diversification of all of these plant-eating insects.

Carotenoids are a diverse class of colorful molecules functioning in photosynthesis, vision, camouflage, and sexual attraction, made by bacteria, fungi, and plants. The only animals that can make them are aphids, mites that spin protective silk webs called spider mites, and gall midges, which are small flies. Other animals must obtain them through their diets. Aphids acquired the genes for them from fungi 270 mya, possibly from infection by them (Altincicek et al., 2012). Aphids can be red or green, depending on the carotenoids in them. Green aphids are susceptible to parasitic wasps; red ones to ladybird beetles. Thus, the carotenoids affect the color of the aphids and what predators they have, and this is subject to natural selection. So HGT has greatly affected the phenotype, including the color and other traits, and evolution, of aphids.

Diatoms are a major group of one-celled phytoplankton. The first genome sequence of a diatom, *Thalassiorsira pseudonana*, which has no common name, was completed in 2004 (Armbrust et al., 2004). This genome consists of DNA from the genomes of a number of species, illustrating symbiotic coevolution, gene transfer between species, and that species can be made up of the genomes of several other species. This diatom genome is a combination of components inherited from: the cell of its ancestor; components of the genomes of the nucleus, chloroplast, and perhaps mitochondria of a red alga that lives symbiotically inside it; a minority of components from HGT from other eukaryotic cells; and a minority of components from HGT from prokaryotic cells. Thus, diatoms are chimeras—organisms that are combinations of more than one organism. Diatoms live in freshwater and soils as well as the sea, have about 10,000 species, number in the trillions, generate about 20% of the oxygen produced annually on Earth, regulate CO_2 in the air and silicon concentrations in rivers and oceans, contribute nearly half of the organic material found in the oceans, and are an important component of the phytoplankton that are the base of the sea's major food webs. So HGT and symbiosis had a profound effect on the planetary system and greatly increased biodiversity by playing essential roles in the evolution of diatoms. Diatoms are not unique. All species are chimeras built with DNA from a number of different taxa.

HGT has rendered the tree of life metaphor obsolete. It must be replaced with a metaphor of a mosaic web, with cross branches that are horizontally connected, in addition to their standard vertical branching pattern. Classification of organisms, the branching and relationships of species, and indeed all taxa of organisms, such as genera and families, and evolutionary theory must be revised to take into account that genes jump between species and all higher taxonomic levels, even between domains, the highest group of organisms (i.e., genes even jump between bacteria and vertebrates, including humans).

As much as 40% of the human genome has moved around during human evolution (Goodsell, 2006).

Following is a discussion of some broad categories of horizontal gene transfer. They are bacterial conjugation, gene transfer agents, transformation, transduction, and transposable elements (including integrons and gene cassettes).

9.5 BACTERIAL CONJUGATION

Bacterial conjugation is the transfer of DNA from a donor bacterial cell to a recipient bacterial cell when cells touch each other, or by a pilus, a structure that connects the two cells. The genetic material does not go through the pilus.

Commonly the donor cell provides a plasmid or transposable element to the recipient cell. A transposable element is a segment of DNA that jumps to other parts of the genome; it will be discussed later in this chapter. Bacterial conjugation is not sexual recombination. The transfer of DNA in bacterial conjugation can increase genetic variability, increase DNA content, and/or result in beneficial adaptations, in the recipient bacterium. Examples of adaptations that may be acquired through bacterial conjugation include the ability to utilize new chemicals, antibiotic resistance, and tolerance to toxins. Some elements transferred via conjugation can be detrimental and can be viewed as bacterial parasites.

9.6 GENE TRANSFER AGENTS

Gene transfer agents (GTAs) are virus-like particles that contain DNA and that are produced by some bacteria and archaea. They package and carry random DNA segments in the host bacteria or archaea that can be passed by HGT to a recipient cell, generally of the same species (see, for example, Maxmen, 2010). These DNA segments become incorporated into the genome of the recipient cell. The genes that code for GTAs are derived from bacteriophage DNA that has integrated into a host chromosome. These integrated phages often acquire mutations that make them defective and unable to produce more phage. Many bacterial genomes contain one or more integrated phages that are defective due to mutations they have undergone. GTAs leave their host cell by bursting and killing it. Some GTA systems appear to be young, while others have been maintained for many millions of years. GTAs can add variability, which allows responses to changes in the environment and can lead to new adaptive functions. The DNA segments they carry to new bacteria or archaea have been subject to natural selection for long time periods, so have a higher probability of being adaptive than random mutations. They have evolved within prokaryotic genomes for long time periods, so will have a high probability of being compatible with the recipient's genome. However, they have a cost to the host population, mostly because they kill the cells they leave. Also, although they have a higher likelihood than random mutations of being adaptive, most new genetic changes they create will be deleterious. Only a few will be adaptive, and still fewer will lead to evolutionary breakthroughs.

9.7 TRANSFORMATION

Transformation is the uptake and incorporation of foreign genetic material from the environment through the cell membrane by an organism. The genetic material travels through the environment. It comes from a donor cell that the recipient cell does not touch. Viruses do not carry the genetic material. It is fairly common in bacteria. It can be the source of new genetic material for the recipient, and can increase its DNA content. It is a source of genetic variability for natural selection to act on. Transformation provides bacteria a way to adapt quickly to a changing environment, and can aid the spread of antibiotic resistance (Engelmoer et al., 2013). It can facilitate adaptation by combining more than one beneficial mutation into a single individual (ibid.), much like sexual reproduction can.

Phage can actually promote transformation in bacteria. Phage interacts with plasmids in bacteria on a massive scale. Keen at el. (2017) found that a subset of the natural phage population, which they call "superspreaders," releases substantial amounts of intact, transformable DNA in the form of plasmids when these phage burst and kill bacterial cells, promoting HGT by transformation. There are two phage superspreaders of the intestinal bacterium, *Escherichia coli*, called SUSP1 and SUSP2. They liberated four evolutionarily distinct plasmids with equal efficiency, including two close relatives of prominent plasmids that carry genes for antibiotic resistance in natural environments. And SUSP2 helped the extensive horizontal transfer of antibiotic resistance genes in communities of soil bacteria from Maryland and Wyoming. What is more, the addition of SUSP2 to cultures of *Escherichia coli* resistant to the antibiotic kanamycin and *Bacillus* sp. (*Bacillus subtilis, B. circulans,* and *B. coagulans*) bacteria sensitive to it resulted in roughly 1,000-fold more kanamycin-resistant *Bacillus* sp. bacteria than arose in control cultures that had no phage. Thus, the phage helped the plasmids spread resistance to the antibiotic from the *Escherichia coli* bacteria to the *Bacillus* sp. bacteria. This suggests that phage superspreaders may play key roles in microbial ecology and evolution.

9.8 TRANSDUCTION AND VIRAL INCORPORATION OF HOST GENES

Transduction is the transfer of DNA from one organism to another by a lysogenic virus. Recall that a lysogenic virus integrates into the host DNA without harming the host. In transduction, the virus then breaks away from the host's genome, taking some host DNA with it. It then moves to another host cell and integrates into its DNA. It thus gives the DNA of its original host to the new host. Transduction can occur between cells of the same or different species. It does not require physical contact between the donor and recipient hosts. Unlike in transformation, the DNA is resistant to breakdown by the enzyme that degrades DNA. DNA is often transferred between bacteria by transduction. It is a source of variability, novel DNA, and increase in DNA content. It is likely important in increasing species diversity, evolutionary rates, and evolutionary breakthroughs. Transduction can transport novel genes with new functions to species when it crosses species boundaries, profoundly affecting host evolution. It is common and important in evolution. Anderson (1970) summarized the extensive evidence that viral transduction is a key mechanism for transporting segments of DNA across species and phylum barriers, and that much evolution depends largely on this transfer. (The phylum is one of the highest levels of classification of organisms.)

Transduction is a form of symbiotic genetic coevolution (see Chapter 9). The virus benefits by living off the host. The host, often a bacterium, can potentially benefit in its evolution by acquiring novel DNA. The variability of the host population increases as a result of transduction. The recipient host gains DNA, increasing the size of its genome, and potentially its complexity. Although transduction can harm the recipient host, it could benefit it. Viruses can transfer DNA between essentially any two taxa, including between domains, such as between bacteria and animals. There are many cases of viruses shuttling prokaryotic genes into eukaryotes. Transduction can potentially result in the evolution of novel adaptive forms and functions, and evolutionary breakthroughs. This can lead to the host receiving the DNA carried by transduction by the virus being in a new adaptive zone, resulting in the availability of many new niches to it, which leads to diversification, greatly increasing the number of species of the host receiving the DNA via transduction.

Payet and Suttle (2013) found that marine phages were more likely to become lysogenic, rather than produce large numbers and burst and kill the host cell, when their hosts were in an environment depleted of nutrients. This is likely an adaptation by the phage to only produce large numbers and kill the host when there are sufficient nutrients to support large host populations for the phage. After all, the phage needs large numbers of bacterial hosts to support large numbers of phage. Lysogenic phages can modify or bring new properties to their host (Menouni et al., 2015). They can increase the ability of the bacteria to survive in harsh environments (ibid.). They can provide some bacteria with resistance mechanisms or other advantages that enhance their chance of survival, sometimes completely altering the host genome (ibid.). These properties of the phage are adaptive for it, because it aids the phage's environment (its host). This illustrates a principle in evolution called levels of selection, since there is selection on two different levels: directly on the phage and on its host. *The significance of this to the ABH is that lysogeny, which is necessary for transduction, is adaptive for the phage: Selection on the phage can lead to symbiotic coevolution whereby the phage sometimes becomes lysogenic. Lysogenic phages are much better for the host than virulent phages that produce great numbers and burst and kill the host cell. Lysogeny can result in transduction, which can lead to any or all of the following: increased variability of the host population (allowing the host to adapt o environmental changes), an increase in host DNA content, the acquisition of novel DNA segments or even novel genes and functions by the host, and key innovations in the host. The new adaptations and functions and key innovations can put the host into a new adaptive zone with many new unoccupied niches, allowing the host to diversify into many new species. The virus can also pick up host genes, increasing its variability, which helps it adapt to changes in its environment. This can also lead to beneficial adaptations in the virus. The adaptation of lysogeny in phage led secondarily to the possibility of novel adaptations and diversification into many new species in bacteria by providing bacteria with new DNA and a source of genetic variability.*

The phage WO was influential in the evolution of the genome of *Wolbachia*, a bacterium which it infects. It was found in 89% of the sampled *Wolbachia* genomes. Phage WO has caused recent HGTs by transduction between *Wolbachia* in two host species that *Wollbachia* infects: a fruit fly and a small wasp that is parasitic on flies. The transfer of DNA by viral transduction between different *Wollbachia* in the same fly or wasp supports a general mechanism for genetic exchange in which genetic material moves horizontally between bacteria that infect the same environment (host). The recombination rate in certain regions of the DNA of this phage is exceptionally fast. Together, this genetic evidence indicates that phage WO is a widespread source of genetic variability in the *Wolbachia* and potentially its invertebrate hosts (Bordenstein and Wernegreen, 2004). The researchers state that gene transfer by bacteriophages could drive significant evolutionary change in the genomes of bacteria that live inside host cells that are typically considered highly stable.

A phage that kills *Wolbachia* has a family of genes that codes for enzymes that pierce and make holes in bacterial cell walls. This gene family has integrated independently into genomes across diverse environments in diverse forms of life, including viruses, archaea, bacteria, fungi, plants, and insects. It likely fights bacterial pathogens in the latter three (Metcalf et al., 2014). Archaea and bacteria that have it likely employ it to fight bacterial competitors. Carrying the gene family is adaptive for any lysogenic virus that carries it because it is beneficial to the host of the virus, for the reasons listed in this paragraph.

Viruses can capture genes beneficial to them from their hosts. Unexpectedly, it was found that viruses often have genes once thought to be restricted to their hosts that are important to their host's metabolism. Cyanophages are phages that infect cyanobacteria, which are bacteria that can carry out photosynthesis. Cyanophages themselves have genes for photosynthesis (Lindell et al., 2004; Millard et al., 2004; Sullivan et al., 2006), and act as genetic reservoirs for their hosts, generating diversity (Zeidner et al., 2005; Sullivan et al., 2006). Viral photosynthesis genes are functional, for they produce both the important molecule mRNA (Lindell et al., 2005; Clokie and Mann, 2006) and protein (Lindell et al., 2005) when they infect cells. (For what follows in this and the next paragraph, see Rohwer and Thurber, 2009, and references therein.) A mere two genera of cyanobacteria account for about 25% of global photosynthesis. Marine cyanophages that infect these two genera also have their own photosynthesis genes. This is adaptive for the phage because the cyanophage shuts down most of its host's DNA activity and production of proteins, which means that the phage cannot use its host cell's energy to produce more phage. So the phage photosynthesis genes allow the phage to generate the energy necessary for viral reproduction. A genus of tiny marine cyanobacteria called *Prochlorococcus* has specific ecological types that live at different depths and are adapted to the different light and nutrient levels found at each depth. The evolution of these ecological types seems to have been aided by phage, which act as a reservoir for genes that are horizontally transferred between hosts via transduction, allowing switching of ecological niches by bacteria. Sharon et al. (2007) showed that *about 60% of genes that encode key photosynthesis proteins (psbA genes) in global ocean surface water are of phage origin. Sometimes over 50% of all cyanobacterial plus phage psbA gene expression is of viral origin*, highlighting the contribution of viruses to photosynthesis and oxygen production (Sieradzki et al., 2019). It is further estimated that *approximately 10% of total global photosynthesis could be performed as a result of genes from phage!*

Other cyanophages have genes that appear to be to be remnants of horizontal gene transfer from cyanobacteria to varying degrees (Chen and Lu, 2002; Sullivan et al., 2005). Some of these genes, such as carbon metabolism genes as well as the photosynthesis ones, occur in many different phage lineages, suggesting they were transferred from cyanobacteria to phages independently many times. They play critical roles during infection, probably enhancing some biochemical processes for the phage. There are other marine phages that also have genes thought to be remnants of HGT from their hosts. The genes captured by phages from their hosts are often beneficial to them.

Transduction has likely played a major role in evolution and the history of life. It may have played a key role in the evolution of oxygenic photosynthesis, including bringing the two photosystems together to complete its evolution. Transduction was likely important in the evolution of many species and in some macroevolutionary breakthroughs and innovations. Each of these breakthroughs would have been followed by diversification into many new species.

Even without transduction, viruses have had a profound effect on the evolution of all cellular life, with a good deal of symbiotic coevolution (see Chapter 8). About 8% of the human genome was built from viruses (Lander, 2001). A large percentage of the DNA of other mammals and many other, in fact, I think, probably all, taxa also came from viruses.

9.9 TRANSPOSABLE ELEMENTS

The next mechanism by which the genome promotes variability, diversity, and evolution by HGT that I will discuss is the transposable element (TE), which is a mobile DNA segment that can move to another part of the genome of the same organism, to another organism of the same species, or to another species. It can sometimes carry with it some extra DNA that is not normally part of it from the genome it is moving from. Movement of TEs is important in creating genetic variability within species, giving them the ability to adapt to changing environmental conditions (Reznikoff, 2003). Movement of TEs is common in prokaryotes (Touchon and Rocha, 2007), and often serves as a mechanism to transfer genes between bacterial species (Frost et al., 1985). *TEs are very common, frequently occupying large portions of a given genome. They are the single most abundant entity of large eukaryotic genomes. They make up more than 85% of the maize genome* (Schnable et al., 2009), *almost 50% and at minimum 45% of that of humans* (Lander et al., 2001; Deininger and Batzer, 2002), *40% of the mouse genome* (Waterston and Pachter, 2002), *and almost half of that of the average mammal* (Percharde et al., 2018). *Genes coding for enzymes that help the movement of TEs to another part of the genome are quite widespread in the genomes of most species and are the most abundant genes known* (Aziz et al. 2010). That enzymes that help TEs move evolved and are common is evidence that TEs were selected for and for my hypothesis that there is selection for a favorable amount and type of variability of the genome.

TEs provide a source of genetic variability on which selection can act. Their effect on evolution can be both immediate and lasting (for reviews, see Oliver and Greene, 2009; Zeb, et al., 2009; Cordaux and Batzer, 2009; and Schaack et al., 2010). Their primary significance to adaptive evolution seems to be their ability to induce changes in the regulation of genes (Sanger, et al., 1977; Olson, et al., 1986), or the coding potential of genes (ibid.; Coulson, A., et al., 1986) without destroying existing gene functions. Changes in the regulation of genes can cause large, rapid changes in organisms that preserve the compatibility of the different parts of the genome.

TEs are among the most dynamic components of the genome. Even identical twins differ genetically due to TEs. Fruit flies are well studied, and a high percentage of known TEs occurred in them. If the fruit fly genome is typical for multicellular organisms for the number of TEs it contains, there are a great many TEs in the many other groups of organisms that have not been studied thoroughly. The rate of successful transposition (movement) of TEs was measured using a TE called the Ty1 element in brewer's yeast using several assumptions, and was found to be about once every few months to once every few years (Paquin and Williamson, 1984), which is very often on an evolutionary time scale. TEs have jumped from one animal phylum to another 12 known times.

Koonin (2016) outlined how viruses and TEs played key roles in major transitions in evolution, including the origin of protocells, the prokaryotic cell, the eukaryotic cell, multicellularity, and cellular defense systems. He presents convincing evidence that these were accomplished by a combination of selective pressure on the hosts from parasitism by the viruses and TEs and cooperation between parasite and host, and by the donation of DNA and from virus and TE to host as well as the co-opting of viruses and TEs to adaptive functions by the hosts.

Berkemer and McGlynn (2020) analyzed thousands of family trees derived from the comparison of DNA similarity data from thousands of microorganisms. They found that before the three domains of life separated, evolution was much faster, mutation rates much higher, and there were many more TEs moving between organisms than today. TEs were thus very important in evolution during the beginning phase of life.

It is astonishing that nearly 50% of the human genome is made up of TEs acquired from outside of it (Lander et al., 2001; Cordaux and Batzer, 2009). It is possible that at least some of this DNA is from symbiotic microbes in the microbiome via HGT to human cells (see Dunning Hotopp et al., 2007). Eukaryotes are truly genetic mosaics of DNA combined from different types of organisms. It is true that much of this DNA is thought to be selfish, parasitic DNA with no function to the eukaryote, but some TEs may have been instrumental in creating new functions (Oliver and Greene, 2009).

Nekrutenko and Li (2001) examined 13,799 human genes and found that 533 cases, just under 4%, helps development. (Most DNA in organisms does not code for proteins.) They found that TE integration into the DNA often has an effect on gene function. They found two mouse genes whose coding regions consist largely of TEs, suggesting that TE insertion might create new genes. They stated that there is increasing evidence for an important role of TEs in gene evolution. They also stated that many TEs are specific to groups of organisms, which implies their integration into coding regions could accelerate the splitting of species into new species.

Usually in nature, DNA codes for RNA, which then codes for a protein. But sometimes DNA can code for RNA, which then codes for DNA. Transposable elements, which are also called transposons, that code in this way are coding backward, or retro, compared to the usual way, so are called retrotransposons. They do exist (Kaessmann, et al., 2009). We now know they are widespread. Studies of retrogenes that code from DNA to RNA and then from RNA back to DNA revealed the first new gene that affects behavior, in the fruit fly (ibid.). Amazingly, analyses of chromosomal gene movement patterns via RNA-based TEs that duplicate genes have allowed new perspectives on the evolutionary biology and origin of human sex chromosomes (ibid.). I infer from this that TEs may have been involved in the evolution of human sex chromosomes. Since DNA coding for RNA is called transcription, RNA coding for DNA is called reverse transcription or retrotranscription.

DNA has many sequences of the four bases A, C, G, and T that repeat themselves over and over. Most of these sequences in humans, about 45% of the genome, are derived from TEs. They are often referred to as interspersed repeats (Smit, 1999; Prak and Haig, 2000). In addition, a good deal of the genome is probably derived from ancient TE copies that have diverged too far to be recognized as TEs.

In mammals, nearly all TEs are one of four types. These are long interspersed elements (LINEs), short interspersed elements (SINEs), long terminal repeat (LTR) retrotransposons, and DNA TEs. The first three are retrotransposons, while DNA TEs are not. Thus, LINEs, SINEs, and LTR retrotransposons first copy themselves from DNA to RNA, and then from RNA back to DNA by reverse transcription. The newly created DNA then inserts itself in another part of the genome. The original TE remains at its original site and a copy of it goes to a new site in this copy-and-paste mechanism. This increases DNA content and is important in evolution. DNA TEs are cut-and-paste types, exiting the DNA and then inserting themselves elsewhere in it. Some insert nonspecifically to any target site, while others bind to specific sequence targets. When a DNA transposon exits a gene, the resulting gap tends to be repaired incorrectly, resulting in many copies of the same sequence. So they also increase DNA content. And sometimes DNA transposons do not simply jump to a new area of the genome, but replicate and move to a new site, increasing DNA content. All four types increase the genetic variability upon which natural selection acts while maintaining the integration and self-compatibility of the genome. All can potentially result in beneficial adaptations and even evolutionary breakthroughs. As of the time of the work of Lander et al. (2001), LINEs, SINEs, LTR retrotransposons, and DNA transposon copies made up 20%, 13%, 8%, and 3% of the human genome's sequence, respectively.

LINEs are among the most successful entities in eukaryotic genomes. In humans, they are about 6,000 bases long. When a LINE RNA undergoes reverse transcription, this often does not proceed all the way to the end, resulting in many shortened, nonfunctional insertions of DNA.

The LINE machinery is believed to be responsible for the majority of reverse transcription in the genome, including the production of processed pseudogenes (Esnault, C., et al., 2000). A pseudogene is a DNA sequence that resembles a gene and was formerly a gene, but has been mutated into an inactive form with no function over the course of evolution. Of the three LINE families in the human genome, only LINE1 (L1) is still active. It is the only autonomously active human TE. It jumps frequently, and makes up 17% of our genome. There are about 950,000 copies of it in our genome, but only about 100 of them are complete and jump. It is abundant in the genomes of almost all mammals.

L1 and Alu, two elements actively increasing in number, make up about 25% of the human genome and contribute extensively to disease (Deininger and Batzer, 2002). Ancient elements that have diverged beyond recognition have also undoubtedly contributed to the composition of mammalian genomes and chromosomes.

There is evidence that some activity of L1 is necessary for brain function, notably of the hippocampus, a brain structure that has a major role in learning and memory, and of the frontal cortex, the brain region especially important for planning appropriate responses to both external and internal stimuli (Erwin, 2014). L1 is highly expressed in early development. In mice, it is needed to develop past the two-cell stage of development. It represses the gene Dux, a master activator in the two-cell stage. This is important because sometimes genes need to be turned off for normal development to proceed. It promotes synthesis of a certain important type of RNA called ribosomal RNA, and plays a key role in coordinating the development of embryonic stem cells, which are cells in early development that can develop into any type of cell in the organism. Working with other factors, it helps regulate the production of RNA, the ability of undeveloped cells to develop into different types of mature cells, and embryonic stem cell self-renewal (Percharde et al., 2018). Nigumann et al. (2002) stated that it has two regulatory regions. They showed that it may be involved in control of a number of cellular genes. *Thus, L1 jumped around the genome of human ancestors and landed in key places, building genes for brain development, synthesis of RNA, embryonic stem cell development, and more, and helped build genes to develop past the two-cell stage in mice.* It clearly helped with large beneficial adaptations. These adaptations would have led to diversification into many new species. Thus, L1 is a great example of how the genome's behavior affects evolution, evolutionary breakthroughs, and diversification into new species.

The L1 sequence has moved from humans to the genome of the bacterium that causes gonorrhea, *Neisseria gonorrhoeae* (Anderson and Seifert, 2011). This shows that HGT can occur from a human host to a bacterium, which has important implications for the coevolution of humans and microbes.

SINEs are thought to use the LINE machinery for moving around genomes. They are short, approximately 100–400 bases long, and do not code for any proteins. The only active SINE in the human genome is what is called the Alu element. There is a SINE that codes for an RNA in the brains of rodents. It is specific to the brain. Some RNA carries the DNA code to direct the manufacture of proteins. This does not do that. But it is involved in fine-tuning the regulation of the production of proteins in the brain (Robeck et al., 2016). Absence of it is associated with alterations in major excitatory transmissions of the rodent brain, and behavior that is harmful to the organism performing it (Briz et al., 2017). SINES are another example of the behavior of the genome affecting evolution and adaptation, and, as a result, promoting diversification.

Long terminal repeat (LTR) retroposons are autonomous and are flanked by long repeating sequences several hundred DNA bases long at each end that are identical to each other. The repeating DNA sequences at each end of an LTR retrotransposon are called long terminal repeats (LTRs) without the word retrotransposon. LTR retrotransposons contain all the elements needed to regulate the activity of their DNA. A few hundred genes use terminators donated by LTR retroposons to stop their coding of RNA. Other genes are regulated by elements derived from repeat elements (Brosius, 1999). These represent examples of how TEs have made creative contributions to the genome.

LTRs may flank both ends of a series of genes, pseudogenes, or a retrovirus integrated into its host cell's genome. In all of these cases they form a retrotransposon. They may also flank both ends of an endogenous retrovirus. Some retrotransposons lack LTRs; all retroviral DNA segments have them. Copies of segments of DNA flanked by LTRs an often be found hundreds or thousands of times in one genome.

Cohen et al. (2009) summarized known examples of LTRs that function as alternative regulators of genes in humans, and the evidence that LTRs were repurposed to cause genes to be expressed significantly differently than they were before the LTRs were present. They said the following. Changes in gene regulation are thought to be major drivers of evolution, with the creation of new regulatory regions likely being instrumental in contributing to the number of species in vertebrates. Transposable elements are recognized as important in gene regulation. There are specific examples of mammalian genes regulated by regions of DNA provided by TEs, including endogenous retroviruses. Studies have revealed hundreds of additional instances where this is likely to be the case. LTRs of retroviruses naturally contain abundant regulatory signals, so endogenous retrovirus LTRs may regulate mammalian genes, and it is reasonable that repurposing of an LTR module that regulates genes provides opportunities for evolution of new patterns of gene regulation. New patterns of gene expression can arise as a result of LTR usage, but this is relatively rare and largely restricted to the placenta. The LTR often seems to be a minor, alternative gene regulator, probably exerting a subtle overall effect on gene expression. We can see from this that transposable elements are important in gene regulation, and that gene regulation is important in evolution. Hence, transposable elements are important in evolution. When they cause evolutionary breakthroughs, this is followed by diversification, increasing the number of species. LTRs are more subtle in their effects on gene expression and hence evolution, but still have some importance on evolution, adaptation, and diversification.

In summary, *LTR retrotransposons have caused some adaptive evolution by providing regulators of genes*. LTR retrotransposons are another example of how the genome's behavior promotes variability, adaptive evolution, and diversification without disrupting the genome, and they provide further evidence that selection optimizes the amount and type of genetic variability.

Alu elements are the most abundant TEs in humans. They are derivatives of endogenous retroviruses (see Chapter 8). They cannot move on their own anymore, and can jump only when accompanied by L1. They are about 300 DNA bases long. There are over a million copies of them dispersed throughout the human genome (Szmulewicz, Novick, and Herrera, 1998). These retrotansposons make up about 10.7% of the human genome. They are estimated to have been inserted in 75% of all human genes (Kim et al., 2004). They originated from the fusion of two distinct DNA segments over 100 mya. Highly conserved in primate genomes, they originated in the genome of a common ancestor of rodents and primates (Kriegs et al., 2007). Most human Alu elements are in the corresponding positions in other primate genomes, but about 7,000 Alu DNA segments are unique to humans (Chimp. Consortium, 2005). Alu mutations are frequently confined to regions in the DNA that do not code for RNA or proteins, so they have little to no effect on the organism (Lander, 2001). Nonetheless, Alu elements have caused changes in control regions of primate DNA, and added DNA content (Shen et al., 2011). Though they cause several diseases and cancer when they land in the wrong area, they also regulate the expression of certain genes that are active in and affect specific tissues, and are involved in the activity of genes that do code for RNA near them (Britten, 1996). They were shown to provide microRNA target sites to numerous transcripts (Smalheiser and Torvik, 2006). MicroRNA is a type of RNA that regulates the expression of genes. The great abundance of Alu sites, their regulation of genes specific to certain tissues, and their ability to provide microRNA targets suggests they are involved in the establishment and extension of complex regulatory networks. Thus, Alu repeats have played a major role in multiple ways in the evolution of the genomes of taxa from rodents to humans. They evolved to be particularly important in regulating genes. Analysis of their organization indicates that they are selected for and benefit their human hosts (Lander, 2001). They are another example of how the behavior of the genome affects evolution, adaptation, and diversification.

Looking at TEs, Britten (2010) found the human genome makes 655 differentperfect full-length matches. This is because the human genome contains many active Alu TEs that have inserted themselves at various places in the DNA relatively recently. There are several types of these. They have many effects, including changes in gene expression, increases in recombination of DNA in chromosomes, and unequal exchange of DNA in chromosomes during this recombination. (In recombination, two similar chromosomes exchange their DNA. When this occurs, sometimes one chromosome receives more DNA than it gives, resulting in an unequal exchange of DNA.)

Rapid evolution of the human line, including brain size, probably occurred during the time the Alus were inserted into human DNA (ibid.). Alus probably underlie rapid human lineage evolution (ibid.). Britten's proposed model is that the many Alu TEs were involved in the striking evolution of the human line. He found ten examples in eukaryotes of sequences of DNA bases derived from past insertions of repetitive DNA or TEs into genomes that participate in the regulation of expression of a useful gene. This shows that segments of repetitive DNA or transposable elements have been inserted in the past in regions of the DNA near genes, been preserved, sometimes modified by natural selection, and now affect regulation of the adjacent gene, turning it on or off. Two of his examples involved the LTRs of mammalian endogenous retroviruses. Another two involved regulation of genes by repeated inserts of segments of DNA in sea urchin genomes. He said that there are now six published examples in which Alu sequences have been inserted long ago into human gene regions, were modified, and now are central in regulation or enhancement of gene activity, and that the number of published examples of Alu sequences affecting gene control is growing rapidly and is likely to continue to do so. He stated, "Taken together, all of these examples show that the insertion of sequence elements in the genome has been a significant source of regulatory variation in evolution." By sequence elements, he meant segments of DNA.

Pace et al. (2008) concluded from their work that DNA TEs have contributed significantly to shaping and diversifying the genomes of multiple mammal and other tetrapod species. Tetrapods are a taxon of four-limbed animals that includes amphibians, reptiles, dinosaurs, birds, mammals, and some others. They identified a set of DNA TE families called SPACE INVADERS (SPIN), which are 2,900 DNA bases long in the genomes of the mouse, rat, bushbaby (a group of small nocturnal primates native to Africa), little brown bat, tenrec (a group of small, spiny mammals found only on Madagascar), possum, anole lizard (lizards that can change color from green to brown and back), and African clawed frog. These gene colonizations all occurred at about the same evolutionary time period, 15–46 mya. They included some of the largest bursts of DNA transposable element activity in any species: almost 100,000 SPIN copies per genome in the tenrec. The process also led to the emergence of a new gene in the rat-and-mouse lineage derived from the gene for an enzyme that moves the SPIN TE from one location in the DNA to another.

One likely result of the incorporation of TEs in the DNA that codes for the vertebrate immune system was a genome capable of the production of a great diversity of antibodies to fight many different diseases. After jumping to an ancestor of jawed vertebrates 550 mya, a DNA TE gave rise to enzymes that splice (cut) DNA, allowing the assembly of genes for antigen receptors by a type of recombination called V(D)J recombination, which is critical to the vertebrate immune system (Huang et al., 2016). Antigen receptors are proteins on the surface of some types of white blood cells that these cells use to identify the target that the immune response is supposed to be directed at. Hence, antigen receptors help stimulate the immune response. In fact, the immune system's V(D)J recombination system operates by a mechanism similar to that of some TEs. These enzymes that splice DNA generate antibodies and other immune proteins that recognize and destroy an ever-changing variety of infectious organisms. Thus, *TEs that were basically parasites were co-opted to be a key part of the immune system.* This allowed great diversification of jawed vertebrates into myriad species. Today, there are about 60,000 species of them, which is about 99% of all living vertebrates. This is another example of how the genome's behavior aids adaptation, evolution, and diversification into many species.

Movement of TEs by retroposition, meaning by DNA coding for RNA and then RNA coding back to DNA, is an efficient mechanism for transporting coding segments of DNA around the genome with the result that they sometimes become associated with and regulated by new regulatory regions of DNA, and also for moving new regulatory segments of DNA into the genome with the result that they sometimes end up controlling genes that they had not controlled before (Brosius, 1999). Brosius (1999) argued that the relationship of TEs to their hosts is more like symbiosis than parasitism, and that host defense is often lenient, tolerating or even supporting TEs. So we can see that TEs can change the regulation of genes profoundly. Changes in the regulation of genes can lead to large changes in the phenotype that maintain the integration and self-compatibility of the genome and phenotype, and this can result in large evolutionary adaptive breakthroughs and key innovations, as we will see later in this chapter. As previously stated, the evolutionary breakthroughs are typically followed by diversification into several new species.

In fact, TEs can cause macroevolutionary leaps by altering genetic regulation. TEs are abundant in mammalian genomes and have potentially contributed to their hosts' evolution by providing novel regulatory or coding sequences. Inactive TE remnants have been transformed into small and large RNAs that regulate genes, enhancing or dampening protein production by genes and gene activity. Qin et al. (2015) found at least 409 human microRNAs that have been made from old TEs, indicating that TEs play important roles in the making of human microRNAs. They found that the proportions of microRNAs derived from TEs in humans are more than other vertebrates, especially nonmammal vertebrates. MicroRNAs are involved in RNA silencing and regulation of gene expression. Feschotte (2008) expanded earlier models and argued convincingly that TEs have been a rich source of material for the assembly and tinkering of eukaryotic gene regulatory systems. TEs contain their own regulatory regions that regulate themselves. These can be repurposed for use in gene regulation in the host.

Transcription factors are proteins that regulate the coding and hence production of RNA from DNA, usually enhancing it. They do this by binding to DNA. The RNA produced then goes on to code for proteins, so DNA indirectly codes for proteins through RNA. So transcription factors regulate the production of proteins by regulating genes. *Transcription factors work in complex networks that would be difficult to evolve by the gradual accumulation of many point mutations.* Retrotransposons have many sites where transcription factors can bind to them, and thus turn them on, causing them to produce more of themselves. If they produce too many copies, this is bad for the cell, especially since they are likely to move into a part of the cell that causes them to disrupt the cell's activity. So cells have evolved mechanisms to shut them down and make them dysfunctional. But dysfunctional retrotransposons are DNA segments that can provide raw material for evolution. They are regions of DNA that can evolve and become DNA regions where transcription factors can bind and regulate genes. This can cause the appearance of new adaptations and functions. It can promote evolution and adaptation by causing more of a protein to be produced. Many sites important for the regulation of human and mouse genes may have evolved from TEs, including dysfunctional retrotransposons. Bourque et al. (2008) showed that TEs played a large role in the evolution of DNA regions where transcription factors can bind to the DNA and regulate genes; that is, they played a role in sites for regulation of the genome, for turning up the production of proteins by genes, or turning it down. Such additions to the repertoire of gene regulation can cause rapid macroevolutionary breakthroughs and major evolutionary transitions. These tend to lead to diversification and speciation into many new species—an increase in biodiversity.

Repurposed retrotransposons called coordinators regulate gene activity in chimpanzees and humans, and are responsible for much of the differences in the faces of the two species (Prescott et al., 2015). *Many TEs inserted themselves into human ancestors after divergence of humans from chimpanzees. More than 1,000 regulatory switches found only in humans have TE remnants* (Glinsky, 2016). Thus, TEs played a key role in the evolution of humans and their divergence from chimps by acting as gene regulators.

Jordan et al. (2003) found that almost 25% of regions in DNA that are involved in turning genes on in humans that they analyzed contain sequences of bases derived from TEs. Regions involved in the simultaneous regulation of multiple genes also contain many TE-derived sequences. These two findings indicate that TEs have contributed a great deal to the evolution of both gene-specific and global patterns of human gene regulation.

Van de Lagemaat et al. (2003) found evidence that TEs affect the expression of many genes by donating signals for their regulation. They found that classes of genes that recently increased in size (i.e., in numbers of genes), such as those involved in responses to external stimuli and immunity, have a great deal of TEs, while highly conserved genes that do not change or evolve very much and have basic functions in development or metabolism lack TEs. Their findings support the view that TEs have played a significant role in the evolution and diversification of mammalian genes. The main point of their findings is that TEs are important in the evolution of genes and of their regulation. Genes that change, adapt, and evolve have many TEs, and genes that do not change or evolve very much do not have many TEs. Changes in genes with basic functions would tend to kill the organism; these genes need to be conserved and not change. Thus, it is adaptive that such genes do not have TEs, which would change them.

Many genes, enough to code for 223 proteins, apparently jumped from bacteria to vertebrates. They have only been detected in bacteria and vertebrates. Most jumped to the human genome, for they are easily detected directly there. No doubt, there are many other genes that jumped from bacteria to other vertebrate species, but have not been detected as yet. There is no reason to think that genes in bacteria jump to humans more readily than to other vertebrates. Orthologs of many genes are in other vertebrates. (Orthologs are genes in different species that evolved from a common ancestral gene by speciation.) At least 113 of these genes that jumped to vertebrates are widespread among bacteria. These genes largely code for enzymes that act within cells, rather than outside of them in places like the bloodstream. This is uncharacteristic of evolutionary innovations specific to vertebrates, which tend to be primarily proteins that act outside of cells. Many of the proteins that TEs code for have established or likely functions in the stress response or the response to toxic substances. Thus, at least some of the genes seem to be involved in important physiological functions and apparently were maintained or even increased in frequency because of their adaptive advantage or advantages. Similar horizontal transfers of genes occurred from bacteria to the worm genome (ibid.).

Forty-seven human genes have been recognized as likely derived from TEs (Smit, 1999; Jurka and Kapitonov, 1999; Lander et al., 2001). These include the genes that code for parts of a protein complex that plays important roles in the rearrangement and recombination of the genes that themselves code for important proteins used in the immune system (Smit, 1999; Jurka and Kapitonov, 1999). Thus, as discussed earlier, TEs were important in the evolution of the immune system. It is awe-inspiring to think that our immune system would be much less effective and would not exist anywhere close to how it does if it were not for TEs.

Mammals such as humans, bears, and rodents have placentas and carry their embryos in a uterus until it is born. They are called placental mammals. They are different from marsupial mammals, which carry their young in a pouch, and monotreme mammals, which lay eggs. In fact, the uterus is a defining character of placental mammals. Amazingly, *the evolution of the uterus appears to have been facilitated by the integration of transposable elements into regions in the genome regulating the gene that codes for prolactin.* Prolactin is a hormone that enables mammals (and birds) to produce milk. It has many other functions, being influential in over 300 separate processes in various vertebrates, including humans. Astonishingly, the evidence is that this jump of TEs that helped form the uterus happened several times independently in many different families of mammals. Indeed, the relevant TEs contain segments of DNA involved in gene regulation that enable the prolactin gene to become expressed in uterine cells (Lynch et al., 2011; Emera et al., 2012). Thus, *the integration of TEs, possibly from other taxa, into the mammalian genome was key and crucial for the evolution of placental mammals and the uterus.* Since prolactin has so many functions, changes in its regulation will have profound effects on the phenotype. Also, see Chapter 8 on the role of viruses in the evolution of live birth in mammals.

Since TEs produce deleterious effects by causing mutations in individuals (Ostertag and Kazazian, 2001), and the vast majority of TEs are deleterious, cells have evolved a number of mechanisms to curtail TE proliferation. Bacteria actually delete genes at high rates to remove TEs and viruses. And organisms have defenses that inactivate TEs, such as chemical alteration of the genome. Chemical modifications of DNA can make transposase, an enzyme that binds to the end of a transposon and catalyzes its movement to another part of the genome, unable to reach certain areas of the DNA. Hosts also use what is termed RNA interference, in which small RNAs shut down TEs. (RNA interference is also used in the regulation of gene expression in eukaryotes.) Strangely, production of high quantities of transposase also decreases overall TE activity, and cells use this technique as well to thwart TEs. Thus, there are many inactive TEs in many species. Large portions of the genomes of many species are TEs, and the vast majority, often easily over 90% of these, are inactive. For example, in vertebrate cells, nearly all the more than 100,000 DNA TEs per genome are associated with genes that code for inactive enzymes that would function to move them around the genome if they were functional. And the activity of TEs in the human line has declined greatly over time. Human DNA TEs seem to have become almost fully inactive. In fact, the vast majority of TEs in the human line are inactive and were formerly functioning.

Some deactivated DNA TEs become segments of DNA within genes that never get coded into proteins (called introns). Many of the inactive regions of the genome that they help create are potential information for evolution. That is because they could potentially become new genes that are coded into proteins or regions that regulate genes. These occasionally can lead to macroevolutionary jumps and key innovations. Thus, the interaction between TEs and defensive responses by their host cells to control them can promote adaptive evolution, potentially sometimes in large leaps. Some TEs generated large families of genes often associated with speciation events. Thus, *although dead TEs comprise what some call junk DNA, they are useful for evolutionary potential.*

TEs add genetic variability, which is usually deleterious, but can occasionally be adaptive (Gonzalez and Petrov, 2009). TEs can occasionally cause evolutionary leaps to new adaptive forms, whether they jump within an organism, between organisms of the same species, or between species. Since genomes are integrated systems, most TEs will land in areas where they will be deleterious, often disabling genes. But TEs are genes or segments of DNA that underwent evolution and the test of natural selection for eons, and in some cases have served some adaptive function, and are not random changes in the genome. They are segments of DNA that are fit and were compatible with the genome they were originally in, and coadapted with it. They thus have a much higher probability of being compatible with and beneficial to the area of the same genome or the new genome they integrate into, and of leading to an adaptive breakthrough, than random mutations. The accumulation of TEs landing in the same general area of the genome can create novel genes or regulatory regions that produce radically new, adaptive features that allow the species to occupy a new adaptive zone. They make the acquisition of novel traits in one or a few generations possible. Once this results in a new adaptive zone, diversification into many new species can follow as a result of the availability of the myriad of new niches available.

Accumulation of TEs adds DNA to the genome, enlarging it. Retrotransposons, which are copy-and-paste TEs, increase DNA content by duplicating themselves before moving to another location. DNA (cut-and-paste) TEs can add DNA content by the mechanism explained earlier. The increase in DNA from either type of TE can increase complexity and occasionally cause an adaptive innovation. Again, this can result in many new available niches, and hence diversification into many new species.

There is substantial evidence that organisms co-opted TEs for new functions in their cells. Transposable element genes have been repeatedly captured by their hosts during evolution in most eukaryotic lineages. *They have taken on the following functions: they protect eukaryotic species against infections, are necessary for reproduction of eukaryotic organisms, allow the replication of their chromosomes, control cell proliferation and death, and are necessary for plant development* (Volff, 2006). This is a spectacular amount of evolutionary innovation from but one form of HGT. This shows that TEs were crucial in the evolution of many important functions. They represent another example of the importance of the behavior of the genome in promoting evolution, adaptive breakthroughs and key innovations, and diversification.

In many eukaryotic genomes, but especially prevalent in higher plants, demonstrated notably in maize, rice, and rockcress, TEs have combined with fragments of DNA in the genome, forming DNA segments that are hybrids from two different sources. These have been captured, rearranged, and increased in number in plant genomes over millions of years. *There are over 3,000 such hybrid TEs in rice, containing fragments derived from more than 1,000 cellular genes.* These TEs frequently contain fragments from multiple regions of the DNA that are fused to form segments of DNA that code for proteins. *The acquisition of gene fragments by these TEs might represent an important new mechanism for the evolution of genes in higher plants* (Jiang et al., 2004, and references therein). That is exciting.

Species being physically close to each other seems to allow HGT between them. And viruses and parasites may aid the spread of TEs across widely diverged species (Schaack et al., 2010). The retrotransposon BovB, a LINE about 3,200 DNA bases long, has jumped between species of vertebrates at least nine times, going to lizards and snakes, egg-laying mammals, marsupial mammals, ruminants, and various African mammals, probably using reptile ticks and/or poxviruses for movement between different species (Piskurek and Okada, 2007; Gogolevski et al., 2008; Walsh, 2013). There are 83 species of poxviruses (smallpox is one disease they cause), and they infect many groups of vertebrates.

Gilbert et al. (2010) presented evidence that interactions between small parasites that are invertebrates and their vertebrate hosts have promoted the horizontal gene transfer of four transposable element families between invertebrates and vertebrates. One of these cases is the parasitic kissing bug, which feeds on the blood of mammals, birds, and reptiles, and transmits trypanosomes, which are parasites that live within cells and cause Chagas disease, a serious tropical disease. The kissing bug's genome has four distinct TE families that invaded the genomes of a diverse, but overlapping, set of vertebrates. This bug's TEs are about 98% identical. They are similar to those of two of the South American mammalian hosts that the bug commonly uses, the opossum and squirrel monkey. They also found a TE family in the great pond snail, a very widespread intermediate host for numerous flatworms that parasitize diverse vertebrates. An intermediate host is an organism that provides a home for the sexually immature form of a parasite and is required by the parasite to undergo development and complete its life cycle. It is distinct from the primary host, so the parasite has two different species of host. This TE family's ancestral DNA base sequence is nearly identical to and clusters with those found in mammals of the Eastern Hemisphere. The kissing bug may have transferred the TEs to the vertebrates directly from exchange of blood or saliva during feeding or through the trypanosomes, and the pond snail to vertebrates through the flatworm parasites. These findings support a hypothesis that host-parasite interactions facilitate HGT among animals (Loreto et al., 2008; Houck et al., 1991). And the large amount of DNA generated by the amplification of the horizontally transferred TEs supports the idea that the exchange of genetic material between hosts and parasites influences the evolution of both their genomes.

P elements are TEs that colonized new genomes by moving from species to species in the fruit fly about 50 years ago, likely by the mite, *Proctolaelaps regalis* (no common name), which is parasitic on fruit flies. Within 50 years, P elements spread to every population of *Drosophila melanogaster* from *Drosophila willistoni* (Daniels, et al., 1990; Engels, 1992). These are two species of fruit fly.

Rolling-circle transposable elements, or Helitrons, are a newly recognized group of TEs abundant in the genomes of zebrafish, invertebrates, and plants. Two rolling circle TEs have entered the genome of the little brown bat, and amplified within the last approximately 30 to 36 million years, and now have copy numbers of about 15,000 and over 100,000, making up at least 3% of its genome (Pritham and Feschotte, 2007; Ray et al., 2008). They are restricted to the lineage of one family of bats called vesper bats. The time of the amplification coincides with very high speciation rates in vesper bats, which is in fact one of the highest speciation rates in mammals. Pritham and Feschotte (2007) say: "Given the demonstrated propensity of rolling circle transposons to mediate the duplication and shuffling of host genes in bacteria and maize, it is tempting to speculate that the massive amplification of Helitrons in vesper bats has influenced the evolutionary trajectory of these mammals." This would no doubt include the coincident high speciation rate. This is another example of the behavior of the genome influencing evolution and increasing species number.

Some rolling circle TEs in maize contain segments of different genes, supporting the idea that these TEs have a role in the shuffling of parts of genes. This also supports the idea that these TEs are involved in the evolution of new proteins. This is because genes code for proteins, so shuffling parts of genes and hence bringing parts of genes together that were previously not close to each other creates new proteins. These rolling circle TEs are plant DNA segments that are hybrids composed of DNA segments from various parts of the genome, like those discussed earlier. There is evidence they actively code for RNA. *Rolling circle TEs in maize seem to continually produce new nonautonomous TEs responsible for the duplication and insertion of gene segments into new locations in the genome and for the unprecedented diversity of genes of maize. The maize genome is continually changing, for TEs continue to change both the parts of the maize genome that have genes and the parts that do not have them, profoundly increasing the genetic diversity of maize* (Morgante, et al., 2005).

Many small RNAs that protect cells by keeping TEs from moving were originally TEs. Organisms repurposed them into molecules that protect cells from TEs, which, as I stated, are usually bad for the cell. Thus, TEs can take on a number of functions for the organism they are in.

In bacteria, some TEs also contain integrons, which allow bacteria to evolve quickly through the acquisition, accumulation, and differential expression of new genes. Integrons are segments of DNA that can capture and express genes from other sources. The integron has what is called a recombination site, where gene cassettes may be inserted. Integrons often acquire gene cassettes and exchange them with other integrons. A gene cassette is a segment of DNA that can move from one place to another in the genome and that possesses usually only one gene, since they tend to be only 500–1,000 DNA bases long. They are not always incorporated into an integron, but sometimes exist freely as circular DNA.

They can jump within an organism's genome or to that of another organism. They can carry antibiotic resistance genes. There are over 40 known antibiotic resistance genes on cassettes. Some cassettes have virulence genes. Most genes in integrons have not yet been identified. Integrons do not occur in eukaryotes.

The peppered moth (*Biston betularia*) is usually gray and rests on the gray trunks of trees, causing it to be camouflaged. A black color variant increased in number in Great Britain because air pollution blackened trees and allowed this form to be better camouflaged than the gray form in polluted areas. This black form was caused by the insertion of a large TE into the gene called cortex (van't Hof, et al., 2016). The evolution of this black form is an iconic example of natural selection discussed in biology textbooks. It is interesting that such a well-cited example of natural selection is the result of a TE.

Some TEs increase their rates of moving from one place to another in the genome when they are under stress (Strand and McDonald, 1985). Increasing genetic variability and the mutation rate under stressful conditions is highly adaptive, for stress implies an environment unfavorable to the organism. Genetic variability allows evolution to adapt to the unfavorable environment, to evolve to a form better adapted to the environment, making it less stressful. Increasing rates of TE movement under stress increases the organism's variability and hence ability to adapt when it most needs these attributes, which is when the environment is unfavorable.

The retrovirus MER1 first infected primate ancestors of humans about 45 to 60 mya, and apparently used interferon gamma in replicating itself. Interferon gamma is a chemical, a cytokine, that is critical for innate and adaptive immunity against infections by viruses, some bacteria, and protozoa. The primate hosts evolved in response, with the result that the remnant of MER1 became an inactive TE. Our primate ancestor converted the inactive TE, making it part of its immune system. It now helps the immune system fight infections, and is part of a system that is alerted by interferon gamma to invasions by pathogens. Humans have hundreds of copies of this TE scattered in their genomes, and MER1 relatives are in lemurs (small primates found only in Madagascar), bats, even-toed hoofed mammals, and carnivores. Thus, retroviruses shaped the evolution of a major branch of immunity, the interferon response. Retroviruses have dispersed many segments of DNA that turn on genes when interferon binds to them in several different mammalian genomes. And their elements are involved in the regulation of essential immune functions in humans. These regulatory DNA sequences now constitute a dynamic reservoir of DNA regions that turn on immune genes when induced to do so by interferon. These retroviruses have fueled genetic innovation in mammalian immune defenses (Chuong et al., 2016).

The bicolored deceiver (*Laccaria bicolor*) is a fungus that lives in association with tree roots and is symbiotic with trees. The tree gives it carbohydrate from the tree's photosynthesis, and the fungus expands the ability of the tree's roots to obtain water and nutrients. The DNA of this fungus shows an expansion of several multigene families, and many repeated DNA sequences and TEs, indicating adaptation to symbiosis with plant roots was aided by all of these (Martin et al., 2008). Plant-fungal symbiosis is thought to have been key to plants colonizing land. It is essential for the health of plants and for forests to thrive.

9.10 TRANSPOSABLE ELEMENTS AFFECT GENOME SIZE, INFLUENCING THE PHENOTYPE AND INCREASING BIODIVERSITY

TEs are selfish DNA. Even though they have a greater probability than random mutations of causing adaptive mutations, they usually have no effect or a deleterious one. Any changes to a well-adapted, coadapted, integrated genome will have a high probability of being deleterious to the organism and genetic system. That is why TEs are called selfish DNA. They proliferate and increase in number selfishly at the expense of the organism. They do not do this consciously, so the metaphor with selfishness is not valid in every aspect. It does mean that there are levels of selection. Selfish DNA segments such as viruses and TEs compete for space in the genome and represent one level of selection. The genome or the organism is another level. Selfish DNA (and RNA), such as viruses (some viruses have RNA) and TEs, are in conflict with the genome and organism; what is adaptive for one level of selection may be harmful to the other. That is why genomes and organisms have evolved ways to delete or silence TEs. Yet, the selfish DNA and RNA segments cannot harm their hosts too much, or they will be eliminated, since they are dependent on their hosts; their hosts are their habitats. If they help their hosts, they are better off, and could even diversify. As such, they undergo group selection. Selfish genetic elements are material for evolution of both new genes and regions that regulate genes. They could cause an existing gene to be more active, causing an adaptive change in the organism. These kinds of changes help the selfish segment by helping its host, and could allow diversification of both host and selfish element. Thus, TEs, viruses, and the like are not entirely selfish. They are both harmful and helpful to the host.

Some taxa have evolved mechanisms to limit their TEs more than others. Salamanders have less effective mechanisms to limit TEs, so have much more of them, and hence much larger genomes than other groups of animals. This is possible because salamanders have lower metabolisms and slower growth rates, and move around less than other kinds of animals. So they can tolerate more TEs. Once they acquired more TEs, they became more sluggish. This allowed more tolerance for TEs, so less selection against them, so yet more TEs, in a positive feedback loop. Furthermore, when a genome has more TEs, additional TEs can enter the host genome where there are existing TEs with a higher probability. Entering the host genome in areas where there are clusters of TEs is less likely to harm the host. This is another positive feedback loop resulting in more TEs. The Neuse River waterdog (*Necturus lewisi*) is a salamander with the largest genome of any four-legged animal, at over 38 times the size of the human genome. As a result, it needs large cells, and its cells are up to 300 times larger than those of a lizard, bird, or mammal. It has to spend a great deal of energy when it duplicates its chromosomes when its cells divide. So it does not have the energy to undergo metamorphosis and stays a larva that lives in the water and breathes with gills for its entire life. It breeds as a larva, and never changes into an adult that lives on land, as the typical amphibian does. It is sluggish and sits still and waits for prey to come its way to eat, again because it lacks energy because of its huge genome. Because of the energy requirements of their large genomes, many salamanders have weak hearts with thin walls. And with a great deal of DNA, cells are larger and fewer. The portion of the brain involved in vision in salamanders was elucidated by Roth et al. (1997). This system has small numbers of large neurons, a low degree of morphological differentiation among neurons, and a small proportion of nerve fibers with the myelin sheath in the optic nerve. The myelin sheath is the protective layer that wraps around the nerve fibers to aid in insulating nerve cells, and to increase the number of electrical signals being transferred. Development is slow, and the brains do not develop very well. Brains of intelligent animals have large numbers of nerve cells. All of these features of the salamander brain indicate that they are not very intelligent. But with large genomes, cells do not differentiate as rapidly, and maintain their potential for regeneration; thus, salamanders have a spectacular ability for regeneration (Sessions and Wake, 2021). They can even regenerate parts of their brains. Thus, the large genomes affect evolution and many aspects of the phenotype of various salamander species. Birds, on the other hand, need high metabolism for flight, and so underwent strong selection to control TEs, and so do not generally have large genomes. The genome of the American robin (*Turdus migratorius*) is less than half the size of the human genome. Bats fly and need high metabolisms, and thus have the smallest genomes of the mammals. The mountain grasshopper (*Podisma pedestris*), with a low metabolism, has a genome over five times the size of that of the human. Thus, TEs have acted as ecosystem engineers, profoundly and significantly affecting natural selection on their hosts, the evolution of their hosts, and their hosts genotypes and phenotypes. In so doing, they have increased the number of niches and lifestyles in their hosts, and hence the number of species and higher taxa. Selfish DNA elements such as TEs have increased biodiversity tremendously by influencing genome size and the factors it is related to. For a good summary of selfish DNA, genome size, and their effects on evolution and phenotypes, see Fox (2022).

9.11 THE MICROBIOME CAN SUPPLY GENES TO ITS EUKARYOTIC HOST

Chapter 7 pointed out that there is a tremendous diversity of viruses, bacteria, archaea, fungi, and invertebrates and their genes in the microbiome in and on all eukaryotic species. Any of these small organisms has DNA segments and genes with the potential to jump into the genome of its eukaryotic host. Most of these transfers of DNA from the microbiome would be disadvantageous to the host. However, at times these DNA transfers could bring new genes and new functions, and provide adaptive changes and even large evolutionary breakthroughs to the eukaryotic host.

9.12 EXON SHUFFLING AND PROTEIN DOMAINS

DNA codes for RNA, and RNA codes for proteins. It is by this method that DNA indirectly codes for all the proteins in an organism. Proteins are the major molecules that determine the characteristics of an organism. Not all of the RNA that DNA codes for is used to code proteins. Genes are made of DNA. The part of the gene that gets coded into protein by the RNA is called an exon. The part of the gene that codes for the RNA that does not get coded into the protein is called an intron. In a process called exon shuffling, two or more exons from different genes can be brought together, or an exon can be duplicated, making two copies of it. This creates new exon-intron structures. Transposable elements can cause exon shuffling. So can unequal genetic recombination from sexual reproduction, and recombination between chromosomes that are not of the same pair. Related to this is the shuffling of protein domains. Proteins are made up of modules called protein domains. A domain is a part of a protein that acts independently of the rest of the protein and acts as a module that fits together with other protein domains like the plastic bricks of a Lego set. Domains fit together to make proteins. Many proteins consist of several domains. Some proteins are made up of but one domain. Many domains together can form multidomain and multifunctional proteins with a vast number of possibilities (Chothia, 1992). One domain may appear in several different proteins. One domain may be in different sets of protein families. Using domains as modules makes a great many different proteins possible. Domains can be put together in many different combinations to make a great number of different proteins. Domains generally vary in length from about 50 to 250 amino acids. Amino acids are the molecules that are the building blocks from which proteins and protein domains are built. Domains often form functional units. Molecular evolution uses domains as modules that can be shuffled into different arrangements to create different proteins with different functions. For example, the protein pyruvate kinase contains three different domains connected by several amino acid chain linkers.

Although additional domains specific to vertebrates will be found, at this point it appears that few new protein domains have been invented in vertebrates. Most protein domains trace at least as far back as a common animal ancestor. There seem to be only about 30,000–40,000 protein-coding genes in the human genome. Genes tend to code for protein domains, not entire proteins. Humans appear to have only about twice as many genes as worms or flies, yet are much more complex.

A major explanation for this is that there is substantial innovation in the creation of new vertebrate proteins, and that human genes are spread out over much larger areas of the genome and create many different proteins by putting together different combinations of protein domains. There are not very many more protein domains in humans than other worms or flies, species, but there are many more proteins and hence greater complexity because the domains are put together in many different ways to produce many more proteins than other species. Novel linear arrangements of domains within proteins abound in humans. There is more alternative splicing,[4] generating a larger number of protein products (Lander et al., 2001). New proteins tend to be created by shuffling pre-existing domains rather than the invention of novel domains. This could well result in about five times as many primary protein products in the human as in the worm or fly.

New architectures can be created by adding or deleting domains, inverting parts of chromosomes, moving segments of DNA between genes, movement of TEs (including between species), unequal genetic recombination between chromosomes in sexual reproduction, recombination between chromosomes that are not of the same pair in sexual reproduction, and slippage of DNA polymerase during replication. DNA polymerase is the enzyme that replicates DNA, making two molecules of DNA from one. Each of these mechanisms can result in new proteins from already existing domains and from exon shuffling. Again, this is using existing DNA segments that have passed the test of natural selection over long time periods. Large segments of DNA and long chains of amino acids are maintained, keeping integration and compatibility of the genome and between protein domains intact. The domains are shuffled around into new combinations to produce novel proteins and adaptations, without losing the integration and compatibility of the parts of the organism and genetic system.

Apic et al. (2001) found that two-thirds of proteins in unicellular organisms and more than 80% in animals are proteins with more than one domain. Ekman et al. (2005) concluded that 40% of prokaryotic proteins and about 65% of eukaryotic proteins have two or more domains. Both these studies show significantly higher percentages of proteins with two or more domains in more complex than in simpler organisms. Many domains in proteins with two or more domains in eukaryotes exist as autonomous proteins in prokaryotes (Davidson et al., 1993), implying at least some eukaryotic proteins with two or more domains evolved from proteins with only one domain. For example, vertebrates have a protein made up of three different domains. In yeast, two of these domains make up one protein and the third domain is a separate protein. In bacteria, each domain is a separate protein (Henikoff et al., 1997). Thus, it appears that single domains combined into multiple domain proteins with novel functions as more complex organisms evolved from simpler ones. Proteins with more than one domain were favored by selection for novel functions, and lineages diverged from each other partly because each lineage evolved different combinations of domains. Species can speciate from one to two species by this mechanism. This is another mechanism by which the genome promotes variability, evolution, evolutionary innovation, and an increase in biodiversity.

9.13 NEW GENES CAN ORIGINATE FROM NONCODING DNA AND NONCODING RNA

New genes can be formed from noncoding DNA. Formation of new genes requires only two features: the generation of a DNA segment that codes for a protein domain and the creation of a site on the DNA to regulate this segment. Levine et al. (2006) reported the origin of five new genes in the fruit fly *Drosophila melanogaster* genome from noncoding DNA. Zhou et al. (2008) found that noncoding DNA sequences play an unexpectedly important role during the origin of new genes in the *Drosophila melanogaster* species complex, and are responsible for 11.9% of the new genes in it. They also found that movement by retrotransposons had generated about 10% of the new genes, and that about 30% of the new genes in the species complex recruited various DNA sequences and formed chimeric (mixed) gene structures. They estimated that the rate of origin of new genes in the species complex is 5 to 11 genes per million years. Also, in bacteria, lysogenic phages act as buffer zones that tolerate genetic variability, increasing the probability of the formation of new genes (Ramisetty and Sudhakari, 2019). Additionally, An et al. (2023) showed how genes important in human brain development and other functions were created from DNA that coded for RNA that was not coded into proteins. The RNA was not coded because it could not leave the nucleus of the cell. A mutation in the DNA caused it to code for the RNA in an altered form that could leave the nucleus and be coded into a protein. In this way, 74 new genes were created, two involved in brain function.

9.14 DUPLICATION OF DNA SEGMENTS

Another built-in tendency in the genomes of organisms to promote evolution and diversification is segmental duplication, the doubling of a segment of typically 10,000 to 300,000 DNA bases into two identical or nearly identical copies. This can happen by unequal genetic recombination during sexual reproduction, with one chromosome giving some of its DNA to the other, which results in one chromosome missing DNA and the other with two copies of the segment the donating chromosome is missing. But segmental duplication can occur by other mechanisms as well. Sometimes the locations of donor and recipient regions are not adjacent, indicating they did not arise by unequal recombination. Segmental duplication sometimes results in the duplication of a gene. At least 3.3% of the human genome consists of DNA segments created by duplication (Lander et al., 2001). Ohno (1970) wrote an entire book demonstrating the importance of gene duplication in evolution.

Complete eukaryotic genome sequences have shown that gene duplications are very common, and large-scale gene duplication events have been associated with important macroevolutionary breakthroughs, transitions, and major leaps in development (Maere et al., 2005). Each of these would have been followed by diversification, creating several new species (Maere et al., 2005). Duplication of DNA segments is much more frequent in humans than in yeast, fly, or worm (Lander et al., 2001). About 60% of protein families are more numerous in the human than in yeast, mustard, worm, or fly (ibid.), indicating that gene duplication has been a major evolutionary force in the human evolutionary line.

Scientists have applied principles involved in the study of populations and DNA base sequence analyses to calculate that, on average, genes duplicate at the rate of about 1% per gene per million years (Lynch and Conery, 2000). The most central regions and regions closest to the ends of human chromosomes have many large recent duplications of DNA segments from other parts of the genome.

Evolutionary novelties can arise through gene duplication as follows. After the gene duplicates into two genes, one gene continues its normal function and codes for a protein domain, while the other does not code for a protein domain and has no function, being what is called a pseudogene. The pseudogene changes and mutates. The functioning gene keeps the needed function going, allowing the pseudogene to mutate many times into many different maladaptive forms. The pseudogene undergoes many unfavorable mutations until it eventually has a lucky adaptive mutation, resulting in an adaptive innovation. Without the functional gene producing a functional protein domain and hence keeping the function occurring during the many unfavorable mutations in the pseudogene, the pseudogene would cause harm or death to the organism carrying it. Redundancy in the genome in the form of two copies of the same gene covers the loss of function of the pseudogene until it mutates to a novel form with a new adaptive function.

Following are some examples of this mode of evolution by gene duplication. Trypsin and chymotrypsin are enzymes that break down proteins so the organism can digest them. Since they each attack different areas of the proteins they digest, an organism needs both of them to effectively digest proteins. Since they would kill the cells of the pancreas that make them, they are made as longer molecules, trypsinogen and chymotrypsinogen, which are broken down to trypsin and chymotrypsin, respectively, when they are safely out of the pancreas. The amino acid sequence of the two molecules and other evidence shows that trypsinogen duplicated and the duplicate gene evolved into chymotrypsinogen. Interestingly, this happened independently in vertebrates and at least some invertebrates. For example, it happened independently in cows and butterflies.

Microtubules are ancient, major components of the cytoskeleton, which is the "skeleton" of a cell. They are found in all eukaryotic cells, and they are involved in cell division, cell movement, transport of materials within the cell, and maintenance of the cell's shape. Actomyosin is important in muscle contraction, and is made up of the two proteins, actin and myosin. It is clear that actin came from a duplicate of the gene for one of the protein domains that is involved in the formation of microtubules. Thus, muscle contraction is possible only because of duplication of a gene or genes for microtubules.

Hemoglobin is a protein in red blood cells that carries oxygen to the body's organs and tissues, and carries carbon dioxide from the organs and tissues back to the lungs. It is made up of subunits called alpha, beta, gamma, and delta chains. It is most efficient when it exists as two alpha chains combined with two beta chains, two alpha and two gamma chains, or two alpha and two delta chains. It started out with only the beta chain, so was less efficient at carrying oxygen. The beta chain gene duplicated and created a pseudogene that eventually evolved into the alpha, gamma, and delta chains, and then these combined as described above. The alpha chain appeared long before the emergence of the first mammals.

Analysis of the genome of the opium poppy (*Papaver somniferum*) reveals that an ancient duplication of a DNA segment occurred 110 mya, followed by fusion of two genes. The gene from the fusion is responsible for a shape shift in some molecules, leading those molecules along the pathway toward morphine-like molecules. Then, a duplication of the whole genome occurred approximately 7.8 mya. Some of the resulting pseudogenes evolved novel functions, leading to the production of morphine by the poppies. Therefore, gene duplication and fusion of genes led to the evolution of specialized metabolic products in the opium poppy (Guo, 2018).

Duplications of DNA segments can create entirely new gene families. In humans, this includes small families, such as the five related genes of the beta-globin cluster of hemoglobin, some of which were just discussed, on chromosome 11, and large ones, such as the receptor gene clusters for smell, which together contain nearly 1,000 genes and pseudogenes (Lander et al., 2001).

9.15 REPETITIVE DNA SEQUENCES

Differing from DNA segments that are duplicated, which are long and can be genes, there are repetitive sequences, which are segments of DNA that occur in multiple copies throughout the genome. In many species, a large part of the genome is highly repetitive. In the human genome, over two-thirds of the sequence consists of repetitive sequences (de Koning et al., 2011). There are two different types of repetitive DNA base sequences. One type consists of repeated sequences adjacent to each other in tandem, the other of repeats dispersed throughout the genome. Many repeated sequences were evolutionary derived from viral infections; that is, from viruses (Villarreal, 2005).

Approximately half of the human genome has no known function or origin. Work by de Koning et al. (2011) demonstrated that approximately half of this mysterious part of our DNA is comprised of repetitive or repeat-derived sequences, which are most likely dominated by TEs. Their work suggested that a large majority of the human genome sequence—over 66 to 69%—may be comprised of repetitive and repeat-derived DNA segments. Their work also indicated that standard methods for detecting repetitive sequences are likely missing large numbers of TE fragments. In the case of a type of TE that is a SINE called a mammalian-wide interspersed repeat, the researchers predicted from their findings that half of the sequence that is likely present in the genome has not been detected. Their work suggests that the human genome consists of substantially more repetitive sequences and TEs have played a much larger role in shaping the history and content of the human genome than researchers previously believed.

In many species, a significant fraction of the genome is highly repetitive DNA. Genomes can contain a large quantity of repetitive DNA base sequences, far more than that portion of the genome occupied by protein-coding genes (reviewed in Gregory and Herbert, 1999 and Hartl, 2000). While less than 2% of the human genome is coded into proteins (Int. Human Genome Seq. Consort., 2004), as pointed out above, repeat sequences may comprise over two-thirds of the human genome (de Koning et al., 2011). Repeat sequences have significantly altered the genome, often adaptively, modifying and shuffling existing genes and creating entirely new genes. Genome size and complexity of species in different taxa do not correlate perfectly well. One reason for this is some simpler species have more repetitive DNA sequences. For this and other reasons, one can conclude that repetitive sequences were (and still are) involved in the evolution of the genome.

9.16 DNA BASE REPEATS

DNA increasing its length through base repeats is another way the behavior of the genome can cause genetic variability, adaptive macroevolution, and diversity. These differ from repetitive DNA sequences in that they are often just a few different bases that are repeated. A species of cellular slime mold, *Dictyostelium discoideum,* which has no common name, is a soil amoeba that feeds on microorganisms such as bacteria. It has the huntingtin gene, which has many functions (Zuccato and Cattaneo, 2016; Sardo et al., 2012, and references therein). This gene: (1) causes the unicellular slime mold cells to aggregate into a many-celled pseudoplasmodium, an organism that crawls and is largely like a slug; (2) controls when the unicellular cells reproduce; and (3) causes them to move toward food. This gene has accumulated triplet repeats of the three DNA bases C, A, and G over evolutionary time. Echinoderms (group that includes sea stars, sea cucumbers, and sea urchins) have two CAG triplets in nonneural tissues. Thus, the DNA where these CAGs are has the base sequence CAGCAG. Amphioxus, a fish-like invertebrate, has two CAGs in neural tissues that are thought to be important in forming its primitive brain. CAGs increase in number as neural complexity in vertebrates increases, with four in zebrafish, six in possums, seven in mice, fifteen in cattle, and eighteen in pigs. It varies from seven to sixteen in monkeys, and from eight to thirty-five in humans. CAG repeats are involved in the formation of the neural tube, the structure that forms the early brain and spine. People with more CAG repeats tend to be more intelligent. But too many such triplets can lead to Huntington's disease, a rare inherited disease that causes the progressive breakdown and degeneration of nerve cells in the brain. In fruit flies, the huntingtin gene has no CAG triplets, and the gene regulates cell division. All of this shows that a gene can be largely conserved and change its function radically, and that addition of DNA repeats can produce profound evolutionary change and increases in complexity. This allows entering new adaptive zones with many unoccupied niches, and hence diversification into many new species. This is another example of how behavior of the genome can lead to evolutionary breakthroughs and increases in biodiversity.

9.17 POLYPLOIDY

Polyploidy is a condition in which an organism has at least double its normal number of chromosomes. Chromosome number can increase in just one generation as a result of faulty cell division. The chromosomes double in number as occurs normally, but then do not separate into two cells, as is normal, and which would cut the number of chromosomes in half and restore their original number. Chromosome duplication may occur more than once in a single species, causing species with more than twice the original chromosome number. The organism in a state of polyploidy is called a polyploid, and polyploidy can also mean the process that produces a polyploid. The production of polyploids generally results in the production of a new species, since polyploids almost always cannot interbreed with their ancestors.

Autopolyploids are polyploids derived from one species. Allopolyploids are polyploids derived from two different species that hybridize with each other. The production of polyploids provides the raw material for great bursts of innovation by duplicating and allowing the divergence of entire pathways. It allows an entire genome to function while another entire duplicate set of genes evolves as pseudogenes through unfavorable pathways until some of them hit on new, adaptive functions. It is the same principle as gene duplication with a pseudogene able to drift through many unfavorable mutations because a good copy of the gene is functioning, but with an entire set of genes functioning; i.e., an entire functioning genome. At the same time, there is an entire genome of pseudogenes able to mutate to unfavorable forms without harming the organism, until some of them acquire adaptive mutations. So there are many more possibilities for one or more pseudogenes having a lucky mutation that leads to an evolutionary breakthrough than if there is only one duplicated gene. Ohno (1970) suggested polyploidy played a major role in evolution.

Polyploidy increases diversity beyond just species number, since most polyploids display novel variation or differences in structure from their parental species. This may help them to use new niches that are different from the niches of their parental species. This could help them survive by reducing competition with their parental species. The differences polyploids have from their parental species can result from higher amounts of proteins produced because there are at least twice as many genes, the coming together of different gene regulation systems if the parental species are two different species, or rearrangements of chromosomes, all of which affect which genes are present and the regulation of these genes (Osborn, et al., 2003; Chen and Ni, 2006; Chen, 2007; Albertin, et al., 2006).

Polyploidy is common outside the animal and plant kingdoms. Some bacteria and archaea are polyploid. Some diatoms, which are common unicellular algae in soil, freshwater, and the sea, apparently are polyploid. There is evidence for a polyploidy-producing event in the ancestry of yeast, and that this was followed by the loss of over 90% of the newly duplicated genes (Wolfe and Shields, 1997). Diverse species of fungus have had past and recent gene doublings, and this has often been followed by diversification in their phenotypes. Some brown algae seem to be polyploid.

Polyploidy is pervasive in plants, where it is estimated that 30–80% of living species are polyploid (Meyers and Levin, 2006). Wood et al. (2009) found that about 15% of the times flowering plants and 31% of the times ferns divide into new species, polyploidy occurs. They found that polyploids do not diversify into new species at a greater rate than plants with the original chromosome number, but after polyploid formation, considerable and sometimes very rapid changes in genome structure and the amount that different genes are expressed have occurred (Matzke et al., 1999). The coast redwood (*Sequoia sempervirens*) is a hexaploid (meaning it has six pairs of chromosomes), with 66 chromosomes (Xu et al., 2002). Aquatic plants include a large number of polyploids (Les and Philbrick, 1993). Evidence suggests that genome multiplication occurred more than one time in the evolution of the flowering plant, thale cress (*Arabidopsis thaliana*) (Blanc, 2000). Three duplications of the entire genome of this plant have been directly responsible for over 90% of its increase in chemicals that regulate its genes and in genes involved in its development in the last 350 million years (Maere et al., 2005). Additionally, this genome showed a great deal of rearrangement, with a patchwork of duplicated regions that indicated insertion of DNA segments, tandem duplication of DNA, inversion of DNA segments, exchange of DNA segments between different chromosomes, and deletion of DNA segments (Blanc, 2000).

Comprehensive analyses of plant genomes whose DNA had been sequenced showed doubling of the genomes of the ancestors of both living seed plant and flowering plant species were crucial in the evolution of these two groups (Yuannian et al., 2011). Recall the small tree with no common name found only on the main island of New Caledonia in the South Pacific, *Amborella trichopoda*, one of the two oldest lines of flowering plants. It is in fact the flowering plant species that diverged from the rest of the flowering plants earlier than all or almost all of the other flowering plants diverged from each other. Said another way, it is one of the earliest branches of the evolutionary tree of flowering plants. Its genome provides an exceptional reference for inferring features of the first flowering plants. The *Amborella* Genome Project (Albert et al, 2013) sequenced its genome, showing that *the first flowering plant appeared as a result of an angiosperm ancestor doubling its entire genome about 200 mya, and that the ancestral flowering plant was a polyploid with a large constellation of both novel and ancient genes that survived to play key roles in flowering plant biology.* The additional genetic material gave the plants the potential to evolve new, never-before-seen structures, such as flowers. Additionally, based on an analysis of the entire genome of thale cress, De Bodt et al. (2005) state there is compelling evidence that *flowering plants underwent two whole-genome duplications during their early evolutionary history.* These events were crucial for the creation of many important developmental and regulatory genes found in living flowering plant genomes. They argue that these ancient polyploidy events might have also had an important role in the origin and diversification of flowering plants.

A genome duplication occurred in a plant species about 319 mya that led to the appearance and then diversification of plants that reproduce by producing seeds. It occurred in a plant species again about 200 mya, and this preceded the diversification of flowering plants. These genome doublings resulted in the diversification of regulatory genes important to seed and flower development, suggesting that they were involved in key innovations that contributed to the evolution, appearance, and rise, followed by diversification and eventual dominance, of seed plants and then flowering plants. It is fair to conclude that polyploidy was crucial in seed plant and then flowering plant evolution, both of which resulted in great diversification. The appearance of flowering plants also led to the diversification into many species of these plants and their symbiotic allies that they coevolved with: fungi in their roots, animals that disperse their seeds, animals that pollinate them, and others. After each of these diversifications, other species evolved that benefitted from the many new species of flowering plants and their symbiotic allies that evolved during each diversification. For example, when a new flowering plant species evolved and a butterfly species coevolved with it and became its pollinator, a species of bird could have evolved that benefitted from the new butterfly species by eating it.

Polyploidy occurs in many diverse taxa of animals. A few hundred cases of polyploidy are known in reptiles, amphibians, fish, crustaceans, insects, and other invertebrates. Polyploidy is more common invertebrates, such as flatworms, leeches, and brine shrimp, than in vertebrates (Otto and Whitton, 2000). In fish, there are polyploids in the salmon family (Salmonidae) and many in the carp and true minnow family (Cyprinidae) (Leggatt and Iwama, 2003). Some polyploid fish species have 400 chromosomes (ibid.). It is common in amphibians. The African clawed frog (*Xenopus laevis*) has several different species with as many as 12 sets of chromosomes (ibid.). All-female species of polyploid lizards that lay fertile eggs without mating with males are numerous.

9.18 CREATION OF NEW SPECIES BY HYBRIDIZATION BETWEEN TWO DIFFERENT SPECIES

Hybridization between two different species can also create new species. This sometimes involves the formation of polyploid species, but this is not always the case. This mechanism is not primarily the result of behavior of the genome, but the behavior of organisms. Nevertheless, it must be discussed because the genome is involved and it often involves polyploidy. Speciation by interspecific hybridization is considered fairly common, if not widespread, in the plant world (Arnold, 1997). It is important in some plant groups (Linder and Risenberg, 2004). Plants form polyploids more easily than animals, so have more polyploid species that are the result of hybridization between two different species. It is estimated that 2–4% of all flowering plants and 7% of all fern species resulted from polyploid hybridization between two different species (Otto and Witton, 2000). It is quite common among oaks, especially in the white oak group. There are even several hybrids between oak species that differ from each other above the species level. Also, hybrid speciation *without* an increase in chromosome number created many sunflower species (Rieseberg, 2003).

In animals, the lonicera fly (*Rhagoletis mendax* × *Rhagoletis zephyria*) is a species created by hybridization between the blueberry maggot (*Rhagoletis mendax*) and the snowberry fruit fly (*Rhagoletis zephyria*). Most ornithologists believe the great skua (*Stercorarius skua*) is a hybrid species between the pomarine skua (*Stercorarius pomarinus*) and one of the northern skua species (*Stercorarius sp.*) (Furness and Hamer, 2003). The Clymene dolphin (*Stenella clymene*) is a hybrid of the spinner dolphin (*Stenella longirostris*) and striped dolphin (*Stenella coeruleoalba*) (Amaral et al., 2014).

If there is a group of several rapidly diverging species, they can sometimes form multiple hybrid species, giving rise to a species complex. An example of this is several closely related but morphologically different genera of cichlid fishes in Lake Malawi in Africa (Genner and Turner, 2011). Cichlid fishes are a highly diverse family of fishes that includes the important food fish, tilapia (which is actually the common name for many species), and these popular aquarium fish: oscars (*Astronotus ocellatus*), angelfish (genus *Pterophyllum*), and discus (genus *Symphysodon*). Experts think many of the species of the duck genus *Anas*, which contains mallards and teals, are hybrid species; many of the different species in this genus can breed with each other and produce fertile offspring.

9.19 DELETION OF SEGMENTS OF DNA

Deletion of segments of DNA is another mechanism for the evolution of new genes and functions, and has occurred numerous times in the evolution of genes. It is another way the genome tends to increase variability and biodiversity. Examples of deletions are found in the evolution of the genes for three of the four hemoglobin protein chains. Recall that hemoglobin evolved by gene duplication. Deletions occurred in the evolution of each of the genes for the alpha, gamma, and delta hemoglobin chains.

A high prevalence of deletion mutations in a regulatory gene that turns other genes on has resulted in the loss of pelvic fins in some three-spined stickleback (*Gasterosteus aculeatus*) (fish) populations (Chan et al., 2010). The pelvic fins in fish are the paired fins at the bottom of the fish that are homologous with (match) the hind limbs of a four-legged vertebrate animal such as a horse.

Androgens are a group of hormones that play a role in male traits and reproductive activity. A receptor is a molecule in cell membranes which responds specifically to a particular chemical that binds with it, causing changes in the organism. Loss of control genes for androgen receptors caused the human penis to lack spines, which are present on the penis of monkeys and the chimpanzee (Reno et al., 2013; McLean et al., 2011). Loss of spines is correlated with longer copulation time. This could have been coupled with tighter bonding during intercourse; long-term, faithful pair bonds; and monogamy in human ancestors. The male hunted and gathered food and brought it to the female and young in human ancestors. This is associated with the three traits just mentioned. Fossil evidence suggests the reason that natural selection strongly favored walking upright on two legs in the evolution of the human line was that it freed the hands for the male to carry food to his mate and offspring (Lovejoy, 1981, 1988). Freeing the hands due to upright walking also allowed human ancestors to make and use tools and manipulate the environment, selecting for greater intelligence because brain power would have allowed more and better use of the hands for these functions. Cooperation and the providing of resources such as food by the male, evolving in conjunction with monogamous pair bonding between the male and female, could have lengthened the period of dependency of the offspring on the parents, making the time of learning as a juvenile longer. This would have made intelligence more favorable and adaptive, and caused natural selection to favor it. All of these traits could have evolved together. So, in humans, loss of spines on the penis due to gene deletions could have evolved together with a suite of traits that led to macroevolutionary change and humans being highly intelligent.

A deletion in a regulatory region that activates a gene that destroys excess primate neurons during embryonic development may help account for why human brains are much larger and have many more neurons than chimpanzees (McLean et al., 2011). *The gene that destroys the neurons is not activated because part of the region that activates it was deleted. This is one reason why humans have more neurons than chimpanzees. Deletion of a regulatory region that activates the gene involved in the growth of the bones of the toes helped us evolve our upright gate* (Indjeian, et al., 2016). *Upright walking is easier with shorter toes. It is miraculous that deletions in the human genome are important in the intertwined evolution of intelligence, pair bonding, male provisioning, walking upright on two legs, and having hands free to use.*

9.20 REGULATORY GENES AND EVOLUTION

Genes are arranged in a hierarchical system. Regulatory or control genes control the genes that code for RNA and protein domains, switching them on and off. A mutation causing a small change in the DNA base sequence of a master control gene can have a very large effect on an organism's phenotype, including structure, color, and behavior. This is another mechanism by which a mutation, even one with a large effect, can occur and yet maintain the compatibility, coordination, and integration between the genes and different parts of the phenotype. The compatibility and integration of the genotype and phenotype have a high probability of being maintained in mutations of control genes because such mutations tend to cause the entire organism to change in a coordinated fashion, such as an increase in size by the same proportional amount in all parts of the organism. The mutation thus has an increased chance of being adaptive compared to mutations in structural genes (genes that do not regulate other genes). A large phenotypic change can occur in one generation. For a hypothetical example, a mouse could have a mutation in a regulatory gene, causing it to produce more growth hormone, which would cause the entire organism to increase in size, having all body parts enlarged equally without disruption of the system. This way, a mouse could hypothetically evolve into a rat.

The ancestor of vertebrates is apparently a sea squirt, also called a tunicate (phylum Chordata, subphylum Urochordata). This is an invertebrate animal that cannot move and is attached to rocks or other hard surfaces on the seafloor as an adult. It is shaped like a flexible, vertical tube, and obtains its food by filtering small organisms out of the seawater. It has a larva called an ascidian tadpole that is shaped somewhat like a tiny eel, and has a head, nervous system, and tail, and can swim. The larva swims around until it finds a favorable place to settle on and change into the adult sea squirt on the seafloor. The evidence supports the hypothesis that in at least one ascidian tadpole, there was a mutation in a control gene that caused the larva to never change into an adult, but rather be able to breed as a free-swimming larva (Delsuc et al., 2006, and references therein). This is called neoteny. Neoteny is the sexual maturity of an animal while it is still in a larval state, or the retention of juvenile features in the adult animal. The neotenic, reproductive ascidian tadpole gave rise to the ancestor of fish and to the vertebrate line. This regulatory gene mutation achieved a huge maroevolutionary breakthrough while conserving the integration and coordination of both the genotype and phenotype of the organism. All systems and parts remained integrated and compatible with each other, because the ascidian tadpole remained intact, with the only changes being that it did not change into an adult sea squirt and was able to reproduce. The large macroevolutionary breakthrough allowed the reproductive ascidian tadpoles that were our vertebrate ancestors to enter into a new adaptive zone. As is the case with any macroevolutionary breakthrough leading to the entry into a new adaptive zone, it was followed by tremendous diversification into many new species. In this case, it was the tremendous number of species in the early vertebrate line. This shows how a small mutation that causes a tiny change in the DNA *in a control gene* can rapidly cause a large, coordinated adaptive change and evolutionary breakthrough and transition in the organism.

Neoteny has been proposed in the evolution of humans, manifested in the slowing or delaying of body development compared to nonhuman primates, resulting in a large head, flat face, and relatively short arms, although humans also have nonneotenic traits, such as relatively large noses. Thiessen (1997) claims neoteny increased as early primates evolved into later forms, and that primates have been evolving flatter faces. Gould (2008) thinks humans have retained the original juvenile features of our ancestors to adulthood. Jones (1995) posited that human evolution's trend toward neoteny may have been caused by natural selection because men preferred neotenous, juvenile facial traits in women, and chose these women, selecting for these traits. The resulting neoteny in male faces was a by-product of this. Shea (1989) presented an opposing view, claiming mistakes of fact and interpretation have led authors to conclude neoteny is important in human evolution. If neoteny played a part in human evolution, changes in control genes would likely have been important in this. I will now propose a new hypothesis, that the mutation in the control gene was caused by a virus or virus-like element.

In addition to a regulatory gene increasing or decreasing how much RNA or protein a structural gene produces, the output from a gene can increase by gene duplication. Both of these mechanisms can greatly alter the phenotype in an adaptive way without disrupting the integration or coadapted nature of the genome and organism, leading to an evolutionary breakthrough.

Finally, a study of 12 animal species showed that after genes duplicate, they often change their expression pattern (Kryuchkova-Mostacci and Robinson-Rechavi, 2016). This can happen by the gene being expressed in different tissues and thereby adopting new roles.

9.21 EXAPTATION AND EVOLUTION

It is not uncommon for a gene, protein, or structure to change function in evolution, causing a change in the phenotype, sometimes a large one. An existing gene, protein, or structure can be altered, causing it to take on a new function. In this process, called *exaptation*, selection acts to create a novel function. Evolution acts by building on to what is already present in the organism. Exaptation can occur by changing the function of a structure or behavior. If it occurs by switching the function of a gene, it is another example of how the nature of the genome can promote evolution, variability, and diversity. Exaptation can lead to new available niches, and thus diversification.

The FLO/LFY genes, called PpLFY1 and PpLFY2, regulate the first cell division and then the development of a moss called the spreading earthmoss (*Physcomitrium patens*). The same FLO/LFY genes underwent exaptation and in flowering plants are master regulatory genes for the development of flowers, without regulating cell division (Tanahashi, 2005).

Prolactin, a protein hormone, is thought to be related to regulation of water and salt balance in fish. It inhibits some species of salamander from changing from larvae into adults, causing them to breed as larvae and not undergo metamorphosis (Gona and Etkin, 1970). It promotes milk production in mammal mothers. It has evolved to take on several other functions in additional cases of exaptation. It is influential in more than 300 different processes in various vertebrates, including humans (Bole-Feysot, 1998). The hormone has important cell cycle-related functions in growth, differentiation, and countering programmed cell death. It is essential in metabolism, development of the pancreas, and regulation of the immune system. As a growth factor, it influences the formation of cellular components of the blood and the formation of new blood vessels. It is involved in the regulation of blood clotting. It has also been modified to different forms in different taxa. All these functions of prolactin occurred by exaptation.

Microtubules are proteins that have been modified a number of times in evolution, taking on different functions by exaptation. There is a microtubule-like structure in the bacterium *Bacillus thuringiensis* (which has no common name) involved in the separating of its small circular pieces of DNA called plasmids after they duplicate into two pieces (Jiang et al., 2016). Microtubules play a key structural role in the flagella and cilia of eukaryotes. Flagella are slender, microscopic, threadlike structures that enable many protozoa, bacteria, sperm, etc., to swim. They are like little whips attached to the cell. They generally attach to unicellular life forms, and there is generally only one or two per cell. Cilia are like little hairs on cells that are present in high numbers, and beat back and forth. They act as sensory organelles or to make the cell move. They can also help sweep food into some organisms. Microtubules form part of the cytoskeleton and provide structure and shape to eukaryotic cells. The microtubule cytoskeleton aids in the transport of material within cells, accomplished by motor proteins that move on the surface of the microtubules. The cytoskeleton formed by microtubules is necessary to an organism's development; this has been shown in the fruit fly (van Eeden and St. Johnston, 1999) and mammals (Beddington and Robertson, 1999). The cellular cytoskeleton can also influence gene regulation (Rosette and Karin, 1995). Microtubules are essential in the development of the nervous system in higher vertebrates (Tucker, 1990). Microtubules act both to restrain cell movement and to establish its directionality in the developing organism. They play major roles in cell division, and are key in separating chromosomes during cell division in eukaryotes. In vertebrates, they are used in the cilia of cells called epithelial cells that form the covering of internal and external surfaces of the body; the tail of sperm; and important parts of nerve cells. And as discussed earlier in this chapter, actin, used in muscle contraction, is derived from microtubules by gene duplication. Thus, microtubules have undergone exaptation many times, taking on many new functions.

Phenotypic structures can undergo exaptation with great effects on evolution. Theropods are the group of dinosaurs that includes *Tyrannosaurus rex*. They were covered with feathers. These functioned to keep them warm and almost surely by males to attract females and intimidate rival males. When birds evolved from small theropods, feathers were exapted for use in flight. Arms of these small theropods also underwent exaptation at the same time that feathers did. They were originally used as arms are usually used. Then when they had feathers, they were used as nets to catch insects to eat. This helped them develop into wings. The wings were used for gliding before they were used for flight. This is a classic example of exaptation. The exaptation was accompanied by an evolutionary breakthrough and transition, with evolution into a new adaptive zone. Then the birds diversified into many new species, taking advantage of the many new unoccupied niches available to them. *Essentially all steps of the evolutionary process from prokaryotes to humans involved evolving new structures, proteins, genes, and functions by modifying existing structures, proteins, and genes. Thus, essentially all steps in evolution involve exaptation.*

9.22 SPECIAL CHROMOSOMES PROMOTED SONGBIRD DIVERSIFICATION

About half of all bird species are songbirds, which are a group of birds that have a unique voice box and exceptional ability to sing. There are about as many songbird species as mammal species. So they have very high biodiversity. All songbirds, but no other birds, have a strange extra chromosome, the germ-line restricted chromosome (GRC), present only in sperm and egg cells (Torgasheva et al., 2019). GRCs differ considerably between species, even closely related ones, suggesting the chromosome has evolved rapidly in different songbird lines since first appearing in the common ancestor about 35 mya. Some GRC genes in zebra finches (genus *Taeniopygia*; there are only two species) make RNA and proteins in the ovaries and testes of adult birds, hinting they may guide sperm and egg development. In other bird species, they seem to be involved in the development of the embryo. It appears the GRC functions only during sperm and egg development and early development, potentially causing great variability within and between populations, leading to speciation if the right selective forces are present. It is possible that the great diversity of songbirds can be at least partly explained by the role played by GRCs in promoting differences between populations and making them unable to breed with each other; i. e., causing them to speciate.

9.23 CONCLUSION

Genomes have a built-in tendency to promote variability in populations, adaptive and innovative evolution, and diversification, creating new species. They are not isolated entities, but linked to other genomes of the same and different species in their ecosystem. Just as no organism or species exists as an autonomous unit ecologically, there is no organism or species that exists as an isolated, autonomous unit evolutionarily or genetically. And no gene exists in isolation, or evolves without an interconnection with other genes in its genome, in the genomes of other organisms, and of other species. All organisms and species are intimately linked ecologically, evolutionarily, and genetically in one evolving, inter-connected system. The genes, genomes, and species of this system coevolve and exchange genetic material. The tree of life metaphor is obsolete, and must be replaced with a model whereby the branches are linked horizontally as well as vertically. DNA jumps between organisms, including between different species, even different kingdoms and domains. Viruses play an important role in this; they transport DNA between organisms by transduction. Genomes are built from DNA segments from many different species. There is selection for an optimal amount and type of genetic variability. The type of genetic changes favored by natural selection are those that do not disrupt the coadapted self-compatibility of the genome and phenotype. Natural selection favors the donating, receiving, and exchanging of entire DNA segments that have undergone natural selection and are compatible with the genomes they are incorporated into. These DNA segments may be transferred between organisms, species, and higher taxa. These DNA segments undergo the test of natural selection within their genomes for long time periods, with the result that they become fit, and coadapted and compatible with the genetic systems they originally reside in, so have a higher probability of being adaptive and coadapted with the genetic systems they are moved to. Genetic recombination, bacterial conjugation, transformation, gene transfer agents, transduction, and transposable elements all result in a much higher probability of an adaptive change than random mutations. This is what I mean by there is selection toward an optimal type of variability. This selection on the nature of variability, favoring relatively large segments compatible with the genome receiving them, is a mechanism for macroevolutionary breakthroughs. Any given horizontal gene transfer or similar alteration of the genome receiving the genetic material will be deleterious with high probability. However, these events often occur in species with many individual organisms, and can occur at any time over very long time periods, meaning there is a very large number of trials. The enhanced likelihood of compatibility of the novel genetic material with the recipient genome and the large number of trials means that the probability of an adaptive change, even a large one, is reasonably high over a sufficient time interval with large populations. In the case of sexual recombination, the probability of a genetic change that increases the organism's fitness is especially high, because entire chromosomes and segments of chromosomes that are already fit and coadapted to each other are shuffled around, creating genetic variability for natural selection to act on while maintaining the compatibility and coadaptedness of the genome. Changes in genetic regulation have a higher probability of causing large, adaptive phenotypic changes than random alterations to the DNA. This is because they can cause large alterations to the phenotype without disrupting the coordinated, coadapted system of the organism. Gene duplication and polyploidy allow genes to mutate randomly over long time periods without hurting the organism until an adaptive breakthrough occurs because there is always a functional gene carrying out the needed function for the organism. Sexual recombination, gene duplication, transposable elements, gene transfer agents, integrons, transduction, polyploidy, and the hierarchical nature of the genome are all mechanisms by which the behavior of the genome can cause rapid macroevolutionary breakthroughs, because they can cause changes, even large ones, that maintain the coordination, compatibility, and coadapted nature of the phenotype and genotype. These breakthroughs are essentially always followed by tremendous diversification into many new species. Then additional species evolve that exploit the niches created by the species that resulted from this diversification.

NOTES

1 Paralogous genes are genes with a similar structure at different locations in the genome, indicating that they diverged from a common ancestral gene; they evolve by a gene duplication event.
2 A point mutation is a mutation in only one base in the DNA.
3 A taxon (plural is taxa) is a group of organisms forming a unit for the purpose of classification, such as a species, genus, family, kingdom, domain, etc. The adjective is taxonomic.
4 Splicing is done to RNA. In RNA splicing, a newly made RNA molecule (that has recently been coded from DNA) is transformed into a mature RNA molecule by the removal of introns and joining together of exons. The resulting RNA molecule then codes for a protein.

REFERENCES

Acuña, R., et al. (March, 2012). Adaptive horizontal transfer of a bacterial gene to an invasive insect pest of coffee. *PNAS USA*. 109 (11): 4197–202. Bibcode:2012PNAS..109.4197A. doi: 10.1073/pnas.1121190109. PMC 3306691. PMID 22371593.

Aguilera, A. & Gomez-Gonzalez, B. (2008). Genome instability: A mechanistic view of its causes and consequences. *Nature Revs. Genet.* 9 (3): 204–17. doi: 10.1038/nrg2268.

Albert, V. A., et al. (2013). The *Amborella* genome and the evolution of flowering plants. *Science* 342 (6165): 1241089.

Albertin, W., et al. (2006). Numerous and rapid nonstochastic modifications of gene products in newly synthesized *Brassica napus* allotetraploids. *Genet.* 173 (2): 1101–13. PMC 1526534 . PMID 16624896. doi:10.1534/genetics.106.057554.

Altincicek, B., Kovacs, J. L., & Gerardo, N. M. (23 April, 2012). Horizontally transferred fungal carotenoid genes in the two-spotted spider mite *Tetranychus urticae*. *Biol. Letters* 8 (2): 253–7. doi:10.1098/rsbl.2011.0704. PMC 3297373. PMID 21920958.

Amaral, A. R., et al. (2014). Hybrid speciation in a marine mammal: The Clymene dolphin (*Stenella clymene*). PloS ONE 9 (1)): e83645.

An, N. A., Zhang, J., Mo, F. et al. (2023). De novo genes with an lncRNA origin encode unique human brain developmental functionality. *Nature, Ecology & Evolution* 7: 264–78. https://doi.org/10.1038/s41559-022-01925-6.

Anderson, M. T. & Seifert, H. S. (2011). Opportunity and means: Horizontal gene transfer from the human host to a bacterial pathogen. *mBio* 2 (1). doi: 10.1128/mBio.00005-11.

Anderson, N. G. (26 Sept., 1970). Evolutionary significance of virus infection. *Nature* 227: 1346–7. doi: 10.1038/2271346a0.

Apic, G., et al. (July, 2001). Domain combinations in archaeal, eubacterial and eukaryotic proteomes. *Journ. of Molecular Biol.* 310 (2): 311–25. doi: 10.1006/jmbi.2001.4776. PMID 11428892. S2CID 11894663.

Armbrust, E. V., et al. (2004). The genome of the diatom *Thalassiorsira pseudonana*: ecology, evolution, and metabolism. *Science* 306: 79–86.

Arnold, M. L. (1997). *Natural Hybridization and Evolution*. Oxford Univ. Press, New York, NY.

Auton, A., et al. (2012). A fine-scale chimpanzee genetic map from population sequencing. *Science* 336 (6078): 193–8. PMC 3532813. doi: 10.1126/science.1216872.

Aziz, R. K., et al. (2010). Transposases are the most abundant, most ubiquitous genes in nature. *Nucleic Acids Research*. 38 (13): 4207–17. doi: 10.1093/nar/gkq140. PMC 2910039. PMID 20215432.

Banks, D.J., Beres, S. B., & Musser, J. M. (Nov., 2002). The fundamental contribution of phages to GAS evolution, genome diversification and strain emergence. *Trends in Microbiol.* 10 (11): 515–521. Abstract available at http://dx.doi.org/10.1016/S0966-842X(02)02461-7.

Bar, D. (16 Feb., 2011). Evidence of massive horizontal gene transfer between humans and *Plasmodium vivax*. *Nature Proceedings* doi: 10.1038/npre.2011.5690.1.

Baudat, F., et al. (2010). PRDM9 is a major determinant of meiotic recombination hotspots in humans and mice. *Science* 327 (5967): 836–40. doi: 10.1126/science.1183439.

Beddington, R. S. & Robertson, E. J. (Jan., 1999). Axis development and early asymmetry in mammals. *Cell* 96 (2): 195–209. doi: 10.1016/S0092-8674(00)80560-7. PMID 9988215. S2CID 16264083.

Berkemer, S. J. & McGlynn, S. E. (21 April, 2020). A new analysis of archaea-bacteria domain separation: variable phylogenetic distance and the tempo of early evolution. *Mol. Biol. and Evol.* doi: 10.1093/molbev/msaa089.

Bézier, A., et al. (2009). Polydnavirus hidden face: The genes producing virus particles of parasitic wasps. *Journ. of Invertebrate Pathol.* 101 (3): 194–203.

Blanc, G., et al. (2000). Extensive duplication and reshuffling in the *Arabidopsis* genome. *Plant Cell* 12: 1093–102.

Blanc, G., et al. (17 Sept., 2010). The *Chlorella variabilis* NC64A genome reveals adaptation to photosymbiosis, coevolution with viruses, and cryptic sex. *Plant Cell*. doi: https://doi.org/10.1105/tpc.110.076406.

Blankenship, R. (27 March, 2017). How cyanobacteria went green. *Science* 355 (6): 1372–3.

Blankenship, R. E. (1992). Origin and early evolution of photosynthesis. *Photosynth. Res.* 33: 91–111.

Bole-Feysot C., et al. (June, 1998). Prolactin (PRL) and its receptor: actions, signal transduction pathways and phenotypes observed in PRL receptor knockout mice. *Endocrine Reviews* 19 (3): 225–68. doi: 10.1210/edrv.19.3.0334. PMID 9626554.

Bolhuis H., et al. (Jan., 2010). Horizontal transfer of the nitrogen fixation gene cluster in the cyanobacterium *Microcoleus chthonoplastes*. *ISME Journ.* 4 (1): 121–30.

Bordenstein, S. R. & Wernegreen, J. J. (1 Oct., 2004). Bacteriophage flux in endosymbionts (*Wolbachia*): Infection frequency, lateral transfer, and recombination rates. *Molecular Biol. and Evol.* 21 (10): 1981–91. https://doi.org/10.1093/molbev/msh211.

Bourque, G., et al. (2008). Evolution of the mammalian transcription factor binding repertoire via transposable elements. *Genome Research* 18 (11): 1752–62.

Britten, R. J. (Sept. 3, 1996). DNA sequence insertion and evolutionary variation in gene regulation. *PNAS USA* 93 (18): 9374–7. ISSN 0027-8424. PMC 38434. PMID 8790336.

Britten, R. J. (16 Nov. 2010). Transposable element insertions have strongly affected human evolution. *PNAS USA* 107 (46) 19945–8. https://doi.org/10.1073/pnas.1014330107.

Briz, V., et al. (17 Aug., 2017). The non-coding RNA BC1 regulates experience-dependent structural plasticity and learning. *Nature Comms.* 8, Article number: 293.

Brosius, J. (1999). Genomes were forged by massive bombardments with retroelements and retrosequences. *Genetica* 107: 209–38. https://doi.org/10.1023/A:1004018519722.

Chan, Y. F., et al. (2010). Adaptive evolution of pelvic reduction in sticklebacks by recurrent deletion of a Pitx1 enhancer. *Science* 327 (5963): 302–5.

Chen, F. & Lu, J. (2002). Genomic sequence and evolution of marine cyanophage P60: A new insight on lytic and lysogenic phages. *Applied and Environmental Microbiol.* 68 (5): 2589–94.

Chen, Z. J. (2007). Genetic and epigenetic mechanisms for gene expression and phenotypic variation in plant polyploids. *Ann. Rev. of Plant Biol.* 58: 377–406. PMC 1949485 Freely accessible. PMID 17280525. doi: 10.1146/annurev.arplant.58.032806.103835.

Chen, Z. J. & Ni, Z. (2006). Mechanisms of genomic rearrangements and gene expression changes in plant polyploids. *BioEssays* 28 (3): 240–52. PMC 1986666. PMID 16479580. doi: 10.1002/bies.20374.

Chimpanzee Sequencing and Analysis Consortium (Sept., 2005). Initial sequence of the chimpanzee genome and comparison with the human genome. *Nature* 437 (7055): 69–87. PMID 16136131. doi: 10.1038/nature04072.

Chothia, C. (June, 1992). Proteins. One thousand families for the molecular biologist. *Nature* 357 (6379): 543–4. Bibcode:1992Natur.357..543C. doi: 10.1038/357543a0. PMID 1608464. S2CID 4355476.

Chou, S., et al. (5 Feb., 2015). Transferred interbacterial antagonism genes augment eukaryotic innate immune function. *Nature* 518: 98–101. doi: 10.1038/nature13965.

Chuong, E. B., et al. (4 Mar., 2016). Regulatory evolution of innate immunity through co-option of endogenous retroviruses. *Science* 351 (6277): 1083–6.

Clokie, M. R. J. & Mann, N. H. (2006). Marine cyanophages and light. *Environmental Microbiol.* 8 (12): 2074–82.

Cohen, C. J., et al. (15 Dec., 2009). Endogenous retroviral LTRs as promoters for human genes: A critical assessment. *Gene* 448 (2): 105–14. https://doi.org/10.1016/j.gene.2009.06.020.

Cordaux, R. & Batzer, M. A. (2009). The impact of retrotransposons on human evolution. *Nature Reviews Genet.* 10: 691–703.

Coulson, A., et al. (1986). Toward a physical map of the genome of the nematode *Caenorhabditis elegans*. *PNAS USA* 83: 7821–25.

Crisp, A., et al. (2015). Expression of multiple horizontally acquired genes is a hallmark of both vertebrate and invertebrate genomes. *Genome Biol.* 16: 50. doi: 10.1186/s13059-015-0607-3. PMC 4358723 . PMID 25785303.

Daniels, S. B., et al. (1990). Evidence for the horizontal transmission of the P transposable element between *Drosophila* species. *Genet.* 124: 339–55.

Davidson, J. N., et al. (March, 1993). The evolutionary history of the first three enzymes in pyrimidine biosynthesis. *BioEssays* 15 (3): 157–64. doi: 10.1002/bies.950150303. PMID 8098212. S2CID 24897614.

De Bodt, S., Maere, S., & Vandepeer, Y. (Nov., 2005). Genome duplication and the origin of angiosperms. *Trends in Ecol. & Evol.* 20 (11): 591–7. PMID 16701441. doi: 10.1016/j.tree.2005.07.008. https://doi.org/10.1016/j.tree.2005.07.008.

Deininger, P. L. & Batzer, M. A. (2002). Mammalian retroelements. *Genome Research* 12: 1455–65. doi: 10.1101/gr.282402.

de Koning, A. P., Gu, W., Castoe, T. A., Batzer, M. A., & Pollock, D. D. (1 Dec., 2011). Repetitive elements may comprise over two-thirds of the human genome. *PLoS Genetics* 7 (12): e1002384. doi: 1 0.1371/journal.pgen.1002384. PMC 3228813. PMID 22144907.

Delsuc, F., Brinkmann, H., Chourrout, D., et al. (2006). Tunicates and not cephalochordates are the closest living relatives of vertebrates. *Nature* 439: 965–8. https://doi.org/10.1038/nature04336.

Duchesne, L. C. & Larson, D. W. (Apr., 1989). Cellulose and the evolution of plant life. *BioScience* 39 (4): 238–41.

Dunning Hotopp, J. C., et al. (21 Sept., 2007). Widespread lateral gene transfer from intracellular bacteria to multicellular eukaryotes. *Science* 317 (5845): 1753–6. doi: 10.1126/science.1142490.

Ekman, D., et al. (April, 2005). Multi-domain proteins in the three kingdoms of life: orphan domains and other unassigned regions. *Journ. of Molecular Biol.* 348 (1): 231–43. doi: 10.1016/j.jmb.2005.02.007. PMID 15808866.

Emera, D., Casola, C., Lynch, V. J., Wildman, D. E., Agnew, D., & Wagner, G. P. (2012). Convergent evolution of endometrial prolactin expression in primates, mice, and elephants through the independent recruitment of transposable elements. *Molecular Biol. and Evol.* 29: 239–47.

Engelmoer D. J. P., et al. (Nov., 14, 2013). Conservative sex and the benefits of transformation in *pneumoniae*. *PLOS PATHOGENS*. https://doi.org/10.1371/journal.ppat.1003758.

Engels, W. R. (Oct., 1992). The origin of P elements in *Drosophila melanogaster*. *Bioessays* 14 (10): 681–86. https://doi.org/10.1002/bies.950141007\.

Erwin, J., A., et al. (2014). Mobile DNA elements in the generation of diversity and complexity in the brain. *Nature Rev. Neuroscience* 15 (8): 497–506.

Esnault, C., et al. (2000). Human LINE retrotransposons generate processed pseudogenes. *Nature Genet.* 24: 363–67.

Feschotte, C. (1 May, 2008). Transposable elements and the evolution of regulatory networks. *Nature Rev. Genet.* 9: 397–405.

Fox, D. (Feb., 2022). Genetically bloated beasts: The surprising persistence of salamanders is forcing us to rethink evolution. *Scientific American* 326 (2): 40–9. Secondary literature.

Frost, L. S., et al. (1985). Mobile genetic elements: the agents of open source evolution. *Nat. Rev. Microbiol.* 3: 722–32.

Furness, R.W. & Hamer, K. (2003). Christopher Perrins, ed. *Firefly Encyclopedia of Birds.* Firefly Books. pp. 270–3, Richmond Hill, ON, Canada. ISBN 1-55297-777-3.

Gasmi, L., et al. (2015). Recurrent domestication by Lepidoptera of genes from their parasites mediated by Bracoviruses. *PLOS Genet.* 11 (9): e1005470.

Genner, M.J. & Turner, G.F. (Dec., 2011). Ancient hybridization and phenotypic novelty within Lake Malawi's cichlid fish radiation. *Mol. Biol. and Evol.* 29: 195–206. (Published online). PMID 22114359. doi: 10.1093/molbev/msr183.

Gilbert, C., et al. (29 April, 2010). A role for host-parasite interactions in the horizontal transfer of DNA transposons across animal phyla. *Nature* 464: 1347–50.

Glinsky, G. V. (2016). Mechanistically distinct pathways of divergent regulatory DNA creation contribute to evolution of human-specific genomic regulatory networks driving phenotypic divergence of *Homo sapiens. Genome Biol. and Evol.* 8 (9): 2774–88.

Gogolevski, K. P., et al. (2008). Bov-B-mobilized SINEs in vertebrate genomes. *Gene* 407: 75–85.

Gona, A. G. & Etkin, W. (June, 1970). Inhibition of metamorphosis in *Ambystoma tigrinum* by prolactin. *General and Comparative Endrocrinol.* 14 (3): 589–91. https://doi.org/10.1016/0016-6480(70)90042-0.

Gonzalez, J. & Petrov, D. A. (15 Dec., 2009). The adaptive role of transposable elements in the *Drosophila* genome. *Gene* 448 (2): 124–33. https://doi.org/10.1016/j.gene.2009.06.008.

Goodsell, D. (Dec., 2006). "Transposase". Molecule of the Month. *Protein Data Bank.* doi: 10.2210/rcsb_pdb/mom_2006_12.

Gould, S. J. (2008). A biological homage to Mickey Mouse. *Ecotone* 4 (1): 333–40.

Grassi, L., et al. (1 May, 2012). Large-scale dynamics of horizontal transfers. *Mobile Genetic Elements* 2 (3): 163–7. https://doi.org/10.4161/mge.21112.

Green, B. R. (2007). Evolution of light-harvesting antennas in an oxygen world. In Fawkowski, P. G. & Knoll, A. H. *Evolution of Primary Producers in the Sea.* Elsevier Academic Press, Cambridge, MA.

Gregory, T. R. & Hebert, P. D. (1999). The modulation of DNA content: proximate causes and ultimate consequences. *Genome Res.* 9: 317–24.

Guo, L. et al. (19 Oct., 2018). The opium poppy genome and morphinan production. *Science* 362 (6412): 343–7. doi: 10.1126/science.aat4096.

Hall, C., Brachat, S., & Dietrich, F. S. (June, 2005). Contribution of horizontal gene transfer to the evolution of *Saccharomyces cerevisiae. Eukaryotic Cell* 4 (6): 1102–15. doi: 10.1128/EC.4.6.1102-1115.2005. PMC 1151995. PMID 15947202.

Han, Y., et al. (Published online: 16 Nov., 2017). Timely trigger of caterpillar zombie behaviour: temporal requirements for light in baculovirus-induced tree-top disease. https://doi.org/10.1017/S0031182017001822.

Hartl, D. L. (2000). Molecular melodies in high and low C. *Nature Rev. Genetics* 1: 145–9. https://doi.org/10.1038/35038580.

Hashimoto, T. et al. (2016). Extremotolerant tardigrade genome and improved radiotolerance of human cultured cells by tardigrade-unique protein. *Nature Communications* 7: 12808.

Hehemann, J.-H., et al. (2010). Transfer of carbohydrate active enzymes from marine bacteria to Japanese gut microbiota. *Nature* 464: 908–12.

Henikoff, S., et al. (Oct., 1997). Gene families: The taxonomy of protein paralogs and chimeras. *Science* 278 (5338): 609–14. doi: 10.1126/science.278.5338.609. PMID 9381171.

Ho, M. W. (2007). *Genetic Engineering Dream or Nightmare? The Brave New World of Bad Science and Big Business.* Third World Network, Gateway Books, MacMillan, Continuum, Penang, Malaysia; Bath, UK; Dublin, Ireland; New York, USA. http://www.i-sis.org.uk/genet.php. ISBN-13: 978-9839747300. ISBN-10: 1858600510.

Hohmann-Marriott, M. F. & Blankenship, R. (2011). Evolution of photosynthesis. *Annu. Rev. Plant Biol.* 62:515–48.

Hoover, K., et al. (9 Sept., 2011). A gene for an extended phenotype. *Science* 333 (1401): p. 1401. doi: 10.1126/science.1209199.

Houck, M. A., et al. (1991). Possible horizontal transfer of *Drosophila* genes by the mite *Proctolaelaps regalis. Science* 6: 1125–9.

Huang, S., et al. (2016). Discovery of an active RAG transposon illuminates the origins of V(D)J recombination. *Cell* 166 (1): 102–14.

Indjeian, V. B., et al. (2016). Evolving new skeletal traits by cis-regulatory changes in bone morphogenetic proteins. *Cell* 164 (1): 45–56.

International Human Genome Sequencing Consortium. (2004). Finishing the euchromatic sequence of the human genome. *Nature* 431: 931–45. https://doi.org/10.1038/nature03001.

Jeffreys, A. J., et al. (Oct., 2001). Intensely punctate meiotic recombination in the class II region of the major histocompatibility complex. *Nat. Genet.* 29 (2): 217–22. PMID 11586303. doi: 10.1038/ng1001-217.

Jiang, N., et al. (30 Sept., 2004). Pack-MULE transposable elements mediate gene evolution in plants. *Nature* 431: 569–73.

Jiang, S, et al. (March, 2016). Novel actin filaments from *Bacillus thuringiensis* form nanotubules for plasmid DNA segregation. *PNAS USA* 113(9):E1200–5.Bibcode:2016PNAS..113E1200J. doi: 10.1073/pnas.1600129113. PMC 4780641. PMID 26873105.

Jones, D. et al. (1995). Sexual selection, physical attractiveness, and facial neoteny: Cross-cultural evidence and implications (and comments and reply). *Current Anthropol.* 36 (5): 723–48. doi: 10.1086/204427.

Jordan, I. K., et al. (Feb., 2003). Origin of a substantial fraction of human regulatory sequences from transposable elements. *Trends in Genetics* 19 (2): 68–72. https://doi.org/10.1016/S0168-9525(02)00006-9.

Jurka, J. & Kapitonov, V. V. (1999). Sectorial mutagenesis by transposable elements. *Genetica* 107: 239–48. doi: 10.1016/S0168-9525(99)01777-1. PMID 1043119.

Kaessmann, H., et al. (2009). RNA-based gene duplication: mechanistic and evolutionary insights. *Nature Rev. Genet.* 10: 19–31.

Kay, E., Vogel, T. M., Bertolla, F., Nalin, R., & Simonet, P. (July, 2002). In situ transfer of antibiotic resistance genes from transgenic (transplastomic) tobacco plants to bacteria. *Appl. Environ. Microbiol.* 68 (7): 3345–51. doi: 10.1128/aem.68.7.3345-3351.2002. PMC 126776 . PMID 12089013.

Keen, E. C., et al. (2017). Novel "superspreader" bacteriophages promote horizontal gene transfer by transformation. *mBio* 8 (1): e02115–16. doi: 10.1128/mBio.02115-16. ISSN 2150-7511.

Kim, T.M., et al. (2004). Periodic explosive expansion of human retroelements associated with the evolution of the hominoid primate. *Journ. Korean Med. Sci.* 19: 177–85.

Kirsch, R., et al. (Nov., 2012). Combining proteomics and transcriptome sequencing to identify active plant-cell-wall-degrading enzymes in a leaf beetle. *BMC Genomics* 201213: 587. doi: 10.1186/1471-2164-13-587.

Kirsch, R., et al. (Sept., 2014). Horizontal gene transfer and functional diversification of plant cell wall degrading polygalacturonases: Key events in the evolution of herbivory in beetles. *Insect Biochem. and Mol. Biol.* 52: 33–50. doi: 10.1016/j.ibmb.2014.06.008.

Koonin, E. V. (19 Aug., 2016). Viruses and mobile elements as drivers of evolutionary transitions. *Philosoph. Trans. of the Royal Soc. B: Biological Sciences* 371 (1701). https://doi.org/10.1098/rstb.2015.0442.

Koonin, E. V., et al. (2001). Horizontal gene transfer in prokaryotes: Quantification and classification. *Annual Review of Microbiol.* 55 (1): 709–42. doi: 10.1146/annurev.micro.55.1.709. PMC 4781227. PMID 11544372.

Kriegs J. O., et al. (April, 2007). Evolutionary history of 7SL RNA-derived SINEs in Supraprimates. *Trends Genet.* 23 (4): 158–61. PMID 17307271. doi: 10.1016/j.tig.2007.02.002.

Kryuchkova-Mostacci, N. & Robinson-Rechavi, M. (Dec., 2016). Tissue-specificity of gene expression diverges slowly between orthologs, and rapidly between paralogs. *PLOS Computational Biology* 12 (12): e1005274. doi: 10.1371/journal.pcbi.1005274. PMC 5193323. PMID 28030541.

Kubyshkin, V., et al. (Feb., 2018). On universal coding events in protein biogenesis. *Bio Systems* 164: 16–25. doi: 10.1016/j.biosystems.2017.10.004. PMID 29030023.

Kurosawa, K., et al. (2008). Rhodostreptomycins, antibiotics biosynthesized following horizontal gene transfer from *Streptomyces padanus* to *Rhodococcus fascians*. *Journ. of the Amer. Chem. Soc.* 130 (4): 1126–7.

Kyndt, T., et al. (2015). The genome of cultivated sweet potato contains *Agrobacterium* T-DNAs with expressed genes: an example of a naturally transgenic food crop. *PNAS USA* 112 (18): 5844–9.

Lander, E. S., et al. (International Human Genome Sequencing Constortium). (15 Feb., 2001). Initial sequencing and analysis of the human genome. *Nature* 409 (6822): 860–921. doi: 10.1038/35057062.

Leggatt, R. A. & Iwama, G. K. (2003). Occurrence of polyploidy in the fishes. *Reviews in Fish Biol. and Fisheries* 13 (3): 237–46. doi: 10.1023/B:RFBF.0000033049.00668.fe. S2CID 11649126.

Les, D.H. & Philbrick, C.T. (1993). Studies of hybridization and chromosome number variation in aquatic angiosperms: Evolutionary implications. *Aquatic Bot.* 44 (2–3): 181–228. doi: 10.1016/0304-3770(93)90071-4.

Levine, M.T., Jones, C. D., Kern, A.D., Lindfors, H. A., & Begun, D.J. (June, 2006). Novel genes derived from noncoding DNA in *Drosophila melanogaster* are frequently X-linked and exhibit testis-biased expression. *PNAS USA* 103 (26): 9935–9. Bibcode: 2006PNAS.103.9935L. doi: 10.1073/pnas.0509809103. PMC 1502557. PMID 16777968.

Li, F., et al. (2014) Horizontal transfer of an adaptive chimeric photoreceptor from bryophytes to ferns. *PNAS USA* 111 (18): 6672–7.

Lindell, D., et al. (2004). Transfer of photosynthesis genes to and from *Prochlorococcus* viruses. *PNAS USA* 101 (30): 11013–8.

Lindell, D., et al. (2005). Photosynthesis genes in marine viruses yield proteins during host infection. *Nature* 438 (7064): 86–9.

Linder, C.R. & Risenberg, L.H. (22 June, 2004). Reconstructing patterns of reticulate evolution in plants. *Amer. Journ. of Bot.* 91 (10): 1700–8. doi: 10.3732/ajb.91.10.1700.

Loreto, E. L., et al. (2008). Revisiting horizontal transfer of transposable elements in *Drosophila*. *Heredity* 100: 545–54.

Lovejoy, C. O. (Jan., 1981). The origin of man. *Science* 211 (4480): 341–50. doi: 10.1126/science.211.4480.341.

Lovejoy, C. O. (Nov., 1988). Evolution of human walking. *Scientific Amer.* 259 (5): 118–25. Secondary literature.

Lynch, M. & Conery, J. S. (2000). The evolutionary fate and consequences of duplicate genes. *Science* 290 (5494): 1151–5.

Lynch, V. J., Leclerc, R. D., May, G., & Wagner, G. P. (2011). Transposon-mediated rewiring of gene regulatory networks contributed to the evolution of pregnancy in mammals. *Nature Genet.* 43: 1154–9.

Maere, S., et al. (2005). Modeling gene and genome duplications in eukaryotes. *PNAS USA* 02 (15): 5454–9.

Martin, F., et al. (2008). The genome of *Laccaria bicolor* provides insights into mycorrhizal symbiosis. *Nature* 452 (7183): 88–92.

Matzke, M. A., et al. (25 Aug., 1999). Rapid structural and epigenetic changes in polyploid and aneuploid genomes. *BioEssays*. https://doi.org/10.1002/(SICI)1521-1878(199909)21:9<761::AID-BIES7>3.0.CO;2-C.

Maxmen, A. (30 Sept., 2010). Virus-like particles speed bacterial evolution. *Nature*. doi: 10.1038/news.2010.507.

McKenna, D., et al. (2016). Genome of the Asian longhorned beetle (*Anoplophora glabripennis*), a globally significant invasive species, reveals key functional and evolutionary innovations at the beetle-plant interface. *Genome Biol.* 17 (1): 227.

McLean, C. Y., et al. (2011). Human-specific loss of regulatory DNA and the evolution of human-specific traits. *Nature* 471 (7337): 216.

Melamed, D., et al. (2022). De novo mutation rates at the single-mutation resolution in a human HBB gene-region associated with adaptation and genetic disease. *Genome Research* 32 (2). doi: 10.1101/gr.276103.121.

Menouni, E. R., et al. (2015). Bacterial genome remodeling through bacteriophage recombination. *FEMS Microbiol. Letters* 362: 1–10. doi: 10.1093/femsle/fnu022.

Metcalf, J. A., et al. (2014). Antibacterial gene transfer across the tree of life. *Elife* 3: e04266.

Meyers, L. A. & Levin, D. A. (2006). On the abundance of polyploids in flowering plants. *Evolution* 60 (6): 1198–206. https://doi.org/10.1554/05-629.1.

Millard, A., et al. (2004). Genetic organization of the psbAD region in phages infecting marine *Synechococcus* strains. *PNAS USA* 101.30 (2004): 11007–12.

Monroe, J.G., Srikant, T., Carbonell-Bejerano, P., et al. (2022). Mutation bias reflects natural selection in *Arabidopsis thaliana*. *Nature* 602: 101–5. https://doi.org/10.1038/s41586-021-04269-6.

Morgante, M., et al. (31 July, 2005). Gene duplication and exon shuffling by helitron-like transposons generate intraspecies diversity in maize. *Nature Genet.* 37: 997–1002.

Myers, S., et al. (Aug., 2006). The distribution and causes of meiotic recombination in the human genome. *Biochem. Soc. Trans.* 34 (Pt. 4): 526–30. PMID 16856851. doi: 10.1042/BST0340526.

Nekrutenko, A. & Li, W.-H. (1 Nov. 2001). Transposable elements are found in a large number of human protein-coding genes. *Trends in Genet.* 17 (11): 619–21. https://doi.org/10.1016/S0168-9525(01)02445-3.

Nigumann, P., et al. (5 May, 2002). Many human genes are transcribed from the antisense promoter of L1 retrotransposon. *Genomics* 79 (5): 628–34. https://doi.org/10.1006/geno.2002.6758.

Ohno, S. (1970). *Evolution by Gene Duplication*. Springer-Verlag, New York, NY; Heidleberg, Germany; Berlin, Germany.

Oliver, K. R. & Greene, W. K. (2009). Transposable elements: powerful facilitators of evolution. *Bioessays* 31: 703–14.

Olson, M. V., et al. (1986). Random-clone strategy for genomic restriction mapping in yeast. *PNAS USA* 83: 7826–30.

Osborn, T. C., et al. (2003). Understanding mechanisms of novel gene expression in polyploids. *Trends in Genet.* 19 (3): 141–7. PMID 12615008. doi: 10.1016/S0168-9525(03)00015-5.

Ostertag, E. M. & Kazazian, H. H. Jr. (Dec., 2001). Biology of mammalian L1 retrotransposons. *Annual Review of Genetics* 35: 501–38. https://doi.org/10.1146/annurev.genet.35.102401.091032.

Otto, S. & Witton, P.J. (2000). Polyploid incidence and evolution. *Ann. Rev. of Genet.* 34: 401–37. PMID 11092833. doi: 10.1146/annurev.genet.34.1.401.

Pace, J. K II, et al. (4 Nov., 2008). Repeated horizontal transfer of a DNA transposon in mammals and other tetrapods. *PNAS USA* 105 (44) 17023–8. https://doi.org/10.1073/pnas.0806548105.

Paquin C. E. & Williamson, V. M. (5 Oct., 1984). Temperature effects on the rate of Ty transposition. *Science* 226 (4670): 53–5. doi: 10.1126/science.226.4670.53. PMID 17815421.

Payet, J. P. & Suttle, C. A. (2013). To kill or not to kill: the balance between lytic and lysogenic viral infection is driven by trophic status. *Limnol. Oceanog.* 58: 465–74.

Percharde, M., et al. (12 July, 2018). A LINE 1-nucleolin partnership regulates early development and ESC identity. *Cell* 174: 1–15.

Pierce, S. K., et al. (2003). Horizontal transfer of functional nuclear genes between multicellular organisms. *The Biol. Bulletin* 204 (3): 237–40.

Piskurek, O. & Okada, N. (2007). Poxviruses as possible vectors for horizontal gene transfer of retroposons from reptiles to mammals. *PNAS USA* 104: 12046–51.

Prak, E. L. & Haig, H. K. Jr. (2000). Mobile elements and the human genome. *Nature Rev. Genet.* 1: 134–44.

Prescott, S. L., et al. (2015). Enhancer divergence and cis-regulatory evolution in the human and chimp neural crest. *Cell* 163 (1): 68–83.

Pritham, E. J. & Feschotte, C. (Feb. 6, 2007). Massive amplification of rolling-circle transposons in the lineage of the bat *Myotis lucifugus*. *PNAS USA* 104 (6): 1895–900. https://doi.org/10.1073/pnas.0609601104.

Qin, S., et al. (June, 2015). The role of transposable elements in the origin and evolution of MicroRNAs in human. *PLOS ONE*. https://doi.org/10.1371/journal.pone.0131365.

Quispe-Huamanquispe, D. G., et al. (24 Nov., 2017). Horizontal gene transfer contributes to plant evolution: The case of *Agrobacterium* T-DNAs. *Front. in Plant Science* 8: 2015. doi: 10.3389/fpls.2017.02015. PMC 5705623. PMID 29225610.

Ramisetty, B. C. & Sudhakari, P. A. (2019). Bacterial "grounded" prophages: Hotspots for genetic renovation and innovation. *Frontiers in Genetics* 10: 65. doi: 10.3389/fgene.2019.00065. PMC 6379469. PMID 30809245.

Ray, D. A., et al. (2008). Multiple waves of recent DNA transposon activity in the bat, *Myotis lucifugus*. *Genome Res.* 18: 717–28.

Raymond, J., et al. (2002). Whole-genome analysis of photosynthetic prokaryotes. *Science* 298 (5598): 1616–20.

Reno, P. L., et al. (2013). A penile spine/vibrissa enhancer sequence is missing in modern and extinct humans but is retained in multiple primates with penile spines and sensory vibrissae. *PLOS ONE* 8 (12): e84258.

Reznikoff, W. S. (March, 2003). Tn5 as a model for understanding DNA transposition. *Molecular Microbiol.* 47 (5): 1199–206. doi: 10.1046/j.1365-2958.2003.03382.x. PMID 12603728.

Rice, D. W., et al. (2013). Horizontal transfer of entire genomes via mitochondrial fusion in the angiosperm *Amborella*. *Science* 342 (6165): 1468–73.

Rieseberg, L. H., et al. (2003). Major ecological transitions in wild sunflowers facilitated by hybridization. *Science* 301 (5637): 1211–6.

Robeck, T., et al. (28 June, 2016). BC1 RNA motifs required for dendritic transport in vivo. *Scientific Reports* 6, Article number: 28300.

Rohwer, F. & Thurber, R. G. (2009). Viruses manipulate the marine environment. *Nature* 459 (7244): 207.

Ropars, J., et al. (5 Oct., 2015). Adaptive horizontal gene transfers between multiple cheese-associated fungi. *Current Biol.* 25 (19): 2562–9. http://dx.doi.org/10.1016/j.cub.2015.08.025.

Rosette, C. & Karin, M. (March, 1995). Cytoskeletal control of gene expression: Depolymerization of microtubules activates NF-kappa B. *Journ. of Cell Biol.* 128 (6): 1111–9. doi: 10.1083/jcb.128.6.1111. PMC 2120413. PMID 7896875.

Roth, G., Nishikawa, K. C., & Wake, D. B. (1997). Genome size, secondary simplification, and the evolution of the brain in salamanders. *Brain, Behavior and Evolution* 50: 50–9. https://doi.org/10.1159/000113321.

Salzberg, S.L., et al. (22 Feb., 2005). Serendipitous discovery of *Wolbachia* genomes in multiple *Drosophila* species. *Genome Biol.* 6 (R23). https://doi.org/10.1186/gb-2005-6-3-r2353-6.

Sanger, F., et al. (1977). Nucleotide sequence of bacteriophage Φ X174 DNA. *Nature* 265: 687–95.

Sardo, V. L., et al. (2012). An evolutionary recent neuroepithelial cell adhesion function of huntingtin implicates ADAM10-Ncadherin. *Nature Neurosci.* 15 (5): 713–21.

Schaack, S. et al. (Sept., 2010). Promiscuous DNA: horizontal transfer of transposable elements and why it matters for eukaryotic evolution. *Trends in Ecol. and Evol.* 25 (9): 537–46. https://doi.org/10.1016/j.tree.2010.06.001.

Schnabe, P. S., et al. (2009). The B73 maize genome: complexity, diversity, and dynamics. *Science* 326: 1112–5.

Schwartz J. A., et al. (2014). FISH labeling reveals a horizontally transferred algal (*Vaucheria litorea*) nuclear gene on a sea slug (*Elysia chlorotica*) chromosome. *Biol. Bulletin* 227: 300–12.

Seaborg, D. (2022). *How Life Increases Biodiversity: An Autocatalytic Hypothesis.* CRC Press/Taylor & Francis. Boca Raton, FL; London, UK; New York, NY.

Sengelov, G, et al. (2001). Effect of genomic location on horizontal transfer of a recombinant gene cassette between *Pseudomonas* strains in the rhizosphere and spermosphere of barley seedlings. *Current Microbiol.* 42: 160–7.

Sessions, S. K. & Wake, D. B. (June, 2021). Forever young: Linking regeneration and genome size in salamanders. *Developmental Dynamics* 250 (6): 768–78. https://doi.org/10.1002/dvdy.279.

Sharon, I., Tzahor, S., Williamson, S., et al. (9 Aug., 2007). Viral photosynthetic reaction center genes and transcripts in the marine environment. *The ISME Journ.* 1: 492–501. https://doi.org/10.1038/ismej.2007.67.

Shea, B. T. (1989). Heterochrony in human evolution: The case for neoteny reconsidered. *Amer. Journ. of Physical Anthrop.* 32 (S10): 69–101. doi: 10.1002/ajpa.1330320505.

Shelomi, M., Danchin, E. G. J., Heckel, D., Wipfler, B., Bradler, S., Zhou, X., & Pauchet, Y. (May, 2016). Horizontal gene transfer of pectinases from bacteria preceded the diversification of stick and leaf Insects. *Scientific Reports* 6: 26388. doi: 10.1038/srep26388. http://dx.doi.org/10.1038/srep26388.

Shen, S., et al. (2011). Widespread establishment and regulatory impact of Alu exons in human genes. *PNAS USA* 108 (7): 2837–42. PMC 3041063. PMID 21282640. doi: 10.1073/pnas.1012834108.

Sieradzki, E., Ignacio-Espinoza, J.C., Needham, D., et al. (12 March, 2019). Dynamic marine viral infections and major contribution to photosynthetic processes shown by spatiotemporal picoplankton metatranscriptomes. *Nature Communications* 10 (1169). https://doi.org/10.1038/s41467-019-09106-z.

Simakov, O., et al. (26 Nov., 2015). Hemichordate genomes and deuterostome origins. *Nature* 527: 459–65. doi: 10.1038/nature16150.

Smalheiser, N.R. & Torvik, V.I. (2006). Alu elements within human mRNAs are probable microRNA targets. *Trends Genet.* 22: 532–6.

Smit, A. F. (1 Dec., 1999). Interspersed repeats and other mementos of transposable elements in mammalian genomes. *Current Opinion in Genet. and Development* 9 (6): 657–63. https://doi.org/10.1016/S0959-437X(99)00031-3.

Strand, D. J. & McDonald, J. F. (1985). Copia is transcriptionally responsive to environmental stress. *Nucleic Acids Research* 13 (12): 4401–10. doi: 10.1093/nar/13.12.4401. PMC 321795. PMID 2409535.

Sullivan, M. B., et al. (2005). Three *Prochlorococcus* cyanophage genomes: signature features and ecological interpretations. *PLOS Biol.* 3 (5): e144.

Sullivan, M. B., et al. (2006). Prevalence and evolution of core photosystem II genes in marine cyanobacterial viruses and their hosts. *PLOS Biol.* 4 (8): e234.

Syvanen, M. (Jan., 1985). Cross-species gene transfer; implications for a new theory of evolution. *Journ. Theor. Biol.* 112 (2): 333–43. doi: 10.1016/S0022-5193(85)80291-5. PMID 2984477.

Szmulewicz, M. N., Novick, G. E., & Herrera, R. J. (1 June, 1998). Effects of Alu insertions on gene function. *Electrophoresis* 19 (8–9): 1260–4. ISSN 0173-0835. PMID 9694261. doi: 10.1002/elps.1150190806.

Tanahashi, T., et al. (2005). Diversification of gene function: homologs of the floral regulator FLO/LFY control the first zygotic cell division in the moss *Physcomitrella patens*. *Development* 132: 1727–36; doi: 10.1242/dev.01709.

Thiessen, D.D. (1997). *Bittersweet Destiny: The Stormy Evolution of Human Behavior.* Transaction Publishers, Piscataway, NJ.

Torgasheva, A. A., et al. (11 June, 2019). Germline-restricted chromosome (GRC) is widespread among songbirds. *PNAS USA* 116 (24): 11,845–50.Touchon, M. & Rocha, E. P. C. (2007). Causes of insertion sequences abundance in prokaryotic genomes. *Mol. Biol. Evol.* 24: 969–81.

Tucker, R. P. (1990). The roles of microtubule-associated proteins in brain morphogenesis: A review. *Brain Research Reviews* 15 (2): 101–20. doi: 10.1016/0165-0173(90)90013-E. PMID 2282447. S2CID 12641708.

Turley, C. (12 Sept., 2001). Manna from heaven: Algal-bacterial coupling in the deep-sea. *Symp. of Soc. for General Microbiol.*, held at Univ. of East Anglia, Norwich, UK.

van de Lagemaat, L. N., et al. (1 Oct., 2003). Transposable elements in mammals promote regulatory variation and diversification of genes with specialized functions. *Trends in Genetics* 19 (10): 530–6. doi: https://doi.org/10.1016/j.tig.2003.08.004.

van Eeden, F. & St. Johnston, D. (Aug., 1999). The polarisation of the anterior-posterior and dorsal-ventral axes during *Drosophila* oogenesis. *Current Opinion in Genetics & Development* 9 (4): 396–404. doi: 10.1016/S0959-437X(99)80060-4. PMID 10449356.

Van Houte, S., et al. (5 Jan., 2015). Baculovirus infection triggers a positive phototactic response in caterpillars to induce 'tree-top' disease. *Biol. Letters* 10:20140680.

van't Hof, A. E., et al. (1 June, 2016). The industrial melanism mutation in British peppered moths is a transposable element. *Nature* 534: 102–5. https://doi.org/10.1038/nature17951.

Villarreal, L. P. (2005). *Viruses and the Evolution of Life.* ASM Press, Washington, DC. ISBN 978-1-55581-309-3.

Volff, J.-N. (Sept., 2006). Turning junk into gold: domestication of transposable elements and the creation of new genes in eukaryotes. *BioEssays* 28 (9): 913–22. https://doi.org/10.1002/bies.20452.

Walsh, A. M., et al. (2013). Widespread horizontal transfer of retrotransposons. *PNAS USA* 110 (3): 1012–6.

Waterston, R. H. & Pachter, L. Mouse Genome Sequencing Consortium (2002). Initial sequencing and comparative analysis of the mouse genome. *Nature* 420 (6915): 520–62. ISSN 0028-0836. https://resolver.caltech.edu/CaltechAUTHORS:20170309-090859678.

Watson, R. A. & Szathmáry, E. (2016). How can evolution learn? *Trends in Ecol. & Evol.* 31 (2): 147–57.

Watson, T. (15 Nov., 2012). Bdelloids surviving on borrowed DNA. *Science/AAAS News.*

Wolfe, K. H. & Shields, D. C. (1997). Molecular evidence for an ancient duplication of the entire yeast genome. *Nature* 387, 708–13.

Woloszynska, M., et al. (Nov., 2004). A fragment of chloroplast DNA was transferred horizontally, probably from non-eudicots, to mitochondrial genome of *Phaseolus. Plant Molecular Biol.* 56 (5): 811–20. doi: 10.1007/s11103-004-5183-y. PMID 15803417. S2CID 14198321.

Wood, T. E., et al. (2009). The frequency of polyploid speciation in vascular plants. *PNAS USA* 106 (33): 13875–9. Bibcode: 2009PNAS.10613875W. JSTOR 40484335. PMC 2728988 Freely accessible. PMID 19667210. doi: 10.1073/pnas.0811575106.

Xu, X., et al. (2002). Origins of polyploidy in coast redwood (*Sequoia sempervirens* (D. DON) ENDL.) and relationship of coast redwood to other genera of Taxodiaceae. *Silvae Genetica* 51: 2–3.

Yuannian, J., et al. (5 May, 2011). Ancestral polyploidy in seed plants and angiosperms. *Nature* 473: 97–100. doi: 10.1038/nature09916.

Zeb, D. H., et al. (2009). Transposable elements and an epigenetic basis for punctuated equilibrium. *Bioessays* 31: 715–26.

Zeidner, G., et al. (2005). Potential photosynthesis gene recombination between *Prochlorococcus* and *Synechococcus* via viral intermediates. *Environmental Microbiol.* 7 (10) (2005): 1505–13.

Zhou, Q., Zhang, G., Zhang, Y., Xu, S., Zhao, R., Zhan, Z., Li, X., Ding, Y., Yang, S., & Wang, W. (Sept., 2008). On the origin of new genes in *Drosophila. Genome Research* 18 (9): 1446–55. doi: 10.1101/gr.076588.108. PMC 2527705. PMID 18550802. (Say all authors names three times in a row without a mistake and the author will send you a free copy of this book).

Zuccato, C. & Cattaneo, E. (Aug., 2016). The Huntington's paradox. *Sci. Amer.* 56–61 (Secondary Literature).

10 Altruism and Cooperation within Populations Are Often Adaptive, and Maintain Populations and Diversity

Darwin's theory of natural selection states that evolution favors those that survive and breed, passing on their genes to future generations. A simplistic interpretation of this would predict that individuals of the same population would tend to compete with each other for limited resources and mates, that most interactions between organisms of the same population and species would be antagonistic, and that organisms would tend to behave selfishly toward each other. Selfish organisms would tend to survive and breed better than the altruists, so a genetic disposition for selfishness would be selected for. Selfish behaviors within a population would tend to increase the probability it would go extinct, decreasing biodiversity in two ways. First, decreasing the number of populations within a species tends to decrease the genetic variability of the species. Biodiversity consists of variability within species as well as number of species. Second, loss of populations increases the likelihood of extinction of a species. Thus, a simplistic interpretation of evolutionary theory conflicts with the Autocatalytic Biodiversity Hypothesis (ABH)/Pachamama Hypothesis.

The purpose of this chapter is to demonstrate that the behavior of organisms toward others in their population and species maintains biodiversity, including both genetic variability within populations and the number of species; that this is accomplished through altruism and cooperation; and that these behaviors are adaptive. The altruistic and cooperative behaviors prevent the decrease in size and loss of populations and the extinction of species. But they do not necessarily in themselves cause speciation and hence do not normally necessarily act as a mechanism to increase the number of species. Rather, they for the most part maintain biodiversity.

This chapter will not argue that organisms do not display selfish behaviors. They clearly do. However, organisms often display cooperative and altruistic behavior, sometimes even at great cost to their survival and ability to directly pass on their genes. Altruism, cooperation, and avoidance of physical battle decrease the probability that populations and species will go extinct from competition for resources or mates, from antagonistic and selfish behaviors, or from fighting. In fact, altruistic and cooperative behaviors are consistent with modern evolutionary theory, and this chapter aims to demonstrate that, showing that the evolution of altruistic behavior supports the ABH. One can call the altruistic and cooperative behaviors a form of ecosystem engineering if one accepts such a broad definition of the term. If not, one can call them social engineering.

Eusocial insects, such as some bees and ants, display tremendous altruism. Honey bees are a prime example. The worker bees are all female. Worker bees vibrate their wings to warm the hive on cold days, often staying out at the periphery, where it is cold, risking death. They sting to defend the hive from invaders. If the invader is a mammal, the stinger is left in the invader, and disembowels the bee. This makes the sting more effective, but kills the bee. They are sterile, and do not reproduce or directly pass on their genes; the queen bee is the only female that reproduces. Workers thus sacrifice directly passing on their genes.

Sacrificing breeding, energy, time, and one's life are all done routinely in the working caste of all of the most highly social insects. Highly social ants show extreme altruism. For example, one to eight individuals of the Brazilian ant species, *Forelius pusillus* (it has no common name), kick sand on the entrance to the nest at the end of the day to close it from the outside, protecting it, with the result that they cannot enter the nest, and will die that night (Tofilski et al., 2008). The behaviors described in this and the previous paragraph are beneficial for the colonies, and involve a great deal of altruism and cooperation. But it is hard to understand in a limited interpretation of Darwinian natural selection how members of an entire caste could give up reproduction and often sacrifice their lives.

10.1 KIN SELECTION

One mechanism that at least partially accounts for the altruistic behavior of highly social bees and ants, and some other animals, is kin selection. In kin selection, organisms can pass on their own genes by helping related individuals, such as siblings, pass on genes that they have in common. Hamilton (1963; more importantly, 1964) formalized this concept, showing the mathematics. His basic idea is that the more related the altruistic organism is to the individual of its species that it is helping, the more likely the altruistic behavior will be favored by natural selection. The greater the benefit to the individual receiving the altruistic act, and the lower the cost to the altruist as a result of performing the act, the more natural selection will favor the act. In highly social ants and bees, such as the western honey bee (*Apis mellifera*), the queen stores sperm and can choose to release it to fertilize an egg

DOI: 10.1201/9781003246640-10

before laying it, or lay an unfertilized egg without releasing any sperm. If she fertilizes the egg, it will have two copies of each chromosome, and develops into a female. If she does not fertilize the egg, it will have one copy of each chromosome, and it becomes a male. In species of highly social ants and bees with only one queen in each colony, the result of this system of reproduction is that females share 75% of their genes with each other and with the queen. In typical species with normal systems of reproduction such as mammals and birds, siblings have 50% of their genes in common with each other and with their parents. The females are called workers and do the work of the hive. This system of reproduction allows the workers to succeed in an evolutionary sense even while giving up their own direct reproduction, because the workers can pass on their genes through the queen by helping her produce more offspring. Kin selection accounts for a great deal to all of the altruism and cooperation of highly social insects. Kin selection is so strong in some species of ants that they will bite through nylon to free other ants tied up in nylon thread, but only if they are related to them.

When in danger of drowning, a colony of fire ants (genus *Solenopsis*) work cooperatively to build rafts (Mlot et al., 2011). Thousands of them pile together and cling to one another in every way—legs on legs, jaws on legs, jaws on jaws—often surviving this way for days. The rafts have physical properties that make them float and protect the ants from the water. The ants start as a sphere. The top ants crawl down to the water and grab onto water-level ants. The next layer crawls to the edge, and so on. That is, the ants take turns being in the safer places and more dangerous places. Thus, they display altruism. This behavior evolved because these ants come from regions of Brazil and Argentina that often flood. Other ant species collectively make structures with their bodies, but only fire ants make rafts. No ants are in charge in building these rafts. In this communal organization, the many act as one, giving an example of a superorganism, which will be discussed later in this chapter. However, the construction process relies on the trapping of ants at the raft edge by their neighbors, suggesting that some "cooperative" behaviors may rely upon coercion.

The naked mole-rat (*Heterocephalus glaber*) is a rodent with almost no hair and wrinkly, flesh-colored skin and two large protruding front teeth (Figure 10.1). It is highly social, with a social structure similar to highly social bees. It lives in colonies averaging 75 to 80, but up to 300, in complex systems of underground burrows up to two to three miles in length, in the dry tropical grasslands of East Africa. The mole-rats dig the burrows. Only one female, the queen, and one to three males, reproduce. The queen is the largest member of the colony. Workers are sterile, with the smaller ones maintaining the nest and doing the food-gathering, and the bigger ones guarding the colony against attack. This system can be partly explained by kin selection. Individuals called helpers that help raise offspring often help raise their nephews and nieces. But genetic evidence has shown that sometimes the helpers are not related to those they are helping. In this case, kin selection does not explain the cooperative behavior, so alternative explanations have been proposed. These include group selection and advantages to the helpers. Both of these explanations will be discussed in more detail later in this chapter, although not with respect to mole rats.

FIGURE 10.1 The naked mole-rat (*Heterocephalus glaber*), here isolated on white to show it clearly, is a highly social, cooperative mammal that shows tremendous altruism within its colony. This is only partially explained by kin selection.

Shutterstock Photo ID: 1667001955

Title: Naked Mole-rat, hairless rat, isolated on white.
Photo Contributor: Eric Isselee

Marmosets, small primates that resemble small monkeys and live in tropical forests of the New World, stay at home past sexual maturity to help care for their younger siblings. This can be accounted for by an unusual form of kin selection. They pick up their siblings' cells by sharing a blood supply through fused placentas when in the womb. All marmosets in the same uterus share a common placenta. They end up as chimeras—composites with some of their own cells and some of their sibling's cells. Ross et al. (2007) found that all 17 organs in marmosets that they looked at had another marmoset's cells and DNA, and even tissues that form the sperm and egg were mixtures of this nature. Most eukaryotes have two copies of each gene. Each copy or form of the gene is called an allele of the gene. The two alleles of a gene can be identical to each other or they can be different from each other. There can be several alleles of a gene in a population, even though any given organism normally has one gene and thus two alleles of any given gene. The researchers found that chimeric marmosets often transmit their sibling's allele, acquired in utero, to their own offspring, instead of transmitting their own allele to their offspring, so an individual that transmits sperm or eggs to an offspring is not necessarily fully the genetic parent of that offspring. The result of all this is that siblings are more related to each other on average than the usual 50% of genes shared in common by most siblings. So marmosets can pass on more of their own genes by caring for their nephews and nieces than uncles and aunts of typical species can.

A subordinate wild turkey (*Meleagris gallopavo*) will help his brother put on an impressive courtship display that directly helps only the dominant brother (Krakauer, 2005). Since they are siblings, this is kin selection.

Sperm cells compete with each other to swim and fertilize the egg. But some human sperm cells sacrifice themselves, exuding their enzymes that break down the egg's membrane prematurely, helping others penetrate the egg cell. Since sperm cells from the same male share on the average a large percentage of their genes, this altruistic sacrifice is kin selection. In the wood mouse (*Apodemus sylvaticus*), sperm aggregate and cooperate, forming moving trains to better fertilize the egg, and some sperm sacrifice themselves, also by prematurely exuding enzymes that break down the egg's membrane, improving the ability of one of the sperm in the train to enter the egg (Moore et al., 2002).

Plants display kin selection. Unrelated American searockets (*Cakile edentula*), a type of plant, competed for soil nutrients by aggressive root growth when grown in the same pot, while related ones did not (Dudley and File, 2007). Adult trees slow their growth when saplings are present, aiding them in surviving. This would be kin selection sometimes, but since seeds are often dispersed far from the parent tree by wind and animals, the saplings would in many cases be unrelated to the adult tree, in which case this may be a form of group selection not fully understood.

Diatoms are single-celled and reproduce by dividing into two cells during their vegetative phase, making clones of genetically identical cells. There are records of freshwater diatoms infected by a fungus dying more quickly than they normally would from the infection with the result that the fungus cannot complete its life cycle and infect other diatoms—altruism that protects their nearby genetically identical kin from infection by the fungus (Canter and Jaworski, 1979).

10.2 THE GREEN BEARD EFFECT

The green beard effect occurs when the carriers of an allele or alleles of a gene or genes recognize the same allele(s) in others who are not necessarily kin, and as a result act altruistically toward them. It involves altruism between organisms with shared alleles expressed by specific phenotypic traits, such as color or height. This could delineate a group of organisms in a population, the members of which selectively display greater cooperation toward each other than toward the rest of the population, forming a clique that can be beneficial to its members who are not necessarily kin. An example of the green beard effect has been found in nature: Workers in colonies of the red imported fire ant (*Solenopsis invicta*) selectively kill queens that share only one copy of the allele called Gp-9 with them and spare those queens that share both copies with them (Keller and Ross, 1998).

10.3 WARNING OTHERS MAY BE KIN SELECTION OR A SIDE EFFECT OF SELFISHNESS

Certain species of birds, squirrels, monkeys, and other organisms warn others of predators with alarm calls. Since this could draw the predator's attention to the callers and allow the predator to know of their presence and location, it could potentially be costly and dangerous to them. Kin selection accounts for alarm calls to some extent. Yellow-bellied marmots (*Marmota flaviventris*) emit warning calls when predators are nearby more often when they are near their offspring and other relatives than when near unrelated individuals. Male prairie dogs (genus *Cynomys*) also call more often in the presence of their kin. But warning calls also evolved due to other factors than kin selection. Belding's ground squirrels (*Urocitellus beldingi*) warn other squirrels of ground predators with multiple-note trills that increase their susceptibility to these predators (Sherman, 1977). They use a single-note whistle when flying hawks appear. Females do not call more in the presence of kin, even their offspring. The males call more often when far from cover and close to the hawk. Sherman (1985) concluded the calls increase the chance the calling squirrels will escape from hawks, and these calls cannot be explained by kin selection. Some argue that warning calls are at least sometimes selfish behaviors that decrease the chance that the caller will be eaten, and that they help other members of the caller's population as a side effect. The most likely way this would be the case is the call could inform the predator that it has been detected, so it will not waste energy pursuing the caller. This is supported by the fact that birds that emit warning calls are attacked less often by predatory birds than birds that do not give warning calls. Second in importance is the calls may frighten off the predator, helping the caller escape. Some think the warning calls may attract predators of the predators, aiding the caller in escaping. This probably happens less often than the first two explanations occur. Some, perhaps all, species throw their voices like ventriloquists, and thus do not significantly increase the chances the caller will be detected by the predator. Thus, the cost to the caller may be insignificant.

Vervet monkeys (*Chlorocebus pygerythrus*) of Africa are thought to have up to 30 alarm calls. Following are three distinct alarm calls that we know the meaning of (Seyfarth et al., 1980). The call for a leopard (*Panthera pardus*) causes monkeys to run up into trees. The call for pythons causes the monkeys to stand on two feet, and look for the snake, and move as far as away from it as necessary for safety. The call that means "eagle" causes monkeys to look up, and scurry down trees and along the ground into the bush.

Gunnison's prairie dogs (*Cynomys gunnisoni*) have distinct alarm calls for hawks, coyotes, domestic dogs, and humans. African elephants (genus *Loxodonta*) have a special warning call they emit when they encounter bee hives. Richardson's ground squirrels (*Urocitellus richardsonii*) of North American prairies have an alarm call audible to human ears and an ultrasonic one inaudible to humans (Wilson and Hare, 2004). The ultrasonic call allows the caller to warn other squirrels without revealing its location to a predator that cannot detect ultrasonic calls.

Some pheromones warn of predators. Minnows and catfish release alarm pheromones when injured, causing nearby fish to hide in dense schools near the bottom. Alarm pheromones in fish may have evolved by kin selection, but it is not fully understood what caused their evolution.

10.4 RECIPROCAL ALTRUISM

Another mechanism of social engineering by which altruism and cooperation between members of a group can evolve is reciprocal altruism, whereby an organism helps another organism at an expense to itself, with the expectation that the beneficiary will reciprocate in the future (Trivers, 1971; Stephens, 1996). For it to spread in the population, the chances of the individuals involved meeting again must be high, and the interactions must be repeated over a long period. The altruists must have sufficient intelligence or other mechanisms for detecting cheaters who receive altruism but do not reciprocate. They also must be able to remember acts of altruism and cheating.

A likely example of reciprocal altruism is found in the common vampire bat (*Desmodus rotundus*) (Wilkinson, 1988), a social (or ecosystem) engineer that feeds on blood, mainly of mammals (including humans), and ranges from Mexico to Brazil, Chile, and Argentina. These bats share food with others in their colony (Wilkinson, 1984). They can only survive approximately two nights without a blood meal, and cannot be assured of finding one every night. Some bats come home hungry, while others arrive bloated with blood. If a bat fails to find a meal, it begs another bat for food, and the other bat may regurgitate a small amount of blood, which sustains the beggar. Amazingly, a study found donor bats approach starving bats and initiate the sharing of food themselves, even more often than recipients approached the donors, showing that begging is not necessary for the altruism and that bats do not share food just to prevent the harassment of begging (Carter and Wilkinson, 2013a). The best predictor of food given from one bat to another was the recipient bat having given food to the donor bat in the past, and it was 8.5 times more important than relatedness. A full 64% of the sharing pairs were not kin, approaching the 67% expected if kin selection was not the explanation. Together with past work, these findings support the hypothesis that food sharing in vampire bats is not explained solely by kin selection or harassment, and requires reciprocal altruism for a comprehensive explanation.

This food-sharing is considered by scientists as true reciprocal altruism, because reciprocity has been shown to be more important than kin selection in the evolution and persistence of this behavior (ibid.). The bats benefit from forming cooperative, reciprocal relationships between both relatives and nonrelatives (ibid.). The data confirm that bats that have behaved altruistically in the past get helped in the future by those they have benefitted. However, since all instances of feeding are between members of the same group, who are on average cousins, some have argued that this is kin selection, which tends to be hard to distinguish from reciprocal altruism. And it has not been shown that previously nonaltruistic bats are refused meals when they need them. Thus, Carter and Wilkinson (2013b) also argued that although correlational evidence indicates a role for fitness benefits that are both direct (reciprocal altruism) and indirect (kin selection), unequivocally demonstrating reciprocal altruism in vampire bats still requires testing whether and how bats respond to those who do not reciprocate in the sharing of meals. But the evidence is largely consistent with reciprocal altruism being at least a partial explanation for food-sharing, since such sharing is routine between nonkin. It appears that when all the evidence is weighed, reciprocal altruism is necessary for a full explanation of vampire bat food-sharing.

Vampire bats are the only bat species known that adopt baby bats in the colony that have lost their mothers. The baby's mother does not need to be the adopting bat's sister. They also do social grooming, usually between females and their offspring, but also between unrelated adult females, which is associated with food-sharing and social bonding (Wilkinson, 1986).

Another example seems to involve emotional reward. Green wood hoopoes (*Phoeniculus purpureus*) (Figure 10.2) are large tropical African birds that massage each other with their beaks (Radford, 2012, and references therein). Both masseuse and recipient benefit with lower stress hormone levels. After receiving massage-like grooming, the recipient reduces its activity levels and relaxes into what looks like a happy stupor for a noticeable period. Subordinate birds seem to enjoy massages the most when they are given by their superiors in the hierarchy. This could be because they are more stressed, so it is particularly relaxing to be massaged by a threatening, dominant bird, since it means they are accepted. The groomer might also benefit because it is trading grooming for food, tolerance, or help from the recipient at a later time, including during conflicts within and between groups. And there is

FIGURE 10.2 Beautiful tropical green wood hoopoes (*Phoeniculus purpureus*) of Africa massage each other with their beaks, resulting in stress hormone reduction in both the masseuse and the recipient, in an intriguing example of reciprocal altruism. This picture shows three of them in Okavango, Moremi, Botswana, Africa.

Shutterstocktock Item ID: 1108992665

Title: Green Wood hoopoe, *Phoeniculus purpureus*, bird family in the nature habitat. Animals sitting in the tree trunk, one bird fly. Wildlife scene from nature, Okavango, Moremi, Botswana.

Photo Contributor: Ondrej Prosicky

some evidence that the masseuses actually experience lower long-term stress levels than the recipients, and that grooming others might even be self-rewarding (ibid.). Other bird species do this, some with other species, which would be symbiosis when both species groom the other species. Studies show these findings on the effects of grooming apply to primates as well (e.g., Dunbar, 2010). The reciprocal altruism is from one bird trading a massage for a future massage or other benefit from the recipient, or because the massage itself benefits both masseuse and recipient, or both of these reasons.

Tree stumps can survive without leaves by grafting their roots onto the roots of neighboring trees of the same species, obtaining water and resources from them. The trees whose roots are grafted onto benefit by getting a larger root system to access more nutrients and water. This reciprocal altruism supports the shift from the perception of trees as individuals to viewing forest ecosystems as superorganisms. Superorganisms will be discussed later in this chapter.

Cooperative breeding is another form of altruism and cooperation primarily explained by reciprocal altruism. It is a hierarchical breeding system with a dominant breeding pair that has offspring, and nonbreeding, subordinate helpers that help raise other's young. The helpers provide care for the breeding pair's offspring (Lukas and Clutton-Brock, 2012). A helper may be defined as an individual who feeds or otherwise aids another organism of its species that is neither its mate nor its dependent offspring. Helpers are saddled with costs, including fewer offspring of their own, obtaining food for and feeding young that are not their offspring, and the need to guard the offspring of others. Kin selection accounts for a large part of why cooperative breeding evolved, but not all of it. Helpers benefit in other ways than passing on their genes through their relatives, including greater time to hunt for food than if they raised their offspring on their own (Hatchwell, 2009), protection from predators, and inheritance of territory. Also, they gain parenting skills that they will need if they reproduce later in life.

Breeders benefit mainly from producing more offspring, which have better survival rates, but also can spend less time caring for offspring and guarding their territory. They do have the costs of the time and energy expended for the suppression of subordinates from breeding. The group augmentation hypothesis states that increasing the size of the group because of the addition of helpers and increased breeding success aids in the survival of both breeders and helpers, and in helper's future breeding success (Mares et al., 2012). The benefits must outweigh the costs for both breeders and helpers for the system to evolve. Studies have shown kin selection is important in driving cooperative breeding in birds in general (Dijk et al., 2014). Based on the benefits I listed, it is apparent that reciprocal altruism is important as well. The two explanations are not mutually exclusive. The best explanation is likely a combination of kin selection and reciprocal altruism. It is probable that the relative importance of the two explanations varies with the species and situation.

Cooperative breeding is most common in birds, where it occurs in about 8% of species worldwide (Jetz and Rubinstein, 2011), but it also occurs in other groups, including insects, such as carpenter bees. Examples of birds that do it are the common moorhen (*Gallinula chloropus*), acorn woodpecker (*Melanerpes formicivorus*), red-cockaded woodpecker (*Leuconotopicus borealis*), and apostlebird. Red-cockaded woodpeckers live in groups of three to four birds, with only one breeding pair. Helper birds take turns sitting on and incubating the eggs that the breeding female lays, and help raise the young. Though the young females will leave the group to look for a new group without a dominant female before the next breeding season, the young males stay with the group which they were born into and become helper birds.

The largest and most spectacular of any bird nest is made by the sociable weaver (*Philetairus socius*) of Southern Africa. It builds large compound community grass nests where many birds raise young. The nests house over a hundred pairs of birds (Covas, 2002), with several generations at a time. Up to over half the birds do not breed in a given season, and nearly all breeding pairs are aided by helpers, who help in both caring for younger siblings and unrelated young (ibid.). So this appears to involve both kin selection and reciprocal altruism.

White-fronted bee eaters (*Merops bullockoides*), birds that inhabit the African savannah, have helpers that assist breeding pairs in raising their young, protecting young from spitting cobras, foraging tirelessly for bees, and delaying having their own young. Generally, the helpers are close relatives, and kin selection is the explanation (Emlen and Wrege, 1988). But not always. Sometimes the helper sneaks in and lays an egg when the parents are off foraging, and then volunteers to help rear what is really her own offspring, along with the other offspring in the nest. There is a case in which a male bird who previously failed in his attempt to reproduce volunteered to help his former mate and her mate raise their young. He was not related to these parent birds. The offspring later helped him raise his young, interacting with him as if they were related. This is reciprocal altruism. The younger bird cannot tell that the helper is not its relative, and seems to have a genetic program to treat a bird that raises it as kin, which it often is. But this program can be exploited by the bird that helps raise it, such that the younger bird repays it.

Cooperative breeding occurs in mammals, such as the sugar glider (*Petaurus breviceps*). It occurs in meerkats (*Suricata suricatta*), members of the mongoose family that live in Africa. In these groups, the level of help from helpers is not correlated with their relatedness to the young they are helping rear (Clutton-Brock et al., 2000). So this is reciprocal altruism.

10.5 GROUP SELECTION

A form of natural selection favoring altruism that is not universally accepted, called group selection, was proposed by Wynne-Edwards (1962; 1986). He asserted that alleles can increase in frequency or even go to 100% in a population or species because of the benefits they bestow on the entire group, even if the behaviors of their bearers are to an extent altruistic and self-sacrificing and the alleles decrease the fitness of the individuals that bear them. He proposed that groups with individuals who limited their breeding had less of a tendency to destroy their environment from overpopulation and eating out their food supply. Those more altruistic groups would outcompete groups of selfish individuals that did not control their reproduction. Groups with more altruists and whose individuals cooperate with each other persist and reproduce by splitting into additional such groups, while groups of selfish individuals tend to survive and reproduce (split into two groups) less. Therefore, groups with a great proportion of altruists outcompete groups of selfish individuals. It works best when the competing groups are isolated and there is very little to no migration between them.

There are examples of group selection in nature. Parasites are often selected to have less of a negative health impact on their host, and to have less of a tendency to kill it. Parasites with a lower negative impact on their host allow it to live longer, giving the parasites more time, and hence a better chance, to infect a new host than parasites that quickly kill their host. Mosquito-borne viruses of rabbits are an example of this. If the reduced negative effect on the rabbit is accomplished by slower breeding by the virus, the group selection involves an "altruistic" sacrifice by the slower-breeding viruses. Groups with more of the slower-breeding viruses will be selected for. This does happen.

Microbiomes undergo group selection favoring those microbe populations that aid their host, or at least do not harm it. When the microbes aid their host, they help their habitat, allowing them to persist and thrive. There may be myriads of such group-selected microbial populations and species, but we will not know their number and prevalence until more study of microbiomes is done. The microbes may be aiding the host and cooperating with each other due to group selection on a *community* of microbes of many species in the microbiome, but this is not known. Much research needs to be done in this area.

Group selection was once viewed favorably by many, but later went out of favor because, it was argued, selfish genes and individuals would outcompete altruistic genes and individuals within the group, eliminating the chance of altruistic groups persisting and being favored by selection (Williams, 1966, © renewed 1992). Later, group selection was revived and arguments emerged in its favor (Wilson and Wilson, 2007, 2008). Wilson and Sober (1994) argued that rejection of group selection was a mistake, and proposed Multilevel Selection Theory, which states that there are units and levels of selection that interact with each other. They may be in conflict with each other, and may work in concert to enhance fitness. The gene is the lowest unit, or level, of selection. The next level up is the cell, then the organism, then the population, then the species, then the ecosystem. Competition between groups must outweigh that between individuals within groups for altruistic genes and traits beneficial to the group to spread by group selection (O'Gorman et al., 2008). The group is usually the population. If any level of selection dies out, all the levels below it will necessarily also die off. So, if the group dies off, all the individuals in it will die off. This must be remembered by those who are skeptical of group selection.

Some experimental work implies group selection occurs. For example, Muir (2009), comparing egg productivity in hens, showed that individual selection led to an exceptionally aggressive strain that launched many fatal attacks on other hens after only six generations, to the detriment of the group. Rauch et al. (2003) demonstrated a host-parasite situation in which more transmissible variants continually arise and grow rapidly for several generations, but die out before dominating the system due to the depletion of susceptible hosts, which hurts the group.

Wilson (2005) argued that kin selection alone can no longer be considered the major cause of highly social systems in animals such as ants and bees. There are several species of extremely social animals that do not have the system of reproduction I described earlier that exists in such highly social insects as honey bees. These include insects, such as termites, and naked mole rats. So the system that causes a relationship of three-quarters between siblings and between mothers and daughters is not required for the evolution of highly social species. This suggests that kin selection alone might not necessarily always account for the evolution of highly social species, and group selection may sometimes be a necessary partial or whole explanation for it.

And the altruism in termites can be extreme. As workers of *Neocapritermes taracua* (no common name), a termite species in French Guiana, near Brazil, age, their jaws wear out, and they cannot chew wood to help the colony as well as they could when they were younger. They develop a deep pouch filled with a blue toxin, made more potent by mixing with their saliva. The older they are, the bigger the pouch. When they fight termites from other colonies, the pouch explodes, and kills both the older termite and its opponent. This is suicide bombing that is an altruistic self-sacrifice that benefits the colony. However, Foster et al. (2006) discussed Wilson's ideas and concluded that kin selection remains the key explanation for the evolution of altruism in eusocial insects.

Kin selection, reciprocal altruism, and possibly group selection are all mechanisms by which natural selection favors organisms that are altruistic and cooperate. This maintains and sometimes increases biodiversity.

10.6 KIN SELECTION WITH POSSIBLE GROUP SELECTION

Following are some examples that can be accounted for by kin selection, but in which group selection cannot be ruled out as a possible additional factor. Unicellular organisms, such as bacteria, some green algae, and protozoa, often kill themselves when their death is imminent, in a process called programmed cell death. Cell activity shuts itself down in steps, methodically ending one function after another. If the organism did not do this, and died without a systematic shutdown, its enzymes would degrade other living cells, its wastes would poison them, and it would change the acidity to a level unfavorable to the organisms near it. By killing itself, the cell leaves reusable nutrients for other cells to use. Cells in the environment of cells that committed suicide by programmed cell death grow better than cells in the environment of cells that died without programmed cell death. Programmed cell death is altruism. Unicellular organisms often reproduce by splitting in two, producing clones with genetically identical individuals. They often disperse short distances at most, so the cell that kills itself is usually genetically the same as the cells near it in its colony. However, this cellular suicide can happen when unrelated cells are nearby, so group selection may play a role. Yet this could be explained by positing that the behavior was solely selected for by kin selection and coincidentally occurs when non-relatives are nearby, in which case group selection is not a required explanation.

Plants that are attacked by herbivores produce natural pesticides—chemical defenses against the animals that chew on them. And they send signals to other plants, warning them of the presence of the herbivores (Walters, 2017; Bruin et al., 1995). Once the receiving plants detect the warning chemicals, they produce their own chemical defenses against any animals that might attack them before this happens. The warning chemicals can travel to unrelated plants. Sometimes these signals even travel between species. Kin and/or group selection might help explain why the sending plant sends the chemical warning signals if the plant receiving the signals is the same species as the plant sending it. Group selection might explain this when the plant that gets warned is not a relative. However, group selection might not be required as an explanation, because the warning chemicals might simply be a secondary effect of kin selection. That is, the unrelated plants could be picking up the warning chemicals that the sending plants are sending to their kin. Individual selection explains why plants receiving the signals would evolve to pick up the signals and respond with a chemical defense. When warning signals are sent from one species to another, it is commensalism or symbiosis.

10.7 UNEXPLAINED ALTRUISM

Altruism can result from a number of causes, and we do not know all of them. *Solenosteira macrospira*, a marine whelk, which is a type of sea snail, has no common name. It lives in the tidal mudflats of Baja California. The male carries the eggs on his back, and on average only 24% of the hundreds of eggs are his. Some carry the eggs fertilized by as many as 25 other males (Kamel and Grosberg, 2012). Carrying this extra weight is costly to the male, causing it to lose weight. It helps the eggs, as the males provide a place for the eggs to develop on, and it keeps the eggs from drying out. When the snails mate, the highly promiscuous female glues her eggs to the male's back, including some fertilized by other males. This male altruism may be the best of limited options the male has, or perhaps it demonstrates to females that the male is a good father, encouraging them to mate with him in the future. But the function is not currently known.

10.8 COOPERATION AND ALTRUISM CAN BE TRANSMITTED CULTURALLY

Many years ago, a troop of 62 savanna baboons (*Papio cynocephalus anubis*) in Kenya underwent a cultural shift (Sapolsky and Share, 2004). It was ruled by aggressive, despotic, belligerent, dominant males that were strong enough to fight with and defeat a neighboring baboon troop. The reward of this victory was waste food at a tourist lodge garbage dump. The meat they ate as a result was tainted with bovine tuberculosis, which soon killed all these aggressive, dominant males. The peaceful, subordinate males, females, and young did not eat the bad meat, and survived. This resulted in a profound cultural and behavioral transformation to pacifism, relaxation of the hierarchy, and a willingness to use affection and mutual grooming rather than threats, violence, and bites. Amazingly, this troop has maintained its cooperative nature for decades, even though new males from elsewhere have replaced the male survivors of the epidemic. Baboon females stay in their troop of birth, but males leave at puberty for new troops. The persistence of the peaceful ways suggests that the resident baboons have been teaching the peaceful customs to newcomers. This is cultural transmission of cooperation and altruism.

10.9 CONFLICT RESOLUTION WITHOUT FIGHTING

Cooperation is displayed in dispute resolution in many animal species. Many species of animals, especially the males, have conflicts over territory and/or females. Many species have the capability of inflicting severe damage to their rivals, or even killing them. But they do not generally cause injury or death. The rivals generally fight following strict rules. For example, bighorn sheep (*Ovis canadensis*) have massive horns and elk (*Cervus canadensis*) have large antlers. Males of both bighorn sheep and elk could seriously injure or kill their opponent in a fight if they struck it on its side. But they generally play by rules, and fight head-to-head, with the result that the winner is generally determined with neither combatant being injured. Rattlesnakes of the genus *Crotalus* could bite and inject venom into their adversaries, killing them. Yet they fight ritualistically following rules by elevating themselves vertically, with their heads high off the ground (Figure 10.3). They each attempt to push the opponent toward the ground from above, pressing on its back. The loser gets pressed toward the ground more often. They avoid biting and envenomating each other. The winner is stronger and a more skilled fighter. Male Mexican beaded lizards (*Heloderma horridum*) are venomous and have sharp claws, and wrestle for several hours for mates and territory. They do not bite each other or use their claws. Generally, neither fighter is hurt. The one who is on top more of the

FIGURE 10.3 Western diamondback rattlesnakes (*Crotalus atrox*) in a ritualistic fight, following strict rules in which these venomous snakes do not bite each other. The contest determines the stronger, more skilled fighter without causing injury or death. The loser leaves, and the winner gets to mate with any available females. Game theory suggests that this evolved by Darwinian selection at the level of the individual, and that group selection is not needed to explain its evolution.

Shutterstock Item ID: 443745355

Title: Western diamondback rattlesnakes fighting for right to mate.
Photo Contributor: Gary McAlea Photography

time wins, and the loser leaves. Like with rattlesnakes, the winner is stronger and a more skilled fighter.

Mantis shrimp (order Stomatopoda) are a group of marine crustaceans that are not true shrimp, but are related to and somewhat resemble them. They have large front appendages suited for grasping their prey. In the family gonadactylidae, these appendages are thick and hard, and can smash and stun prey. The peacock mantis shrimp (*Odontodactylus scyllarus*) (Figure 10.4) generates an extremely fast strike that requires major energy storage and release (Patek et al., 2004). Cox et al. (2014) measured the speed of the fastest known mantis shrimp, the purple spot mantis shrimp (*Gonodactylus smithii*), and found that it reached 30 meters (about 98.4 feet) per second at accelerations of 150,000 meters (about 492,126 feet) per second. These mantis shrimp could easily kill an opponent with a direct hit. They can inflict serious damage on animals considerably bigger than themselves. Some species require natural burrows in the sea for their survival. They cannot dig these burrows, and they use them for protection and homes. There are fewer of these cavities than there are individuals seeking them, so there is intense competition for them. Yet, the animals have evolved a system of signaling, communication, and rules that determines the winner while preventing the contestants from fighting and injuring one another.

FIGURE 10.4 The peacock mantis shrimp (*Odontodactylus scyllarus*) of the family gonodactylidae can inflict tremendous damage or even death to its opponents with its raptorial appendages in fights for limited natural holes on the seafloor, but settles disputes for these holes peacefully by elaborate communication. Game theory suggests that this can be explained by individual selection, with no need to invoke group selection.

Shutterstock Item ID: 257451349

Title: Vividly colored Peacock Mantis Shrimp on a black sandy sea bed.
Photo Contributor: Richard Whitcombe

Thus, the general principle is that most organisms play by rules, limit contests to ritualized combat, and settle disputes without fighting. Hence, large numbers of individuals are not killed in fights. This rule is not universal. There are instances of cheating, violent fights, injury, and death. But they are the exception in the vast majority of species.

If the animals fought violently instead of using ritualized combat, signaling, and well-followed rules, the population would be greatly reduced or could even become extinct. Group selection is not the main explanation for ritualized combat and the avoidance of violent fighting, even though this is good for the population. There is a simpler explanation. This process has been analyzed by use of the mathematical system known as game theory, which determines the best strategy an organism could use when involved in situations having conflict of interest with other organisms (Maynard Smith,

1982, and references therein). Maynard Smith compared the strategies animals use in fighting to determine the best strategy. He listed three behaviors in fighting: conventional fighting, which means snarling, growling, bluffing, etc.; escalation or dangerous fighting, which is actual contact fighting; and retreating. From these, one can discern and name five combat strategies. The first is used by the *dove*, who fights conventionally, and retreats if the opponent escalates. The *hawk* escalates immediately and keeps this up regardless of the opponent's actions. The *bully* escalates at the beginning, but retreats if its opponent retaliates. The *retaliator* fights conventionally, and escalates only if its opponent escalates. The *prober* fights conventionally at first, but escalates if the opponent fights conventionally. It turns out the most effective strategy, the one that brings the highest long-term payoff, called an evolutionary stable strategy, is that used by the retaliator. Since the animal that uses this strategy does exactly what its opponent does, it is called tit for tat. The reason aggression and escalation are not effective is that they risk retaliation by the opponent, and hence injury or death. This is why animals generally follow rules when they fight; do not cheat; use a great deal of communication, ritual, and bluff; and as a general rule determine the stronger, better fighter without dangerous combat, bloodshed, injury, or death. It is in the interest of each combatant to play by the rules and curb its aggression. It is not the case that animals never cheat or break the rules, or that there is never bloodshed in animal combat; these phenomena happen. Some species have fights that result in injury or death. But ritualized combat is the rule and explainable by standard Darwinian selection at the level at the individual. Invoking group selection is not necessary. The population evolves toward the retaliator strategy that plays by rules and avoids dangerous fighting in favor of ritualized combat and communication because that is the most effective and stable strategy for the individual.

The null hypothesis—that biodiversity would be just as high as observed in nature if the mechanisms discussed in this chapter that select for altruism and cooperation between organisms of the same population did not exist—can be rejected. If these mechanisms did not exist, organisms within the same population would act selfishly and aggressively, and, as a result, populations would be in constant conflict, decrease in size, and often go extinct. Therefore, *the mechanisms that select for altruism and cooperation, such as kin selection, reciprocal altruism, group selection, and perhaps other such mechanisms are powerful agents for maintaining biodiversity*. However, they do not by themselves as a rule increase biodiversity; they generally merely maintain it.

10.10 THE SUPERORGANISM

A population may function as a superorganism, a social unit of many individual organisms of the same species that work together, in some respects as one organism. Division of labor is highly developed and specialized, and each individual cannot survive on its own. Examples include colonies of highly social ants, bees, termites, and naked mole rats, and many human groups. It can produce results governed by the collective group; that is, what one might say the group collectively "wants," such as bees choosing a nest site. Some think the superorganism has a group brain, also called swarm intelligence, meaning the organisms effectively function like the nerve cells of a unified "brain." Here "intelligence" emerges from the interaction of many simple components. Many individuals working together as one can achieve spectacular accomplishments, and act similar to a highly intelligent organism.

Individual fish of many species are not very smart, but by following simple rules, a school of fish can often outwit even the most cunning predators, by using a group brain. Such rules include swim close, but not too close, to each other; stay lined up; and avoid unattractive stimuli.

Superorganisms might result from kin selection, reciprocal altruism, group selection, or any combination of these. The superorganism maintains populations and species richness, but does not necessarily increase biodiversity.

REFERENCES

Bruin, J., et al. (1995). Do plants tap SOS signals from their infested neighbours? *Trends in Ecol. & Evol.* 10 (4): 167–70.

Canter, H. M. & Jaworski, G. H. M. (1979). The occurrence of a hypersensitive reaction in the planktonic diatom *Asterionella formosa* Hassell parasitized by the chytrid *Rhizophydium planktonicum* Canter Emend., in culture. *New Phytologist* 82: 187–206.

Carter, G. G. & Wilkinson, G. S. (2013a). Food sharing in vampire bats: reciprocal help predicts donations more than relatedness or harassment. *Proc. Royal Soc. B. Biological Sci.* 280: 2012–573. doi: 10.1098/rspb.2012.2573.

Carter, G. G. & Wilkinson, G. S. (2013b). Does food sharing in vampire bats demonstrate reciprocity? *Communicative and Integrative Biol.* 6 (6): e25783. doi: 10.4161/cib.25783.

Clutton-Brock, T. H., Brotherton, P. N. M., O'Riain, M. J., Griffin, A. S., Gaynor, D., Kansky, R., Sharpe, L., & McIlrath, G. M. (2000). Contributions to cooperative rearing in meerkats. *Animal Behav.* 61: 705–10.

Covas, R. (2002). Life-history evolution and cooperative breeding in the Sociable Weaver. PhD Thesis. Percy Fitzpatrick Institute—Univ. of Cape Town, Cape Town, So. Africa.

Cox, S. M., et al. (7 Feb., 2014). A physical model of the extreme mantis shrimp strike: Kinematics and cavitation of Ninjabot. *Bioinspiration and Biomimetics* 9: 016014 (16 pp). doi:10.1088/1748-3182/9/1/016014.

Dijk, R. E., Kaden, J. C., Argüelles-Ticó, A., Dawson, D. A., Burke, T., & Hatchwell, B. J. (Sept., 2014). Cooperative investment in public goods is kin directed in communal nests of social birds. *Ecol. Letters* 17 (9): 1141–8. https://doi.org/10.1111/ele.12320.

Dudley, S. A. & File, A L. (2007). Kin recognition in an annual plant. *Biol. Letters* 3 (4): 435–8.

Dunbar, R. I. M. (2010). The social role of touch in humans and primates: behavioural function and neurobiological mechanisms. *Neuroscience & Biobehavioral Reviews* 34 (2): 260–68.

Emlen, S. T. & Wrege, P. H. (1988). The role of kinship in helping decisions among white-fronted bee-eaters. *Behavioral Ecol. and Sociobiol.* 23 (5): 305–15.

Foster, K. R., et al. (Feb., 2006). Kin selection is the key to altruism. *Trends in Ecol. and Evol.* 21 (2): 57–60. PMID: 16701471. doi: 10.1016/j.tree.2005.11.020. Epub 20 Dec., 2005.

Hamilton, W. D. (1963). The evolution of altruistic behavior. *Amer. Nature* 97: 354–6. doi: 10.1086/497114.

Hamilton, W. D. (1964). The genetical evolution of social behavior. *Journ. of Theoretical Biol.* 7 (1): 1–52. doi: 10.1016/0022-5193(64)90038-4.

Hatchwell, B. J. (2009). The evolution of cooperative breeding in birds: kinship, dispersal and life history. *Philos. Trans. of the Royal Soc. of London, Series B, Biol. Sci.* 364: 3217–27.

Jetz, W. & Rubinstein, D. R. (2011). Environmental uncertainty and the global biogeography of cooperative breeding in birds. *Current Biol.* 21: 1–7.

Kamel, S. J. & Grosberg, R. K. (2012). Exclusive male care despite extreme female promiscuity and low paternity in a marine snail. *Ecol. Letters* 15 (10): 1167–73. 2 https://doi.org/10.1111/j.1461-0248.2012.01841.x.

Keller, L. & Ross, K. G. (6 Aug., 1998). Selfish genes: a green beard in the red fire ant. *Nature* 394: 573–5. doi: 10 (1038): 29064.

Krakauer, A. H. (2005). Kin selection and cooperative courtship in wild turkeys. *Nature* 434 (7029): 69–72.

Lukas, D. & Clutton-Brock, T. (2012). Life histories and the evolution of cooperative breeding in mammals. *Proc. of the Royal Soc. B. Biol. Sci.* 279: 4065–70. https://doi.org/10.1098/rspb.2012.1433.

Mares, R., Young, A. J., & Clutton-Brock, T. H. (2012). Individual contributions to territory defense in a cooperative breeder: weighing up the benefits and costs. *Proc. of the Royal Soc. B. Biol. Sci.* 279: 3989–95. 2https://doi.org/10.1098/rspb.2012.1071.

Maynard Smith, J. (1982). *Evolution and the Theory of Games.* Oxford Univ. Press, Oxford, UK.

Mlot, N. J., Tovey, C. A., & Hu, D. L. (2011). Fire ants self-assemble into waterproof rafts to survive floods. *PNAS USA* 108 (19): 7669–73.

Moore, H., et al. (1 Mar., 2002), Exceptional sperm cooperation in the wood mouse. *Nature* 418: 174–7. doi: 10.1038/nature00832.

Muir, W. M. (2009). Genetic selection and behaviour. (Meeting Abstract). *Canadian Journ. of Animal Sci.* 89 (1): 182. doi: 10.1037/1089-2699.12.1.17.

O'Gorman, R., Wilson, D. S., & Sheldon, K. M. (2008). For the good of the group? Exploring group-level evolutionary adaptations using multilevel selection theory. *Group Dynamics: Theory, Research, and Practice* 12 (1): 17–26.

Patek, S. N., Korff, W. L., & Caldwell, R. L. (2004). Deadly strike mechanism of a mantis shrimp. *Nature* 428 (6985): 819–20. doi:10.1038/428819a. PMID 15103366.

Radford, A. N. (2012). Post-allogrooming reductions in self-directed behaviour are affected by role and status in the green woodhoopoe. *Biol. Letters* 8 (1): 24–7.

Rauch, E. M., Sayama, H., & Bar-Yam, Y. (2003). Dynamics and genealogy of strains in spatially extended host-pathogen models. *Journ. of Theoretical Biol.* 221: 655–64.

Ross, C. N., et al. (2007). Germ-line chimerism and paternal care in marmosets *(Callithrix kuhlii)*. *PNAS USA* 104 (15): 6278–82.

Sapolsky, R. M. & Share, L. J. (2004). A pacific culture among wild baboons: its emergence and transmission. *PLOS Biol* 2 (4): e106.

Seyfarth, R. M., Cheney, D. L., & Marler, P. (14 Nov., 1980). Monkey responses to three different alarm calls: Evidence of predator classification and semantic communication. *Science* 210 (4471): 801–3. doi: 10.1126/science.7433999.

Sherman, P. W. (23 Sept., 1977). Nepotism and the evolution of alarm calls. *Science* 197 (4310): 1236–53. doi: 10.1126/Science.197.4310.1246.

Sherman, P. W. (Oct., 1985). Alarm calls of Belding's Ground Squirrels to aerial predators: Nepotism or self-preservation? *Behav. Ecol. and Sociobiol.* 17: 313–23.

Stephens, C. (1996). Modeling reciprocal altruism. *British Journ. for the Philos. of Sci.* 47 (4): 533–51.

Tofilski, A., et al. (2008). Preemptive defensive self-sacrifice by ant workers. *Amer. Naturalist* 172 (5): E239–E243.F

Trivers, R. L. (1971). The evolution of reciprocal altruism. *Quarterly Rev. of Biol.* 46: 35–57.

Walters, D. (2017). *Fortress Plant. How to Survive When Everything Wants to Eat You.* Oxford Univ. Press, Oxford, UK.

Wilkinson, G. S. (1984). Reciprocal food sharing in the vampire bat. *Nature* 308: 181–4. https://doi.org/10.1038/308181a0.

Wilkinson, G. S. (1986). Social grooming in the common vampire bat, *Desmodus rotundus. Animal Behav.* 34 (6): 1880–9.

Wilkinson, G. S. (1988). Reciprocal altruism in bats and other mammals. *Ethol. and Sociobiol.* 8: 85–100.

Williams, G. C. (1966, copyright renewed 1992). *Adaptation and Natural Selection.* Princeton Univ. Press, Princeton, NJ; Oxford, UK. New paperback, ISBN 978-0691-18286-5.

Wilson, D. R. & Hare, J. F. (2004). Animal communication: ground squirrel uses ultrasonic alarms. *Nature* 430 (6999): 523.

Wilson, D. S. & Sober, E. (1994). Reintroducing group selection to the human behavioral sciences. *Behav. and Brain Sciences* 17 (4): 585–654.

Wilson, D. S. & Wilson, E. O. (2007). Rethinking the theoretical foundation of sociobiology. *Quarterly Rev. of Biol.* 82 (4): 327–48.

Wilson, D. S., & Wilson, E. O. (2008). Evolution "for the Good of the Group": The process known as group selection was once accepted unthinkingly, then was widely discredited; It's time for a more discriminating assessment. *Amer. Scientist* 96 (5): 380–9.

Wilson, E. O. (2005). Kin selection as the key to altruism: its rise and fall. *Social Research* 72 (1): 159–66.

Wynne-Edwards, V. C. (1962). *Animal Dispersion in Relation to Social Behaviour*. Oliver & Boyd, Edinburgh, UK.

Wynne-Edwards, V. C. (1986). *Evolution through Group Selection.* Blackwell Scientific Pubs., Palo Alto, CA. Cloth, ISBN 0-632-05139- X; paper, 0-632-01541-1.

11 Network Theory Models and Empirical Evidence Support the Thesis That Coevolved Diversity and Connectance Correlate with Ecosystem Stability, Productivity, Resilience, and Persistence

The Autocatalytic Biodiversity Hypothesis (ABH)/Pachamama Hypothesis is a comprehensive hypothesis. The Gaia Hypothesis is an idea proposed by James Lovelock and Lynn Margolis that proposes that organisms made Earth's life-friendly atmosphere of high oxygen and favorable levels of other gases (Lovelock, 1972; Lovelock and Margulis, 1974; Lovelock, 1979). The ABH states that this is true, but life increased biodiversity in many other ways as well. Thus, the Gaia Hypothesis is a subset of the ABH. The ABH incorporates the Gaia Hypothesis. Natural selection is one of the mechanisms by which the ABH operates. The ABH is profound and supported by the evidence. It states that organisms increase biodiversity by a number of mechanisms and that there is one unifying mechanism, which is that all species are ecosystem engineers that have a net positive effect on the species they interact with, their ecosystem, and biodiversity, under natural conditions over sufficient time. There is currently no unifying mathematical model that explains the ABH. Such a model would be desirable. I hope such a model comes about at some point. For the present, network theory is one field of mathematics that is especially relevant to the hypothesis. This chapter discusses how network theory largely supports the hypothesis.

11.1 LIFE PROMOTES REGULATION OF EARTH'S SYSTEM

The Gaia hypothesis proposes that the Earth system self-regulates by negative feedback, keeping Earth in a habitable state, and that this is done by life. Lenton (2002) looked at the effect of life on the state of the Earth and its response to perturbations and drivers of the climate, arguing that that life has not survived for over 3.8 billion years purely by chance, but that Earth with its life has mechanisms for regulation. It is more resilient and resistant to many perturbations, though not to all of them, as a result of life. Life extends the time that Earth is habitable. Theoretical principles indicate that the interaction of life with the physical-chemical environment makes regulation of the system likely (ibid.). There seems to be an inherent tendency for properties that regulate and stabilize the Earth to accumulate and strengthen as life evolves (ibid.). Lenton and van Oijen (2002) suggested that the Earth's biological system, which they refer to as Gaia, is self-organizing. Their simple cellular automata model exhibited emergent self-regulation as a consequence of feedback coupling between life and its environment. They state: "Large perturbations have occasionally suppressed life but the system has always recovered without losing the capacity for large-scale free energy capture and recycling of essential elements." This is consistent with the ABH. Other chapters in this book, especially Chapters 2 and 3, as well as the companion book to this one (Seaborg, 2022), discuss many of the ways by which life regulates the Earth system.

11.2 NETWORK THEORY

Network theory is a powerful way to study complex ecosystems. Ecological networks are being used to study diversity, complexity, and stability issues. They describe interactions between species, such as predators and prey, symbiosis, and competition, in a community or ecosystem (Pascual and Dunne, 2006). They can be represented as a set of points characterizing the species, connected by a set of links, which represent the interactions among each ordered species pair (Newman, 2010). The interaction could be a predator and its prey, or competition, for example. Complex network models are composed of a set of compartments, describing either species or broader functional groups, and a set of links representing interactions, or energy or biomass flows among compartments. Network theory has proven effective at throwing light on the ongoing debate as to whether diversity, connectance, and complexity tend to promote stability, resilience, and persistence in, or destabilize, ecosystems, a question highly relevant to the ABH. Connectance is the connections between species; it is the links or relationships between species. Symbiosis, competition, and predators and their prey are connections or links between species that are forms of connectance. Complexity is used to mean number of species in the ecosystem and/or connectance between species. It can be viewed as a combination of the two. Stability of an

DOI: 10.1201/9781003246640-11

ecosystem can be viewed as its ability to return to its original functioning state after a stress (Johnson et al., 1996; Worm and Duffy, 2003; Dunne et al., 2005; Hooper et al., 2005; Kondoh, 2005; Loreau and De Mazancourt, 2013). It can be considered the ability of an ecosystem to maintain or return to its level of biodiversity, even when acted on by a deleterious outside force. Of course, a deleterious outside force would tend to cause it to lose species, so stability can be viewed as entailing the ability to keep from losing species or to regain lost species. Thus, it can be viewed as the ability of an ecosystem to recover from a disturbance or loss of diversity. Examples of outside forces or disturbances are large storms, meteor strikes, volcanism, unusually large fires, and human disturbances such as introduction of exotic species and the cutting of large numbers of trees. Resilience has largely the same meaning as stability. Persistence is the ability of the ecosystem to persist through time. Network theory gives a quantitative foundation to the widely held perception of the interconnectedness of nature (Fath and Patten, 1999). Biodiversity refers to the variability among living organisms. For this discussion, it is the number of species in a community or ecosystem. Thus, a community or ecosystem with many species has higher biodiversity than one with few species. It can also be referred to as diversity or species diversity. In this chapter, I will use biodiversity, diversity, and species diversity as synonyms. The diversity-stability hypothesis states that species diversity increases the stability of the community—that diversity stabilizes the community—by dampening environmental fluctuations between species. That is, the higher the number of species in a community, the more stable and resistant to change or collapse it is. The phrase often used for this is *diversity breeds stability*. The alternative hypothesis is that most species are functionally redundant and have no effect on stability of the community. My view is that the connections—the relationships—between species are as important as the number of species in determining a community's stability.

Before the term network became popular, Odum (1953) and Elton (1958) concluded that diversity breeds stability based on repeated observations that simple communities on land with few species were characterized by more violent fluctuations in population density than diverse land communities with many species. For example, agricultural monocultures (monocultures are ecosystems with just one species, such as farms with only one crop growing) are unstable and readily invaded, disrupted, and caused to crash by plant-eating insects, and destroyed by fungi, other pathogens, and inclement weather. Monocultures are unstable and cannot long exist without constant human support. Outbreaks of plant-eating insects occur readily in boreal forests, which occur in the high, cold latitudes and have relatively few species, but not in tropical rainforests, which are highly diverse. The frequency of invasions by insects that eat trees and other plants and of fungi that attack plants is higher in simple island communities of low species diversity than more diverse mainland communities. Disruption by species that are not native to the community is much easier in simple ecosystems with few species than complex ones with high species diversity, such as rainforests. Similarly, an animal with a diverse microbiome is more resistant to pathogens. Elton (1958) concluded complex, diverse communities, with many predators and parasites, prevented populations from undergoing explosive growth and were more resistant to invasions. MacArthur (1955) concluded that stability increases as the number of links between species increases, and that stability is more easily accomplished in more diverse communities of species. He thereby linked community stability with both increased links between species in the food web and increased species diversity. A link is a relationship between two species, such as a predator and its prey, two competitors, or two species that are symbiotic with each other. An ecosystem with many predator species is more stable, all else being equal, because the redundancy stabilizes the system. That is, if the main predator species declines or is killed off entirely, other species of predators can increase their populations, take on the (formerly) main predator's role, and control the prey populations. Large whales such as the blue whale (*Balaenoptera musculus*) eat a small type of crustacean called krill, keeping their populations stable and preventing them from depleting the phytoplankton. Without whales or other predators controlling them, krill populations would explode and the krill would deplete the supply of phytoplankton, and many ocean food webs would crash catastrophically, causing a tremendous decrease in biodiversity in the sea. When the populations of large krill-eating whales were severely lowered due to whaling, some species of seals and penguins increased in numbers and increased their consumption of krill, keeping krill levels relatively stable and preventing them from exploding. So the krill did not deplete the phytoplankton, and the ocean ecosystem remained stable and intact. This redundancy and high diversity of predators allowed the ecosystem to persist without experiencing a catastrophic loss of species. A system with only one predator would not be able to replace the lone predator if it should severely decrease in number or die off. If this happened, the ecosystem would crash, with a great loss of diversity.

The concepts of trophic and trophic level must be defined here for the reader unfamiliar with these concepts. Trophic refers to the feeding and nutrition relationships of different species in a food web. The trophic level of a species is its position in a food web. A food web starts at trophic level 1 with primary producers such as plants that obtain their energy from sunlight via photosynthesis and nutrients from the soil. Level 2 is herbivores that eat plants, level 3 is carnivores that eat herbivores, and level 4 is carnivores that eat other carnivores and perhaps herbivores. The last level consists of the decomposers, such as vultures, fungi, and bacteria, which recycle the nutrients back into the soil, making them available to plants again. Ecological communities with higher biodiversity form more complex and branched trophic food webs.

When the gray wolf (*Canis lupus*) thrived in the Greater Yellowstone ecosystem, its main prey, mule deer (*Odocoileus hemionus*), white-tailed deer (*Odocoileus virginianus*), and elk (*Cervus canadensis*), moved from one area to another to eat, because if they stayed in one place all the time, wolves could learn where they always were and easily find them. Once wolves were killed off by humans and no longer existed in the ecosystem, deer and elk no longer feared staying in the same place to eat, and so ate in one area until it was denuded of its trees and other vegetation before moving to a new area to eat. Also, without wolves to eat them, the populations of deer and elk exploded. With larger numbers, they collectively ate much more. The result of these two effects was that the trees that they ate, mainly aspen (genus *Populus*) and willows (genus *Salyx*), as well as other vegetation they ate, had population crashes. This led to the decline in populations or complete loss of many species of mammals, birds, reptiles, amphibians, insects, and other invertebrates that were dependent on or benefitted from aspens or willows and other vegetation in the forests. Deer and elk had population crashes because they did not have enough aspen, willow, and other trees to eat. Trees keep river banks stable, so when they were removed, banks of rivers eroded and soil ran into the rivers, causing plant and invertebrate species that live on river banks to decline or die off, and fish, invertebrates, and species that live in rivers to decrease in numbers or die out because of the sedimentation in their water. Without trees to provide cooling shade on river banks, water temperatures increased, and aquatic invertebrate and fish diversity plummeted. North American beavers (*Castor canadensis*) need aspen, among other trees, so they disappeared in many areas. So species dependent on the pond habitat that they create died off. Many species benefit from beavers. This includes various species of mammals, birds, reptiles, amphibians, fish, insects, other species of invertebrate, and plants that live in ponds. Many scavengers that benefit from wolf kills also diminished when wolves were removed. So there was a tremendous decrease in the number of species. This catastrophic loss of species as a result of the removal of a key species in the food web from an ecosystem is called a trophic cascade. When the wolf was reintroduced to the Greater Yellowstone ecosystem, there was a great recovery of all impacted species. So the wolf was crucial to the survival of many species, and maintained high diversity in the ecosystem. Predators often maintain high species diversity by this mechanism. A similar trophic cascade occurred when the mountain lion (*Puma concolor*) was eliminated or greatly reduced in many parts of its range. A number of species of shark maintain diversity by the same mechanism. This shows the interconnectedness of species and the importance of this connectance. It demonstrates the inter-dependence of species.

However, May (1972, 1973) used an oversimplified model to attempt to mathematically demonstrate that network stability *decreases* with increased diversity and connectance. Bear in mind that connectance between species is correlated with diversity, and related to the ABH is the importance of connectance (relationships) between species, so the relationship of connectance to stability is relevant to the ABH. A positive correlation would support the ABH, while a lack of a correlation or a negative one would be counter to it. Some authors did work that supported May's view (Pimm, 1979, 1980; Chen and Cohen, 2001; Gross et al., 2009; Allesina and Tang, 2012). Allesina and Tang (2012) found that connectance negatively affects the local stability of symbiotic community networks. Some of the theoretical attempts at justification of this view are: in complex, diverse systems, more can go wrong; complexity enables the effects of disturbances, such as species loss or species invasion, to spread and amplify through the network; time lags in response to changes in the ecosystem can destabilize diverse systems; and if there are many connections between species, any given species is more likely to be connected to and benefit other species, so its loss is more likely to negatively impact other species. Thus, Vieira and Almeida-Neto (2015) found that extinction cascades, where the extinction of one species leads to the extinction of others, were more likely to happen in highly connected communities.

However, most of the evidence supports the view that connectance is correlated with stability, persistence, and resilience. De Angelis (1975) showed stability can increase with increasing connectance if certain criteria are met. Haydon (1994) used more realistic assumptions than May, and found stability increases with connectance. Dunne et al. (2002) and Dunne and Williams (2009) also found that connectance enhances network stability. But most importantly, Yodzis (1981) discovered that the networks were stable with a much higher probability when one uses elements in accord with actual food web patterns in nature, as opposed to random ones. A fallacy of some who dismiss the diversity-stability hypothesis is their failure to consider that species in ecosystems are not thrown together randomly. If 95% of the species of an ecosystem are herbivores, it is unstable even at very high diversity. But the species of natural ecosystems coevolve. This results in relative numbers of species of plants, herbivores, predators, species in symbiotic relationships, and so on that are close to optimal. For example, there are never ecosystems with only predators, which would be unstable. Ecosystems in nature have fewer predators than their prey, with the result that predators do not deplete their prey. They have a highly favorable ratio of species at different levels in the food web: of predators to prey, of herbivores to plants, and so on. And there are many symbiotic relationships, which helps keep the ecosystem stable. The species in any given ecosystem do not come together randomly. Thus, the bulk of the evidence supports the thesis that *coevolved* diversity and *coevolved* connectance both generally lead to ecosystem stability.

Koh et al. (2004) applied a mathematical model informed by real data to assess species coextinction. Coextinction is the tendency that, if a species goes extinct, other species will go extinct because they are dependent on it. For example, if a species of insect eats only one species of plant and that plant species goes extinct, the insect will also go extinct. The scientists examined the coextinction levels between host species and the species that benefit from them for a broad range of coevolved systems between species: fig trees and the wasps that pollinate them, animals and their parasites, butterflies and the plants their caterpillars eat, and ant butterflies whose caterpillars are raised by ants and their host ants. The latter is symbiosis because the ants also benefit because the caterpillar makes a nutritious liquid that feeds the ants. Applying a method based on the number of host species per species they benefit, they estimated that 6,300 of the species that are benefitted are coendangered with host species currently listed as endangered, indicating that the loss of a species can lead to extinction of a great number of other species. They concluded that current extinction estimates need to be recalibrated by taking species coextinctions into account. This shows the tremendous connectance and interdependence of species and the high coextinction of species. A high degree of connectance, interdependence, and coextinction of species is evidence for the ABH because it shows that it is common for species to help each other, and even for species to be dependent on each other. Species help each other by being ecosystem engineers.

Coextinction and the effects of loss of taxa on other taxa need more study. As pointed out in Chapter 7, two species chytrid fungi are currently killing amphibians all over the world in great numbers, sending a large number of species extinct. Zipkin et al. (2020) found that the loss of amphibian species in Panama due to this fungus caused a decline of their snake predators. After the mass mortality of amphibians, the snake community contained fewer species and was more uniform across their study site, with several species in poorer body condition, despite no other systematic changes in the environment. With less amphibian prey, snake species number declined from 30 to 21. This finding, which is not surprising, is an example of high connectance as well as coextinction in nature.

Interconnectedness, coextinction, and what is called quasi-stable states are illustrated in a catastrophe that occurred in northern California's kelp forests. What some people call starfish are correctly called sea stars because they are not fish. The sunflower sea star (*Pycnopodia helianthoides*), a major predator of sea urchins, was driven locally extinct from the kelp forests of Northern California due to human-induced warming of the ocean leading to sea star wasting disease, starting in 2014. Sea star wasting disease is a disease of sea stars and some animals related to them that starts with the emergence of lesions, followed by body fragmentation and death. Without predators, the purple sea urchin (*Strongylocentrotus purpuratus*) population exploded. They are voracious herbivores of kelp, and they decimated the kelp forests. This is a naturally dynamic system that has shown great resiliency to extreme environmental challenges in the past, but the loss of the sea star predators caused the ecosystem to collapse (McPherson et al., 2021). The kelp forests were simultaneously subjected to a marine heatwave and El Niño event combined with the sea urchin population explosion. Bull kelp (*Nereocystis luetkeana*), the dominant canopy-forming kelp species north of San Francisco Bay, thrives when strong upwelling of cold, deep water brings nutrients to the surface along the California coast. The warm water brought by marine heatwaves and El Niño events suppressed coastal upwelling, resulting in low nutrient conditions in which kelp grows poorly. Upwelling is the movement of water from deeper to shallower and surface waters. The upwelled water brings great quantities of nutrients with it to the surface and shallow waters, making the nutrients available to phytoplankton, allowing them to grow in great numbers. Phytoplankton are the basis, the lowest level, of the ocean's food webs. So upwelling is crucial to a healthy ecosystem in the sea. Kelp also does better when water temperatures are not too high, apart from nutrient content. An unusual marine heatwave started in the Northeast Pacific in 2014 and spread down the West Coast in 2015. At approximately the same time, a powerful El Niño event developed, transporting warm water up the California coast from the south. The warm water coincided with an increase in sea urchin numbers along the North Coast. The area covered by kelp forests along 350 kilometers (about 217.5 miles) of northern California coastline dropped by more than 95% from 2014 to 2019, with just a few small, isolated patches of bull kelp remaining, as a result of a combination of the sea urchin predation, warm water, and low nutrients. Kelp forests declined in southern California waters, but to a lesser extent than in the north. In northern California waters, there are no other sea urchin predators, such as sea otters (*Enhydra lutris*), which keep patches of kelp forest healthy in some southern California waters such as Monterey Bay. In these southern areas, the sea otters replaced the sea stars as predators of the sea urchins, allowing the kelp to thrive and the ecosystem to persist and be stable. In southern California, the main kelp species is giant kelp (*Macrocystis pyrifera*). In this catastrophe, there were two alternative quasi-stable states of the ecosystem. A quasi-stable state is a state that is mostly, but not entirely, stable. It can be thrown into an alternative state by outside disturbances such as those caused by human destruction of the environment and disruption of the climate. The general rule is that in nature, the natural quasi-stable state is more diverse, more complex, has more information, and has higher primary productivity (in this case, primary productivity is growth of the kelp). In this situation, the natural state is the high kelp, high sea star, moderate sea urchin quasi-stable state. This state was replaced by the quasi-stable state of many sea urchins, few sea stars, and almost no kelp, which is much less diverse, complex, and information-rich, and of lower primary productivity. The reason that the quasi-stable state with many sea urchins has low diversity, complexity, information, and productivity is that there are so many species dependent on kelp that decrease or die off when the kelp is reduced or killed. This state is an unnatural one induced by human intervention. The existence of two alternative quasi-stable states is common in nature. The predominate quasi-stable state tends to be natural, and higher in diversity, complexity, and information. The alternative state is commonly unnatural, human-induced, and of lower diversity, complexity, and information.

Dunne et al. (2002) found robustness (the quality of an ecosystem being in good condition) increases with food-web connectance, but appears independent of species diversity. Stability is enhanced when species at a high level on the food web feed on multiple prey species and when species at an intermediate level on the food web are consumed by multiple predator species. They found food webs experience "rivet-like" thresholds beyond which they display extreme sensitivity to removal of highly connected species. Higher connectance delays the onset of this threshold, so enhances stability. This also supports the hypothesis that connectance, or many links between species, causes stability. Highly connected communities were also shown to be persistent (James et al., 2012) and resilient (Okuyama and Holland, 2008). Thébault and Fontaine (2010) showed that a high number of species promotes both the resilience and persistence of symbiotic communities.

Damschen (2019) was able to reduce the likelihood that plant species would go extinct by about 2% per year by connecting patches of habitats of pine savanna. Connecting the patches also increased the likelihood of colonization of the patches by plants by about 5% per year. After 18 years, connected patches had 14% more species than unconnected patches. This shows that connecting habitat patches increases stability and diversity. Since connecting habitat patches connects species, this implies that connectance promotes stability and diversity.

Insects exemplify the importance of connectance effectively. They are ecologically important enough and provide so much evidence for the ABH/Pachamama Hypothesis that an entire chapter could be devoted to them. But the same general concepts that such a chapter would describe are covered elsewhere in this book, so the main points will be summarized briefly here in lieu of adding an entire chapter. Over 40% of insect species are threatened with extinction over the next few decades due to human activities (Sánchez-Bayoa and Wyckhuysbcd, 2019). This will have catastrophic consequences to ecosystems and biodiversity. Insects make up perhaps about 70% of all animal species, and potentially 90% of them (Erwin, 1997). Plant-eating insects are an important food source for praying mantises, wasps, spiders, scorpions, centipedes, many other invertebrates, frogs, toads, salamanders, caecilians (limbless, wormlike amphibians that live mostly in the tropics), lizards, some snakes, turtles, birds, echidnas (spiny, egg-laying mammals that live in Australia), anteaters, rodents, and numerous other animals. Some plants consume them to supplement their nitrogen. About 60% of birds rely on insects as a food source (ibid.). In Australia, insectivorous birds that now lack their insect prey are preying on other birds (ibid.). Insect-eating bats are especially dependent on them.

Insects are ecosystem engineers. They pollinate 80% of wild plant species. Ants are important seed dispersers. Termites and ants till and aerate the soil and provide many other ecosystem services. Predatory insects regulate the populations of various animal species; herbivorous insects regulate plants; fungus-eating ones control fungi populations; and insects that carry diseases such as malaria by mosquitoes regulate many vertebrate species. Their coevolution with plants and with their predators has led to many evolutionary innovations in insects, plants, and their predators. Dung beetles, some termites, and some flies decompose and recycle. Western honey bees (*Apis mellifera*) produce honey, a food source for many animals. Some insect species transport other species from place to place; for example, some small mites hitch rides on insects. Insects are hosts to countless viruses, bacteria, archaea, fungi, and invertebrates in their microbiomes. Ecosystem services provided by wild insects have been estimated at $57 billion annually in the United States alone. Insects have greatly increased biodiversity by their own species number, and by providing numerous ecosystem services since they appeared 400 mya. They also exchange DNA by horizontal gene transfer with other taxa, promoting the evolution of the taxa and of insects themselves. All of these positive effects on biodiversity and other species provided by insects are due to their connectance to the species they aid. The loss of even 10% of insect species would lead to an enormous extinction of species that are either directly or indirectly dependent on them. Entire ecosystems would collapse. Insects thus illustrate the importance of connectance to ecosystem stability, diversity, resilience, and persistence.

Fowler (2009) showed that increasing the number of species and the connectance between the species results in an increased probability of local stability in communities with competition, when some species would show unstable dynamics in the absence of competition. Here, adding more species or links and increasing the competitive negative feedbacks in the network contributes to stability. He also showed that an increase in network connectance and number of competitive links stabilizes oscillations. (Too much oscillation can drive populations to local extinction.) This is interesting, since it demonstrates that stability increases with the number of links, even if they are negative (representing competition).

Melian et al. (2009) combined symbiotic and competing interactions of herbivores in a model of such ecological networks, showing that *species persistence is increased by the correlation between strong species dependences and the ratio of the total number of symbiotic to competitive interactions per species*. This shows that *the nature of the connections between species affects persistence*. Mougi and Kondoh (2012) showed *increasing complexity, defined as number of species and connectance, leads to increased community stability*.

McNaughton (1977) said that the diversity-stability hypothesis appears widely misunderstood by ecologists. He pointed out that fluctuations in the abundances of species with different adaptive modes could stabilize a community in an environment that varies. He said that the available empirical evidence supports the hypothesis at the primary producer level. He stated that this includes "data from an experimental perturbation of successional old fields and data on the impact of environmental fluctuations on properties of grasslands in the Serengeti-Mara ecosystem of Tanzania and Kenya." He added: "The data on diversity-stability relationships in plant communities indicate that the traditional verbal model is considerably more robust in application than recent 'more rigorous' mathematical models." By traditional verbal model, he means the diversity breeds stability hypothesis.

Tilman (1999) used theory and experiments to explore the effects of diversity on stability, productivity, and susceptibility to invasion. A model of resource competition predicted that increased diversity causes community stability to increase, but population stability to decrease. He stated that community stability depends on ecosystem productivity increasing with diversity. He said that models of competition predict, and field experiments confirm, that greater plant diversity leads to greater primary productivity. Further, lower levels of available limiting resources at higher diversity should decrease an ecosystem's susceptibility to invasion. He concluded that biodiversity should be included with species composition, nutrient supply, climate, and disturbance as a major regulator of population and ecosystem structure and dynamics. Tilman and Downing (1994) described a long-term grassland study (Tilman, 1987; Tilman, 1988) that showed that primary productivity in plant communities with higher diversity is more resistant to, and recovers more fully from, a major drought. Each additional species lost from the grassland community had a progressively greater impact on resistance to drought. Lehman, Tilman, and Gaines (2000) showed that three markedly different models of multispecies competition all predict that greater diversity increases the stability over time (defined as the ratio of mean abundance to its standard deviation) of the entire community, decreases the temporal stability of individual populations, and increases community productivity. These effects of diversity on stability of the community and populations are in agreement with Tilman's (1999) model (see above). As diversity increases, the temporal stability of entire communities is predicted to increase fairly linearly, without clear saturation (Lehman, Tilman, and Gaines, 2000). The model predicts that species composition will be as important as diversity as far as effects on community stability and productivity are concerned. More diverse communities have greater temporal stability because of higher productivity at higher diversity, competitive interactions, and statistical averaging (ibid.).

Continuing this line of inquiry, Tilman et al. (2014) found that species diversity is a major determinant of ecosystem stability, productivity, resistance to invasion, and nutrient flow. They pointed out that hundreds of studies spanning terrestrial, aquatic, and marine ecosystems show that high-diversity mixtures are about twice as productive as monocultures of the same species, and this difference increases through time. These impacts of higher diversity have many causes, including the fact that different species are complementarity to each other, greater use of limiting resources in diverse communities, nutrient-cycling feedbacks that increase nutrient stores and supply rates over the long term, and decreased herbivory and disease. These experimentally observed effects are consistent with predictions based on a variety of theories, all of which have trade-off-based mechanisms that allow long-term coexistence of many different competing species. Diversity loss can have an effect as great as, or greater than, the effects of elevated CO_2, drought, nitrogen addition, and other environmental disrupters, with respect to ecosystem stability, productivity, resistance to invasion, nutrient flow, and the survival of the species. Of course, the impact of each of these factors depends on how severe the factor is.

As discussed in Chapter 6, Simkin et al. (2016) measured nitrogen levels in over 15,000 forest, woodland, brush, and grassland sites, and found human-induced high nitrogen levels have caused a loss of plant diversity. Species richness increased at low deposition levels, but 24% of sites had lower species richness, which decreased with increased nitrogen input. Grasslands, shrublands, and woodlands were susceptible to species losses at lower nitrogen loads than forests, which are more diverse than each of these three ecosystems, supporting the thesis that diversity breeds stability and resilience.

A study showed that diversity promotes resilience in ecosystems by showing that loss of diversity is causing loss of resilience in the Amazon Basin. It also illustrates a quasi-stable state, and how natural quasi-stable states created by nature and life are more diverse and complex than when humans convert them to other quasi-stable states. The Amazon is being converted by humans from a high-diversity and high-complexity quasi-stable state of rainforest to a quasi-stable state lower in diversity and complexity of savannah. Bouton et al. (2022) applied established indicators to quantify changes of Amazon resilience, using satellite data from 1991 to 2016, which include measurements of the density of life in the forest. About one-fifth of the Amazon has already been destroyed, mainly from human activities, such as logging, mining, damming it, and planting crops. Their study led them to warn that the Amazon rainforest is approaching a threshold beyond which over half of it could be transformed into savannah in only decades. This would convert the Amazon rainforest into an environment like an African savannah with grass sparsely populated with trees. Three-quarters of the Amazon Basin has been losing resilience against droughts and other adverse weather events since the early 2000s. This loss of resilience makes it recover more slowly from such events as droughts, and is significantly greater than the loss of biomass. Deforestation and climate change are the main drivers of this decline. Increase in the length of the dry season and frequency of droughts due to deforestation and climate change may already have pushed the Amazon close to a critical threshold of rainforest dieback. If too much resilience is lost, dieback is inevitable even without deforestation. Many interlinked factors, including cutting and burning of forest, degradation, droughts, fires, and climate change, could combine to reduce resilience and trigger the crossing of a tipping point in the Amazon rainforest. The loss of rainforest would mean billions of metric tons[1] of CO_2 would be released into the atmosphere. For comparison, humans produce an estimated 36 billion metric tons of carbon each year. This would also reduce Earth's ability to recycle CO_2, and accelerate global climate change. There would be a positive feedback whereby global warming causes rainforest death, and loss of rainforest accelerates climate change. And loss of rainforest affects the global atmospheric circulation, affecting climate in places far from the Amazon.

Much research has demonstrated that ecosystems with more species are more efficient at removing nutrients from soil and water than are ecosystems that are less species-rich (Spehn,

2005; Cardinale et al., 2006; Bracken and Stachowicz, 2006). Species with similar niches in the same ecosystem partition niches, which means they avoid competition by using slightly different resources. For example, in two bird species, one might eat slightly larger seeds than the other. Or two species of phytoplankton might live at different depths in the ocean, receiving different amounts or intensities of light. Cardinale (2011) showed that niche partitioning among species of algae can increase the uptake and storage of nitrate, a nutrient pollutant of global concern. Nitrate uptake rates increased linearly with the number of species, and were driven by niche differences among species. So higher diversity resulted in more niche partitioning, and was correlated with better use of nitrate as a nutrient. This decreased the nitrate in the water, so the ecosystem was healthier. Low levels of nitrate are helpful to an ecosystem, since it is a nutrient. But nitrate becomes a toxic pollutant at high levels. This shows that communities with more species and more niche partitioning are better at taking advantage of the niche opportunities in an environment, allowing diverse systems to capture a greater proportion of biologically available resources such as nitrogen. *Communities that are more diverse and have more niche partitioning are more efficient at utilizing nutrients. And biodiversity may help to buffer natural ecosystems against the ecological impacts of pollution due to an excess of nutrients.*

Maestre et al. (2012) found in a global study in dryland ecosystems, which cover about 41% of the Earth's land surface, that *plant species diversity enhanced the ability of ecosystems to maintain multiple functions, such as carbon storage, productivity, and the buildup of nutrient pools. They found plant diversity is key to a healthy ecosystem* and a buffer against desertification in Samburu National Reserve, Kenya. A wide variety and a high number of plant species was even more important than other factors, such as annual rainfall and microbes in the soil. Loss of biodiversity reduces the services the ecosystem can provide.

Fath and Patten (1999) looked at the influence of indirect flows in networks and showed that network organization is, on the whole, more symbiotic when indirect flows are taken into account than is apparent from direct interactions alone. They stated that using network analysis, objects can be studied as part of a connected system and the indirect effects can be identified and quantified. This provides a quantitative foundation to the interconnectedness of nature.

11.3 A NEW PROPOSED PRINCIPLE OF ECOLOGY AND EVOLUTION: LINKS BETWEEN SPECIES ARE FUNDAMENTAL

I will now propose a new hypothesis of ecology and evolution, which I formulated and call the principle of positive interactions and ecosystem stability. It states that *an ecosystem will evolve to the maximum possible number and proportion of positive connections, since this is its most stable state. The number and proportion of positive interspecific relationships is a primary factor in promoting stability in a natural, coevolved ecosystem. The more positive interspecific relationships in a natural ecosystem, the more stable it is*. Stating it in a succinct phrase: *positive connections breed stability*, or, if you prefer, *positive links breed stability*. Interspecific interactions, relationships, links, connections, and connectance all have very similar meanings here. Thus, one could state this principle as follows: *High positive connectance is a primary factor determining ecosystem stability*. Recall that I define stability as the ability of an ecosystem to maintain or return to its level of biodiversity, even when acted on by an outside force. If an outside force caused a loss of diversity, this generally means an ability or tendency of the ecosystem to return to its previous or baseline level of diversity. The principle says that stability is correlated with a greater proportion and number of positive interactions between species. Since a greater number of interactions requires a greater number of species, a corollary of this proposed principle is that ecosystems evolve toward the greatest possible number of species, because this is their most stable state. That is, ecosystems maximize biodiversity. But the original and more important proposed principle is that ecosystems tend toward higher positive connectance and maximize it. *The number of links are at least as important as the number of species in determining an ecosystem's stability.*

Predator-prey interactions are symbiotic, and positive for both predator and prey, *when the population and species are considered*, in a natural ecosystem. Predators have deleterious effects on the organisms in the prey population that they eat, but positive effects on the ones they do not eat. The predator regulates the prey population, preventing it from exploding and depleting its food source and thus declining precipitously or going locally extinct. Thus, the predator and prey population have positive effects on each other, are symbiotic, and have a connectance that is positive in both directions under normal circumstances.

A high number of positive interspecific interactions is the most stable state of an ecosystem, and ecosystems evolve toward many positive interspecific interactions. A higher number of positive interactions between species requires a higher number of species, so high species diversity is correlated with stability, and ecosystems evolve toward high species number in their evolution toward a greater number of positive interactions between them. These two tendencies occur not by interecosystem group selection, since entire ecosystems with less positive connections do not constantly become extinct and outcompeted by ecosystems with more positive links. Rather, each organism in a population and species, and each population and each species in an ecosystem, undergoes natural selection, and this results in the ecosystem being more diverse and with more positive links, because this is a state of higher stability. Ecosystems naturally evolve toward the most stable state. Remember that closely associated species in a natural ecosystem have undergone coevolution with one another. Of course, there are limits to the number of species and links an ecosystem can have. It is limited by space, nutrients, and its energy source, which could be, for example, sunlight or the heat of a volcanic vent. The number of species and links varies from one ecosystem to another. Rainforests have more species and connections than deserts. And the number and percentage of links that are positive is limited, because a certain amount of competition will inevitably be present. However, *every ecosystem will evolve to the state with the highest possible number of positive links because this is its most stable state*. Thus, *ecosystems maximize biodiversity and the number and proportion of positive interactions between species in them*.

11.4 CONCLUSION

There is not a full consensus on the relationship between diversity, complexity, and connectedness on the one hand and stability on the other, but the bulk of the evidence supports, and the majority of researchers believe, that they are positively correlated. And biodiversity is correlated with ecosystem connectance, productivity, resilience, and resistance to invasion. It is correlated with good nutrient flow and efficient use of resources. Stability may indeed decrease with increasing diversity if species are randomly thrown together to build an ecosystem. However, the main point concerning the diversity-stability hypothesis is that *coevolved* diversity breeds stability, and coevolved diversity is the only kind that occurs in nature. An ecosystem with a million species of predators and no plants or herbivores is unstable and will disappear in a short time. However, this ecosystem would never occur in nature, where the species in ecosystems coevolve. The ecosystems that exist under natural conditions are not random associations of species, but coevolved ecosystems with near-optimal ratios of primary producers, herbivores, first-level carnivores, top carnivores, decomposers, species that consume at multiple trophic levels, parasites, and hosts. Such ecosystems are very stable. And this considers only trophic interactions. Ecosystems are further stabilized by increasing the numbers of symbiotic and commensal species. Recall that biogeochemical cycles such as the nitrogen cycle are stable, diverse, mutualistic closed loops (Chapter 6), and that all eukaryotes are complex, diverse ecosystems of interacting symbiotic and commensal species (Chapter 7). These systems are stable largely because of their high diversity and large number of symbiotic and commensal interactions. The evidence in this chapter indicates that there is a positive correlation between biodiversity and stability, connectance, resilience, persistence, productivity, resistance to invasion, and efficient nutrient flow and use of resources, in natural ecosystems. Complexity and connectivity have similar positive correlations with stability, resilience, and persistence. These correlations are consistent with the ABH.

A reasonable hypothesis is that ecosystems tend to evolve toward the highest possible biodiversity and positive connectance and more stable states.

NOTE

1 A metric ton is 1,000 kilograms, or 2,204.6 pounds.

REFERENCES

Allesina, S. & Tang, S. (2015). The stability-complexity relationship at age 40: a random matrix perspective. *Population Ecol.* 57: 63–75.

Boulton, C. A., Lenton, T. M., & Boers, N. (7 March, 2022). Pronounced loss of Amazon rainforest resilience since the early 2000s. *Nature Climate Change* 12: 271–8. https://doi.org/10.1038/s41558-022-01287-8.

Bracken, M. E. S. & Stachowicz, J. J. (2006). Seaweed diversity enhances nitrogen uptake via complementary use of nitrate and ammonium. *Ecology* 87: 2397–403.

Cardinale, B. J. (2011). Biodiversity improves water quality through niche partitioning. *Nature* 472: 86–9.

Cardinale, B. J., et al. (2006). Effects of biodiversity on the functioning of trophic groups and ecosystems. *Nature* 443: 989–92.

Chen, X. & Cohen, J. E. (2001). Global stability, local stability and permanence in model food webs. *Journ. of Theor. Biol.* 212: 223–35.

Damschen, E. I. (27 Sept., 2019). Habitat connectivity enhances diversity. *Science* 365 (6460): 1478–80. doi: 10.1126/science.aax8992.

De Angelis, D. L. (1975). Stability and connectance in food web models. *Ecology* 56: 238–43.

Dunne, J. A. & Williams, R. J. (2009). Cascading extinctions and community collapse in model food webs. *Philos. Trans. Royal Soc. B. Biol. Sci.* 364: 1711–25.

Dunne, J. A., et al. (10 July, 2002). Network structure and biodiversity loss in food webs: robustness increases with connectance. *Ecol. Letters* 5 (4): 558–67. https://doi.org/10.1046/j.1461-0248.2002.00354.

Dunne, J. A., et al. (2005). Modelling food-web dynamics: complexity-stability implications. In: Belgrano, A., Scharler, U. M., Dunne, J. A., & Ulanowicz, R. E. (eds.), *Aquatic Food Webs: An Ecosystem Approach.* Oxford Univ. Press, Oxford, UK, pp. 117–29.

Elton, C. S. (1958). *Ecology of Invasions by Animals and Plants.* Chapman and Hall, London, UK.

Erwin, T. L. (1997). Biodiversity at its utmost: Tropical forest beetles. In: Reaka-Kudla, M. L., Wilson, D. E., & Wilson, E. O. (eds.). *Biodiversity II.* Joseph Henry Press, Washington, DC, pp. 27–40.

Fath, B. D. & Patten, B. C. (March, 1999). Review of the foundations of network environ analysis. *Ecosystems* 2: 167–79. https://doi.org/10.1007/s100219900067.

Fowler, M. S. (2009). Increasing community size and connectance can increase stability in competitive communities. *Journ. Theor. Biol.* 258: 179–88.

Gross, T., et al. (2009). Generalized models reveal stabilizing factors in food webs. *Science* 325: 747–50.

Haydon, D. (1994). Pivotal assumptions determining the relationship between stability and complexity: an analytical synthesis of the stability-complexity debate. *Amer. Nat.* 144: 14–29.

Hooper, D. U., et al. (2005). Effects of biodiversity on ecosystem functioning: a consensus of current knowledge. *Ecol. Monogr.* 75: 3–35.

James, A., et al. (2012). Disentangling nestedness from models of ecological complexity. *Nature* 487: 227–30.

Johnson, K. H., et al. (1996). Biodiversity and the productivity and stability of ecosystems. *Trends in Ecol. and Evol.* 11: 372–7.

Johnson, S., et al. (16 Dec., 2014). Trophic coherence determines food-web stability. *PNAS USA* 111 (50): 17923–8. arXiv: 1404.7728. doi: 10.1073/pnas.1409077111. Bibcode: 2014 PNAS..11117923J.

Koh, L. P., et al. (10 Sept., 2004). Species coextinctions and the biodiversity crisis. *Science* 305 (5690): 1632–4. doi: 10.1126/science.1101101.

Kondoh, M. (2005). Is biodiversity maintained by food-web complexity? The adaptive food-web hypothesis. In: Belgrano, A., Scharler, U. M., Dunne, J. A., & Ulanowicz, R.E. (eds.), *Aquatic Food Webs: An Ecosystem Approach.* Oxford Univ. Press, Oxford, UK, pp. 130–42.

Lehman, C. L. & Tilman, D., & Assoc. Ed.: Gaines, S. D. (2000). Biodiversity, stability, and productivity in competitive communities. *Amer. Nat.* 156: 534–52.

Lenton, T. M. & van Oijen, M. (29 May, 2002). Gaia as a complex adaptive system. *Philosoph. Transactions of the Royal Society B. Biol. Sciences* 357 (1421). https://doi.org/10.1098/rstb.2001.1014.

Lenton, T. M. (2002). Testing Gaia: The effect of life on Earth's habitability and regulation. *Climatic Change* 52: 409–22. https://doi.org/10.1023/A:1014201801949.

Loreau, M. & de Mazancourt, C. (2013). Biodiversity and ecosystem stability: a synthesis of underlying mechanisms. *Ecol. Letters* 16: 106–15.

Lovelock, J. (1979). *Gaia. A New Look at Life on Earth.* Oxford Univ. Press, Oxford, London.

Lovelock, J. E. & Margulis, L. (1974). Atmospheric homeostasis by and for the biosphere: the Gaia hypothesis. *Tellus. Series A. Stockholm: Internat. Meteorol. Inst.* 26 (1–2): 2–10. Bibcode: 1974Tell...26....2L. DOI: 10.1111/j.2153-3490.1974.tb01946.x. ISSN 1600-0870.

Lovelock, J. E. (1972). Gaia as seen through the atmosphere. *Atmospheric Environment.* 6 (8): 579–80. Bibcode: 1972 AtmEn...6..579L. DOI: 10.1016/0004-6981(72)90076-5.

MacArthur, R. H. (1 July, 1955). Fluctuations of animal populations and a measure of community stability. *Ecology* 36 (3): 533–6. https://doi.org/10.2307/1929601.

Maestre, F. T., et al. (13 Jan., 2012). Plant species richness and ecosystem multifunctionality in global drylands. *Science* 335 (6065): 214–8. doi: 10.1126/science.1215442.

May, R. M. (1972). Will a large complex system be stable? *Nature* 238: 413–4.

May, R. M. (1973). *Stability and Complexity in Model Ecosystems.* Princeton Univ. Press, Princeton, NJ.

McNaughton, S. J. (May-June., 1977). Diversity and stability of ecological communities: A comment on the role of empiricism in ecology. *Amer. Naturalist* 111 (979). doi: https://doi.org/10.1086/283181.

McPherson, M. L., Finger, D. J. I., Houskeeper, H. F., et al. (2021). Large-scale shift in the structure of a kelp forest ecosystem co-occurs with an epizootic and marine heatwave. *Communications Biol.* 4, Article number: 298. https://doi.org/10.1038/s42003-021-01827-6.

Melian, C. J., et al. (2009). Diversity in a complex ecological network with two interaction types. *Oikos* 118: 122–30.

Mougi, A. & Kondoh, M. (2012). Diversity of interaction types and ecological community stability. *Science* 337: 349–51.

Newman, M. (2010). *Networks: An Introduction.* Oxford Univ. Press, Oxford, UK.

Odum, E. P. (1953). *Fundamentals of Ecology.* Saunders, Phila., PA.

Okuyama, T. & Holland, J. N. (2008). Network structural properties mediate the stability of mutualistic communities. *Ecol. Letters* 11: 208–16.

Pascual, M. & Dunne, J. A. (eds.) (2006). *Ecological Networks: Linking Structure to Dynamics in Food Webs.* Oxford Univ. Press, Oxford, UK.

Pimm, S. L. (1979). Complexity and stability: another look at MacArthur's original hypothesis. *Oikos* 33: 251–7.

Pimm. S. L. (1980). Food web design and the effect of species deletion. *Oikos* 35: 139–49.

Sánchez-Bayoa, F. & Wyckhuysbcd, K. A. G. (April, 2019). Worldwide decline of the entomofauna: A review of its drivers. *Biol. Conservation* 232: 8–27. https://doi.org/10.1016/j.biocon.2019.01.020.

Seaborg, D. (2022). *How Life Increases Biodiversity: An Autocatalytic Hypothesis.* CRC Press/Taylor & Francis. Boca Raton, FL; London, UK; New York, NY.

Simkin, S. M., et al. (12 April, 2016). Conditional vulnerability of plant diversity to atmospheric nitrogen deposition across the United States. *PNAS USA* 113 (15) 4086–91. https://doi.org/10.1073/pnas.1515241113.

Spehn, E. M., et al. (2005). Ecosystem effects of biodiversity manipulations in European grasslands. *Ecol. Monogr.* 75: 37–63.

Thébault, E. & Fontaine, C. (2010). Stability of ecological communities and the architecture of mutualistic and trophic interactions. *Science* 329: 853–6.

Tilman, D. & Downing, J. A. (27 Jan., 1994). Biodiversity and stability in grasslands. *Nature* 367: 363–5. https://doi.org/10.1038/367363a0.

Tilman, D. (1988). *Plant Strategies and the Dynamics and Structure of Plant Communities.* Monogr. in Population Biol., Vol. 26. Princeton Univ. Press, Princeton, NJ.

Tilman, D. (1999). The ecological consequences of changes in biodiversity: a search for general principles. *Ecology* 80: 1455–74. https://doi.org/10.1890/0012-9658.

Tilman, D. (Sept., 1987). Secondary succession and the pattern of plant dominance along experimental nitrogen gradients. *Ecol. Monogr.* 57 (3): 189–214. https://doi.org/10.2307/293708017.

Tilman, D., et al. (Nov., 2014). Biodiversity and ecosystem functioning. *Ann. Rev. of Ecol., Evol., and Systematics* 45: 471–93. https://doi.org/10.1146/annurev-ecolsys-120213-091917.

Vieira, M. C. & Almeida-Neto, M. (2015). A simple stochastic model for complex coextinctions in mutualistic networks: robustness decreases with connectance. *Ecol. Letters* 18: 144–52.

Worm, B. & Duffy, J. E. (2003). Biodiversity, productivity and stability in real food webs. *Trends in Ecol. and Evol.* 18: 628–32.

Yodzis, P. (1981). The stability of real ecosystems. *Nature* 289: 674–6.

Zipkin, E. F. (14 Feb., 2020). Tropical snake diversity collapses after widespread amphibian loss. *Science* 367 (6479): 814–6. doi: 10.1126/science.aay5733.

12 Scientific Laws Indicate the Essential Inevitability of the Chemical Evolution of Life under Favorable Conditions: A Hypothesis That Incorporates the Autocatalytic Biodiversity Hypothesis Postulates a Tendency for Increase in Information in Systems with Favorable Conditions

Because of the laws of physics and chemistry, the evolution of simple chemicals into life is essentially inevitable when there are favorable conditions, such as existed on early Earth. The second law of thermodynamics states that there is a tendency for an increase in entropy, which is disorder, in any closed system. This means order tends to decrease in any closed system. However, as a result of the laws of physics and chemistry, there is a tendency toward an increase in order, complexity, and information Earth, and presumably any planet or moon with conditions favorable to life. This is not a violation of the second law of thermodynamics, since the biosphere of any such planet or moon is not a closed system, but an open system, because it receives energy from an outside source: its star. Of course, in the case of Earth, this source is the sun. In fact, England (2013) showed mathematically that *life tends to result from the second law of thermodynamics*. The essential inevitability of the evolution of life from simple chemicals occurring under favorable conditions is relevant to the Autocatalytic Biodiversity Hypothesis (ABH) because this inevitability is a tendency for information and complexity to increase, just as the tendency for ecosystems to maximize diversity and for life to promote diversity are examples of the tendency of information and complexity to increase under similarly favorable conditions. This is true because biodiversity is a form of information. Thus, the essential inevitability of life to evolve from simple chemicals and the tendency of life to cause an increase in biodiversity are both forms of the tendency of systems to increase in information content under the right conditions and with an energy input.

12.1 PRECURSORS TO LARGE BIOLOGICAL MOLECULES WERE DELIVERED FROM SPACE

Small, simple molecules that were used as building blocks to form the larger molecules used in living organisms, such as amino acids, have been found in space by radio astronomers (Irvine, 1998). These simple molecules include hydrogen sulfide (H_2S), methane (CH_4), hydrogen cyanide (HCN), formaldehyde (HCHO), methanol (H_3COH), and ammonia (NH_3). Several organic molecules, including alcohols, ketones, ethers, and others have also been found, distributed in dark clouds and hot cores in giant molecular clouds (Ohishi, 2008). The amino acid glycine may have been detected, but this is not certain (ibid.). Amino acids are the building blocks for proteins. Amazingly, over 140 molecules have been detected in space (ibid.). Some of the basic molecules necessary for the chemical evolution of life have been found in comets and asteroids, as well as interstellar molecular clouds. These simple molecules are highly reactive and can form more complex molecules when they are mixed. We know they form more complex molecules in space in our solar system because these more complex molecules have been found in meteorites that hit Earth. This includes 10 amino acids that are used in proteins, 60 other amino acids, several nucleobases (chemicals that are part of DNA and RNA), a simple sugar (sugars are in DNA, RNA, and carbohydrates), and fatty acids that can be assembled into the membranes of living cells.

Oba et al. (2022) have found all five of the nucleobases—adenine, guanine, cytosine, thymine, and uracil—used

DOI: 10.1201/9781003246640-12

as the chemical building blocks to make DNA and RNA, in meteorites from space. This adds to the evidence, but does not prove, that these key precursors to life came from space. Some have objected that the nucleobases could be contamination from the soil, but some of the nucleobases were in higher concentrations than in the surrounding soil, and the researchers detected more than a dozen other compounds related to life in the meteorites, but not in the soil. The researchers say that the nucleobases could have been generated by photochemical reactions prevailing in the interstellar medium and later incorporated into asteroids during solar system formation. This study shows that the nucleobase building blocks of DNA and RNA could have come from interstellar space on meteors.

Oba et al. (2019) created an environment similar to molecular clouds that are found between stars. In this simulated environment, some of the nucleobases that are in DNA and RNA were created. They also found amino acids and found that some pairs of amino acids combined together into one molecule with two joined amino acids in this environment. These molecules would have accumulated on the surfaces of mineral dust particles in the molecular clouds. The dust particles would have then come together into asteroids because of gravity, carrying the biological molecules along with them. It is also possible that reactions creating these and even more complex molecules occurred in asteroids early in the history of our solar system.

It is thought that these molecules reached Earth through meteorite impacts and dust from space four bya. These extraterrestrial compounds could well have been part of the primordial soup necessary for the origin of life. Primordial soup is the term for the liquid, probably water, containing various chemicals where life first evolved via chemical evolution. Thus, the main source of amino acids that make up proteins, nucleobases that make up DNA and RNA, simple sugars, fatty acids capable of assembling into the membranes of living cells, and perhaps other biological molecules were very likely delivered to Earth from the solar system outside Earth. In fact, even today organic compounds of the type that life evolved from are falling to Earth from space, mainly on dust particles, at a high rate, although at a lower rate than in the past.

Molecular clouds are composed of the remnants of stars that have gone through their life cycle. Stars larger than our sun explode as supernovas, while others shed their outer layers and collapse into white dwarfs. This produces huge molecular clouds, which can be seen as large dark areas in most galaxies, including the Milky Way, because they obscure the light of the stars behind them. Molecular clouds light years across are composed of dust particles and gas. As the dust aggregates under the force of gravity, new stars and planets form solar systems.

The primary material that formed the inner solar system, which is comprised of the four terrestrial planets and the asteroid belt, is the dust and gas that composed our molecular cloud. Volatile materials were delivered into the inner solar system by comets and interplanetary dust particles near the end of the hot early phase of the formation of the sun and its planets. Once the inner planets were cool enough, they were seeded with complex organic compounds delivered from the latter phase of the solar nebula, which is the disc-shaped cloud of gas and dust left over from the sun's formation from which the planets of the solar system formed (Oró, 1961; Delsemme, 1984; Anders, 1989; Chyba and Sagan, 1993).

12.2 HOW LIPIDS AND AMINO ACIDS ARE FORMED IN MOLECULAR CLOUDS AND POSSIBLY HOW AMINO ACIDS ARE FORMED ON EARTH

Lipids are important biological macromolecules whose functions include storing energy, signaling, and acting as key structural components of cell membranes. They are present in meteorites that are high in carbon, and can spontaneously assemble into membranes. They can be synthesized in the laboratory when ultraviolet radiation impinges on simulated molecular clouds (Dworkin et al., 2001). The dust particles that compose a molecular cloud are coated with a thin film of ice mixed with simple molecules such as methanol (CH_3OH), carbon monoxide (CO), carbon dioxide (CO_2) and ammonia (ibid.). Icy dust grains in the interstellar medium, icy comets, and ice covered meteorites contain simple molecules such as CO_2, water, methane, hydrogen cyanide, and ammonia, and are exposed to complex radiation fields, including UV light. When ultraviolet light from a nearby star is absorbed by the ice, it drives a variety of photochemical reactions that make biologically relevant molecules from such molecules as the simple ones listed in the previous two sentences. The ice in molecular clouds is also exposed to various forms of high-energy radiation and low energy electrons (LEEs) that can drive chemical reactions. Esmaili et al. (2018) simulated astrophysical conditions with the cryogenic (extremely cold) temperatures that occur in molecular clouds in an ultrahigh vacuum and used an electron gun to generate LEEs that irradiated thin films of ice mixed with simple molecules such as methane, carbon dioxide, and ammonia. They therefore simulated the chemistry of the icy dust grains in molecular clouds, comets, and meteorites and the energy they receive. The amino acid glycine was detected in the resulting mixture, showing that the copious secondary LEEs generated by ionizing radiation can drive the synthesis of an amino acid from simple molecular ingredients, none of which possess the carbon-to-carbon (C-C) or carbon-to-nitrogen (C-N) bonds so important in the formation of many of the molecules of life. The UV and LEE experiments demonstrate how life's building blocks could have formed in space and then been delivered to Earth by dust particles, comets, and meteorites.

Miller (1953) and Miller and Urey (1959) decided that a mixture of some of life's precursors—hydrogen gas (H_2), water vapor, methane, and ammonia—were plausible components of Earth's early atmosphere at the time of life's origin. They exposed such a mixture to continuous electrical sparks to simulate lightning, and amino acids were synthesized, including glycine, α-alanine and β-alanine, and possibly aspartic acid and α-aminobutyric acid. However, the current consensus is that at the time of life's origin, the Earth's atmosphere was not composed mainly of these gases, but was mostly nitrogen, with small amounts of carbon dioxide and argon. Stribling and Miller (1987) reported the synthesis of amino acids, but in very low yields, when this atmosphere was exposed to electrical discharge.

The experiments discussed thus far collectively indicate that life's precursor molecules, even amino acids, formed with ease in outer space, and possibly in what may have been the Earth's primordial atmosphere. This supports the idea that there is a tendency for the chemical evolution of life to occur and for information and complexity to increase under favorable conditions.

12.3 SOURCE OF NUCLEOTIDES NOT FULLY WORKED OUT

Nucleobases are nitrogen-containing biological compounds that form nucleosides, which, in turn, are components of nucleotides. Nucleotides are the building blocks that make up the nucleic acids DNA and RNA. They are composed of nucleobases attached to a sugar and a phosphate. The sugar in DNA is called deoxyribose, and the sugar in RNA is ribose. Nucleotides are more complex than amino acids, and their source is still a gap in our understanding of how life began. It is important to know how they were formed to understand how DNA and RNA were formed at life's beginning. Oró and Kimball (1961) showed that adenine could be made from hydrogen cyanide and ammonia in a water solution, and Powner et al. (2009) were able to synthesize a nucleotide in the laboratory. Pearce et al. (2022) pointed out that the basic building blocks of RNA could have been delivered by carbon-rich meteorites, or produced by processes beginning with the synthesis of hydrogen cyanide in the early Earth's atmosphere. Lightning and external chemistry driven by ultraviolet light could produce the hydrogen cyanide. But the full details of how nucleotides could be made in natural conditions in space or on Earth are not fully understood.

Nam et al. (2018) showed a general synthetic path to create ribonucleosides in a microdroplet environment using the magnesium ion (Mg^{+2}) as a catalyst. This suggests a possible scenario for the spontaneous production of the random ribonucleosides necessary to generate the nucleotides necessary to make primitive RNA. This is significant, since many think RNA played a central role in the origin of life, and was the first genetic material, and was later replaced by DNA. And Nam et al.'s work provides support for these hypotheses. Becker et al. (2019) reported that they were able to make nucleosides from small molecules and ribose, the sugar molecule in RNA, driven solely by wet-dry cycles. In the presence of phosphate-containing minerals, nucleotides also formed. The pathway allowed the formation of all the precursor molecules needed to form DNA and RNA. However, while these two studies are steps in the right direction, they involve relatively complex laboratory syntheses. It is not certain that these reactions would actually occur and produce nucleosides in the environment in which life evolved.

Ranjan et al. (2018) modeled Earth at about 3.5 bya, when they and many others think life originated, and looked at geologic records to see the amount and effects of volcanic activity that took place at this time. They found volcanism would have produced tremendous amounts of sulfur dioxide, which would have dissolved in the shallow waters of ponds and entered into in lakes, producing large, even millimolar, concentrations of sulfites and bisulfites. They proposed that, in these bodies of water, chemical reactions of sulfites and bisulfites could have played a role in the making of ribonucleotides, but research is needed to test this hypothesis.

12.4 SOURCE OF EARTH'S WATER

As to how Earth obtained its water, it was thought in the past that bombardments by icy comets from the outer solar system 4.5 to 3.8 bya brought Earth's water, and they likely did deliver some. However, Cleeves et al. (2014) showed that between 30 and 50% of it came from the solar nebula. Since the planets were formed from this, this source of water is millions of years older than the solar system itself. And Hallis et al. (2015) presented evidence that it mostly came from water-soaked dust grains during the Earth's formation. Further, the ocean is not Earth's largest water reservoir. It is estimated that Earth's interior holds several times as much water as is found at the surface. Piani et al. (2020) showed that enstatiite chondrite meteorites, which were originally formed near the center of the solar nebula that created the Solar System, and which are similar to the building blocks that helped form Earth, contain sufficient hydrogen to have delivered to Earth at least three times the mass of water in its oceans, and thus most of Earth's water. This indicates meteors from the inner solar system where the four rocky planets are located likely gave Earth most of the water in its interior. The meteors did not deliver water, but contained enough hydrogen, which could combine with oxygen to to create it. The meteors of the inner solar system likely contributed most of the water to the seas as well. Still, some water seems to have come from meteors and comets from the outer solar system. So dust grains, meteors, and comets delivered Earth's water, mostly before Earth was formed.

Since the water was mostly delivered to the Earth before the debris that formed it came together, Earth's water was originally trapped underground and not available to life aboveground. Volcanoes supplied the heat that formed water vapor that escaped into the atmosphere and eventually condensed and formed a good deal of Earth's aboveground water—lakes, rivers, and the oceans, which were originally fresh water. Without volcanoes, most water would be trapped underground in the crust and mantle. Life aboveground would not be possible, and there would no river, lake, or ocean ecosystems. Together, space debris and volcanoes delivered the water that made life aboveground possible.

The water ultimately formed oceans about five kilometers (about 3.1 miles) deep. The organic compounds and the water, both delivered from space, could well have been part of the primordial soup necessary for the origin of life, once the water had reached the surface.

12.5 VOLCANOES PROVIDED LAND: SOLAR DISTILLATION OF WATER

Volcanoes provided land masses above sea level on the early Earth. They created the land that still exists today. Solar heat evaporated water from the ocean, some of which fell as freshwater on the land, as rain. Rain is essentially distilled water. The metabolic processes required by life today are inhibited by the high salt content of seawater, so it is reasonable to assume that the first forms of life needed fresh water conditions (Damer and Deamer, 2015); hence the need for the distillation and precipitation of fresh water onto volcanic land masses on the early Earth. Organisms in the sea, with its high salt concentration, have evolved mechanisms to pump the salt out, so they can carry out their metabolic processes and survive.

12.6 HOW THE BUILDING BLOCKS OF RNA WERE JOINED TOGETHER TO MAKE RNA AND HOW RNA INCREASED IN COMPLEXITY

Nucleotides, the building blocks of RNA, had to be joined together to produce this biological molecule that is thought to be one of the first molecules of life. Short chains of RNA can be synthesized nonenzymatically under laboratory conditions, but there is no consensus on how these polymers[1] could have been synthesized on the early Earth. Ferris et al. (1989) reported that the formation of short strands of RNA from nucleotides may have been catalyzed by charged clays such as montmorillonite (montmorillonite is a very soft sheet silicate group of minerals). And minerals have been shown to catalyze the joining together of nucleotide-like molecules (Huang and Ferris, 2006). RNA could also have formed in salty ice water. Kanavarioti et al. (2004) found that in the presence of metal ions in dilute solutions, frozen samples of nucleosides form chains up to 11 bases long, suggesting that that ice matrices on the early Earth may have accelerated the formation of nucleoside chains.

Mizuuchi et al. (2022) performed long-term experiments on the evolution of RNA that can duplicate itself using a self-encoded RNA replicase, an enzyme that duplicates RNA molecules. The RNA increased in complexity in the lab, diversifying into several coexisting host and parasite lines, whose frequencies in the population fluctuated at first, but then gradually stabilized. After generations of evolution, the final population, which consisted of five RNA lineages, formed a network of replicating molecules with diverse interactions, including cooperation that helped the replication of all replicating RNA molecules. These results are evidence for the capability of molecules that replicate to spontaneously increase in complexity through Darwinian evolution, which is a critical step for the evolution of life.

12.7 HOW THE FIRST CELLS WERE MADE

Recently, a hypothesis has been put forward that life started in volcanic hot springs (Damer and Deamer, 2020). An origin of life on land provides a solution to how amino acids were connected together to form proteins and how nucleotides were put together to form nucleic acids. The process starts with amino acids, nucleotides, and lipids being synthesized, perhaps in space and delivered to Earth by comets, meteorites, and dust in the solar system, perhaps by volcanic geochemistry on Earth, or by a combination of the two. And amino acids could be formed in the early atmosphere by lightning. The biological molecules then could have been transported to hot springs by rain. As a result, nucleic acids, proteins, and lipids ended up in hydrothermal pools with other compounds and begin to evolve into cells (Damer and Deamer, 2015). As an initial test of this idea, Deamer et al. (2006) added four amino acids, four nucleic acid bases, phosphate, glycerol, and a lipid into a small, boiling spring on Mount Mutnowsky in Kamchatka, Russia. Within minutes, many tiny vesicles containing the compounds Deamer et al. provided appeared around the edges of the pool, demonstrating that primitive "cells" can emerge spontaneously. Then in the laboratory, they mixed nucleotides with lipids and put them through cycles of wetting and drying at acidic pH ranges (pH 3), and elevated temperatures, similar to those in hot springs. *Long chains from 10 to 100 nucleotides in length were formed, demonstrating that nucleotides can spontaneously join together into DNA and RNA! These were later shown to resemble RNA* (DeGuzman et al., 2014; Da Silva et al., 2015). Deamer (2021) then showed *the same reaction that produced the long nucleotide polymers occurred in nature, in an actual natural hot spring in New Zealand.*

The mixtures of organic compounds in thermal pools undergo repeated cycles of three phases: wet, dry, and moist gels. The dry phase sufficiently concentrates the amino acids and nucleotides so that polymerization reactions (reactions that join the amino acids together and the nucleotides together) can form peptides (chains of amino acids not long enough to be called proteins) and nucleic acids (DNA and RNA). But also, Holden et al. (2022) showed that the amino acids glycine or L-alanine bond together at the interface of water droplets and air. This is another mechanism to form proteins from amino acids, without employing a dry phase. In the wet phase, protocells[2] that can encapsulate biological macromolecules spontaneously assemble from lipids (Deamer and Barchfeld, 1982). Cell membranes in life today are composed of double layers of lipids of a type called phospholipids. *When they are extracted from living cells and dispersed in water, phospholipids spontaneously form vesicles by self-assembly. If biological macromolecules like proteins and the nucleic acids DNA and RNA are present during this self-assembly, they become enclosed in membranous spheres to form protocells, which is one of the essential steps required for the origin of life* (ibid.; Deamer, 1985; Deamer, 2019)! The hydrated phase is therefore required for the mixture of lipids and large biological molecules such as proteins and nucleic acids to assemble into protocells. Under certain conditions, protocells can even divide into smaller daughter cells (Luisi, 2016). The hydrated phase stresses the protocells, which undergo natural selection and evolution. *Amino acids readily form peptides when exposed to alternating cycles of drying and wetting* (Rodriguez et al 2015), *and nucleotides also join together into short nucleic acids resembling RNA under alternating dry and wet cycles* (Rajamani et al., 2008; DeGuzman et al., 2014). *If lipids are present, the polymers are encapsulated to form microscopic protocells.*

In the moist gel phase, protocells come together and communicate by exchanging sets of large molecules such as chains of nucleic acids and peptides that can act as catalysts. Those large molecules that aid survival and reproduction during ensuing cycles are selected for. The researchers hypothesize that natural selection began for the first time at this point when the more robust protocells survived long enough to be transported to other hot springs. Some remained in the hot springs, but others migrated to land and the sea. Selection continued until protocells developed a genetic code and ribosomes (factories in the cell that make proteins), evolved the necessary machinery to grow by DNA replication and the manufacture of proteins, and had the ability to divide into daughter cells that could survive and reproduce. This paved the way for the emergence of the first cell, and finally the first microbial mats. Microbial mats are sheets of microorganisms that grow in layers; they are usually bacteria and archaea, but sometimes consist of only bacteria. They are likely the first communities of living organisms. They were more common in the past. All of this shows the tendency of chemicals to evolve spontaneously into life.

When the water from a dilute concentration of protocells evaporates, causing them to concentrate into a film on a mineral surface, the protocells fuse with each other to produce a structure with hundreds to thousands of lipid layers. This would be like a puddle containing protocells evaporating during the time when the first cells were evolving, leaving the many lipid layers on granular mineral surfaces or even sand and pebbles. The result is that large biological molecules resembling proteins, DNA, and RNA get captured between the lipid layers. These large molecules have a variety of potential functions such as storing genetic information (as DNA does) or acting as catalysts to drive chemical reactions of a primitive metabolism (as proteins do).

The proteins, DNA, RNA, and other molecules move around and are mixed in this structure. Upon the addition of water, which would be like rain falling on early Earth, new protocells form, each with different combinations of proteins, DNA, and RNA than the original protocells, as a result of the mixing of these large biological molecules in the structures with many layers of lipids. (The mixing is analogous to sexual reproduction because sexual reproduction also mixes large biological molecules in the form of segments of DNA, producing new combinations of DNA.) Natural selection then acts on the protocells with the new combinations of proteins, DNA, and RNA, as follows: The protocells are exposed to environmental stresses that tend to disrupt fragile combinations of the biological molecules, while more robust combinations survive. This process of creating protocells with new combinations of proteins and nucleic acids and natural selection on them is repeated for generations, until the first cells evolve.

Each protocell is a microscopic experiment in which varying mixtures of large biological molecules such as proteins, DNA, RNA, and lipids are tested for their ability to evolve biologically useful functions. We do not know yet what these functions might be, but two of the simplest are stabilization of the cell, as the cytoskeleton (the cell's "skeleton") does in life today, and production of microscopic pores that provide access to nutrients in the environment. The most important function would likely be for some of the large biological molecules to be able to use DNA to make new DNA and RNA and to guide the production of proteins. At that point, the most primitive form of cellular life will have emerged on the early Earth.

Hanczyc et al. (2007) demonstrated that a solution containing RNA, fatty acids, and clay produces structures that contain clay, which acts as a surface that promotes biochemical reactions, and RNA, enclosed within a membrane much like that around protocells. They showed that mineral surfaces such as those provided by clay can bring together and organize key components in the chemical evolution of life. They demonstrated that a diverse array of mineral particles could have been incorporated in vesicles and acted as catalysts that promote reactions that produce molecules of life there.

The above studies provide clear evidence that there is a tendency for RNA and proteins to be formed from basic chemistry and incorporated into lipid protocells as a result of physical and chemical laws. Favorable conditions for this to happen would have been present in the combination of space and the early Earth. They also could have occurred on other planets and moons in our solar system, and likely occurred or are occurring on planets and moons outside our solar system.

Although the evolution of self-replicating DNA, protein synthesis, and cell division have yet to be demonstrated, researchers believe it is only a matter of time, because this is fully plausible within known physical and chemical laws. They are already part way there. For instance, RNA could have been the first molecule of life, since it can act both as a genetic material and as an enzyme (Szostak et al., 2001, and references therein). This has led many to believe that RNA may have been important in the evolution of systems that can replicate themselves before they evolved to be living cells (see Neveu et al., 2013). There is a consensus that DNA evolved later as a more stable form of genetic material. The fact that DNA is more stable than RNA would have led to selection for it to replace RNA as the genetic material.

There are other models of life's origin, such as in deep-sea vents. Because I want to be brief, I have presented what I consider to be the most plausible model. At any rate, my argument that life will inevitably evolve as a result of physical and chemical laws under the conditions of outer space and early Earth is supported by all of the proposed models.

It is even possible that life on Earth started on Mars and was delivered to Earth on asteroids. Life cannot evolve until its planet cools down enough for water to exist as a liquid. Evidence is mounting that Mars formed and cooled down before Earth, and that it had methane, a necessary ingredient for the chemical evolution of life. And we now know that microbial life can travel great distances through the Universe in asteroids and other such debris. More research is needed before we will know definitively whether or not Earth life originated on Mars. If it is true, it is still consistent with the inherent tendency of life to evolve under favorable conditions.

12.8 LIFE'S EARLIEST FOSSILS ARE CONSISTENT WITH THE INEVITABILITY OF LIFE'S ORIGIN

Djokic et al. (2017) found evidence that supports the hot springs model for the origin of life discussed above in the Dresser Formation of Western Australia, which has an estimated age of about 3.48 bya. They found geyserite, sinter terracettes, and mineralized remnants of ancient hot spring pools and vents with fossil evidence of the existence of microbes, indicative of the earliest life on land, and even fossil stromatolites. Geyserite, or siliceous sinter, is a form of opaline silica that is often found as crusts or layers around hot springs and geysers. Sinter is a hard deposit containing silica or calcium carbonate precipitated from mineral springs. A terracette is a landform consisting of a hillside ridge arranged as part of subparallel step-like sequences of such ridges. Terracettes occur on step hillsides and are regularly spaced. Stromatolites are layered sedimentary formations formed by photosynthetic microorganisms, such as cyanobacteria, sulfate-reducing bacteria, and proteobacteria. Though living ones exist today, many are fossils, including those found in the Dresser Formation. Schopf (2002) observed microfossils of bacteria in the same region. The hot springs model is also supported by DNA sequencing evidence that points to the original prokaryotic life form as a dweller of hot springs adapted to very high temperatures (Di Giulio, 2008).

Schopf et al. (2018) examined the earliest fossils of cells, which are about 3.465 billion years old. They consist of eleven specimens of five taxa of prokaryotic filamentous cellular microfossils from the Apex chert of northwestern Western Australia. Earlier, there was dispute as to whether these were fossils of microbial life or nonbiological deposits. The researchers confirmed that these were indeed fossils of microbes. They found the fossils contained the earliest known archaea. One was an archaeal methane producer, two others were methane consumers, and two of the species were primitive photosynthesizers. Thus, life was already diverse 3.465 bya. And the microbes existed when there was very little oxygen present in Earth's atmosphere. The findings further suggest that methane-cycling communities of methane-producing methanogens and methane-consuming methanotrophs were a significant component of Earth's early biosphere. The methanogens would have supported the methanotrophs.

Bell et al. (2015) analyzed small fragments captured and preserved in minerals called zircons from Western Australia. What they saw was consistent with a biological origin. They suggested from this analysis that *life could have originated 4.1 bya, 300 million years earlier than previously thought, shortly after Earth formed 4.54 bya and oceans formed 4.4 bya, and prior to the massive bombardment of the Earth 3.9 bya.*

The work of Schopf et al. and Bell et al. strongly indicates that chemical evolution of life began a geologically very short period of time after conditions were favorable for it on Earth. Deamer (2019) estimates that it took 500 million years for life to evolve from the earliest primitive RNA-based life to the microbes preserved in the Dresser formation. The complex and varied structures of the fossils at such an early point in Earth's history indicate that life evolved and diversified (as prokaryotes) immediately after it first appeared, and at a rapid pace. Thus, there was a tendency—even an inevitability—for the spontaneous chemical evolution of life and it rapidly becoming complex and diversifying on Earth as a result of chemical and physical laws.

There are a hundred billion stars in a typical galaxy and a hundred billion galaxies in the known Universe, and there is a growing consensus among astronomers that exoplanets are commonplace. Therefore, the increasingly accepted idea that life is common throughout the Universe is very likely valid. *The emergence of life on habitable planets, of which Earth is one, is essentially inevitable.* This is related to the ABH. There is an inherent tendency for information, including diversity, to increase, on planets and moons with the right conditions.

12.9 A GENERAL HYPOTHESIS THAT INCORPORATES THE ABH

One can combine the following four observations into a general hypothesis: (1) The evolution of life is inevitable where there are favorable conditions. (2) The complexity of the most complex taxa at any point in geologic time will increase until the maximum complexity possible is reached, and this is because of life's tendency to fill empty niches, as discussed in Chapter 2 of the companion volume to this book (Seaborg, 2022); this is part of the ABH. (3) Life has an inherent tendency to increase diversity because all species are ecosystem engineers that have a net positive effect on diversity in natural ecosystems over sufficient time. This is the main unifying mechanism of the ABH. There are also other mechanisms by which life causes its own diversification. (4) Thus, under favorable conditions, life will evolve by chemical evolution, increase in the complexity of the most complex taxa until its maximum possible complexity is reached, and increase in diversity. From these observations, I propose the *Information Increase Hypothesis*, which proposes that there is a tendency for information, diversity, and complexity to increase in the right conditions (such as occur on Earth). This a more general hypothesis than the ABH, and incorporates it; that is, the ABH is a subset of this hypothesis. Since there is no increase in diversity during the chemical evolution of life, one can state the hypothesis as follows: *There is a tendency for a system to increase in information and complexity under the right conditions*. Diversity is a form of information, so stating the hypothesis this way includes the increase in diversity after chemical evolution.

This hypothesis applies to planets and moons outside our Solar System in addition to our Solar System. It thus applies to the whole Universe, but in only very limited areas that represent a tiny fraction of it. It does not imply teleology; does not conflict with the second law of thermodynamics; recognizes that catastrophes can happen that for a period of time decrease diversity and information (meteorite strikes, massive volcanism, nearby supernovae, etc.); and recognizes that on some planets and moons, life will only evolve to the prokaryote stage, making it necessary to limit its claims. Further evidence for it will be discussed in the next chapter. It is original because it combines the tendency of life to arise from chemical evolution under the right conditions with the ABH. Many already accept the tendency for chemical evolution to occur, so this alone would not be original.

NOTES

1 Polymers are chains of molecules that are of the same type. Proteins are polymers made up of amino acids. DNA and RNA are polymers of nucleotides.

2 A protocell is a self-organized sphere made of lipids that has some similarities to, but is much simpler than, a living cell. Protocells are proposed as stepping stones toward the evolution and origin of cells.

REFERENCES

Anders, E. (1989). Pre-biotic organic matter from comets and asteroids. *Nature* 342: 255–7.

Becker, S., et al. (4 Oct., 2019). Unified prebiotically plausible synthesis of pyrimidine and purine RNA ribonucleotides. *Science* 366 (6461): 76–82. doi: 10.1126/science.aax2747. PMID 3160430.

Bell, E. A., et al. (24 Nov., 2015). Potentially biogenic carbon preserved in a 4.1 billion-year-old zircon. *PNAS USA* 112 (47): 14518–21. https://doi.org/10.1073/pnas.1517557112.

Chyba, C. & Sagan, C. (1992). Endogenous production, exogenous delivery and impact-shock synthesis of organic molecules: an inventory for the origins of life. *Nature* 355 (6356): 125–32. doi: 10.1038/355125a0.

Cleeves, L. I., et al. (26 Sept., 2014). Nature or nurture for solar system ices? *Science* 345 (6204): 1590–3. doi: 10.1126/science.1258055.

Da Silva, L., Maurel, M. C., & Deamer, D. (2014). Salt-promoted synthesis of RNA-like molecules in simulated hydrothermal conditions. *Journ. Molecular Evol.* 80: 86–97.

Damer B., & Deamer, D. (2015). Coupled phases and combinatorial selection in fluctuating hydrothermal pools: a scenario to guide experimental approaches to the origin of cellular life. *Life* 5: 872–87. doi: 10.3390/life5010872. https://doi.org/10.3390/life5010872.

Damer, B. & Deamer, D. (April, 2020). The hot spring hypothesis for an origin of life. *Astrobiol.* 20 (4): 429–52. doi: 10.1089/ast.2019.2045. ISSN: 1531-1074.

Deamer, D. (2019). *Assembly of Life: Can Life Begin on Earth and Other Habitable Planets?* Oxford Univ. Press, Oxford, UK.

Deamer, D. (10 Feb., 2021). Where did life begin? Testing ideas in prebiotic analogue conditions. *Life* 11: 134. https://doi.org/10.3390/life11020134.

Deamer, D., Singaram, S., Rajamani, S., Kompanichenko, V., & Guggenheim, S. (2006). Self-assembly processes in the prebiotic environment. *Philos. Trans. Royal Soc. Lond. B.* 361: 1809–18.

Deamer, D. W. (1985). Boundary structures are formed by organic compounds of the Murchison carbonaceous chondrite. *Nature* 317: 792–94.

Deamer, D. W. (2019). Personal communication.

Deamer, D. W. & Barchfeld, G. L. (1982). Encapsulation of macromolecules by lipid vesicles under simulated prebiotic conditions. *Journ. Molecular Evol.* 18: 203–6.

DeGuzman, V., Shenasa, H., Vercoutere, W., & Deamer, D. (2014). Generation of oligonucleotides under hydrothermal conditions by non-enzymatic polymerization. *Journ. Molecular Evol.* 78: 251–62.

Delsemme, A. H. (Dec., 1984). The cometary connection with prebiotic chemistry. *Origins of Life* 14 (1–4): 51–60. https://doi.org/10.1023/A:1006574110907.

Di Giulio, M. (7 April, 2003). The universal ancestor was a thermophile or a hyperthermophile: tests and further evidence. *Journ. of Theoretical Biol.* 221 (3): 425–36. https://doi.org/10.1006/jtbi.2003.3197.

Djokic, T., et al. (9 May, 2017). Earliest signs of life on land preserved in ca. 3.5 Ga hot spring deposits. *Nature Comms.* 8, Article number: 15263. doi: 10.1038/ncomms15263.

Dworkin, J. P., et al. (2001). Self-assembling amphiphilic molecules: Synthesis in simulated interstellar/precometary ices. *PNAS USA* 98: 815–9.

England, J. L. (2013). Statistical physics of self-replication. *Journ. of Chemical Physics* 139: 121923. https://doi.org/10.1063/1.4818538.

Esmaili, S., et al. (2018). Glycine formation in CO_2:CH_4:NH_3 ices induced by 0-70 eV electrons. *Journ. of Chem. Physics* 148 (164702). doi: 10.1063/1.5021596.

Ferris, J. P., et al. (March, 1989). Mineral catalysis of the formation of dimers of 5′-AMP in aqueous solution: The possible role of montmorillonite clays in the prebiotic synthesis of RNA. *Origins of Life and Evol. of Biosph.* 19 (2): 165–78.

Hallis, L. J., et al. (13 Nov., 2015). Evidence for primordial water in Earth's deep mantle. *Science* 350 (6262): 795–7. doi: 10.1126/science.aac4834.

Hanczyc, M. W., et al. (Feb., 2007). Mineral surface directed membrane assembly. *Origins of Life and Evol. of Biosph.* 37 (1): 67–82.

Holden, D. T., et al. (3 Oct., 2022). Aqueous microdroplets enable abiotic synthesis and chain extension of unique peptide isomers from free amino acids. *PNAS USA* 119 (42): e2212642119. doi: 10.1073/pnas.2212642119.

Huang, W. & Ferris, J. P. (16 June, 2006). One-step, regioselective synthesis of up to 50-mers of RNA oligomers by montmorillonite catalysis. *Journ. Amer. Chem. Soc.* 128 (27): 8914–9. https://doi.org/10.1021/ja061782k.

Irvine, W. M. (Oct., 1998). Extraterrestrial organic matter: a review. *Orig. Life Evol. Biosph.* 28: 365.

Kanavarioti, A., Monnard, P.-A., & Deamer, D. W. (5 July, 2004). Eutectic phases in ice facilitate non-enzymatic nucleic acid synthesis. *Astrobiol.* 1 (3): 271–81. http://doi.org/10.1089/15311070152757465.

Luisi, P. L. (2016). *The Emergence of Life: From Chemical Origins to Synthetic Biology*, 2nd ed. Cambridge Univ. Press, Cambridge, UK.

Miller, S. L. (15 May, 1953). A production of amino acids under possible primitive Earth conditions. *Science* 117 (3046): 528–29. doi: 10.1126/science.117.3046.528.

Miller, S. L. & Urey, H. C. (1959). Organic compound synthesis on the primitive Earth. *Science* 130 (3370): 245. doi: 10.1126/science.130.3370.245.

Mizuuchi, R., Furubayashi, T., & Ichihashi, N. (2022). Evolutionary transition from a single RNA replicator to a multiple replicator network. *Nature Communications* 13 (1460). https://doi.org/10.1038/s41467-022-29113-x. www.nature.com/naturecommunications.

Nam, I., Nam, H. G., & Zare, R. N. (2 Jan., 2018). Abiotic synthesis of purine and pyrimidine ribonucleosides in aqueous microdroplets. *PNAS USA* 115 (1): 36–40. doi: 10.1073/pnas.1718559115. PMID 29255025; PMCID: PMC5776833. Published online 18 Dec., 2017.

Neveu, M., et al. (April, 2013). The "strong" RNA world hypothesis: Fifty years old. *Astrobiol.* 13 (4): 391–403. Bibcode: 2013AsBio..13..391N. doi: 10.1089/ast.2012.0868. PMID 23551238.

Oba, Y., Takano, Y., Furukawa, Y., et al. (26 April, 2022). Identifying the wide diversity of extraterrestrial purine and pyrimidine nucleobases in carbonaceous meteorites. *Nature Communications* 13 (2008). https://doi.org/10.1038/s41467-022-29612-x.

Oba, Y., et al. (27 Sept., 2019). Nucleobase synthesis in interstellar ices. *Nature Communications* 10, Article number 4413. doi: 10.1038/s41467-019-12404-1.

Ohishi, M. (1 Feb., 2008). Molecular spectral line surveys and the organic molecules in the interstellar molecular clouds. *Proc. of the International Astronomical Union* 4, Symposium S251: Organic Matter in Space, pp. 17–26. doi: https://doi.org/10.1017/S174392130802108X.

Oró, J. (29 April, 1961). Comets and the formation of biochemical compounds on the primitive Earth. *Nature* (190): 389–90.

Oró, J. & Kimball, A. P. (Aug., 1961). Synthesis of purines under possible primitive earth conditions. I. Adenine from hydrogen cyanide. *Archives of Biochem. and Biophys.* 94 (2): 217–27. doi:10.1016/0003-9861(61)90033-9. PMID 13731263.

Pearce, B. K. D., Molaverdikhani, K., Pudritz, R. E., et al. (2022). Towards RNA life on Early Earth: From atmospheric HCN to biomolecule production in warm little ponds. arXiv:2201.00829 [astro-ph.EP], (or arXiv:2201.00829v2 [astro-ph.EP] for this version). https://doi.org/10.48550/arXiv.2201.00829.

Piani, L., et al. (28 Aug., 2020). Earth's water may have been inherited from material similar to enstatite chondrite meteorites. *Science* 369 (6507): 1110–3. doi: 10.1126/science.aba1948.

Powner, M. W., Gerland, B., & Sutherland, J. D. (2009). Synthesis of activated pyrimidine ribonucleotides in prebiotically plausible conditions. *Nature* 459: 239–42.

Rajamani, S., et al. (2008). Lipid-assisted synthesis of RNA-like polymers from mononucleotides. *Origins of Life and Evol. of Biosph.* 38: 57–74.

Ranjan, S., et al. (Aug., 2018). Sulfidic anion concentrations on early Earth for surficial origins-of-life chemistry. *Astrobiol.* 18 (8): 1023–40. http://doi.org/10.1089/ast.2017.1770.

Rodriguez, M., et al. (2015). Formation of oligopeptides in high yield under simple programmable conditions. *Nature Comms.* 6, Article number: 8385. https://doi.org/10.1038/ncomms9385.

Schopf, J. W., et al. (2 Jan., 2018). SIMS analyses of the oldest known assemblage of microfossils document their taxon-correlated carbon isotope compositions. *PNAS USA* 115 (1): 53–8. https://doi.org/10.1073/pnas.1718063115.

Schopf, J. W., et al. (2002). Laser–Raman imagery of Earth's earliest fossils. *Nature* 416 (6876): 73–6. doi: 10.1038/416073a. PMID 11882894.

Seaborg, D. (2022). *How Life Increases Biodiversity: An Autocatalytic Hypothesis*. CRC Press/Taylor & Francis. Boca Raton, FL; London, UK; New York, NY.

Stribling, R. & Miller, S. L. (Sept., 1987). Energy yields for hydrogen cyanide and formaldehyde syntheses: The HCN and amino acid concentrations in the primitive ocean. *Origins Life and Evol. of the Biosphere* 17: 261–73. https://doi.org/10.1007/BF02386466.

Szostak, J. W., Bartel, D. P., & Luisi, P. L. (2001). Synthesizing life. *Nature* 409: 387–90. (One also can find this article on the Szostak Lab publication page.)

13 The Solar System, Sun, Jupiter, Earth's Moon, and Nonbiological Earth All Aid Life, Suggesting the Autocatalytic Biodiversity Hypothesis Does Not Account for All of Earth's Biodiversity

In addition to life, nonbiological forces often promote and maintain biology, biodiversity, and evolution. This means the Autocatalytic Biodiversity Hypothesis (ABH) does not provide a complete explanation of Earth's biodiversity. It is honest and good science to recognize the imperfections of one's hypotheses and to criticize one's own ideas. Hence, this chapter discusses how these nonbiological forces, including the solar system, sun, moon, and nonliving Earth, promote and protect life.

13.1 THE SUN SUPPORTS LIFE

The sun is the major nonbiological promoter of life and biodiversity in the solar system. It is the main source of energy, heat, and light for Earth and its life. Each day, over a million terrawatts, over 7,000 times the global energy use by humans at the time this was written, strike Earth from the sun. This powers photosynthesis, keeps life warm, and drives wind and weather patterns. Skin cells use sunlight to make essential vitamins in humans and animals. The Earth revolves around the sun at a distance that is within a narrow zone that allows life. This has been the case since life started about 4.54 bya. If Earth were a little closer to the sun, it would be too hot for life; if it were a little farther from the sun, it would be too cold for life. The habitable zone is narrow.

Airapetian et al. (2016) presented evidence that Coronal Mass Ejections (CMEs) from the sun initiated reactions that produced nitrous oxide (N_2O), a powerful greenhouse gas that would have greatly warmed the Earth when it was young and the sun's heat output was low. This would have provided heat when life needed it. However, the N_2O would have been in the upper atmosphere, and most greenhouse warming occurs lower in the atmosphere, so a mechanism was needed to lower the altitude of the N_2O to the point where it would have been helpful to life. There is also the problem that CMEs are highly destructive to life and could strip away the atmosphere. Airapetian et al.'s claim is possibly valid. But the major mechanism that warmed the planet when the sun produced less heat was the production of the powerful greenhouse gas, methane, by methanogens, a biological mechanism.

13.2 THE EARTH IS PLACED WELL FOR LIFE

The nonliving Earth also is a system that promotes life, evolution, and diversity. The Earth's distance from the sun, size, tilt, speed of rotation on its axis, and speed of revolution around the sun are all highly favorable to life on Earth. The Earth's interior is very hot because of the remaining heat of formation from when the planet came together from gravitational forces and formed; frictional heating, caused by denser core material sinking to the center of the Earth; and heat from radioactive decay. This heat is crucial to life. It powers volcanoes and deep-sea vents. Hot, molten rock under the Earth's surface and movement of Earth's tectonic plates cause an average of about 30 volcanic eruptions per day worldwide. Volcanoes and deep-sea vents contribute a great deal to life and biodiversity.

13.3 VOLCANOES AND EARTH'S INTERNAL HEAT AID LIFE

Volcanoes were crucial in creating the early atmosphere, emitting water vapor, nitrogen, hydrochloric acid, methane, ammonia, sulfur gases, carbon monoxide, and copious quantities of CO_2. These molecules were important in the chemical evolution of life, and all of them except hydrochloric acid and carbon monoxide are needed by life. The greenhouse gases emitted by volcanoes in this early atmosphere were another source of heat that helped early life to survive when the sun produced much less heat. Again, life (in the form of methanogens producing methane) was the primary source of the heat needed by life at this early time.

Volcanoes provided and still provide minerals and other nutrients to the Earth's surface and soil, and these helped build the soil, although the soil was mainly built by life. Volcanic dust, ash, and rocks decompose into soils, giving them nutrients and an exceptional ability to hold nutrients and water, making them very fertile. Oldoinyolengai is a volcano in Tanzania that has ejected carbonate minerals for millions of years; these are nutrients that fall as part of its ash and feed a rich grassland that supports a diverse ecosystem of herbivores, carnivores,

DOI: 10.1201/9781003246640-13

and other species. This includes zebras, lions, millions of wildebeest, and so on. Galapagos tortoises (*Chelonoidis niger*) supplement their diet with volcanic mineral rocks. Volcanic ash fertilizes the sea, causing phytoplankton blooms, supporting entire oceanic food webs. It also adds nutrients to the seafloor that are used by the life there. The protoplanetary disc and colliding meteors that formed Earth also provided many minerals essential to life.

Volcanoes created most of today's land surface on Earth, both the continents and islands. Volcanoes also build seafloor. Igneous (volcanic) rocks form over 80% of the Earth's surface, both above and below the sea. Kīlauea's eruption in Hawaii in 2018 added almost 300 hectares (741 acres) of seafloor that was colonized by microbes within 100 days. Surtsey is a volcanic island that emerged in 1963 along the midocean ridge near Iceland.

Volcanoes created and still create diverse environments that support life, such as mountains, islands, hot springs, and deep-sea vents. The isolation of some of these habitats often fosters speciation and high numbers of species found only in those specific locations. The Hawaiian Islands were created by volcanoes, and are so far from other land that 90% of their species are found only in Hawaii. They had very diverse ecosystems before human impacts. On these islands, lava flows isolate small patches of forest into tiny ecosystems where insects and perhaps other invertebrates speciate. Volcanic mountains can isolate populations of the same species on either side of them, fostering speciation. Volcanoes create hollow lava tubes that are cave habitats for many species. Often these species live off roots dangling in the cave from trees that grow above the lava tube. Ecosystems of prokaryotes thrive in volcanic hot springs.

The Galapagos Islands and Hawaiian Islands are volcanic islands with unique ecosystems with many species only found on each of those islands because of their isolation. Additionally, populations in various species on different islands within island groups have speciated in both the Hawaiian and Galapagos Islands. For example, the bird genus *Geospiza* is found only on the Galapagos Islands. Populations on different islands have speciated such that there are different species on different islands. (Although called Darwin's finches because of his studies of them, they are actually in the tanager family, Thraupidae.) There are several other cases of speciation of animals and plants between islands on Hawaii and on other volcanic groups of islands.

Volcanism helped create Earth's atmosphere. The atmosphere holds heat and helps generate weather and lightning. It also acts as a protective shield, burning up many meteors, preventing them from crashing into the Earth and harming life. Some of it was created by volcanoes. Nitrogen, 80% of the atmosphere, helps regulate temperature while conserving carbon, protects a little against incoming radiation, and isecessaryy for life. It was stored in the Earth's mantle when the Earth was formed, and escapes to the atmosphere as nitrogen gas by volcanism, primarily at the convergence of the tectonic plates (Mikhail and Sverjensky, 2014). Carbon dioxide is needed for photosynthesis and helped keep the planet warm at life-favorable temperatures before humans added excessive levels of greenhouse gases. Volcanoes exude copious quantities of CO_2. Of course, much of the atmosphere was created by life. Almost all of the O_2 and the protective ozone layer were made by life.

Before the seas existed, Earth's water was trapped underground and not available to life aboveground. Volcanoes supplied the heat that formed water vapor that escaped into the atmosphere and eventually condensed and formed a good deal of Earth's aboveground water—lakes, rivers, and the oceans, which were originally fresh water. Without volcanoes, most water would be trapped deep below the ground in the crust and mantle. Life aboveground would mostly not be possible, and there would no river, lake, or ocean ecosystems. Together, space debris and volcanoes delivered the water that made life aboveground possible to Earth's surface. Space debris brought it to Earth's interior, and volcanoes delivered it to the surface.

Volcanism helped create Earth's atmosphere. The atmosphere holds heat, helps generate weather and lightning, and shields life from harmful radiation, mainly the sun's ultraviolet light. It also acts as a protective shield, burning up many meteors, preventing them from crashing into the Earth and harming life. Some of it was created by volcanoes. Nitrogen, 80% of the atmosphere, helps regulate temperature while conserving carbon, protects a little against incoming radiation, and is necessary for life. It was stored underground in the Earth's mantle when the Earth was formed, and escaped to the atmosphere as nitrogen gas by volcanism, primarily at the convergence of the tectonic plates (Mikhail and Sverjensky, 2014). Carbon dioxide is needed for photosynthesis and helped keep the planet warm enough to be favorable to life before humans added excessive levels of it and other greenhouse gases to the air. Volcanoes exude copious quantities of CO_2. Of course, much of the atmosphere was created by life. Almost all of the O_2 and the protective ozone layer were made by life.

Minerals from beneath the Earth's crust below the seafloor help fuel phytoplankton blooms and support oceanic food webs. These come to the seafloor and seawater nonbiologically by seafloor spreading, deep-sea vents, and deep-sea volcanoes. Many of life's essential trace elements—zinc, copper, cobalt, manganese, and selenium—are provided by plate tectonics, as erosion of the Earth's crust supplies nutrients to the oceans. All of these processes are powered by Earth's internal heat. Deep-sea vents supply nutrients and heat that support rich, diverse ecosystems that are not powered by the sun. The vents supply great quantities of nutrients to life. Cold seawater enters magma chambers through cracks in the vents. Here it is heated to 750°F (about 399°C). The dense, salty water dissolves solid rock under high pressures and temperatures in the vents, where complex chemical reactions form sulfur and iron compounds needed by life. Seawater full of minerals from inside Earth exits the vents. These feed a diverse community of animals that live on the hydrothermal vents. They are also carried away by currents. They circulate through the sea for thousands of years, eventually rising to the surface. The mineral nutrients are used by communities of organisms from deep in the ocean to near its surface as they move through the sea. If it were not for seafloor spreading, erosion of Earth's crust, deep-sea vents, and deep-sea volcanoes, the sea would have the chemistry of a soda lake, like the Great Salt Lake, with much less of a variety of minerals, supporting a much lower biodiversity. The system is recycled, since every few million years, the entire volume of the ocean circulates through this system. Deep-sea vents, deep-sea volcanoes, and seafloor spreading are a major influence on the chemistry of the sea, and provide a great deal of the nutrients for the phytoplankton blooms that support major ocean food webs.

Finally, undersea volcanoes might have exuded CO_2 and perhaps methane, and warmed the Earth, ending the Huronian glaciation and Snowball Earth about 2.1 bya. This is not known for sure.

13.4 UPWELLING IS CAUSED BY A COMBINATION OF NONBIOLOGICAL AND BIOLOGICAL FACTORS AND BRINGS THE NUTRIENTS THAT SUPPORT MANY OF THE OCEAN'S FOOD WEBS TO THE PHOTIC ZONE

Upwelling brings the nutrients supplied by seafloor spreading, deep-sea vents, and deep-sea volcanoes to the surface of the ocean by delivering dense, cooler, usually nutrient-rich water from deeper in the sea to the surface. Warmer, usually nutrient-poor water at the surface is replaced by the nutrient-rich upwelled water. This is caused in large part by wind, water, and the Earth's rotation. It is caused partly by winds blowing across the sea surface, causing a wind-water interaction. Also, the right ambient air and water temperatures are essential for it to happen. This is aided by life's sequestration of carbon, keeping temperatures low enough for upwelling to occur. So upwelling is caused by a combination of nonbiological and biological factors.

Upwelling brings nutrients to where sunlight and CO_2 are available, so phytoplankton can use the nutrients as they photosynthesize. This causes phytoplankton blooms. Billions of plankton bloom and can double in one day, covering hundreds to thousands of square kilometers (1,000 square kilometers is about 386 square miles) of sea surface, and weighing over 800,000 tons. Phytoplankton can cover up to a fifth of the global sea surface, producing about 45 billion tons of biomass per year. Phytoplankton blooms support most sea biomass. A simplified version of the plankton-based food web involves phytoplankton, zooplankton, predatory zooplankton, krill, filter feeders, other invertebrates, small predatory fish, large predatory fish, marine birds, and marine mammals. Upwelling supports this entire food web, a tremendous amount of diversity and biomass.

Blooms can fade after only 72 hours. Many plankton sink, carrying nutrient with them and sequestering carbon. Some minerals are quickly recycled, and some are recycled slowly, again emerging from hydrothermal vents millions of years later. Some nutrients are transferred to land when they travel up food webs from phytoplankton to sea birds, seals, and other organisms that defecate or die on land, and this helps terrestrial ecosystems. Thus, terrestrial ecosystems receive some nutrients from the sea via biological processes.

13.5 DEEP-SEA ERUPTIONS CREATE HABITATS FOR DIVERSE FLOATING COMMUNITIES OF SPECIES

In 2012, the Havre Seamount, in the deep sea in the South Pacific Ocean, erupted in what is estimated to be the largest underwater volcanic eruption ever recorded, and created a pumice raft, a floating platform of millions of different pieces of buoyant volcanic rock. Each rock in a pumice raft attracts and harbors a small community of marine organisms, including algae, barnacles, corals, and others. Some rocks have huge numbers of organisms. They pick these up from ecosystems such as coral reefs. Undersea eruptions producing pumice rafts occur about once every five years. This shows that nonbiological forces such as deep-sea eruptions can create habitats for diverse communities of species. A pumice raft from 2019 from a deep-sea volcano near Tonga traveled far and spread out over more than 1,300 kilometers (about 808 miles) of Australia's east coast from northern Queensland to northern New South Wales. However, pumice rafts can also be destructive by smothering coastlines, thus hurting habitat and decreasing diversity, although this is not a permanent effect.

13.6 PLATE TECTONICS AND CONTINENTAL DRIFT INCREASED DIVERSITY AND AIDED THE EVOLUTION OF UPRIGHT WALKING IN HUMAN ANCESTORS

Earth's internal heat also runs some of its geophysical systems, such as plate tectonics and continental drift. When the continents moved apart, isolation of populations caused a great deal of speciation and diversification, primarily on land, but also in the sea. Plate tectonics put the continents in their favorable positions for currents, temperature regulation, and hence life, as will be discussed in this chapter. And when North America collided with South America, ocean currents changed, making Africa drier and with less rainfall in the region where human ancestors lived. Forests thus shrank and grasslands increased there. Less forest habitat caused some primates to leave the trees and populate the savannah, where they evolved from primates that walked on their hands to human ancestors that walked upright on two feet. This probably would not have happened if our ancestors remained in trees; it likely requires a grassland habitat. This freed the hands and caused natural selection to favor the large brain and high intelligence that *Homo sapiens* evolved.

13.7 THE MAGNETIC FIELD PROTECTS AND IS USED BY LIFE ON EARTH

Earth's internal heat also helps generate the Earth's magnetic field. The magnetic field, generated by the rotation of Earth's molten core, extends far beyond the Earth. The magnetic field blocks the solar wind. The sun sends out a stream of charged particles called the solar wind, which is capable of stripping Earth of its atmosphere. Mars has no magnetic field, so the solar wind stripped much of its atmosphere away, leaving it with a very thin atmosphere. The solar wind would also do severe damage to life if it hit Earth. The magnetic field of our planet protects the planet from the solar wind by deflecting much of it. This has allowed Earth to keep its atmosphere, and hence its high diversity and abundance of life. The magnetic field also protects life from cosmic rays, which constantly come to Earth, and could cause nausea, damage to DNA, cancer, and even death to humans and animals, if Earth were not protected from them. The atmosphere shields Earth from them as well, although much less so than does the magnetic field.

Earth's magnetic field also captures the charged particles from the sun and holds them around Earth in what are called the Van Allen radiation belts, further protecting Earth's atmosphere and biology. The two donut-shaped Van Allen belts, a collection of charged particles held in place by Earth's magnetic field, greatly reduce the impact of the solar wind on the Earth and its atmosphere. Our planet has another barrier protecting it. Baker et al. (2014) found that the plasmasphere, a giant cloud of charged particles filling the outermost region of Earth's atmosphere, scatters electrons, creating an almost impenetrable barrier through which the most energetic Van Allen belt electrons cannot migrate.

Occasionally the sun emits Coronal Mass Ejections (CMEs) that would strip away the atmosphere, and be very destructive to higher life, if Earth lacked protection. Coronal Mass Ejections tear the magnetosphere open, allowing vast numbers of charged particles to breach the magnetic field's first layer. But Earth has a second layer and line of defense, the inner magnetic field, which steers the radiation to the poles. The solar particles get trapped and interact with the atmosphere, causing auroras. The result is that the Earth and life are protected from the CMEs. The magnetic field has protected Earth for billions of years, allowing life to thrive, evolve, and diversify.

Bacteria, termites, ants, bees, molluscs, fish, salamanders, toads, box turtles, sea turtles, some other reptiles, birds, mice, mole rats, bats, whales, and other animals use the Earth's magnetic field. They use it for various reasons, including navigation and migration. The Red Fox (*Vulpes vulpes*) appears to use magnetoreception when predating small rodents.

13.8 BIOLOGICAL AND NONBIOLOGICAL FACTORS COMBINE TO CAUSE THE OCEAN'S CIRCULATION SYSTEM, WHICH GREATLY AIDS LIFE

The winds of the polar jet stream and water of the circumpolar current together form an essentially impenetrable barrier around Antarctica, fully circumnavigating it, and isolating it from the rest of Earth, depriving it of warmth. That is why it is much colder than expected, even taking into account its polar location. During winter, the polar jet stream can reach speeds of over 100 mph. Antarctica thus gained about 5.5 million square miles of ice each winter before humans started warming the planet. A constant gale blows across the Weddell Sea, often freezing it. The Weddell Sea is the Southern Ocean's southward extension with a coastline on East and West Antarctica. When saltwater freezes, tiny crystals grow and knit themselves together. When the crystals bond, they expel salt into the sea. Salt forms brine that drips down long, narrow tubes in the ice as the brine forms. Brine is denser than seawater, so it sinks. The water is also very cold, and this also causes it to sink, since cold water is denser than warm water. The area where this occurs increases as winter progresses and ice increases. Each year, over 200 billion tons of ice form in the Weddell Sea alone, releasing tens of trillions of gallons of brine into the sea.

There is a vast chasm in the sea's bottom two miles deep off Antarctica's coast. *One trillion gallons of salty brine, the equivalent of 500 Niagara Falls, plunge down through the Weddell Sea each hour.* It spreads out to the edge of Antarctica's continental shelf, then falls into the chasm as a vast submarine waterfall, slowly going two miles down. This sinking and outflow from Antarctica drives the salty water toward the equator. All the water in the bottom of every ocean on Earth has its start within six miles of Antarctica. Antarctica produces the coldest, densest water on the planet. This water thus sinks and drives the entire circulation system of the Earth. The frigid, salty water becomes part of a worldwide circulation system, stirring and cooling all the world's seas, driving the ocean's currents. The water from Antarctica moves as currents on a journey that could take a thousand years to complete. As the water from Antarctica migrates to the equator, the cold bottom current mixes with less salty, warmer water, and slowly rises, then joins other currents. It eventually returns south, where it again cools, goes back to Antarctica, freezes, and releases salt, completing the journey. So Antarctica drives the sea's currents, setting up a conveyor belt. The dense, cold Antarctic waters form the basis of global sea circulation. Variations in temperature and salinity cause sinking and rising of water in different areas to drive the conveyer belt. The ocean has a "circulatory system" of currents that move entirely through all of the oceans.

The Arctic aids this oceanic circulatory system. In the Arctic, frigid winter winds freeze the sea. Salt is exuded from Greenland ice into the sea by the same mechanism as in Antarctica. So cold, dense North Atlantic water sinks and drives the current there, moving south. All over the world, slow-moving currents keep sea temperatures within narrow, intermediate ranges, keeping them from temperature extremes. *They regulate average temperatures of seas to within 1°F (about 0.55°C)! This thermal stability aids life all across the globe.*

Oceans and the atmosphere interact with and affect each other, being intimately interconnected. Sea currents affect the atmosphere. The sinking water from Greenland drives the current system that moves south and ultimately drives the Gulf Stream. The Gulf Stream transports warm water from the Gulf of Mexico north to the eastern United States and Europe. It follows the eastern coastlines of the United

States and Newfoundland before crossing the Atlantic Ocean as the North Atlantic Current, and then heads north along Europe. Warm air blows from the sea across the Gulf Stream and warms the eastern United States. Warm air blows from the ocean across the North Atlantic Current as it branches into two currents and warms Europe. Rome is approximately at Chicago's latitude, and would be cold like Chicago and not have its mild Mediterranean climate if it were not for a branch from the Gulf Stream warming Europe. The Kuroshio Current transports heat from southern areas of China north to Japan, also in conjunction with wind blowing over it. Ocean currents distribute heat around the Earth, making the temperature more uniform from one place to another—conditions favorable to life. Without these currents, life would be less diverse, because there would be many more areas with extreme temperatures. Earth's climate is profoundly affected by sea currents, in a way that greatly aids life. The Carolinas are warm partly because of the Gulf Stream, which exists because it takes root in the Arctic. And the transfer of heat and energy by currents allows the African and Asian monsoon system to occur through this transfer's effects on sea surface temperature, the hydrological cycle, atmospheric circulation, and variation in the intertropical convergence zone. The ocean circulation affects wind currents that help deliver the rain of the monsoons. Reliable monsoons are critical to vegetation and the animals it supports in Africa and Asia. Loss of the currents that support them would cause drought and reduced diversity. The constant, dependable circulation of the sea around the Earth—its currents—and the stable temperatures it produces have made the planet much more hospitable for life. Sea currents also aid the movement of countless taxa, including sea turtles, fish and their larvae, invertebrate larvae, and several others. The invertebrate larvae include larvae of coral, crustaceans, echinoderms, and many others.

A large pool of cold water near the seafloor in the Bering Sea that is a by-product of sea ice and that persists through summer is important to the Bering Sea ecosystem, being a refuge where Arctic cod (*Arctogadus glacialis*), hide from predators such as Pacific cod (*Gadus macrocephalus*), which are less cold tolerant. The Arctic cod eat and grow there, and are eaten by polar bears and seals when they leave the refuge. Thus, the pool of cold water allows higher biodiversity to exist.

The currents are favorable to life partly becausee life has sequestered carbon and thus regulated the temperature, making the climate favorable to life-friendly currents. If life did not sequester great amounts of carbon, the Antarctic and Arctic ice caps would be smaller. This would impact Earth's climate. The global average temperature would be higher, so the edges of the Arctic and Antarctic ice caps would melt, sending freshwater into the sea. Being warmer and with a lower salt concentration, the water from both ice caps would be less dense, and so would not sink. It is the sinking of this water that provides the power to drive the currents. Without the sinking of the water, the currents would become weaker or die out completely, and the circulation system of ocean currents would become weaker or stop altogether. As a result, differences in temperature between different areas in the sea would be greater. This would be less favorable for life, and diversity would be lower. Also, currents and wind would not warm Europe, the eastern United States, and southern areas of China north to Japan. This would also make conditions worse for life and thus cause lower diversity. Thus, biology plays a big role in keeping global temperatures low enough for the ocean circulation system to be strong and favor life.

The positions of the continents are also important in ensuring there is a strong global system of currents. However, the continents were in very different positions at various times in the past. They were once more clumped together. Gondwanaland was a supercontinent comprised of South America, Africa, Arabia, India, Antarctica, and Australia that existed from about 550 mya, and began to break up about 180 mya. Its remnants make up about two-thirds of today's continental area. This is not the only time many continents were together as one giant continent. And it is predicted that approximately 200–250 million years from now, the continents will come together again. When the continents are together, the seas are large and continuous. Currents would be different in these situations. They would likely not dampen temperature differences in the sea or aid life as effectively. So currents aid life at the present time, but this might not have been the case at times in the past when the continents were in different positions than they are now. It also might not be the case in the future after the continents move to new, different positions. At present, life and nonbiological factors work together to cause the ocean's currents to cause biodiversity to be high, but this was not always, and will not always be, the case.

13.9 NONBIOLOGICAL TRANSPORT OF NUTRIENTS BETWEEN ECOSYSTEMS

Dust blown from the Gobi Desert provides nutrients, mainly phosphorus, to Sierra Nevada ecosystems in California. This accounts for the ability of otherwise phosphorus-poor soil to support giant sequoia trees. Some 18–45% of the Sierra Nevada's dust comes from Asia, and a smaller amount comes from California's Central Valley (Aciego, 2017). Erosion would deplete the nutrients of many Sierra Nevada forests if not for these inputs. Winds blow fossil diatoms in the Sahara Desert across the Atlantic Ocean to the Amazon, Caribbean Islands, and Everglades, nourishing these ecosystems. Since the diatoms were once living, this involves biology along with the nonbiological process. The Great Plains of North America are rich in minerals supplied by erosion from the Rocky Mountains, allowing the growth of grasses and the diverse prairie ecosystems there. The Ganges River washes iron and other nutrients from the Himalayas to the Ganges delta in Bangladesh, greatly enhancing diversity in that region. Such cryptic transport of minerals creates fertile soil that allows plants and their ecosystems to thrive, and helps support food webs all over the planet. The Earth's systems are profoundly interconnected. Organisms are not isolated from processes that occur far away in time and space. The land, sea, air, and life are interconnected; constantly feedback on one another; coevolved; and work to increase diversity and aid life. There are many other cases of geological and other nonliving factors working with life with a resultant increase in diversity.

13.10 LIGHTNING HELPS LIFE

Earth's lightning is constantly visible from space. There are 40 lightning strikes per second, and three million per day, on Earth. The sun's heat evaporates water. The water vapor thus formed creates on the order of 40,000 thunder clouds daily, planetwide. Rising columns of moist air driven by heat from the sun generate powerful up-drafts that cause water vapor to rise. In clouds high in the atmosphere, where it is cold, ice particles form. As ice and water droplets collide at high speeds, vast charges of static electricity build up. An average thunder cloud has enough energy to power a city the size of Denver for ten hours. In time, the charge builds up to the point where air molecules are torn apart, and a lightning bolt five times the temperature of the sun's surface discharges. The vast majority of the atmosphere's nitrogen exists as molecular nitrogen, N_2, not useable by life. Lightning dissociates N_2 into two nitrogen atoms. Some of the nitrogen atoms so formed combine with molecules of oxygen (O_2) to form nitrate (NO_3-), making 13,000 tons of it per day, and between 5 and 10 billion kilograms of it per year. Nitrate is useable by life. It is carried to the ground in rain, which delivers it to the soil, where plants and microbes can use it. This also removes it from the air, so this greenhouse gas does not heat the Earth to temperatures unfavorable to life. High atmospheric oxygen created by life is needed for this process. Evapotranspiration by trees adds a large quantity of the water vapor needed to create the lightning necessary for converting nitrogen to a form useable by life. So some of the nitrogen useable by life is produced by physical forces and life working together. Of course, most of the nitrogen useable by life is produced by nitrogen-fixing bacteria in the soil.

Lightning is also helpful to life because it causes fires. Fires fertilize the soil, increase the growth of vegetation, and have many other positive effects on life and diversity (see Chapter 3).

13.11 THE MOON AIDS LIFE AND MAY HAVE CAUSED MAJOR EVOLUTIONARY BREAKTHROUGHS

The moon was created about 4.5 bya, about 30–50 million years after the formation of the Solar System. When the Earth was very young, before the moon existed, the length of a day on Earth—the time for one full rotation of the planet—was about six hours. The moon's gravity generates tides by pulling hardest on the part of Earth closest to it. The moon revolves around the Earth slower than the Earth rotates on its axis. Hence, the moon slows the Earth's rotation down. The sun has a similar effect, though smaller, because it is so far away. The two together add 1.7 milliseconds to the length of a day each century. Without the moon, the length of each day on Earth would be much shorter, with about six hours of light alternating with a like number of hours of dark. Recall from Chapter 2 that the longer days created by the moon's pull allowed more photosynthesis and a greater production of oxygen. Apart from this, we do not know if six-hour days in the absence of the moon would result in a less diverse planet, or whether life would have adapted to six-hour day-night cycles, evolving to be as diverse as the biosphere became with its longer days.

The moon and sun cause the tides of the ocean. This has a profound effect on life. Life has adapted to the tides, and many species synchronize their breeding with the tides. These include the California grunion (*Leuresthes tenuis*), a fish that spawns on beaches in southern California and Baja California from two to six nights after the new and full moon starting soon after high tide. Corals cannot move, and synchronize the annual release of their sperm and eggs with the phase of the moon and water temperature. Entire colonies of coral release their sperm and eggs on the same night. This is adaptive because it maximizes the probability that their sperm and eggs will find each other. In addition, the intertidal zone is a diverse habitat with many species adapted to the rising and falling of the tides.

The moon affects the activities of many species. The moon's light helps nocturnal animals see. It guides moths and helps nocturnal predators hunt. Many of the 450 species of migrating birds use the moon and stars to navigate, and sea turtles locate nesting beaches by this means. (Birds and sea turtles also use the Earth's magnetic field to navigate.) Dung beetles (superfamily scarabaeoidea) steer straight at night as they roll their dung balls by using polarized moonlight. Some lunar light scatters off gas molecules in the atmosphere and becomes polarized. They also use the Milky Way. In the morning and afternoon, they use the sun, and at midday, they use the wind (Dacke et al., 2021). They dance at the start of each trip (ibid.). Wildebeest stay in safer areas during new moons because lions are less visible at those times. Many other animals and even plants use the moon to adaptively guide their behavior. Light pollution from humans is adversely affecting countless animal species that use moonlight. For example, it makes it harder for sea turtles to find their nesting beaches.

The moon was much closer to the Earth in the geologic past. It therefore had a larger gravitational effect, making the high tides much higher and the low tides much lower. This would probably have stranded marine fungi higher on land and potentially for longer periods than today, allowing longer times for selection for terrestrial adaptations, promoting the evolution of land fungi. The moon and tides would have likely done the same thing to the algae that evolved symbiotically with fungi on land, and eventually evolved into land plants. Sea invertebrates might have been stranded higher and longer on land than if they had been so stranded today as well, although this would have been later in Earth's history, so the high tides would have been lower than at the time fungi and algae were stranded. Over many generations, selection could have favored adaptations in these invertebrates for living on land as they were washed and stranded on the shore by the tide for long time periods because of the higher high tides than today's. The same principle might apply to the evolution of amphibians from fish. The preamphibian fish might have had to survive and breathe out of water for a longer time because it was washed higher on the shore due to the higher tides than occur today. This could have selected for the evolution of lungs. Some would die, some would adapt to life on land. Of course, it is possible the precursors to land invertebrates and amphibians intentionally walked onto land for short periods, instead of being washed up. Or it could have been a combination of the two. At any rate, the moon very likely played a role in the evolution of terrestrial fungi and plants, and possibly invertebrates and vertebrates.

When the solar system was young, there were many more meteors and comets that could have potentially hit the Earth, causing the extinction of a large number of species. The closer proximity of the moon to the Earth at that time means that the moon may have protected the Earth from strikes from these bodies somewhat at that time, because its gravity could have caused them to crash into the moon or take a course that missed the Earth. It is less important in that role now that it is farther from the Earth and there are fewer meteors and comets.

The axial tilt of the Earth results from the amount of water in the seas, and influences of the sun, other planets, and *especially the moon*. The tilt oscillates between 22.1 and 24.5 degrees, and is currently at a mean of about 23.4 degrees. This is stable and results in seasons in temperate and polar zones, which are favorable to diversity. The moon stabilizes the precession (wobble about an axis) of the Earth. Without the moon, the Earth's precession would oscillate tremendously, wildly, causing extreme climatic fluctuations. The day-night cycle and the seasons would also be more erratic than they are now. This is another means by which the moon greatly aids life, causing increased biodiversity. This stabilization will continue for another two billion years. This is fortuitous, since the moon was formed when a large planetoid collided with the Earth, throwing out a great amount of dust that coalesced into the moon. This is a random event not ordained by physical laws. The many benefits to life that the moon provides are a result of chance, of good luck.

The Earth undergoes precession in a period of about 26,000 years, due mainly to the gravitational pull of the sun and moon. As a result, how far north the monsoons in Africa go changes over a period of 20,000 to 40,000 years. Because of this, the monsoon belt on this continent was about 700 kilometers (about 435 miles) farther north than today about 125,000 years ago. Thus, the Sahara Desert experienced monsoons and abundant rainfall. It was a green ecosystem of trees and grasses. It was passable by humans, and this time corresponds closely to one of the periods when fossil evidence indicates that *Homo sapiens* migrated out of Africa into Europe and Asia, for modern human fossils found outside of Africa have been dated to about 90,000 to 120,000 years ago in the eastern part of the Mediterranean (Hershkovitz, 2018). (There were hominid migrations from Africa to Europe and Asia at other times as well.) Of course, the period of relatively high monsoons did not abruptly end 125,000 years ago, so was still occurring, even if perhaps less vigorously, into the period of migration, and so the times of the conditions that allowed migration overlap with the fossil dates. Hence, *the precession of the Earth may have played a major role in setting the time frame when one of the migrations of humans out of Africa, ultimately colonizing the rest of the world, happened.*

13.12 JUPITER AND SATURN MAY HAVE HELPED MAKE EARTH ABLE TO SUPPORT LIFE

Jupiter is massive; it is two and a half times the mass of all the other planets in the solar system combined. Some think it protects Earth from bombardment by gravitationally directing comets and meteorites away from Earth, but others think it draws comets toward the Earth, and others think that the effects balance each other. It receives about 200 times more asteroid and comet impacts than Earth. Some of the asteroids and comets that hit it would have hit the Earth. There is evidence that Jupiter formed early (Kruijer, et al., 2017) and near the sun (Raymond et al., 2016). Alternatively, it may have formed farther out, and then moved inward, close to the sun. Either way, it may once have been much closer to the sun than it is today. Thus, Jupiter could have removed much of the planet-forming disk of the early solar system, keeping the inner planets limited to four in number and small. This favored life on Earth because if Earth were too big, it could have been drawn too close to the sun to be capable of having life. And limiting the size and number of the other three inner planets prevented potential perturbations in Earth's orbit that could have been unfavorable to life. Jupiter would also have cleared meteors that could have struck Earth at this time. But if Jupiter had stayed close to the sun for too long, it would have removed so much debris that Earth would have been too small to maintain a magnetic field and hence retain its atmosphere and support life. Mars is small and without a magnetic field because Jupiter removed debris that would have accumulated to enlarge Mars, which is only 10% of the size of Earth. It is thought that Saturn moved inward to the region of the four inner planets and became locked in a motion resonance with Jupiter. Saturn pulled Jupiter outward, and they thus both went to their current positions. Saturn prevented Jupiter from scattering too much debris, leaving enough for Earth to grow to a size that can have a magnetic field and hence maintain an atmosphere, and thus sustain life. If this idea holds up, Jupiter and Saturn helped make life on Earth possible.

13.13 THE GALACTIC TIDE MAY HAVE AIDED BIOLOGY

Klacka (2010) presented evidence that the galactic tide could affect the Oort cloud. The galactic tide is a tidal and gravitational force due to the gravitational field of our Milky Way galaxy. The Oort cloud is an area far beyond Pluto that is composed of icy bodies such as comets. This means the galactic tide could perturb the orbits of comets and occasionally cause them to strike the Earth, affecting the frequency of comet strikes on Earth. This is important because comets brought some of the water to the Earth. Therefore, the galactic tide might have aided Earth's biology. The galactic tide might also cause occasional catastrophic impacts by comets on Earth. This research opens the possibility that Earth's biology is part of a larger web, affected not just by the Earth, sun, moon, and Jupiter, but the Milky Way galaxy itself.

13.14 THE POSSIBILITY OF LIFE ON OTHER PLANETS

The idea that the solar system is a life-generating system that also protects life would be supported if life were present now or in the past elsewhere in the solar system. It may have been present in the past or be present now on Mars, Venus, and some moons in our solar system. Although Mars has a thin atmosphere, it may have had prokaryotes in the past, and they could exist below its surface today. Nazari-Sharabian, et al. (2020) state the following in a review of the search for water on Mars. It is widely accepted that Mars had abundant water very early in its history. Long ago, it was periodically wet, with favorable conditions for microbial life on its surface. Today its surface is very dry and inhospitable to life. But NASA found an underground lake in 2016, and Italian scientists found a subglacial lake in 2018. It is still an open question whether Mars harbors life or did so in the past.

Greaves et al. (2020) and Bain et al. (2020) detected phosphine in the middle atmosphere of Venus at concentrations of from 5 to 20 parts per billion, which is thousands of times more than in Earth's atmosphere. This is difficult to explain without a biological source. Still, some researchers suggest that the gas could have resulted from unexplained atmospheric or geologic processes. Many planetary scientists have hypothesized that Venus has prokaryotic life. It once had an atmosphere where life as we know it could have thrived, and had oceans as recently as 700 mya. Today, a cloud layer 31 miles below the top of its atmosphere may reach temperatures as mild as 30°C (86°F), and has a pressure not far from that at ground level on Earth. The microbes may have originated on the surface when it had benign conditions, then ascended to the current life-favorable altitude when the planet's surface became very hot and dry. They would likely have had to survive extreme temperatures and to have evolved to survive in high acidity. But there are prokaryotes on Earth with both of those adaptations.

Jupiter's moon Europa has more liquid water than in all of Earth's seas, although the salty seas lie beneath about ten miles of ice. They are dark, but tidal forces from Jupiter provide heat, and perhaps they even have geothermal vents. The water is interacting with rock, a necessity for chemical evolution. Life could be at the vents (if they exist), below the seafloor, or clinging to the lower surface of the ice layer.

NASA's Cassini spacecraft photographed geysers of frozen water spewing from cracks in Saturn's moon, Enceladus. Scientists think reservoirs of liquid water lie beneath the frozen surface and are warmed by gravitational interactions between Enceladus and other moons around Saturn and from gravitational forces from Saturn. The chemicals needed for the chemical evolution and existence of life might be there too. Life might exist in the underground water.

Titan is Saturn's largest moon. Although its temperature is –179°C (–290.2°F), its atmosphere is thought to have a chemistry similar to primordial Earth's. It has polycyclic aromatic hydrocarbons, and carbon chain anions, which appear to be involved in the production of large complex organic molecules (Desai et al., 2017). These could be biological molecules. However, its atmosphere lacks water vapor. Titan might have a subsurface liquid ocean serving as a biological environment (Grasset et al., 2000; Fortes, 2000). Unlike on the sun-heated Earth and Mars, a major source of heat for the moons of Jupiter and Saturn are tidal forces from the strong gravitational forces of the massive gas giant planets they revolve around.

About 4.5 billion years from now, the sun will become a red giant. Although Earth will have been long dead from excessive solar heat, other bodies will at various future times become better capable of supporting life, as the greater portion of the sun's heat, and the habitable zone, shifts farther out into the solar system (Rushby et al., 2013). In the future, the sun could provide sufficient heat for bodies like Europa, Enceladus, and Titan to support life even if they do not do so now. This would allow evolution, diversification, and some increase in complexity to proceed on one or more of these moons. However, this will not last the billions of years thought to be required for evolution to produce complex multicellular life. Life would be limited to prokaryote-like organisms. Interestingly, Vanderburg et al. (2020) found what is likely a planet approximately the size of Jupiter circling a white dwarf 6,500 light years from Earth, demonstrating that planets can survive the red giant phase of their star, and still survive when the star becomes a white dwarf. The planet occupies an orbit comparable in distance from its star to that of Jupiter.

The argument for a life-friendly solar system may be extended to the Universe, since it too is likely a system that generates life. This would only apply to small areas where conditions are right. Almost all of the Universe is not life-friendly. The percentage of stars with planets with conditions favorable to life may be low, but the great number of stars with planets means the odds are high that life is common in the Universe. The portion of the Universe we have explored for life compared to the entire Universe is so small that it is analogous to a glass of seawater compared to all the world's oceans. If many other stars have planets with diverse ecosystems, the Universe is a system that favors life, although only under very specific conditions that exist in only a very small percentage of it. The laws of physics and chemistry favor the evolution of intelligent life and high biodiversity on planets with the right limited, specific conditions. The number of such planets may be quite large. About 4,100 exoplanets have been found, and it is estimated that about one in five star systems contains a planet that is at least somewhat similar to Earth. That means there are up to tens of billions of such star systems in the Milky Way alone, and that does not include the many moons that could harbor life. And the Milky Way is only one of an estimated 100 to 200 billion galaxies in the Universe.

13.15 ARE THE PARAMETERS ON EARTH CLOSE TO OPTIMAL FOR LIFE? IF SO, IS THIS DUE TO LIFE?

Let us now briefly revisit the question I raised in Chapter 2. It is interesting to ponder how close to optimal for life the Earth's system parameters are, and how much this is due to life vs. nonbiological factors. For instance, do the oxygen, greenhouse gases, and water levels in Earth's atmosphere cause a near-optimal amount of lightning and fires for life? How close to optimal are atmospheric greenhouse gas levels for biodiversity and life? How much are these parameters that are favorable to life due to life vs. nonbiological factors? How close to optimal are Earth's distance from the sun, orbital ellipse, rotation, and tilt, for seasons and climate to maximize diversity? Do the magnetic field, atmosphere, solar wind, CMEs, and DNA's ability to repair damage to it combine to create a mutation rate that is near-optimal for the evolution of complex life and high biodiversity? These are interesting questions that should be explored. Similar questions can be formulated for other parameters. Of course, there is likely a range of highly favorable values rather than one optimal value. These questions are hard to answer because it is hard to measure what the diversity would be if the parameters were altered. There are also other problems in answering these complex problems. But they are worth contemplation and experimentation because they could throw light on the question of how much the solar system and Earth are systems that favor life and maximize biodiversity, as well as how much life's alterations of the planet favor and caused high biodiversity.

13.16 CONDITIONS WERE NOT ALWAYS FAVORABLE TO LIFE ON EARTH AND WILL NOT ALWAYS BE SO

It is important to bear in mind that the conditions on Earth that exist today that are highly favorable to life, created by a combination of biological and nonbiological factors, did not always exist and will not exist indefinitely into the future. And nonbiological factors can create conditions highly unfavorable to life and diversity. From about 4.5 to 3.8 bya, comets and asteroids bombarded Earth extensively and violently. Near the end of this, during a period called the Late Heavy Bombardment, impacts may have increased. These may have provided heat needed for chemical evolution, but later they likely killed off a great many organisms, greatly harming life and diversity. Also, large meteors and comets could hit the Earth at any time, causing mass extinctions. A meteor strike caused the Cretaceous–Paleogene extinction event of approximately 66 mya that caused the extinction of perhaps 75% of Earth's species and killed off the dinosaurs. Additionally, massive volcanism can cause mass extinctions. Volcanism was likely the principal cause of the Permian-Triassic extinction event of about 251.9 mya, the largest mass extinction in Earth's history.

At the beginning of Earth's history, before methanogens heated the planet, Earth was probably frigidly cold because the sun's luminosity was much lower than today. These conditions were highly unfavorable for life. And this chapter previously discussed how the continents in the past were, and in the future will be, in much less favorable positions for high diversity.

The position of the continents may have played a role in a hypothesized Snowball Earth different from the one that occurred 2.4 to 2.1 bya and that some claim was caused by the Great Oxygenation Event (see Chapter 3). This later Snowball Earth was a phase when much of the Earth was frozen, postulated to have lasted from about 750 to about 660 mya. It is not accepted by everyone. If it occurred, it was triggered by one or more nonbiological cause(s). Possible candidates are the eruption of a supervolcano or many volcanoes that exuded particles that blocked sunlight, changes in Earth's orbit, changes in the heat output of the sun, a reduction in the atmospheric levels of CO_2 and/or methane, and less absorption of the sun's heat. The latter two are thought to have been caused by the position of Earth's continents. The continents were mainly in the tropics at the start of this Snowball Earth. Tropical continents receive more rain, causing weathering and discharge of carbon into the sea, sequestering it. This would have sequestered very large amounts of carbon, removing tremendous quantities of CO_2 and/or methane from the atmosphere. Tropical continents are not as dark as the ocean, so reflect more solar heat; most absorption of the sun's heat today occurs in tropical oceans. Ability to absorb solar heat is more important in the tropics than elsewhere. Once it covered a large enough area, the ice would have reflected a great deal, of the sun's heat, further decreasing the temperature, in a positive feedback loop. These are nonbiological causes that made the temperature very cold and unfavorable for life and diversity. The Earth was brought out of the snowball state and warmer temperatures were restored by a combination of biological and nonbiological factors. The CO_2 levels needed to thaw Earth have been estimated to be 350 times today's levels, which is about 13% of the atmosphere (Crowley et al., 2001). During a period of 4 to 30 million years, enough CO_2 and methane were released and accumulated to cause a greenhouse effect that melted the surface ice in the tropics until a band of permanently ice-free land and water developed. This was darker than the ice, so absorbed the sun's heat, and helped thaw the rest of the Earth in another positive feedback loop. The CO_2 and methane were emitted mainly by volcanoes, but also produced by microbes. So the Earth was thawed mainly by nonbiological processes, but with an assist from biology.

The Sahara Desert is nearly a third of Africa. It also had a grassland ecosystem with a much wetter climate than today about 8,000 years ago. At this time, Earth's tilt on its axis changed from around 24.1 degrees to the present-day 23.5 degrees, due to the gravitational influences of other planets. The Sahara received less sunlight as a result, and so the monsoons decreased. Therefore, vegetation decreased. Plants release water vapor into the air and cause rain. Thus, the decrease in vegetation caused rainfall to decrease even more. So conditions became yet drier. Thus, there were even fewer plants, and so yet less rainfall, and so on, in a positive feedback loop that led to the present-day desert. Present-day conditions were reached about 1,100 years ago. Therefore, conditions started becoming much less favorable for life in the Sahara Desert about 8,000 years ago, due to nonbiological factors.

In the distant future, conditions will deteriorate. Random extraterrestrial events pose risks to biodiversity. These include asteroid and comet impacts. A supernova within 100 light years of Earth would be devastating to life here. Predictable, nonrandom variations in the eccentricity, axial tilt, and precession of the Earth's orbit could bring the system far from optimality for life. The Earth's axial tilt may begin to undergo chaotic variations, with changes in the axial tilt of up to 90°, sometime in the next 1.5 to 4.5 billion years (Neron de Surgy and Laskar, 1997).

Also, bear in mind that nonbiological forces in the Solar System and Universe can be destructive and decrease diversity and information. Examples of this are meteorite strikes and massive volcanism on Earth, and nearby supernovae.

As described in Chapter 2, the sun's increased heat output will eventually sequester so much carbon by weathering and stop volcanoes from adding it to the atmosphere, that C4 plants will not be able to carry out photosynthesis by 600 million years from now. And the sun will keep increasing its output until it is too hot for life to exist. Ocean water will evaporate to water vapor, a greenhouse gas, causing a positive feedback loop. Models predict that in about a billion years, this will reach a tipping point whereby the Earth's seas will completely evaporate, when surface temperatures reach 70°C (about 158°F) (Leconte et al., 2013). In approximately 2 to 3 billion years, Earth's magnetic field may cease, and the solar wind will remove the atmosphere. In about 7.5 billion years, the sun will be a red giant that engulfs the Earth and makes it too hot for any life to survive. Thus, the time that the solar system is favorable to life on Earth is limited. However, a life-friendly solar system will have lasted a long time by the time it ends. If we call the start of favorable conditions the time of life's origin on Earth, at about 3.48 bya, and the end the cessation of C3 photosynthesis, in about 0.6 billion years, then the solar system will have been favorable to life on Earth for about 4.08 billion years. That is for Earth alone. But conditions may have been favorable to life on other planets such as Mars before the origin of life on Earth. And when the sun is a red giant, about 7.5 billion years from now, it might support life on moons such as Europa, Enceladus, and Titan for a while. So the time the solar system is favorable to some form of life in some part of it, albeit sometimes only prokaryotic life, could be over 7.5 billion years, although there could be periods when there is no life between periods when there is life somewhere in the solar system.

REFERENCES

Aciego, S. M., et al. (2017). Dust outpaces bedrock in nutrient supply to montane forest ecosystems. *Nature Communications* 8, Article number: 14800. https://doi.org/10.1038/ncomms14800.

Airapetian, V. S., et al. (2016). Prebiotic chemistry and atmospheric warming of early Earth by an active young Sun. *Nature Geoscience* 9: 452–5.

Bain, W., et al. (Oct., 2021). Phosphine on Venus cannot be explained by conventional processes. *Astrophysics* 21 (10): 1277–1304. arXiv:2009.06499 [astro-ph.EP] or arXiv:2009.06499v1 [astro-ph.EP]. http://doi.org/10.1089/ast.2020.2352. Published Online: 7 Oct., 2021.

Baker, D. N., et al. (27 Nov., 2014). An impenetrable barrier to ultrarelativistic electrons in the Van Allen radiation belts. *Nature* 515: 531–4.

Crowley, T.J., Hyde, W.T., & Peltier, W.R. (2001). CO_2 levels required for deglaciation of a "near-snowball" Earth. *Geophys. Res. Lett.* 28 (2): 283–6. Bibcode:2001GeoRL..28..283C. doi: 10.1029/2000GL011836.

Dacke, M., et al. (Jan., 2021). How dung beetles steer straight. *Annual Review of Entomol.* 66:243–56. https://doi.org/10.1146/annurev-ento-042020-102149.

Desai, R. T., et al. (2017). Carbon chain anions and the growth of complex organic molecules in Titan's ionosphere. *Astrophys. Journ. Lett.* 844 (2): L18. arXiv:1706.01610. Bibcode:2017ApJ...844L..18D. doi: 10.3847/2041-8213/aa7851.

Fortes, A. D. (2000). Exobiological implications of a possible ammonia-water ocean inside Titan. *Icarus* 146 (2): 444–52. Bibcode: 2000Icar..146..444F. doi: 10.1006/icar.2000.6400.

Grasset, O., et al. (2000). On the internal structure and dynamic of Titan. *Planetary and Space Science* 48 (7–8): 617–36. Bibcode: 2000P&SS...48..617G. doi:10.1016/S0032-0633(00)00039-8.

Greaves, J. S., Richards, A. M. S., Bains, W., et al. (14 Sept., 2020). Phosphine gas in the cloud decks of Venus. *Nature Astron.* 5: 655–64. https://doi.org/10.1038/s41550-020-1174-4.

Hershkovitz, I., et al. (26 Jan., 2018). The earliest modern humans outside Africa. *Science* 359 (6374): 456–59. Bibcode: 2018Sci...359..456H. doi:10.1126/science.aap8369. PMID 29371468.

Klacka (2010). Galactic tide. arxiv.org/abs/0912.3112.

Kruijer, T. S., et al. (27 June, 2017). Age of Jupiter inferred from the distinct genetics and formation times of meteorites. *PNAS USA* 114 (26): 6712–6. https://doi.org/10.1073/pnas.1704461114.

Leconte, J., et al. (11 Dec., 2013). Increased insolation threshold for runaway greenhouse processes on Earth-like planets. *Nature* 504: 268–71.

Mikhail, S. & Sverjensky, D. A. (19 Oct., 2014). Nitrogen speciation in upper mantle fluids and the origin of Earth's nitrogen-rich atmosphere. *Nature Geoscience* 7: 816–9.

Nazari-Sharabian, et al. (9 May, 2020). Water on Mars—A literature review. *Galaxies* 8 (2): 40. https://doi.org/10.3390/galaxies8020040.

Neron de Surgy, O. & Laskar, J. (Feb., 1997). On the long term evolution of the spin of the Earth. *Astronomy and Astrophysics* 318: 975–89. Bibcode: 1997A&A...318..975N.

Raymond, S. N., et al. (21 May, 2016). Did Jupiter's core form in the innermost parts of the Sun's protoplanetary disc? *Monthly Notices of the Royal Astron. Soc.* 458 (3): 2962–72. https://doi.org/10.1093/mnras/stw431.

Rushby, A. J. (18 Sept., 2013). Habitable zone lifetimes of exoplanets around main sequence stars. *Astrobiol.* 13 (9): 833–49. https://doi.org/10.1089/ast.2012.0938.

Vanderburg, A., Rappaport, S.A., Xu, S., et al. (16 Sept., 2020). A giant planet candidate transiting a white dwarf. *Nature* 585: 363–7. https://doi.org/10.1038/s41586-020-2713-y.

14 Counterexamples to and Arguments against the Autocatalytic Biodiversity Hypothesis and Answers to Them

There are a number of counterexamples to and arguments against the Autocatalytic Biodiversity Hypothesis (ABH). They must be acknowledged for the sake of good, honest science. In this chapter, I describe them and give answers to them as best I can, and assess the current state of these arguments to the best of my capabilities.

When life first evolved, there were likely a number of different genetic codes competing with each other. As a result of natural selection, only one code prevailed. This is a decrease in diversity due to life, due to natural selection. New codes did evolve in mitochondria, chloroplasts, some prokaryotes, and ciliated protozoa, increasing diversity. But the vast majority of life has the one canonical genetic code. However, in the long run, this may have increased diversity by allowing viruses to promote evolutionary innovations in their hosts through transduction, and transposable elements to be shared between species, allowing macroevolutionary breakthroughs. These innovations and breakthroughs would have been followed by diversifications into many new species. The ability to share genes between species may have been a selective factor favoring a near-universal genetic code. If there were several genetic codes, transferring of DNA between species would be more limited than it is with a shared, universal genetic code. This would make evolutionary innovations (and the resulting diversifications) less easy to occur, less probable, and less frequent.

Competition can lead to a decrease in diversity if the dominant form causes extinction of its competitor. Placental mammals eliminated marsupials in most parts of the Earth by competition. However, this is not the rule, since diversity is generally maintained during competition, which can even promote diversity. This is discussed in detail in Chapter 6 of my companion book to this book (Seaborg, 2002).

Most natural selection is normalizing selection, which selects against individuals at the extremes, and favors those that are most like the average of the population. A hypothetical example is a moth that is brown and is camouflaged on brown tree trunks. Predators select out any moths that deviate from the brown color, finding and consuming any that are not brown or even a shade of brown that differs too much from the color of the tree trunk. This type of selection decreases genetic variability within populations and species, and genetic variability is a type of biodiversity. Directional selection selects for an increase in a characteristic, and predators can be the selective agent. For example, lions select for the fastest gazelles, because they can catch and eat any that are not extremely fast. There is a limit as to how far this selection can progress. Gazelles cannot become faster than a maximum possible speed because of physical limits on what muscles and hearts can do. By selecting for only those gazelles near the limit of possible speed, this mode of selection decreases genetic variability in species and populations. This is generally true for this type of selection; it is not limited to lions acting as a selective agent on gazelles. However, sometimes natural selection favors increased genetic variability in species and populations. This is discussed in Chapter 3 of the companion book to this one (ibid.).

Life can decrease diversity, having a negative effect on biology through unfavorable ecosystem engineering. Oxygen produced by cyanobacteria from the Great Oxidation Event (GOE) was a toxic gas that prokaryotes were exposed to, causing a reduction in diversity. Only prokaryotes existed at this time. As discussed in chapter 3, this likely did not lead to a mass extinction.The GOE might have contributed to the Huronian glaciation and Snowball Earth, about 2.4 to about 2.1 bya, because the oxygen produced would have combined with and greatly reduced the amount of the potent greenhouse gas, methane, in the atmosphere. The Huronian glaciation was actually several ice ages, not a single continuous event. Oxygen produced by photosynthesis made the temperature lower and would have caused a drop in planetary diversity. But it may not have been the sole cause of the glaciation. It is not even certain there was a Snowball Earth at this time. There was a loss of diversity, but likely not a mass extinction. Furthermore, the Huronian glaciation may have helped increase oxygen levels. Its removal of methane from the air helped regulate Earth's temperature in the long run, lowering the temperature in the face of a sun that kept increasing its heat output and the addition of greenhouse gases to the atmosphere as a result of volcanism. See Chapter 3 for more details.

It is confusing to some as to whether the addition of nutrients to aquatic ecosystems results in increased or decreased levels of oxygen without some explanation. It is important to explain this at this point, so that the rest of this chapter will be clear. The addition of small-to-moderate quantities of nutrients such as nitrogen and phosphorus to freshwater or marine ecosystems increases phytoplankton growth, but not

DOI: 10.1201/9781003246640-14

excessively. The greater numbers of phytoplankton produce more oxygen through photosynthesis. They also increase atmospheric oxygen by burying reduced carbon when they die and sink. However, the addition of large quantities of nutrients causes tremendously large phytoplankton blooms. They become overpopulated and deplete their nutrient supply, resulting in the death of great numbers of phytoplankton. They are then decomposed by bacteria that obtain their food from them. This decomposition consumes extremely large amounts of oxygen, and produces aquatic ecosystems depleted of oxygen.

Poulton et al. (2021) found that oxygen rose and fell three times between 2.43 and 2.22 bya. The rises were to 10^{-5} of the present atmospheric level. The rises are correlated with glacial periods. The mechanism is that oxygen rose and depleted atmospheric methane, leaving CO_2 and water vapor as the remaining greenhouse gases. And methane is 30 times as powerful a greenhouse gas as CO_2, and a much more powerful greenhouse gas than water vapor. So, a glacial period occurred each time oxygen rose. In seas covered with ice, cyanobacteria, the principal oxygen producers at that time, did not receive sufficient sunlight, and died back. In time, subglacial volcanoes emitted CO_2 and methane, and methanogens produced methane, so Earth warmed up. The ice on the ocean melted, and cyanobacteria received more sunlight. And the CO_2 from volcanoes reacted with rainwater, forming carbonic acid, which dissolves rocks, causing weathering. This delivered more nutrients such as phosphorus into the oceans. So cyanobacteria built their populations up and produced oxygen, methane fell, and glaciers returned. This cycle occurred three times. Then oxygen increased a fourth time, and a tipping point was reached and it did not go down again. Permanent oxygenation of the atmosphere had arrived. No one knows what caused this threshold to be crossed, but theresearchers hypothesize that volcanism provided a sufficiently large influx of nutrients to the seas for cyanobacteria to permanently thrive. At this point, oxygen levels were high enough to permanently suppress methane's influence on the climate, and CO_2 from volcanic activity and other sources became the dominant greenhouse gas that kept the planet warm. They found that this permanent atmospheric oxygenation did not occur until about 2.22 bya (100 million years later had been estimated until their work). This extends the GOE by 100 million years, and means it took 100 million years longer than thought to reach a permanent tipping point of high oxygen. It means oxygen levels fluctuated across the threshold of 10^{-5} for about 200 million years. This work also provides further evidence for a link between oxygenation and massive climate swings. Recall that the sun emitted much less heat at this time. It shows the production of oxygen by life sometimes threw Earth into glacial periods, decreasing diversity. However, in the long run, this reduction of methane made the planet better for life as the sun increased its output and there was a need to cool the Earth. In spite of all of oxygen's negative effects on life, it greatly increased life's diversity and complexity once organisms evolved tolerance for it and then the ability to use it in cellular respiration. It was necessary for the evolution and diversification of complex life, which needed it as an energy source.

Sulfur-reducing prokaryotes reduce elemental sulfur to hydrogen sulfide (H_2S), a gas toxic to most life. When H_2S rises to the stratosphere, it destroys the ozone layer, allowing damaging ultraviolet radiation to reach Earth's surface. And it destroys hydroxyl radicals in the atmosphere; these consume methane. So H_2S can cause higher levels of this potent greenhouse gas, perhaps sometimes exceeding 100 ppm. This raises temperatures to levels unfavorable to most life. H_2S also inhibits the growth of nitrogen-fixing bacteria. Sulfate-reducing microbes also generate methylmercury, which is toxic to the nervous system, as a by-product of their metabolism, through methylation of inorganic mercury. They are the main source of this form of mercury in aquatic systems (Compeau and Bartha, 1985). They employ inorganic sulfur compounds for respiration in the absence of oxygen. Sulfur-reducing prokaryotes and their relatives have greatly diversified. They occur in about 69 genera within nine phyla in the bacteria and 37 genera within two phyla in the archaea (Florentino, 2016). But only a few taxa are true sulfur-reducing microbes, using sulfur reduction as the only or main catabolic reaction. Sulfate reducers occur in deep-sea vents, hot springs, and other extreme environments (ibid.). They are ubiquitous in environments that lack oxygen, where *they have an important role in both the sulfur and carbon cycles* (Muyzer and Stams, 2008). Thus, sulfate-reducers have a negative impact on biodiversity in some respects, but a positive one in others. They are limited to only a few habitats by the high oxygen levels produced by photosynthetic organisms because they cannot grow in the presence of oxygen. Under normal circumstances, their net effect on diversity is to increase it. Note that it is life that limits their abundance and destructiveness and thus protects almost all of the biosphere (by producing oxygen).

Canfield (1998) argued that the ocean was euxinic (lacking in oxygen and high in toxic hydrogen sulfide; the noun is euxinia) from the middle to late Proterozoic eon, which would be from about 1.52 bya to 541 mya. He thus postulates that deep ocean water remained very low in oxygen and high in sulfur long after the GOE. These euxinic seas are called Canfield Oceans. The sulfur-isotope record supports his hypothesis (ibid.). He suggests that deep ocean waters were not rich in oxygen until about 541 mya, in association with a second large oxygenation of the Earth's surface. In Canfield Oceans, which had low oxygen levels, sulfate-reducing microbes would have thrived and produced toxic hydrogen sulfide, which would have built up to high levels. Thus, in postulated Canfield Oceans, life—sulfate-reducing microbes—produced toxic hydrogen sulfide in the absence of oxygen, killing some species off and reducing biodiversity. This could explain why it took so long from the first cells, which appeared between 3.465 and 4.1 bya, to the Cambrian explosion that produced all the major animal groups alive today, of about 541 mya (Ward, 2009). This could be an example of ecosystem engineering that inhibits evolutionary progress and diversity in natural conditions over a long time period if it holds up. It would thus be a counterexample to the ABH.

However, there are some problems with this idea. Canfield says the deep sea was anoxic from about 1.52 until 1.0 to 0.541 bya. Since the GOE occurred about 2.3 to 2.45 bya, the rest of the ocean may have been oxygenated from about 2.45 bya to 541 mya (really until present), except for some relatively short-lived periods of low oxygen. So, it is likely that there was enough oxygen for life to thrive and continue evolving most of this time period in the large expanse of ocean that is not the deep sea. Thus, sulfate-reducing microorganisms were probably limited to the deep sea. How much H_2S they produced, its concentration in waters beyond the deep sea, and how much area beyond the deep sea was free of it are not known. Thus, the extent to which sulfate-reducing microbes adversely affected life, diversity, and evolution beyond the deep sea is not known. These are interesting questions Canfield's ideas raised that need exploring.

Life was mostly unicellular during most of the time that Canfield Oceans are postulated to have existed. Unicellular organisms can evolve quickly and would have adapted to the high H_2S levels to some extent. The Ediacaran animals, the earliest multicellular animals, appeared about 635 mya, and were simple. The first complex animals appeared during the Cambrian explosion, about 541 mya. Thus, Canfield oceans did not have a significant effect on the complex animals that resulted from the Cambrian explosion. Also, the Ediacaran animals were multicellular forms that existed before the Cambrian explosion, from approximately 635 to 541 mya. They survived in spite of the possibility that Canfield oceans existed until up to 541 mya. It is hard to see how these animals could have survived if every part of the seas had high levels of toxic H_2S and very low levels of oxygen.

At any rate, the Cambrian explosion could not have occurred until the ozone layer was built up by life. High levels of ultraviolet (UV) light would have inhibited the evolution of more complex life in the shallow and middle levels of the ocean whether there was high H_2S or not until the ozone layer was created.

Also, recall that sulfate-reducing bacteria aided the GOE and oxygenation of the atmosphere. And zooplankton outcompeted sulfate-reducing bacteria, controlling their numbers, although when this started is not clear. See Chapter 3 for details on these last two sentences. Also, sulfate-reducing microbes are part of the sulfur cycle and help other microbes in that cycle, acting in commensalism and symbiosis with them. Sulfate-reducing bacteria are also in symbiotic consortia with methanotrophs and filamentous sulfur bacteria, mainly in low-oxygen marine sediments, reducing the amount of methane released into the air, generally without increasing toxic H_2S. This lowers atmospheric temperature to the benefit of life. These consortia have many species and are globally distributed. In some marine areas, the process produces H_2S, but it is used by commensal filamentous sulfur bacteria, and by animals, such as clams and tube worms, through the help of their symbiotic microbes. These animals have symbiotic sulfide-oxidizing bacteria, allowing them to benefit from the H_2S. And the consortia aid many other species by producing nutrients for other prokaryotes and some animals. See Chapter 2. Even more, symbiotic sulfide-oxidizing bacteria that live in clams and tube worms consume and thus reduce the amount of H_2S, and provide nutrition to their hosts, while receiving a home from them. They benefit from sulfide-reducing microbes, which help and are part of this multispecies symbiotic system. Additionally, sulfate-reducing bacteria add to diversity in the deep sea because they consist of 60 genera containing 220 known species. All this supports the ABH.

Since sulfate-reducing microbes cannot grow in the presence of oxygen, they usually grow only in limited areas in the deep sea, since most other areas on Earth are high in oxygen because of its generation by photosynthesis and the burial of reduced carbon by life. But if there is an event that limits oxygen, they can increase. During the largest mass extinction in Earth's history, the Permian-Triassic mass extinction of about 251.9 mya, severe depletion of oxygen seems to have occurred with the result that sulfate-reducing microbes became the dominant force in oceanic ecosystems, producing copious amounts of H_2S, contributing to the mass extinction. In this extinction, there was over-riding of the negative feedback that stabilized oxygen levels, producing a very low 15% oxygen level in the air, as well as extremely low oxygen in the seas. Massive volcanism caused high CO_2, which heated Earth and the seas. The hotter water is, the less oxygen it absorbs, so the oceans had less dissolved oxygen. Additionally, warmer seas can decrease the heat difference between the poles and equator, which decreases ocean currents that transport oxygen throughout the sea, making at least some areas of the oceans even more oxygen-depleted. And many organisms in the sea died as a result of the lack of oxygen in the ocean. The resulting dead organisms further depleted oxygen as they decomposed, because decomposition of organisms utilizes and hence depletes oxygen. The seas therefore became very low in oxygen. Meyer et al. (2011) studied limestones in south China to show that high populations of phytoplankton, when oxygen was high enough again for them to grow, delayed recovery, as bacteria decomposed these phytoplankton after they died and thus consumed and depleted the oxygen dissolved in the sea. Depletion of oxygen allowed sulfate-reducing bacteria to increase greatly. So hydrogen sulfide levels went up, killing animal life. Hydrogen sulfide also destroys the ozone layer, so damaging ultraviolet radiation increased. When dead organisms were under water where there was no oxygen, microbes called methanogens that can only grow in the absence of oxygen and that produce methane as a waste product, used them as nutrient sources, decomposing them, producing methane, which further heated the atmosphere and depleted oxygen. The atmosphere became dominated by hydrogen sulfide and ultra violet radiation, which is harmful to life, and methane. So life combined with nonbiological factors to play a part in this mass extinction. Of course, the nonbiological cause of heat from greenhouse gases from massive volcanism was the main cause of the extinction that led to other problems such as the ones just discussed.

As it ascends into the atmosphere, hydrogen sulfide destroys chemicals that destroy methane. The result is higher levels of this potent greenhouse gas, perhaps exceeding 100 parts per million, a very high level. It also inhibits the growth of bacteria that convert nitrogen in the air to a form life can use. The low-oxygen conditions and high levels of hydrogen sulfide caused the die- offs of many kinds of animals.

Rothman et al. (2014) postulated that the Permian-Triassic mass extinction was caused by a genus of methanogenic archaea, *Methanosarcina*, which could not grow in the presence of oxygen, that acquired a new metabolic pathway via gene transfer at the time, enabling them to efficiently metabolize acetate into methane. That would have caused a sharp increase in methane in the oceans and atmosphere. Additionally, massive volcanism might have exuded large quantities of nickel, which is a scarce cofactor for enzymes involved in methane production. This also would have allowed methanogens to produce more methane. The additional methane resulting from either or both of these mechanisms would have increased temperatures to the point where they were unfavorable for life, decreasing diversity. If true, this is also counter to the ABH. It is not known whether this hypothesis is valid; it needs further testing.

Evidence indicates that life combined with nonbiological factors to reduce diversity by creating seas depleted of oxygen and high in hydrogen sulfide a number of times in the Mesozoic era (252-66 mya) (Leckie et al., 2002; Takashima et al., 2006), including the Posidonienschiefer event (about 183 mya), Selli event (about 120 mya), Paquier event (about 111 mya), and Bonarelli event (about 94 mya) (see Jenkyns, 2010). Volcanoes exuded large quantities of CO_2 into the air. Ocean crust production added CO_2 in at least some cases. Methane-producing sources other than life added methane. The greenhouse gases caused higher temperatures, which evaporated more water, causing more rain, which increased weathering, bringing more nutrients to the sea. More warmth also caused sea levels to rise because water expands when heated and the hotter water melts the polar ice caps. This brought yet more nutrients to the ocean because when the sea submerges land it did not previously cover, it receives nutrients from it. Submarine volcanism increased nutrients, mainly iron. The increased nutrients from all these sources were exceptionally large, and so caused massive blooms of phytoplankton and bacteria, most notably cyanobacteria. When these organisms died, the decomposition of the dead organic matter by microbes consumed large amounts of oxygen, leading to severely low oxygen levels (eutrophication). Higher temperatures also decreased the amount of oxygen that dissolved in seawater. The poles can heat up more than the tropics when temperatures increase, so the heat gradient between the poles and equator would have decreased, decreasing ocean currents that transport oxygen throughout the sea, making the oceans even more depleted of oxygen. This low-oxygen state of the seas is called an oceanic anoxic event. When the seas became extremely low in oxygen, sulfate-reducing bacteria were able to increase, producing large quantities of H_2S, killing large numbers of organisms. Volcanoes also add small amounts of H_2S. In areas that had excessively high nutrient input from rivers that led to plankton blooms followed by die offs that depleted oxygen, and where the geography favored density stratification that made it difficult for the mixing of the water layers, the sea could have readily evolved from poorly oxygenated to depleted of oxygen and ultimately euxinic conditions.

During the Bonarelli event, oxygen-depleted environments expanded from the photic zone of the shallow sea to a depth greater than 3,500 meters (11,483 feet) in the Atlantic Ocean, possibly resulting in about 20% of marine organisms becoming extinct in various habitats within an interval of less than one million years. While life did not initiate any of the above extinction events, and nonbiological factors played important roles, life played a large role in all of them. Sulfate-reducing prokaryotes may have worsened some or all of these events because they could expand in oxygen-depleted conditions and produce H_2S. Yet, in every case, even in the Permian-Triassic mass extinction, life eventually restored oxygen to life-favorable levels. This inhibited sulfate-reducing microbes and H_2S production. Interestingly, Leckie et al. (2002) said plankton experienced accelerated rates of *both* speciation and extinction at or near the major Cretaceous oceanic anoxic events, indicating it is not clear these events led to decreases in biodiversity. And they said the oceanic anoxic events were short-lived episodes. There is no modern analog for global ocean oxygen depletion, so models of these events are difficult to fully test. So we do not know how serious these events were or how large a role life played in them. These important questions need further study.

The Ordovician period ended with a series of extinctions that occurred about 447–444 mya and collectively make up the second largest of the five major extinction events in Earth's history in terms of percentage of genera that became extinct. Complex multicellular organisms lived in the ocean at that time. Almost 85% of animal species and 49–60% of animal genera went extinct. There were two major pulses of extinctions. The cause of the first pulse is hotly debated. The most commonly accepted hypothesis is that the extinctions were caused by the onset of cold conditions, then an ice age, that ended the long, stable greenhouse conditions that had been typical of the Ordovician. The Late Ordovician glaciation was preceded by a fall in atmospheric CO_2 from 7,000 ppm to 4,400 ppm (Young et al., 2010). This led to glaciation, since the sun gave off much less heat at this time. Phytoplankton could have bloomed and sequestered carbon when they died and sank to the seafloor. But the main cooling mechanism was likely the expansion of primitive nonvascular land plants for the first time. These plants do not have true roots, but they have root-like structures that accelerated chemical weathering. By this mechanism, nonvascular plants may have sequestered sufficient atmospheric CO_2 to trigger the glaciations at this time (Lenton et al., 2012). Nonvascular plants have no sophisticated vascular system to transport nutrients and water. They include mosses, hornworts, liverworts, and some algae, especially green algae. Sea levels dropped as water was taken up by growing glaciers. The shallow seas disappeared, eliminating habitats. Seas on continents also dried up, eliminating these important habitats. The seas returned with diminished life, and withdrew again with the next glaciation phase, eliminating niches in the shallow seas again. The lower sea levels also changed ocean currents, disrupting climate and habitats. The idea that the ecosystem engineering of nonvascular land plants under natural conditions over a relatively long time period was a major factor causing this mass extinction is counter to the ABH. However, it has also been hypothesized that the glaciation was caused by an increase in volcanism that deposited new silicate rocks, which drew CO_2 out of the air as they were weathered. Volcanoes can emit cooling sulfur aerosols to the atmosphere, and this too could have occurred. Heavy weathering from the uplifting of the Appalachian Mountains in the United States during the Late Ordovician also sequestered large quantities of CO_2. Additionally, the supercontinent Gondwana drifted over the South Pole, and ice caps formed on it; this lowered Earth's temperature.

A minority hypothesis postulates that the first extinction pulse was caused by a ten-second gamma-ray burst that could have destroyed the ozone layer and exposed life on land and in the upper portions of the seas to ultraviolet (UV) radiation, killing much of it. The net effect of the UV radiation is to cool the stratosphere more than it warms the troposphere. So the UV could have initiated global cooling (Melott et al., 2014). On the other habnd, a recent study suggested the first extinction pulse was caused by volcanism, which induced global warming and low oxygen levels, rather than cooling and glaciation (Bond and Grasby, 2020).

Marine life diversified some, and gained some of its former diversity during the cold period, and a new cold-water ecosystem, the "Hirnantia biota," was established (Sheehan, 2001). This decreased the net loss of diversity from the first pulse.

It is most likely that cooling, not warming, was the cause, and it appears nonbiological factors were partly responsible for the mass extinction. But a large amount of the cooling was due to life in the form of nonvascular plants. However, the decrease in greenhouse gases caused by nonvascular land plants aided life and diversity over a long time frame because the sun continually became hotter and volcanoes continuously added greenhouse gases to the atmosphere over time. The decrease in greenhouse gases also aided life and increased diversity by regulating the acidity of both freshwater and oceanic environments in the long run. The net, overall effect over a long time frame of the lowering of greenhouse gases and global temperatures, and the acidity of aquatic ecosystems, due to nonvascular plants, was an increase in diversity, and was beneficial to life. If one considers the entire time nonvascular land plants have been on Earth from their appearance until the present time, their actions have increased diversity much more than they have decreased it.

The second extinction pulse happened after warm conditions returned. It is associated with extreme worldwide conditions of low oxygen and high sulfur levels. Retreating glaciers could have exposed land to weathering, causing a great flow of phosphates into the ocean. This would cause extreme cyanobacteria blooms, followed by eutrophication. The worldwide distribution of black shales in the late Hirnantian stage, which is the final stage of the Ordovician period and lasted from 445.2 to 443.8 mya, is indicative of a global oxygen depletion event (Melchin et al., 2013). The oxygen-depleted oceans seem to have allowed sulfate-reducing microbes to grow and produce hydrogen sulfide (Zou et al., 2018). The geologic record in China indicates that this second extinction pulse was correlated with intense euxinia (ibid.). On a global scale, euxinia was probably one or two orders of magnitude more prevalent than in the modern day. The depleted oxygen and high hydrogen sulfide levels resulted from a combination of both nonbiological factors and life. The biological factors are cyanobacteria and the bacteria that decomposed them. Thus, life helped cause this second extinction pulse of the Ordovician period. However, cyanobacteria have aided life and increased diversity far more than they have hurt life and decreased diversity. They were the first to evolve photosynthesis, the first to fix nitrogen, are a food source at the base of ocean food webs, and provide other ecosystem services. Therefore, when unusual nonbiological conditions such as retreating glaciers occur, cyanobacteria can be one of the causative agents of a mass extinction, but they usually cause diversity to be maintained or increase, and have increased diversity over geologic time much more than they have decreased it. The bacteria that decomposed the cyanobacteria normally help oceanic ecosystems by acting as decomposers, removing dead cyanobacteria and other dead organisms. If they were not decomposed, cyanobacteria (and other species of organisms) would accumulate in great numbers, and become a major problem for marine ecosystems. Bacteria that carry out decomposition also recycle nutrients and make them available to their ecosystems. Thus, these bacterial decomposers have also caused diversity to be maintained and increase far more than they have decreased it over time, consistent with the ABH.

Following the mild climate of the earlier Devonian, the Earth became very cold, and there was widespread glaciation. This greatly contributed to late Devonian mass extinction of about 376-360 mya, in which many species of animals and plants went extinct. It is one of the five mass extinctions in Earth's history. Life seems to have played a major role in it. It occurred when land plants increased their maximum height from 30 centimeters (about 11.8 inches) at the start of the Devonian period, to 30 meters (about 98.4 feet) at the end of the period, due to the evolution of advanced systems for transporting nutrients and water from one part of the plant to another, which permitted the growth of complex branching and rooting systems (Algeo and Scheckler, 1998). Plant roots and their symbiotic fungi weathered silicate rocks, reducing CO_2 in the atmosphere. Plants were also buried in great numbers, decreasing atmospheric CO_2 concentrations from about fifteen to three times present levels, producing vast coal deposits in such places as China, at a time when heat from the sun was much lower than it is today. The development of seeds at this time permitted reproduction and dispersal in any areas which that were not waterlogged, allowing plants to reach and colonize previously unreachable or inhospitable inland and upland areas (ibid.). As a result, plants took on an important role as ecosystem engineers at this time. The Earth's orbit was also mostly farther from the sun, and this seems to have contributed to lowering Earth's temperature, so a nonbiological factor was involved as well.

Forests dominated by *Archaeopteris*, an extinct genus of tree-like plant with fern-like leaves, expanded rapidly during the closing stages of the Devonian period. While early Devonian plants had only primitive root-like structures that penetrated but a few centimeters, *Archaeopteris* had deep root systems that broke up the upper layers of bedrock and stabilized a deep layer of soil. The roots chemically broke down rocks and caused weathering, sequestering carbon and thus lowering temperatures, and releasing nutrients rapidly (ibid.). This could have resulted in a relatively sudden input of high levels of nutrients into rivers, where they may have caused a great growth of freshwater phytoplankton. When these died and were decomposed by bacteria, oxygen would have been consumed in great amounts, causing low oxygen levels in the water. Some of the high levels of nutrients would have been carried to the oceans, causing the same problem there. Reefs at this time were dominated by stromatolites and, to a lesser degree, corals—organisms which that grow well only in low-nutrient conditions. This could have played a large role in the loss of reefs and other groups of organisms at this time. Black shales, which occur when there is severe deficiency of oxygen in the bottom waters of the oceans, correlate well with extinction episodes in the Late Devonian period tropical marine sea bottom, implicating low oceanic oxygen levels more than global cooling as the cause of the extinction (Algeo et al., 1995), although both played a role.

There are alternative hypotheses to the above. A meteorite impact has been suggested, but lacks convincing evidence. A nearby supernova explosion could account for the dramatic drop in atmospheric ozone that could have permitted massive ultraviolet damage to life, triggering the extinction. UV damage to pollen and spores over many thousands of years during this event supports this (Fields et al., 2020). However, the long-lived, extra-terrestrial radioisotopes that would support this supernova idea were not found. Volcanism with the ejection of cooling particles (as opposed to warming CO_2 and methane) has been suggested as a cause of the cooling, and there is some correlation between volcanic regions on the Siberian craton (a major Precambrian tectonic unit in Northern Eurasia) and the extinction (Ricci, et al., 2013). The Earth's Milankovitch cycle seems to have contributed to lowering Earth's temperature as well. A Milankovitch cycle is any slow but regular change in the Earth's orbit around the Sun, and in the tilt of the Earth's axis. This affects how much heat Earth receives from the sun, and thus its temperature. The continued reduction of carbon in the air eventually pulled the Earth out of its Greenhouse Earth state into the Icehouse that continued throughout the Carboniferous and Permian periods.

Hence, a variety of factors may have contributed to the widespread glaciation and global cooling. But weathering by large land plants apparently played a major part in the extinction. This ecosystem engineering by plants contradicts the ABH/Pachamama Hypothesis.

The removal of carbon from the air by vascular land plants lowered Earth's temperature, and thus was helpful to life in the long run as the sun increased its heat output. It also helped regulate the acidity in both freshwater and marine ecosystems over time. The increase in diversity as a result of the sequestration of carbon by land plants lasted a long time and is still ongoing. The depletion of oxygen and glaciation they helped cause were for relatively short times. Vascular land plants increased diversity far more and for a far longer time than they decreased it, by removing greenhouse gases, producing local rainfall, providing shade, manufacturing oxygen, preventing soil erosion, providing food and habitat to animals, and other mechanisms.

Life not does not always act as a carbon sink and can even be a carbon source. Freshwater ecosystems are a major source of methane, contributing 6.5×10^{-10} grams in CO_2 equivalents per year towards global carbon emissions (Emilson et al., 2018), offsetting about 25% of the terrestrial carbon sink (ibid.). The source of the majority of methane emissions are sediments around the shore, where a large amount of plant material is decomposed (ibid.). Methane production is at least 400 times higher from sediments composed of decaying large aquatic plants than sources on land (ibid.) The increased atmospheric methane increases temperature when decreasing it would be favorable to life in face of a sun increasing its output. Of course, life sequesters much more carbon than it adds to the atmosphere.

The calcite in calcium carbonate allows the coccoliths of coccolithophores to scatter a greater quantity of light than they absorb. This results in induced photoinhibition, which is reduction in the photosynthetic capacity of a plant, alga, or cyanobacterium due to too much light. Additionally, a high concentration of coccoliths can lead to an increase in surface water temperature simultaneous with a decrease in the temperature of deeper waters. This causes more stratification in the water column and a decrease in the vertical mixing of nutrients. This makes less nutrients available to life in shallow waters where photosynthesis is possible. This is an example of the effect of a life form's ecosystem engineering decreasing productivity. It is not known if this leads to a decrease in diversity. If so, it is counter evidence to the ABH. However, the net overall effect of coccolithophores and their ecosystem engineering on diversity is to increase it. They are at the base oceanic food webs, sequester a great deal of carbon, produce oxygen, and provide other ecosystem services.

The following conclusions can be drawn. Large macroevolutionary innovations result in new adaptive zones with many new available niches, so are followed by large adaptive radiations, increasing diversity significantly. However, some such breakthroughs can result in ecosystem engineering that is disruptive to the ecosystem. This can alter the environment, making it less favorable to life, decreasing diversity for a period of time. This deterioration of the environment can last a long time, but it has never thus far lasted indefinitely. Life has in all known cases up to this point come up with a mechanism or mechanisms to restore conditions favorable to biology. Moreover, the innovation generally results in diversification of the group with the innovation. The innovation tends to profoundly change the ecosystem in a way that is deleterious to biology at first, but ultimately beneficial to life. The ecosystem engineering that results from the innovation tends to change the environment in a way that increases the diversity of other life forms in the long run.

The innovation of being able to reduce sulfate resulted in the production of toxic hydrogen sulfide, but allowed the sulfate-reducing microbes to use sulfate as an energy source and be involved in symbiotic relationships with other microbes in the sulfur cycle. It allowed great diversification of sulfate-reducing microorganisms. Some became part of symbiotic relationships in multicellular animals. The toxic hydrogen sulfide they produced as a waste product was used by some other microbes. When photosynthesis evolved, life produced enough oxygen that these sulfate reducers were controlled, and could only grow in very limited areas. However, there were times when low oxygen conditions eliminated this control. This was always initiated by nonbiological factors, such as volcanism. It did involve biology, and the hydrogen sulfide production was mostly biological, since most was produced by microbes. But in time, biology has always restored high oxygen levels and the regulation of sulfate-reducing microbes.

When photosynthesis first appeared, it disrupted the environment in two ways. First, oxygen was toxic to a good deal of the prokaryotic life that existed then. This problem was solved because (a) minerals combined with oxygen, keeping its levels low for a long time, giving life time to evolve resistance to oxygen, and (b) all organisms at the time of the GOE were prokaryotes, which can evolve resistance to toxins very quickly. The microbes did evolve resistance to oxygen, and then the ability to use it as an energy source. This allowed multicellular life to evolve and diversify into a tremendous number of new species. The second disruption photosynthesis caused was a freezing of the Earth because oxygen reacted with and removed methane from the air. This resulted in a great loss of diversity for a while. So photosynthesis decreased diversity and was disruptive to the biosphere at first, and this lasted a period of time. But photosynthesis resulted in a huge net increase in diversity, first because the photosynthesizers diversified tremendously because of the adaptation they had, and second because the high levels of oxygen they produced resulted in the evolution of respiration to use this oxygen as an energy source. The oxygen produced also led to the formation of the ozone layer, which was also necessary for the evolution of complex life on land and in the shallow sea. The decrease in temperature caused by the removal of methane from the atmosphere by oxygen, catastrophic at first, in the long run was beneficial to life as the sun's heat increased. When all its effects are considered for the entire time it was present, photosynthesis caused an increase in diversity that far exceeds the loss of diversity it caused.

In the case of plants, the evolution of simple nonvascular plants and later complex vascular plants both resulted in new adaptive zones and great diversification of both groups. The innovations that led to the first nonvascular land plants were disruptive to the biosphere. These plants helped cause an ice age that led to a great loss of diversity. But in the long run, their ecosystem engineering kept temperatures from being too hot as the sun increased its heat output, for a much longer time than the duration of the ice age they helped cause. Carbon sequestration by nonvascular plants also kept the acidity of aquatic systems at life-favorable levels. Their net long-term effect was to make conditions favorable to life and increase diversity.

Vascular plants helped cause conditions depleted of oxygen and perhaps an ice age, a disruptive effect of their macroevolutionary breakthrough. Again, the removal of greenhouse gases, although unfavorable to life at first, in the long run aided life and diversity by making temperatures cooler and better for life for a long time, continuing to the present day. The low atmospheric oxygen levels caused by the plants and the phytoplankton that the plants caused to bloom was another disruption associated with an evolutionary innovation. It was temporary, and vascular plants increased atmospheric oxygen much more and for a much longer time than they caused it to decrease. The many services they provide to the biosphere resulted in an increase in diversity that far outweighs the loss of diversity from the temporary low oxygen levels and low temperatures they helped cause. Phytoplankton too helped increase diversity much more than they decreased it, by producing oxygen, sequestering carbon, and providing an important food source to many animals, over the eons. Both plants and phytoplankton also greatly increased biodiversity by diversifying into many species as a result of their evolutionary innovations.

Macroevolutionary breakthroughs that were disruptive to ecosystems and resulted in ecosystem engineering that decreased diversity and had negative impacts on other species for limited time periods included the evolution of sulfate-reducing microorganisms, photosynthesis, nonvascular land plants, and vascular land plants. They all resulted in adaptive radiations. Life evolved a mechanism to control sulfate-reducing microbes. The other three breakthroughs resulted in organisms that performed ecosystem engineering that was highly favorable to life and diversity in the long run. The result of each of the four was an increase in diversity that is much greater than the loss of diversity each caused. Sulfate-reducing microbes are not an exception to this because they diversified into a large number of species, they are important in the sulfur cycle, they provide niches for and are consumed by other species, they are symbiotic with a number of prokaryotic as well as animal species, and the vast majority of the time they are controlled by photosynthesis and so do not poison the environment by producing high levels of hydrogen sulfide.

14.1 THE MEDEA HYPOTHESIS

Ward (2009) wrote a book counter to the ABH. He calls his book and hypothesis the Medea Hypothesis, in keeping with the naming of hypotheses about life's effects on life after

Greek goddesses. The word "Gaia," used to name the Gaia Hypothesis, means "Good Mother," and she was the Greek goddess of the Earth. Medea was a Greek goddess who killed her children, so was a bad mother. The name Ward gave to his hypothesis is fitting because his hypothesis proposes that life is suicidal and tends to hurt life. I shall answer his major arguments in this section.

He says populations of organisms outbreed their environment. However, in nature, populations are controlled by predators and pathogens. Because of this, organisms do not normally outbreed their environment. There is negative feedback between predators and their prey, and pathogens and their hosts, and this keeps populations from becoming overpopulated and destroying their environment. This is discussed in Chapter 9 of the companion volume to this one (Seaborg, 2022). Ward says life is self-poisoning in closed systems. Earth and the biosphere are not closed systems. They receive energy from the sun. Also, the wastes of any given species are nutrients for other species, and these waste/nutrients are recycled. Recall, for one example, the nitrogen cycle in the soil (Chapter 6). He says species compete for resources, leading to extirpation or emigration of one or more of the species. This can happen, but is not the rule. Competing species partition niches and coexist in nature. Competing species even evolve to specialize on different resources, allowing their coexistence. This is discussed in Chapter 6 of this book's companion volume (ibid.). He claims most biological feedback is positive, although he acknowledges some is negative. In fact, although biologically driven feedback can be positive and destabilizing, this is not the rule and tends to occur only when a nonbiological catastrophe, such as massive volcanism or an asteroid strike, has destabilized the normal stabilizing characteristics of the biosphere. Normally, biological systems consist of stabilizing negative feedback loops. I pointed out stabilizing negative feedback mechanisms involving life with respect to CO_2 and oxygen in Chapters 2 and 3, respectively, although I also pointed out that these feedbacks are weak and can be overridden by catastrophic events. But most importantly, predator-prey and parasite-host systems are stabilized by negative feedback, and this is significant. I discuss this in Chapter 9 in the companion book on the ABH (Seaborg, 2022). And positive feedback between species tends to lead to evolutionary breakthroughs and diversification in natural systems, as is seen in coevolution resulting in symbiosis (ibid., Chapter 4). It can improve the environment for life when biology is involved with it. For example, it was involved in life increasing oxygen levels. He says life will deplete the planet of CO_2, causing an end to life. Life sequestered carbon since its inception and does so today, regulating the temperature to the benefit of life. However, as I pointed out in Chapter 2, life has mechanisms to do this while conserving carbon and carbon dioxide. The sun is constantly increasing its heat output. Because life conserves carbon as it sequesters it relatively slowly, the sun will become too hot for life over 1.7 billion years before life depletes carbon to a catastrophic point. Thus, life will not deplete the carbon supply. He states that organisms that make the environment better for themselves cause destabilizing positive feedback. They do not. They put the environment in a stable state that is better for them and other species. Beavers build dams that convert parts of rivers into ponds, benefiting them and many other species. This is a stable system; they do not continue to change the ecosystem once the pond is created. This is true for elephants maintaining grasslands by trampling and eating small, young trees and all other cases of species improving the habitat for themselves. In these cases, other species also benefit. Often, as in the case of beavers and elephants, diversity is increased because a second habitat is added. A pond is not necessarily more diverse than a river, but a pond and a river together are more diverse than only a river. These points are discussed in Chapter 5 of the companion book.

Ward claims there is what he calls biotic plundering and gives an example of marine plankton depleting their nutrients, dying and sinking. But as discussed in Chapter 8, marine plankton are controlled by viruses that kill vast numbers of them daily, fertilizing the sea. Other plankton use nutrients created by this so-called viral shunt to grow and thrive. Marine plankton are also controlled by the many species that consume them. There are times when marine plankton deplete their resources and even cause eutrophication. But this is not the normal situation, and tends to happen in conjunction with uncommon catastrophic events, such as massive volcanism, or unnatural events, such as human pollution. Similar arguments can be made for other species that might destroy their environment; predators and disease regulate them.

Ward claims the CLAW Hypothesis has it backward, and that dimethyl sulfate (DMS) does not regulate temperature by stabilizing negative feedback, but causes destabilizing positive feedback, making Earth cooler when it is cool and warmer when it is warm. But the references he lists for this do not in fact support this claim. He provides no evidence to back up the assertion. He says that DMS is produced by dust, not phytoplankton, without providing evidence or any references. Moreover, it is now clear that under normal circumstances, the CLAW hypothesis is valid as stated, and life is involved in stabilizing the temperature via DMS and negative feedback.

One of the references he cites concerning DMS, however, makes a point counter to the ABH. The authors, Watson and Liss (1998), do not say that DMS makes Earth cooler when it is cool and warmer when it is warm. However, they argue that in glacial periods, the colder and drier climate has a more vigorous atmospheric circulation. This causes greater ocean circulation responsible for mixing subsurface nutrients, and also influences the transport of nutrients, such as iron, in atmospheric dust. These two sources of increased nutrients cause more phytoplankton growth, which leads to greater cooling by sequestration of CO_2 and increase in DMS. This would make the cold glacial period even colder, hurting diversity, which is counter to the ABH. The problem is that the Earth does not keep getting colder in glacial periods. Eventually, the climate warms and an interglacial period manifests.

Ward says that the Cambrian explosion was associated with a decrease in biomass, but actually a great increase in biomass occurred then (Franck et al., 2006). He attributes a large drop in temperature that occurred at that time to the appearance of Cambrian animals, without giving a mechanism, explanation, or evidence. He cites a reference that does not in fact say that that was the cause. It is hard to believe the rise of the major animal phyla caused a great drop in temperature.

Ward makes a good point that after the asteroid struck Earth and caused the mass extinction of 66 mya, fires emitted vast amounts of soot into the atmosphere, blocking sunlight and causing global cooling for months after the impact, worsening the extinction event. This was possible because of the extensive forests at the time, so the extinction was primarily caused by a nonbiological force, but forests allowed the fires to burn.

Ward discusses eutrophication events, production of toxic hydrogen sulfide by life, evolution of an almost universal genetic code, production of the greenhouse gas methane by methanogens, production of oxygen in the GOE and its toxic and temperature-lowering effects, temperature drops caused by land plants (not least the colonization of land by plants), and Snowball Earths. All of these were discussed in this book, and need not be discussed now.

Ward can be forgiven for many of his erroneous statements in the area of population biology, such as that populations outbreed their environment. He is a paleontologist, not an ecologist.

14.2 HUMAN IMPACTS AND THE AUTOCATALYTIC BIODIVERSITY HYPOTHESIS/PACHAMAMA HYPOTHESIS

One may very well accept the large volume of evidence presented in this book for the ABH/Pachamama Hypothesis, and recognize that organisms are the major engines creating, increasing, and maintaining biodiversity and that ecosystems maximize diversity. And the evidence presented may well have convinced the reader that all species are ecosystem engineers that have a net positive effect on biodiversity, their ecosystem, and other species under natural ecosystems over sufficient time periods. But one might challenge the generality of this with the question: What about humans? Is this species not an exception to this rule? *Homo sapiens* is a species that is an ecosystem engineer with massive negative effects of the biosphere, and that has caused and continues to cause a great decrease in diversity and an extinction crisis.

There are a number of arguments in answer to this. First, the ABH states that all species are ecosystem engineers with positive effects on biodiversity over sufficient time periods *in natural ecosystems. This means in natural situations.* Of course, natural means *without significant human impact.* This means that if human impacts are significant, the system is not natural, and thus the ABH does not apply. It follows from this that the ABH does not apply to humans themselves and their impacts. Humans are the only species that alter the environment profoundly enough and to such an extent as to make them an unnatural influence. Humans are not within the normal range of species the ABH pertains to. Humans evolved an adaptation, a large brain capable of extreme intelligence and innovation, which allowed them to escape the natural controls imposed on all other species. Humans have removed the predators and many of the diseases that regulate their population. At the time of this writing, their population is almost eight billion and increasing. They have invented numerous means of altering the environment to their short-term benefit, and directing unprecedented percentages of resources to themselves. So humans have an adaptation that allowed them to escape the natural controls that ensure a species will have its numbers regulated to the point that it will not generally have a negative impact on biodiversity and its ecosystem. They are the sole exception to the ABH, the only species that it is not true for. They have an environmental impact that is much greater than any other species. In view of this, another way of stating of the ABH is: *All species except humans are ecosystem engineers that have a net positive impact on diversity, their ecosystem, and other species,in natural conditions over sufficient time.* This is a sufficient answer to the criticism, and no further arguments need be given. Yet additional interesting, valid answers to the objection exist, and are worth discussing.

Another answer to the counter-argument is that humans are not destined to be on this planet for much longer if they continue their destructive ways. Nature does not need people, but people need nature. Humanity's destruction of the environment will lead to the extinction of humans if they do not mend their ways. This is happening now, as climate change, the extinction crisis, soil erosion, pollution from plastics and toxins, and other insults humans are inflicting on Earth, and its response to them, are making humanity's ability to obtain food and water, remain healthy, and survive less and less feasible.

Spratt and Dunlop (2007) pointed out that the reports of the Intergovernmental Panel on Climate Change, which significantly informs climate policymaking and the public narrative on it, tend toward reticence and caution, erring on the side of least drama and danger, and downplaying more extreme and more damaging outcomes. They say that at this point this is becoming dangerously misleading, given the seriousness and speed of the global climate impacts. High-impact events once had low probability, but are now becoming significantly more likely. Of particular concern to them is the likelihood of passing critical thresholds that result in step changes in the system, such as melting the polar ice sheets and hence sea level rise; melting the permafrost, which releases more greenhouse gases in a destabilizing positive feedback loop; and releasing carbon from other carbon stores. In these examples, the impacts of global warming are nonlinear, which is much more dangerous than if they were linear. They claim that human civilization will crumble by 2050 if we do not stop climate change at this time. Although the dates of the fall of civilization and of human extinction cannot be known, many other informed scientists share their view that humans will become extinct at some point in the future if we do not make radical changes in our treatment of the Earth.

A thorough study of human impacts found that the Earth may be on the brink of the sixth mass extinction (Harfoot et al., 2021). Mass extinctions can take up to 2.8 million years. Yet the current extinction crisis is happening in the course of a human lifetime. Kolbert (2014) elaborated on this human-induced sixth mass extinction in a book. Since we are dependent on other species to eat, obtain clean water, and survive, a mass extinction jeopardizes our survival as a species.

The number of serious environmental problems humans are creating are not limited to climate change and loss of biodiversity, and are numerous. They include plastic and chemical pollution of air and water, loss of topsoil, overpopulation, causing the increase of pathogens resistant to antibiotics, habitat destruction, eutrophication of large areas of the sea (causing dead zones with very little life), and damming of rivers (including rivers in rainforests, flooding and submerging the forests), to name only some.

We are also affecting our ability to reproduce. Chemical pollution is so bad that male sperm counts are declining to the point where the level of sperm production could be below that needed for successful conception by 2045 (Levine et al., 2017). A meta-analysis found male sperm counts declined by 50 to 60%, dropping on average 1% to 2% per year, between 1973 and 2011, among men unselected by fertility from North America, Europe, Australia, and New Zealand (ibid.). Female reproductive systems are also being adversely affected by chemicals. These problems are discussed in a book that shows how serious the problem is, and that it threatens the survival of the human race (Swan, with Colino, 2021). Clearly, we cannot survive as a species if we cannot reproduce.

One scenario is that disease could greatly decrease the human population. A large pandemic would most likely be the result of our destruction of biodiversity and the environment. There are a number of new, emergent diseases, such as Ebola virus, Marburg virus, and coronaviruses. New pathogens are not limited to viruses. Various animal species have pathogens that can jump to and infect humans. As we encroach on natural ecosystems, we come into contact with these species. Forest fragmentation, agricultural expansion, and livestock production are all bringing humans into closer contact with a variety of other species that could transmit disease to humans. Concentrated livestock production brings together large populations of genetically similar, often immune-suppressed animals that are highly vulnerable to disease outbreaks. Some of these diseases could potentially be passed to humans. Destruction of rainforests is a primary way we come into contact with animals that can spread diseases to us. Forest fragmentation and habitat destruction in Africa are linked to outbreaks of the deadly Ebola virus, which is thought to have come from fruit bats in the rainforest. Bats seem to be particularly effective hosts for viruses that infect humans. Horseshoe bats (family Rhinolophidae) are the family that most commonly carry severe acute respiratory syndrome (SARS)-related coronaviruses. Some horseshoe bat species may carry coronaviruses harmless to them that are pathogenic to their bat competitors and to their predators, such as snakes, in a symbiotic relationship (Villarreal, 2020). The findings of Rulli et al. (2021) indicated that interactions between humans, livestock, and wildlife in China may form hotspots with the potential to increase SARS-related coronavirus transmission from animals to humans. As we encroach on more habitats, we come into contact with more wild species that could give us new diseases that we have not developed an immunity to.

In addition, many diseases carried by mosquitoes and other arthropods are increasing their range as humans warm the planet. All of these diseases will control human population if humans do not decrease their destruction of nature. If disease severely decreases the human population, this would support the ABH, because life would be responding and regulating the numbers of the species that is attacking its biodiversity. Disease alone will not cause extinction of the human race. But a large pandemic could decrease the human population to a very low level and put it in a permanent preindustrial state from which humans cannot recover. Humans would continue to survive, but would never be able to have a technological society again. This would greatly decrease our negative environmental impact, and preserve biodiversity to an unknown extent. For a pandemic to do this, it would have to cause enough mortality to decrease the population to the point that there would not be enough farmers to grow food, shippers to get essential supplies to market, people to run water delivery systems, and so on. And the disease would have to kill off a large number of the people with the knowledge needed to run a technological society.

If humans wake up and change to the point that they have a positive impact on biodiversity, there would be no species that are exceptions to the ABH. On the other hand, if we drive ourselves extinct because of some combination of overpopulation, a mass extinction, habitat destruction, climate change, loss of food production, chemical and/or plastic pollution, lack of potable water, or some other environmental catastrophe of our making, we will cease to have the profound environmental impact we now have. There will be a mass extinction as we fall. But then, without further human destruction occurring, nature, life, and biodiversity will gradually recover. Of course, this will take millions of years. Still, the extinction of humans and the resultant recovery of biodiversity would be consistent with and support the ABH.

Third, nature can respond and solve some of the human-caused environmental problems in time. Prokaryotes and fungi could evolve new forms and even new species with the ability to consume plastics and other toxins humans have added to the environment, creating new breakdown products that other prokaryotes and fungi evolve variants and/or species to consume, in a chain continuing until the plastic or toxin is fully broken down. This would not only help solve the plastic and chemical pollution problem, it would also increase diversity as the plastics and chemicals are utilized as new niches. A new strain of bacterium in the genus *Pseudomonas* can break down polyurethane plastic, a nonbiodegradable plastic that is building up in the environment and causing serious environmental problems. This plastic releases toxic chemicals that would kill most bacteria when it is degraded.

The microbe utilizes energy obtained from the breakdown of the plastic. The bacterium can use the breakdown of polyurethane as a sole source of carbon, nitrogen, and energy. Some fungi can also break down polyurethane. The larvae of the greater wax moth (*Galleria mellonella*) eat and degrade the plastic, polyethylene, which is used in about 40% of all plastic products, with over a trillion plastic bags made from it used annually. In the absence of the moth larvae, this plastic takes an extremely long time to break down, and is a major environmental problem. The larvae break it down to ethylene glycol. In lab tests, 100 worms can consume 92 milligrams of polyethylene in just 12 hours (Bombelli, 2017). The degradation is accomplished by bacteria in the larvae's microbiomes. The bacteria must have evolved this capability in recent times. Microbes and other organisms could also evolve to break down human-made toxic chemicals that are not plastics. Humans could also artificially select for microbes that can break down plastics and toxic chemicals. These examples show nature can respond to human pollution of the Earth. They also illustrate how chemicals, even toxic ones, when in abundance, are resources that organisms will evolve to utilize, solving problems created by toxic chemicals that are too abundant. When there are large amounts of any plastic, chemical, or species (such as an exotic species) in the environment, it is a resource, and selection will favor any microbe or other species that evolves to utilize, consume, and regulate such an abundant resource.

Fourth, one could even argue that the human impact will in the long run enhance diversity even if we do destroy ourselves and cause a mass extinction. Much or all of the anthropogenic increase in carbon in the atmosphere will eventually be sequestered by life and nonbiological processes. Moreover, carbon we add to the atmosphere will be used by life. It will replenish some of the carbon lost to the system by biological and nonbiological sequestration, making it available to life again, helping alleviate the problem of the loss of carbon from the biosphere discussed in Chapter 2. In that chapter, I pointed out that life has largely solved this problem by conservation and storage of carbon, but there may be a point when replenishing some of it is beneficial to life. The destructive addition of some other nutrients such as nitrogen and phosphorus could possibly likewise eventually have positive effects, replenishing nitrogen and phosphorus and perhaps other nutrients buried and lost to the system. Introduced species, so destructive to diversity, could in time coevolve and coadapt with native species. In cases where there are more introduced species added than the native species that they drive extinct, there will be a net gain of species locally, although a net loss globally. And at least some of the introduced species in time could diverge from their parent populations in their land of origin into new species. This would result in an increase in biodiversity globally. In geologic time scales, the result of the great number of species introduced by humans to new areas could possibly be an increase in diversity, even globally. However, this is not a certainty, and we should not introduce any species to new areas with this hope in mind. Introduced species are at present causing the loss of more species than any other cause, except for habitat destruction. (Introduced species will likely soon be surpassed by climate change as a cause of extinction of species.) We should do everything we can to prevent the introduction of exotic species to any habitat, and to control and eradicate them where they already exist.

Finally, there will be an adaptive radiation following the sixth mass extinction, returning diversity at least to nearly the level it reached just prior to the current environmental crisis. Past mass extinctions indicate that there will be probably refugia where some species survive, providing a source for recovery and diversification. The continued increase in diversity since the Permian mass extinction suggests that the diversification after the human-induced mass extinction could continue to a diversity that is above the level that existed just prior to the start of this extinction. But we do not know if the postrecovery Earth will have a little less than, about the same as, or higher biodiversity than the Earth before the human-caused mass extinction. It could have lower diversity. We simply do not know. Also, bear in mind that the recovery will likely take a million or more years.

Of course, I am not advocating the continued destruction of the Earth and biodiversity in the hope that this might result in a long-term increase in diversity. Nor should we add greenhouse gases to the atmosphere to replenish the carbon. There are two problems with advocating such actions. First, it means the extinction of the human race. Second, we will not be the only species to go extinct. If we continue our destructive ways, there will be a mass extinction with a loss of untold numbers of wonderful species of animals, plants, fungi, and other taxa. The Earth could be inhospitable to complex life and have none to very little of it for a long time. And there is no guarantee that our carbon input, introduced species, or a mass extinction will have a net long-term positive effect on life or result in increased biodiversity in the long term. After at least a million years, global biodiversity will in all probability come back to approximately its preindustrial level, but whether it will be lower, the same, or higher than this level after full recovery cannot be predicted.

REFERENCES

Algeo, T.J. & Scheckler, S. E. (1998). Terrestrial-marine teleconnections in the Devonian: links between the evolution of land plants, weathering processes, and marine anoxic events. *Philosophical Transactions of the Royal Society B. Biological Sciences* 353 (1365): 113–30. doi: 10.1098/rstb.1998.0195. PMC 1692181.

Algeo, T.J., et al. (March, 1995). Late Devonian oceanic anoxic events and biotic crises: "Rooted" in the evolution of vascular land plants? *GSA Today* 5 (3).

Bombelli, P. (24 April, 2017). Polyethylene bio-degradation by caterpillars of the wax moth *Galleria mellonella. Current Biol.* 27 (8): R292-R293. https://doi.org/10.1016/j.cu.

Bond, D. P.G. & Grasby, S. E. (18 May, 2020). Late Ordovician mass extinction caused by volcanism, warming, and anoxia, not cooling and glaciation. *Geology* 48 (8): 777–81. Bibcode: 2020Geo....48..777B. doi: 10.1130/G47377.1.

Canfield, D. A. (3 Dec., 1998). A new model for Proterozoic ocean chemistry. *Nature* 396: 450–3. https://doi.org/10.1038/24839.

Compeau, G. C. & Bartha, R. (Aug., 1985). Sulfate-reducing bacteria: principal methylators of mercury in anoxic estuarine sediment. *Applied and Environmental Microbiology* 50 (2): 498–502. doi:10.1128/AEM.50.2.498-502.1985. PMC 238649. PMID 16346866.

Darwin, C. R. (1958). *The Origin of Species*. A Mentor Book from New American Library, New York, NY and Toronto, Canada; The New English Library Limited, London, UK First published as *On the Origin of Species by Means of Natural Selection, or the Preservation of Favoured Races in the Struggle for Life*, 24 Nov., 1859.

Emilson, E. J. S., et al. (4 May, 2018). Climate-driven shifts in sediment chemistry enhance methane production in northern lakes. *Nature Communications* 9, Article number 1801. doi: 10.1038/s41467-018-04236-2.1089969.

Fields, B. D., et al. (18 Aug., 2020). Supernova triggers for end-Devonian extinctions. *PNAS USA* 117 (35): 21008–10. arXiv:2007.01887. Bibcode: 2020PNAS..11721008F. doi: 10.1073/pnas.2013774117. ISSN 0027-8424. PMC 7474607. PMID 32817482.

Florentino, A. P., et al. (2016). Ecophysiology and application of acidophilic sulfur-reducing microorganisms. In Rampelotto, P. H. (ed.). *Biotechnology of Extremophiles: Advances and Challenges. Grand Challenges in Biology and Biotechnology*, vol. 1, pp. 141–75. Springer International Publishing, New York, NY. doi:10.1007/978-3-319-13521-2_5. ISBN 978-3-319-13521-2.

Franck, S., et al. (10 March 2006). Causes and timing of future biosphere extinctions. *Biogeosciences* 3: 85–92. https://doi.org/10.5194/bg-3-85-2006.

Harfoot, M.B.J., Johnston, A., Balmford, A., et al. (2021). Using the IUCN Red List to map threats to terrestrial vertebrates at global scale. *Nature, Ecology, and Evolution* 5: 1510–9. https://doi.org/10.1038/s41559-021-01542-9.

Jenkyns, H. C. (9 March, 2010). Geochemistry of oceanic anoxic events. *Geochemistry, Geophysics, Geosystems* 11 (3). doi: 10.1029/2009GC002788. https://doi.org/10.1029/2009GC002788. ISSN: 1525-2027.

Kolbert, E. (2014). *The Sixth Extinction. An Unnatural History*. Henry Holt & Company, New York, NY.

Leckie, R. M., et al. (Sept., 2002). Oceanic anoxic events and plankton evolution: Biotic response to tectonic forcing during the mid-Cretaceous. *Paleoceanography and Paleoclimatology* 17 (3): 13–1–13–29. https://doi.org/10.1029/2001PA000623.

Lenton, T., Crouch, M., Johnson, M., et al. (1 Feb., 2012). First plants cooled the Ordovician. *Nature Geoscience* 5: 86–9. https://doi.org/10.1038/ngeo1390.

Levine, H., et al. (Nov.-Dec., 2017). Temporal trends in sperm count: a systematic review and meta-regression analysis. *Human Reproduction Update* 23 (6): 646–59. https://doi.org/10.1093/humupd/dmx022.

Melchin, M. J., et al. (2013). Environmental changes in the Late Ordovician-early Silurian: Review and new insights from black shales and nitrogen isotopes. *Geological Society of America Bulletin* 125 (11/12): 1635–70.

Melott, A.L., et al. (2004). Did a gamma-ray burst initiate the late Ordovician mass extinction? *International Journ. of Astrobiology* 3 (2): 55–61. arXiv:astro-ph/0309415. Bibcode: 2004IJAsB...3...55M. doi: 10.1017/S147355040400 1910. S2CID 13124815.

Meyer, K. M., Yu, M. Jost, A. B., et al. (1 Feb., 2011). δ13C evidence that high primary productivity delayed recovery from end-Permian mass extinction. *Earth and Planetary Science Letters* 302 (3–4): 378–84. https://doi.org/10.1016/j.epsl.2010.12.033.

Muyzer, G. & Stams, A. J. (June, 2008). The ecology and biotechnology of sulfate-reducing bacteria. *Nature Reviews Microbiology* 6 (6): 441–54. doi: 10.1038/nrmicro1892. PMID 18461075. S2CID 22775967.

Poulton, S.W., Bekker, A., Cumming, V.M., et al. (29 March, 2021). A 200-million-year delay in permanent atmospheric oxygenation. *Nature* 592: 232–6. https://doi.org/10.1038/s41586-021-03393-7.

Ricci, J., et al. (2013). New 40Ar/39Ar and K-Ar ages of the Viluy traps (Eastern Siberia): Further evidence for a relationship with the Frasnian-Famennian mass extinction. *Palaeogeography, Palaeoclimatology, Palaeoecology* 386: 531–40. doi: 10.1016/j.palaeo.2013.06.020.

Rothman, D.H., et al. (31 March, 2014). Methanogenic burst in the end-Permian carbon cycle. *PNAS USA* 111 (15): 5462–7. Bibcode:2014PNAS..111.5462R. doi: 10.1073/pnas.1318106111. PMC 3992638. PMID 24706773.

Rulli, M.C., D'Odorico, P., Galli, N., et al. (31 May, 2021). Land-use change and the livestock revolution increase the risk of zoonotic coronavirus transmission from rhinolophid bats. *Nature Food*. https://doi.org/10.1038/s43016-021-00285-x.

Seaborg, D. (2022). *How Life Increases Biodiversity: An Autocatalytic Hypothesis*. CRC Press/Taylor & Francis. Boca Raton, FL; London, UK; New York, NY.

Sheehan, P. M. (May, 2001). The Late Ordovician mass extinction. *Annual Review of Earth and Planetary Sciences* 29 (1): 331–64. Bibcode: 2001AREPS..29..331S. doi: 10.1146/annurev.earth.29.1.331. ISSN 0084-6597.

Spratt, D. and Dunlop, I. (7 Sept., 2007). What lies beneath: The scientific understatement of climate risks. *Breakthrough—National Centre for Climate Restoration* (Melbourne, Australia).

Swan, S. H., with Colino, S. (23 Feb., 2021). *Count Down: How Our Modern World Is Threatening Sperm Counts, Altering Male and Female Reproductive Development, and Imperiling the Future of the Human Race*. Simon & Schuster, New York, NY.

Takashima, R., et al. (Dec., 2006). Greenhouse world and the Mesozoic ocean. *Oceanography* 19 (4): 64–74.

Villarreal, L. (2020). Personal communication.

Ward, P. (2009). *The Medea Hypothesis. Is Life on Earth Ultimately Self-Destructive?* Princeton Univ. Press, Princeton, NJ, and Oxford, UK.

Watson, A. J. & Liss, P. S. (29 Jan., 1998). Marine biological controls on climate via the carbon and sulphur geochemical cycles. *Philosophical Transactions of the Royal Society B: Biological Sciences* 353 (1365). https://doi.org/10.1098/rstb.1998.0189. Print ISSN:0962-8436. Online ISSN:1471-2970.

Young, S. A., et al. (15 Oct., 2010). Did changes in atmospheric CO2 coincide with latest Ordovician glacial-interglacial cycles? *Palaeogeography, Palaeoclimatology, Palaeoecology* 296 (3–4): Pages 376–88.

Zou, C. (2018). Ocean euxinia and climate change "double whammy" drove the Late Ordovician mass extinction. *Geology* 46 (6): 535–8. Bibcode:2018Geo....46..535Z. doi: 10.1130/G40121.1.

15 Implications

There are several important implications to the Autocatalytic Biodiversity Hypothesis (ABH)/Pachamama Hypothesis. Most importantly, its statements that organisms create biodiversity, that ecosystems maximize biodiversity, and that there is a tendency toward increasing information (biodiversity) in ecosystems suggest that it would be wise to value biodiversity. There is a tendency in nature to increase and maintain biodiversity, suggesting we should view it as having inherent value. The argument is strengthened when one considers that symbiosis is a fundamental relationship and that all species are interconnected and dependent on each other. This is not to say that nature is conscious and makes value judgements. The increase in diversity happened as a result of natural physical, chemical, and biological principles that are not teleological. However, if one were to fail to respect the law of gravity, he or she would not survive long, and this is analogous to ignoring the tendency toward diversity in ecosystems and the reality of fundamental symbiotic interdependence. We are currently suffering the consequences of not valuing biodiversity.

The biodiversity of our planet provides many valuable services to humans. Organisms sequester carbon, regulating temperature, and acidity in aquatic ecosystems. Plants and other photosynthesizers produce oxygen, and life buries reduced carbon. As a result, life created Earth's high oxygen levels and protective ozone layer. Life provides us with medicines in abundance. Approximately 7,000 drugs used in Western medicine are derived from plants; many others come from animals, coral reefs, fungi, bacteria, and other life. Less than 5% of Amazon plant species have been studied for their potential medicinal benefits (Dunnell, 2018). Yet approximately 25% of all drugs employed in Western medicine are derived from rainforest plants (ibid.). This includes treatments for heart disease, fungal infections, cancer, malaria, brain diseases, and many other ailments. According to the U.S. National Cancer Institute, more than two-thirds of all medicines that can be used to combat cancer are derived from rainforest plants (Beaudry, 2019). One example is: chemicals from the Rosy Periwinkle (*Catharantus roseus*), found only in Madagascar, have increased the chances of survival for children with childhood leukemia from 20% to 80%.

Organisms provide many of the products we use, often daily, such as gums and resins, wood, paper, ingredients in toothpaste, oils, fibers, contraceptives, and biodegradable pesticides. Natural ecosystems are home to many indigenous peoples today, many living in the rainforest, some still uncontacted by us in technological civilizations. Life built the soil we need to grow food. Most of our crops are pollinated by animals, especially the western honey bee, which pollinates crops worth tens of billions of dollars per year.

Nature can teach us a great deal. For example, the bonobo is a chimpanzee that settles disputes without violence, dolphins show a great amount of altruism, and we can learn much about sonar from bats. Organisms provide an endless quantity of fascinating questions for scientific study, both for the pure knowledge of basic science and applications for bettering our lives. Biodiversity and nature are beautiful and give us great joy. It is calming and has been shown to be good for the physical and mental health to be with animals and plants and in nature. It provides economic benefits from tourism. The loss of biodiversity means the loss of all of these services and attributes.

No human problem or cause, no matter how worthy, has any long-term relevance if we do not stop the human-caused extinction crisis and save biodiversity. If the loss of species crosses a critical threshold, civilization will fall, and humans could even possibly become extinct. We cannot cause the loss of great numbers of species without causing other problems. Biodiversity is so fundamental that essentially any problem one can name is exacerbated by its loss and alleviated by its preservation. For example, forests remove CO_2 from the atmosphere through photosynthesis and their burial, alleviating anthropogenic climate change. But if trees are cut or burned, they release CO_2 into the air, exacerbating climate change. Flooding and destroying rainforests with large dams results in the release of methane when the trees are decomposed underwater. Recall that methane is 30 times as powerful a greenhouse gas as CO_2. All the problems of climate change are then exacerbated, from sea level rise to drought. Loss of forests causes local droughts, reducing our capacity to access clean water and to grow food. Removal of natural vegetation causes soil erosion, decreasing our ability to grow food and causing siltation of rivers. Loss of coral decreases fish that many people depend on for a livelihood or food. This list could go on for pages, but I will stop here, since my point has been made. We can see clearly that we need biodiversity for our food, water, and many other amenities. Without sufficient biodiversity, we would have mass starvation and massive deaths from lack of potable water. The ABH is relevant to all of these services that life provides, for all of these services are examples of ecosystem engineering by organisms.

DOI: 10.1201/9781003246640-15

Preserving any species helps solve many other problems and increases the chance humans and other species will continue to survive. Let us take one example, the Couch's spadefoot toad (*Scaphiopus couchii*), which relies on monsoon rains in Arizona to breed in great numbers in August. Human-induced global climate change has stopped the summer monsoons, and the toads are potentially headed to extinction because they cannot breed sustainably. If we made it a priority to save the toads, we would have to reduce fossil fuel emissions and preserve our forests in order to decrease emissions of CO_2 into the air and thus decrease the temperature, which would bring back the monsoons and allow the spadefoot toads to breed and survive. In saving just one species, we would alleviate climate change; increase our chances of continued survival; be able to produce more food; decrease droughts in the southwestern United States and other places; save other species dependent on desert monsoons; increase the amount of water available to people, plants, and wildlife; and solve or alleviate a myriad of other problems. And there is nothing special about the Couch's spadefoot toad with respect to this argument. One could make this argument using any number of other species.

We must give the Earth and biodiversity the highest priority, and put the Earth and biodiversity above short-term human needs. We need biodiversity to survive. Without a minimum required amount of it, we could go extinct. And if a large number of species went extinct, decreasing biodiversity, a great many other species dependent on the species that went extinct would similarly die off. This is coextinction and is discussed in Chapter 11.

It is also true that many solutions to major problems result in the preservation of biodiversity. Stopping the burning of fossil fuels in favor of renewable energy and conservation alleviate climate change, which slows the drying and burning of rainforests caused by increased temperatures. Empowering women and recognizing their equal rights reduces the amount of children they have, reducing population growth, reducing species extinctions. Again, this list could continue for paragraphs. The reader is encouraged to conceive of more examples of how preserving diversity alleviates other problems and vice versa.

There are several human activities that would benefit from the preservation of biodiversity. I will discuss only two, hoping the reader will be imaginative and think of some of the many others. Agriculture would benefit by techniques that preserve the biodiversity of agricultural systems. This includes preserving the biodiversity of the soil. Soils contain an abundance of biologically diverse organisms that provide essential ecosystem benefits, including cycling of nutrients, decomposition of dead organisms, building and maintaining soil structure, carbon sequestration, supporting plants and aboveground ecosystems, and the regulation of populations of soil species. One teaspoon of healthy soil contains billions of soil microorganisms, including bacteria, algae, and fungi. It is the soil life that allows soils to store carbon. Soil stores carbon in large quantities, keeping it out of the atmosphere and yet available to the biosphere. Soil biodiversity reduces the impacts of extreme droughts and floods, which are becoming more common with human-induced climate change. Soil with healthy structure absorbs water during intense rains and holds it during droughts, improving agricultural yields during weather extremes. Ants and earthworms are ecosystem engineers that improve soil structure and make tunnels that allow air, water, and nutrients to travel through the soil.

The United States uses more than one billion pounds of pesticides, which include insecticides, herbicides, and fungicides, every year. U.S. agriculture has become 48 times more toxic to insect life since it started using neonicotinoid pesticides in the 1990s. As little as 0.1% of pesticides interact with their targeted "weed" or "pest." The rest contaminate the soil, water, and air, and have significant negative effects on the biosphere. Pesticides kill or harm soil organisms, including bacteria, soil algae, earthworms, ants, beetles, ground-nesting bees, and other soil invertebrates. They reduce the abundance and diversity of soil organisms, damaging important dynamics in the soil community. Although many soil species are negatively impacted by pesticides, some increase because they have less competitors or predators. As a result, some species that were abundant become rare and vice versa. The result is lower biodiversity, and a less healthy and more poorly functioning ecosystem.

Pesticides often remain in the soil for years to decades after they are applied, continuing to harm soil organisms. Farmers typically use mixtures of multiple pesticides. Washington apples are treated with an average of 51 different pesticides in 6 to 17 applications per year. The continuous use of pesticides does not allow sufficient time for soil ecosystems to recover. Gunstone et al. (2021) found that pesticides of all types pose a clear hazard to soil invertebrates, and negative effects were evident in both lab and field studies, across all studied pesticide classes, and in a wide variety of soil organisms.

Pesticides also reduce the diversity of the aboveground community. Insecticides kill predators of the targeted insects. The targeted insects develop resistance to the insecticides by natural selection. These two factors result in greater numbers of insects that eat the crop being grown. This results in lower crop yields. The ecosystem on the farm with lower biodiversity is less stable and less healthy.

Pesticide companies claim that their products are needed to feed the world's growing population. In fact, pesticides over time reduce crop yields by destroying the soil ecosystems that sustainable food production depends on.

In addition to pesticides, fertilizer harms soil life. Growing monocultures of single crops can cause a loss of soil biodiversity. Growing multiple plant species together and rotating crops can preserve soil biodiversity and improve yields.

Agricultural practices such as composting and cover cropping build healthy soil ecosystems and reduce the need for pesticides. Cover crops are plants that are planted to cover the soil rather than to be harvested. They help with soil quality, fertility, and erosion, and with water retention. They help control pests, weeds, and diseases. They can increase biodiversity and aid wildlife. There are sustainable farming methods that allow the elimination of pesticides and less use of fertilizer. These include integrated pest management and use of the natural predators of the insects that eat our crops. Organic fertilizer is better for soil life. Eliminating pesticides and sustainable agriculture would preserve soil biodiversity. Increased soil biodiversity would improve crop yields and increase the amount of carbon stored in soil.

Another reason to eliminate pesticides is to protect the pollinators of our crops. Many are declining because of pesticides. Many are declining because of habitat destruction and climate change. We need to address these problems and conserve pollinating species to protect our food supply. In the Unites States alone, more than 150 crops depend on animal pollinators, including nearly all fruits and grains.

Agroecology is a method of farming that mimics nature, using biodiversity and the recycling of resources to increase food production, eliminate pesticides and fertilizers, and regulate organisms that attack crops with natural predators and other natural means. Growing multiple crops increases food production. It also increases resilience because one crop could fail in conditions that enable another crop to thrive. Agroecological systems are diverse, and provide habitats for frogs, lizards, birds, and other predators of crop-eating insects, helping control them. Social systems in agroecology are different than those of standard agriculture. Communities work together to share practices that best fit their specific ecological and cultural circumstances. These may vary from one place to another. Agroecology recycles nutrients and mimics nature in that it produces no waste. Manure from animals is used for fertilizer, and nitrogen-fixing bacteria in plants roots also provide nitrogen in a form life can utilize. Crop residues such as straw feed farm animals. Biodiversity is kept high and used to advantage. This system of agriculture increases yields, produces healthier food, reduces pollutants that would run into water and cause eutrophication, and sequesters carbon.

In fact, the ABH implies that, if all species are ecosystems engineers that help other species and increase biodiversity under natural conditions, there is no such thing as a pest under natural conditions. What we consider to be pests only cause problems for us because we have set up conditions that would never occur naturally. Large farms with only one crop would never occur naturally. Nature abhors monocultures and organisms create high biodiversity. Monoculture farms cause unnatural population explosions of taxa such as insects and fungi that exploit them only because we provide these taxa with unnaturally high numbers and concentrations of food.

The second activity that would benefit from a goal of preserving biodiversity is how we manage our forests. There are several ways to remove carbon dioxide from the atmosphere. The most effective one is preserving forests and planting trees that are native to the area. Photosynthesis by trees removes CO_2 from the air, storing it in trunks, branches, leaves, and roots. Today, Earth's forests absorb about 16 billion tons of CO_2 per year. Approximately one-half of this is lost to the cutting and burning of forests and other disturbances. Preserving forests, planting native trees, and allowing degraded forests to regenerate would preserve and restore biodiversity. It also is a very effective method of removing CO_2 from the air, while storing it and keeping it available to the biosphere. Preserving forests also prevents soil erosion, fertilizes the soil, conserves water, generates local rainfall, and provides many other useful services.

There are countless other examples of how protecting biodiversity helps us. It is a good exercise for the reader to think of some.

Deep ecology states that nature, biodiversity, and species have inherent value, and species have an inherent right to exist, that we are ethically obligated to preserve nature and biodiversity, and prevent the extinction of species. Ethics has a subjective aspect to it. But the ABH can be interpreted to support deep ecology because the fact that ecosystems have a tendency toward and maximize diversity could be interpreted to imply that diversity has inherent value. The fact that organisms create biodiversity and that all species are ecosystem engineers that have a net positive effect on biodiversity in natural ecosystems over sufficient time periods can be interpreted to imply that nature, biodiversity, and species have inherent value. Pragmatic environmentalism, the complementary philosophy to deep ecology, states that we should preserve nature, ecosystems, and species because we are dependent on them for our survival and well-being because they provide us with resources and services, such as oxygen, favorable climates, food, water, and so on. The ABH is relevant to pragmatic environmentalism because an idea related to it is that symbiosis and commensalism are fundamentally important. This pertains to humans because of our dependence on other species and biodiversity for our survival and well-being, and because other species make our survival and comfort possible through symbiosis, and commensalism in which we are the beneficiary species. For example, we benefit from organisms that build, fertilize, and aerate the soil; plants that produce oxygen and bury carbon; plants and animals we eat; medicines from rainforests; pollinators of our food crops; and so on. The relationship between deep ecology and environmentalism is complementary. Each is dependent on and suggests and supports the other. If we respect the inherent right of nature and species to exist, we preserve them and so benefit from them and better survive and thrive. If we are serious about preserving nature for our survival, this is best accomplished if we respect nature's right to exist.

Humans are unwittingly performing a scientific test of the ABH, giving overwhelming evidence of its validity. It is estimated that humans are currently causing extinction rates 1,000 times higher than natural background rates of extinction, and future rates are likely to be 10,000 times higher (De Vos et al., 2015). If all species currently deemed threatened become extinct in the next century, future extinction rates will be 10 times recent rates (Pimm et al., 1995). As we cause species to become extinct, several additional species go extinct because they depend on the species we exterminate. This shows that many species are dependent on other species for their survival. Further, we know that the species that were helping other species before they went extinct did so by ecosystem engineering. For example, when a tree species goes extinct, this negatively impacts species in the soil that benefit from its holding the soil in place and preventing erosion; herbivores that feed on the tree; animals and fungi that use the tree as habitat or shelter; symbiotic root fungi; animals that eat fruit it produces; and so on. When an animal species that pollinates a plant species goes extinct, this hurts the plant species, which could then go extinct. Remember from Chapter 11 that Koh et al. (2004) estimated from their study that 6,300 of the species that are benefitted by host species are coendangered with host species currently listed as endangered, suggesting that the extinction of a species can lead to extinction of a great number of other species. Recall also that the loss of amphibian species in Panama caused a decline of their snake predators (Zipkin, 2020). Thus, the extinction crisis is evidence for the ABH. It also shows the importance of symbiosis and commensalism, since species dependent on their symbiotic allies and commensal helpers are negatively impacted if the species that benefit them go extinct. Recall that the importance of symbiosis and commensalism and their tendency to aid diversity are hypotheses related to of the ABH.

Ecosystems consist of a tightly linked web of interconnected species. The loss of a few species causes the loss of more. We are seeing verification of the network theory models that postulate a correlation between diversity, complexity, and connectance on the one hand, and stability, resilience, and persistence on the other, in natural ecosystems, when we cause the extinction of a species, and as a result, several species dependent on it have significant population declines or become extinct via coextinction. This is demonstrated especially well when keystone species are eradicated. I discussed the great loss of diversity when the wolf was exterminated from the Greater Yellowstone ecosystem, and its recovery when the wolf was reintroduced there, in Chapter 9 of the companion book to this one (Seaborg, 2022). When beavers are locally eradicated, the entire pond ecosystem they created can disappear. When large herbivores such as elephants and bison are removed, forests replace grasslands, with a loss of diversity because of the loss on an entire ecosystem. This shows that species help other species exist, and that life maintains biodiversity, both of which are consistent with the ABH. All of the observations in this paragraph are evidence for the ABH provided by an unintentional human experiment that involves the extinction of species.

It is not necessary to list the many ways we can preserve diversity in this discussion. They are well known and one can read about them in several places. But it is important to be aware of the importance of protected areas and indigenous people. Preserves have kept extinction rates of amphibians, birds, and mammals 20% lower than they would have been without them. Nearly 13% of Earth's land is protected in preserves but only 2% of the ocean is in preserves. Two studies have shown the most effective method of preserving rainforest, the ecosystem with the highest diversity on Earth's land, is by helping the indigenous people who live there protect their forest (Nepstad et al., 2006; Porter-Bolland et al., 2011). This preserves the most acres of rainforest and species per dollar, and lasts the longest of any form of protection. Nations with tropical rainforests often either lack the money or the commitment to hire enough guards and pay them sufficiently to protect rainforests in the national parks they create. So exploiters of rainforests are often able destroy the rainforests in these parks, which are in many cases "paper parks"—parks on paper only. As a general rule, the amount of rainforest that can be preserved is greater if one wisely aids responsible indigenous people who live in the rainforest than if a government buys rainforest to establish a national park and hires sufficient guards to protect it, spending the same amount of money on each effort. This is largely because the indigenous people already live in the rainforest and are as a whole motivated to protect it. They are the rainforest's natural guards. They are often struggling to survive, and need help in protecting their rainforest homes from those who would profit by destroying the forests. Rainforests have the highest biodiversity of terrestrial ecosystems. Rainforests with the highest diversity are in indigenous lands, and the highest of those are in the forests of uncontacted tribes. We should use the most effective strategy to preserve diversity. We should protect as much of nature as possible, but should give priority to the hotspots with the highest biodiversity when we need to choose which areas to protect. Hence, protection of indigenous rainforest is a wise strategy for the protection of biodiversity.

By far the most important taxa in terms of numbers of individuals and species, impact on ecosystems, and importance in maintaining and creating diversity are the taxa with small organisms, and taxa humans do not tend to notice or pay attention to. The three most important taxa are, in order of importance for all criteria just listed, with the most important listed first: viruses, prokaryotes, and arthropods, especially insects. Over 90% of all species of life are viruses, over 90% of all species of cellular life are prokaryotes, and over 50% of all eukaryote species are insects. Most life is deep underground and out of the photic zone. The highest animal diversity is in the rainforest canopy, and there are more species of insect there than in any other class of animal. Thus, an implication of the ABH is that we should pay closer attention to the small organisms. We should study viruses, prokaryotes, and insects, both as an effective way to learn more truths in basic science and for applications. Another implication is that we should pay attention to organisms that dwell underground, out of the photic zone. Another is that we should study the rainforest canopy more thoroughly.

Not surprisingly, biodiversity is correlated with ecological importance. Viruses regulate global climate, keep phytoplankton populations healthy via the viral shunt, increase marine unicellular species diversity via kill the winner, regulate populations of all cellular species, promote evolution by passing genes between cellular species, and were important in a number of major macroevolutionary transitions in eukaryotes. Prokaryotes are responsible for the major chemistry that runs ecosystems: anoxygenic and oxygenic photosynthesis, respiration, nitrogen fixation, a great deal of decomposition, and more. They aid multicellular animals and plants (by being in their microbiome) in digestion, immunity, and many other vital functions. They transfer genes to multicellular organisms, aiding their evolution. Insects are the most impactful animals in ecological effects and in generating diversification. They are food sources, pollinators, seed dispersers, decomposers, soil improvers, and regulators of plant populations. Thus, viruses, the most diverse taxon, have the highest ecological impact, and prokaryotes, the second most diverse taxon, have the second highest. Insects are a lower taxonomic level (class) than viruses and prokaryotes, so not comparable. But insects are one of the most diverse classes of animal *and* one of the most ecologically important ones.

Another implication of the ABH is that the microbiome, connections between species, and coevolution lead us to the conclusion that the concept of the individual is not entirely correct, and that, in a real sense, there is no such thing as an individual. Multicellular organisms are ecosystems with diverse microbiomes of viruses, bacteria, archaea, fungi, and invertebrates that they are dependent on and have symbiotic and commensal relationships with. And the organisms and species in the microbiome are connected to each other by symbiotic and commensal relationships. They are dependent on their host, subject to group selection favoring those species that help, or at least do not harm, the host. The species of the microbiome affect the evolution of the host, even providing it with genetic material via horizontal gene transfer. They can promote host speciation. The host is a not an individual, but an ecosystem, a collection of interacting organisms. Viruses can be involved in macroevolutionary transitions in multicellular organisms. And every "individual" is tightly connected to other organisms in its population, and to many other species. It is connected to other members of its species, and to predators, parasites, prey, hosts, competitors, commensals, symbiotic species, and so on. Each of these affects its survival and evolution, and can exchange genes with it. This connection to other organisms of its population and to other species it interacts with holds true whether the "individual" is an animal, plant, fungus, protist, prokaryote, virus, or member of any other taxon. The individual is real and the organism is an individual in that it is a functioning, living unit that has a separateness, and it is more consequential to the individual organism in question than to other organisms if it dies. But the organism is not an individual in the sense that it is a part of a larger system that it cannot be separated from. The relationship between the autonomous self and the whole system is complementary, where self and system are interdependent. Each depends on the other: The individual cannot exist without the system, but the system is made up of, and so needs, the various individuals. Thus, the view that there is no individual at all and the view that there is a real, autonomous individual are both only partly true, and both views are needed for the whole reality to be understood.

The same is true for the gene. No gene exists autonomously or in a vacuum; each gene is connected to and interacts with other genes in its genome, with genes in other organisms of its population, and with genes of other species in its community. Genes even jump from one part of an organism's genetic material to another, from one organism to another, and between species, even between domains. Viruses can give DNA to animals and plants. The gene is real in the same sense that the individual is real, and not real in the same sense that the individual organism does not exist apart from its microbiome and ecosystem. The gene exists because deleting or changing it would affect the organism it is in. It does not exist in the sense that it is meaningless except in the context of its genome, the gene pool of its population, and all the genes of the organisms that the organism with the gene in question interacts with in the ecosystem, directly or indirectly.

Similarly, no species exists in isolation. All species are tightly connected to other species. All species affect the evolution of other species, and the evolution of all species is affected by other species. Sometimes species coevolve. Sometimes, this coevolution can lead to macroevolutionary breakthroughs, such as the evolution of flowers and animal pollination. Yet we need the concept of species for models, and for discussing speciation and interspecific relationships. And species are real in the sense that removing one species affects other species. The reality of individual species and their lack of autonomous existence have the same complementary relationship as do the individual and its ecosystem. Species are complimentary with their ecosystems. Species exist as entities while lacking full existence without other species and their ecosystems. Species need their ecosystems to exist and to define them, and ecosystems are made up of species, so need them to exist.

Another implication of the ABH is the importance of coevolution. Much evolution is coevolution, with feedback between species. Species evolve in a many-stranded, interacting web. *Most evolution between tightly interacting species is coevolution.* This includes herbivores and the plants they eat, predators and their prey, parasites and their hosts, competition between species, and symbiosis. Also, the geology, mineralogy, atmosphere, water, soil, chemistry, and physics of the Earth evolved in a coevolutionary feedback with life.

Another implication is that symbiosis is one of the most important, predominate interspecific interactions. Symbiosis is common, pervasive, organizes ecosystems, and sometimes creates tremendous adaptive radiation into extraordinarily high biodiversity when it first appears and the symbiotic species diversify. This is followed by more diversification when other species evolve to exploit the niches provided by the many newly evolved symbiotic species. Symbiosis has a tendency to evolve under certain conditions. Symbiosis creates and maintains biodiversity. The high diversity we see today would not be possible without symbiosis. There is no known species that could survive without it. Even the mitochondria and chloroplasts of eukaryotes evolved by symbiosis. The many symbiotic relationships between animals and plants show its importance. These include pollination, and seed dispersal. There are many microbe species that are symbiotic with their host and/or with each other in the microbiome of every multicellular species. This suggests we might benefit from working *together* with other species, while preserving them, rather than exploiting them to their extinction or nearly to it. The pervasiveness of symbiosis implies that we would also benefit from working together as a race, with emphasis on cooperation, rather than settling conflicts of interest through violence and war. In addition to symbiosis, the altruism displayed within species re-enforces this idea.

Another implication is that commensalism, too, is very common and important, and structures ecosystems. Commensalism maintains and increases biodiversity. Keystone species increase diversity tremendously, mainly through commensalism, but also via symbiosis when the beneficiary species return the help that they receive from the benefactor species. All species provide niches for at least some other species. There are a tremendous number of microbial species that are commensal with their host and/or with each other in the microbiome of every multicellular species.

An interesting implication from Chapter 12 is that the Universe is primed for life, the scientific laws are precisely right for the appearance of life, and chemical evolution of life is inevitable, under favorable conditions for life, such as are present on Earth. There is a tendency for a local increase in order, complexity, and information without violating the second law of thermodynamics when conditions are right and there is an input of energy from an outside source. In the case of the Earth, the outside source providing the energy is the sun. There is an enormous number of stars in the Universe. Thus, it is inevitable that some stars will be the right size and will have planets the right distance from them and with the right conditions for life to evolve. There are almost certainly many other planets with life, and a large subset of those with complex, multicellular, even intelligent, life. Chapter 13 demonstrates an additional implication, that the solar system creates conditions favorable to life. The sun, Jupiter, the moon, and the nonbiological Earth all contribute to this. While this does not contradict the ABH, it shows factors other than life contribute to biodiversity. It suggests a more general hypothesis than the ABH, which postulates that the Solar System is a system that helps and promotes life, and increases in diversity, information, and order in local areas with the right conditions, such as on Earth. Diversity is a form of information, so when the nonbiological and biological factors of the Solar System promote increases in diversity, they promote an increase in information.

In my view, these are the main implications to the ABH/Pachamama Hypothesis. However, the implications have a subjective aspect and are open to debate. The reader is encouraged to think of his or her own implications of the hypothesis.

CONCLUSION

It is time for an entirely new way of thinking and new model and synthesis for ecology, evolution, and population biology. It is time for a paradigm shift. *Ecosystems maximize diversity, organisms create diversity, and all species are ecosystem engineers that have a net positive effect on biodiversity, other species, and their ecosystems, in natural conditions over sufficiently long time periods.* Symbiosis and commensalism are common, pervasive, structure ecosystems, and maintain and greatly increase diversity. The concept of the autonomous individual is only partially valid. Genes, organisms, and species do not exist or evolve in isolation, but are connected to each other and larger systems. The evolution of all genes, organisms, and species affects and is affected by other genes, organisms, and species. Much of evolution is coevolution, even between life and the nonliving environment. *All evolution between closely interacting species is coevolution.* Ecosystems are holistic, interacting systems. The number and nature of relationships between species are at least as important as the number of species in determining the stability and resilience of ecosystems. Genomes promote variability, adaptive evolution, and evolutionary breakthroughs. Evolutionary breakthroughs lead to diversification. Besides selecting for adaptive traits in the genes and organism, natural selection selects for a favorable amount and nature of variability of the genotype and phenotype.

A paraphrase of Darwin's last sentence in *The Origin of Species* (1858) is as pertinent to the ABH/Pachamama Hypothesis as to evolution by natural selection, and serves as a fitting end to this treatise: There is grandeur in this view of life, where, with its several powers, from so simple a beginning, endless forms most beautiful and most wonderful have been, and are being, evolved.

REFERENCES

Beaudry, F. (ed.) (updated Feb. 18, 2019). Tropical rainforests are nature's medicine cabinet. *Earth Talk*. ThoughtCo, https://www.thoughtco.com (secondary literature).

Darwin, C. R. (1958). *The Origin of Species*. A Mentor Book from New American Library, New York, NY and Toronto, Canada; The New English Library Limited, London, UK.

De Vos, J. M., et al. (April, 2015). Estimating the normal background rate of species extinction. *Conservation Biology* 29 (2): 452–62. https://doi.org/10.1111/cobi.12380.

Dunnell, T. (21 July, 2018). Medicinal plants of the Amazon and their uses in modern medicine. www.savacations.com/medicinal-plants-amazon-uses-modern-medicine/ (secondary literature).

Gunstone, T., Cornelisse, T., Klein, K., et al. (4 May, 2021). Pesticides and soil invertebrates: a hazard assessment. *Frontiers in Environ. Science*. https://doi.org/10.3389/fenvs.2021.643847.

Koh, L. P., et al. (10 Sept., 2004). Species coextinctions and the biodiversity crisis. *Science* 305 (5690): 1632–4. doi: 10.1126/science.1101101.

Nepstad, D., et al. (Feb., 2006). Inhibition of Amazon deforestation and fire by parks and indigenous lands. *Conservation Biol.* 20 (1): 65–73.

Pimm, S L., et al. (21 July, 1995). The future of biodiversity. *Science* 269 (5222): 347–50. doi: 10.1126/science.269.5222.347.

Porter-Bolland, L., et al. (15 March, 2012). Community managed forests and forest protected areas: An assessment of their conservation effectiveness across the tropics. *Forest Ecol. and Management* 268: 6–17. doi: 10.1016/j.foreco.2011.05.034. Journal homepage: www.elsevier.com/locate/foreco.

Seaborg, D. (2022). *How Life Increases Biodiversity: An Autocatalytic Hypothesis.* CRC Press/Taylor & Francis. Boca Raton, FL; London, UK; New York, NY.

Zipkin, E. F. (14 Feb., 2020). Tropical snake diversity collapses after widespread amphibian loss. *Science* 367 (6479): 814–6. doi: 10.1126/science.aay5733.

Glossary

The companion book to this one does not have a glossary. This glossary covers terms in both this book and its companion book.

adaptive zone A broad area of adaptation. A set of related ecological niches. When a species invades a new habitat or evolves a new function, it typically enters a new adaptive zone. This allows it to diversify into many species that then occupy the many new, unoccupied niches available to it. An example of an adaptive zone is all niches for daytime vertebrate flight. When birds evolved, they moved into this adaptive zone and diversified into the many niches in this zone, from niches of hummingbirds to eagles. Some other authors define this term somewhat differently than the way it is used and defined in this book.

allele A form of a gene. A gene may have several different alleles. Diploid organisms have two alleles of each of their genes; these alleles may be identical or different.

amino acid A molecule that is the basic building block of proteins. There are 20 amino acids used by living organisms to make proteins.

anoxygenic photosynthesis A chemical reaction that uses the energy of sunlight to convert hydrogen sulfide and carbon dioxide to carbohydrate (food) and elemental sulfur. It is a form of photosynthesis that does not produce oxygen.

antagonistic coevolution Coevolution in which two species are in conflict, as in predators and their prey, or two species competing with each other. One species develops a defense against the other, the second species responds with a counter to that defense, the first species evolves another adaptation against the second species, and so on. Antagonistic coevolution can also occur between the two sexes within a species.

archaea (**singular**: **archeon**) Prokaryotes that constitute a domain of life, and are not bacteria. The other two domains of life are bacteria and eukaryotes. Some archaea live in extreme habitats, such as hot springs with very high temperatures.

Autocatalytic Biodiversity Hypothesis (abbreviated **ABH**; also called the **Pachamama Hypothesis**) The hypothesis of the author of this book, David Seaborg, that this book presents evidence and arguments for. It states that all species are ecosystem engineers that have a net positive effect on other species, their ecosystem, and biodiversity, in natural ecosystems, over sufficiently long time periods to make a fair assessment of their effects. It also postulates that ecosystems maximize biodiversity. It posits that both symbiosis and commensalism are of fundamental importance in structuring ecosystems and increasing biodiversity. It has similarities to the Gaia Hypothesis and was inspired by it, but is distinctly different from it. It is original, profound, radical, supported by the evidence, scientific, and testable. It does not propose New Age ideas such as the Earth is a conscious entity directing diversity increase. Its originator is firmly opposed to such interpretations of the hypothesis.

bacteriophage (also called **phage**) A virus that attacks bacteria or archaea.

bacterium (plural: **bacteria**) Prokaryotes that constitute a domain of life. Only a very small percentage are pathogenic to other organisms.

biodiversity The variety of life on Earth, in an ecosystem, or in a biological group (taxon). The biological variety and variability of life on Earth. It is a measure of variation at the genetic, population, species, and ecosystem level. It can be approximated as the number of species. One definition of it is the number of species and how equally individuals are distributed between species.

biological pump The sea's biologically driven sequestration (burial) of carbon from the atmosphere and from land runoff, mainly via rivers, to the ocean's interior and seafloor sediments.

biosphere The part of the Earth occupied by life; it is the sum of all Earth's ecosystems.

biota The animal and plant life of a particular region, habitat, or geological period.

bivalve mollusc a mollusc that has a shell of two parts that has a hinge, and that it can open and close; inside the shell is the soft-bodied animal. They are invertebrates. *See* mollusc.

calcium carbonate A chemical compound that contains calcium and carbon, is important in the carbon cycle and hence regulation of the amount of carbon in the biosphere, and is in echinoderms (the group that includes sea stars and sea urchins), coral, the shells of bivalve molluscs, the shells of some phytoplankton, and some other groups of organisms. Its chemical symbol is $CaCO_3$. Limestone is mostly calcium carbonate and was mostly formed by biological organisms, sequestering great quantities of carbon.

Cambrian explosion A rapid diversification of animals during a geologically short interval of time approximately 541 mya, at the beginning of the Cambrian period (this period was about 541 to 485.4 mya) when practically all major animal phyla first appeared suddenly and quickly.

capsid A shell made of protein that surrounds, contains, and protects the genetic material of a virus. The genetic material of a virus maty be DNA or RNA.

carbon The most important element for life. It is in all large biological molecules, including DNA, RNA, proteins, carbohydrates, and fats. It is also in carbon dioxide. Chemical symbol is C.

carbonate A chemical compound that has carbon and oxygen. Many carbonates are used by life.

carbon cycle The biogeochemical cycle by which carbon is exchanged between the biosphere, Earth, water, and atmosphere. It consists of a short-term and a long-term carbon cycle. *See* long-term carbon cycle, short-term carbon cycle.

carbon dioxide A molecule used in photosynthesis and produced in cellular respiration. Animals, including humans, expel it when they breathe. It is the main and most abundant greenhouse gas, causing warming of the planet because people are adding too much of it to the atmosphere. People are also adding too much of it to freshwater ecosystems and the ocean, making both of these aquatic systems too acidic. Chemical symbol is CO_2.

carbon sequestration The removal of carbon from the biosphere and atmosphere and storing it elsewhere, such as in the Earth or under the seafloor. This includes weathering and burial. *See* weathering.

carbon sink An entity that removes more carbon from the biosphere and atmosphere than it releases to them. In the case of the long-term carbon cycle, this carbon is stored in the solid Earth, including the seafloor.

carbon source An entity that adds carbon to the biosphere or atmosphere. This includes volcanoes, movement of the continents over the seafloor's calcium carbonate and organic sediments, and spreading of the seafloor. It also includes the burning of fossil fuels by humans.

catalyst (the verb is **to catalyze**) A substance that increases the rate of a chemical reaction, aids its occurrence, or both. An enzyme is a specific type of catalyst. *See* enzyme.

cell The basic structural and functional unit of life (except viruses). All organisms except viruses are made up of cells.

cellular organism An organism made up of cells. All organisms except viruses are cellular organisms.

chloroplast The organelle of the cell that carries out photosynthesis in organisms that can photosynthesize, such as cyanobacteria, phytoplankton, and plants. Chloroplasts have their own genomes, hence their own DNA.

chromosome The structure in the nucleus of eukaryotic cells that has DNA combined with protein. In prokaryotes, which do not have a nucleus, the DNA scattered throughout the cell can be called the chromosome.

class A taxonomic group of related orders of organisms. The taxon between order and phylum.

coevolution Evolution by positive feedback between two species, in which a change in the first species leads to a change in the second species, which leads to another change in the first species, and so on, until an equilibrium is reached in which there is no net gain in further changes in either species with respect to the traits undergoing coevolution. Coevolution may be positive or symbiotic, as in the evolution of pollination. Alternatively, it may be negative or antagonistic, as in predators and their prey, or two species competing with each other. There is also symbiotic genetic coevolution, which involves the transfer of genes between species. *See* symbiotic coevolution; antagonistic coevolution; symbiotic genetic coevolution.

commensalism (the adjective is **commensal**) A relationship between two species in which one benefits and the other is not affected, being neither helped nor harmed. An example is a hermit crab using the shell of a dead snail. The hermit crab obtains a protective, moveable home, and the snail is not affected because it is dead.

control gene *See* regulatory gene.

diversification An increase in the number of species of a group (taxon) of organisms due to many speciation events. It often occurs when the taxon reaches a new adaptive zone with many new unfilled niches with no competition from other species, and no predators. It also occurs if the taxon invades a new habitat and thus has access to many new, unoccupied niches and a lack of competition from other species, and a lack of predators. Diversification due to these two reasons lasts only until all the new niches are filled and a new equilibrium with the environment is reached. Another term for diversification is adaptive radiation.

DNA The genetic material of all organisms, except for some viruses. DNA codes directly for RNA and indirectly for proteins. In eukaryotes, the majority of it is in the nucleus of the cell, but it is also in the mitochondria and chloroplasts. Prokaryotes do not have a nucleus, mitochondria, or chloroplasts, so the DNA is spread throughout their cell, most of it on one circular chromosome, but some DNA can also be in smaller circular structures called plasmids in prokaryotes only. Some viruses have DNA as their genetic material; some have RNA.

domain A group of related kingdoms. The highest taxonomic group. There are three domains: archaea, bacteria, and eukaryotes.

ecology The study of the relationships of organisms and species to each other and to their environment. This includes humans. Ecology considers organisms at the individual, population, species, community, ecosystem, and biosphere levels. Relationships include herbivore-plant, predator-prey, parasite-host, competition between species and between individuals of the same species, commensalism, and symbiosis.

ecosystem A biological community of interacting organisms and their physical environment. An ecosystem consists of all the organisms and the physical environment with which they interact.

Ediacaran biota The life of the Ediacaran period (about 635–541 mya). It was simple and all of it was in the sea. It included the first multicellular organisms. Many of the animals could not move, and were attached to rocks or the seafloor. Many of the animals were frond-shaped and tubular. They included trilobites and jellyfish. They were not as complex on the average as the animals that followed them, from 541 mya to the present.

endogenous retrovirus (**ERV**) A type of jumping DNA segment in the genome that resembles viruses and can be derived from them. ERVs sometimes get packaged and moved within the genome, with the result that they end up serving a vital role in gene expression and regulation for their host. They comprise up to 5–8% of the human genome. Apparently, some of them evolved from viruses, although some researchers think some viruses evolved from ERVs, since ERVs can mutate and become separate from their cell, and even pathogenic.

enzyme A catalyst that regulates the rate at which chemical reactions of living organisms proceed, generally increasing their rate. Enzymes aid the chemical reactions of organisms. Without enzymes, many of these reactions would not take place at a perceptible rate. Enzymes regulate all aspects of cell metabolism. All enzymes are proteins. *See* catalyst.

et al. From the Latin phrase meaning "and others." A citation reading Jones, V.B. et al. means "Jones, V.B. and others".

eukaryote An organism whose cell or cells have a nucleus enclosed in a membrane. The nucleus contains the cell's DNA. Eukaryotes have organelles enclosed in membranes, such as mitochondria and chloroplasts. Eukaryotes comprise one of the three domains of life, the other two being the bacteria and archaea, which are both prokaryotes.

eusociality (the adjective is **eusocial**) The highest level of social organization in an animal species. It is defined by cooperative care of juveniles (and eggs if the animal is egg-laying), including care of offspring by individuals that do not reproduce; overlapping generations; and a single female or caste producing the offspring. Honey bees, some ant species, and naked moles-rats are eusocial animals.

eutrophication The depletion of oxygen in a freshwater or marine ecosystem that occurs because an excess of nutrients such as phosphorus or nitrogen, are added, leading to a population explosion of phytoplankton, which causes a high population of bacteria that decompose the phytoplankton when they die in great numbers. The bacteria deplete the oxygen when they decompose the phytoplankton. The resulting system has very low oxygen levels, and so cannot support very much life. Eutrophication can lead to the death of many fish and invertebrates.

euxinia (the adjective is **euxinic**) Conditions of depleted oxygen and high levels of toxic hydrogen sulfide (H_2S). Usually refers to an aquatic ecosystem, mainly the ocean.

evolution Change in the characteristics of a gene, genome, trait, organism, population, or species over time. It is mostly the result of natural selection. It is how the great diversity of life came about. Life started as one-celled prokaryotes and evolved to produce viruses, trees, elephants, humans, and the rest of life.

exaptation The change in function of an already-existing gene, protein, structure, or trait due to natural selection. It may or may not involve changing its form. Evolution often works by taking existing genes, proteins, structures, and traits, and building on them, adapting them to new functions. An example of exaptation is the fins of fish used for swimming evolved into legs used for walking when amphibians evolved from fish. Exaptation can include a behavior being changed to a new function.

exon The part of a gene that is coded into a protein domain.

exon shuffling A process whereby two or more exons from different genes are brought together, or an exon is duplicated, making two copies of it. It is a mechanism for forming new genes.

exoskeleton The external skeleton of arthropods.

family A group of related genera. The taxon between genus and order.

feedback A process in which the effect alters its cause. The effect and cause may or may not have causes and effects between them. It makes a circular loop, called a feedback loop. Factor A affects factor B, which affects factor C, which affects factor A. *See* negative feedback, positive feedback.

food web The branching flow of energy in an ecosystem from primary producers, such as plants, to herbivores to carnivores to decomposers. *See* trophic level.

Gaia One of the Greek primordial deities, the personification of the Earth and ancestral mother of all life. The Gaia Hypothesis is named for this goddess.

Gaia Hypothesis The hypothesis of James Lovelock and Lynn Margulis that posits that life made the Earth, especially the atmosphere, favorable to life, and regulates it, keeping it favorable to life, by negative feedback. This hypothesis inspired and influenced the Autocatalytic Biodiversity Hypothesis/Pachamama Hypothesis.

game theory A mathematical method for determining the best strategy to use to win in a conflict. It is more than this, but this definition applies for this book.

gene A basic unit of heredity and a sequence of bases in DNA that codes for the production of either a protein or RNA.

gene transfer agents Virus-like particles that contain DNA and that are produced by some bacteria and archaea. They package and carry random DNA segments in the host bacteria or archaea that can be passed by horizontal gene transfer to a recipient cell. They are thus agents of horizontal gene transfer. The genes that code for gene transfer agents are derived from phage DNA that has integrated into a host chromosome.

genome The complete set of genes or genetic material in a cell or organism.

genotype The complete set of genetic material, the entire genome, of an organism. It is all of the DNA of an organism (or all of the RNA in a virus whose genetic material is RNA). Genotype contrasts with phenotype. *See* phenotype.

genus (plural: **genera**) A group of related species. The taxon between species and family.

green beard effect A phenomenon that occurs when the carriers of an allele or alleles of a gene or genes recognize the same allele(s) in others who are not necessarily kin, and as a result act altruistically toward them. It involves altruism between organisms with shared alleles expressed by specific phenotypic traits, such as color or height.

group brain The "brain" from the collective intelligence of a highly social group or colony of organisms that function as a super organism. The intelligence of the group can exceed that of any individual. Examples where it may occur include highly social bees, ants, naked mole-rats, and humans. It is thought by some, but not all, thinkers to exist.

group selection Natural selection between groups of the same species. The group with more altruists and cooperators out competes the group with more selfish individuals. This controversial idea is not universally accepted or rejected by evolutionary biologists.

hectare A unit of area equal to 10,000 square meters.

horizontal gene transfer (HGT), also called **lateral gene transfer** The lateral movement of genetic material between organisms other than from parent to offspring. It can occur between organisms of the same or different species, even between different domains and between viruses and cellular orgaisms. It can generate variability without disrupting the compatibility of the parts of the organism.

humus The organic component of soil, formed by the decomposition of plant and animal matter by soil organisms.

immunoglobulin A (IgA) An antibody that plays a role in the immune function of mucous membranes. It attacks pathogens, and regulates gut bacteria and maintains the symbiotic balance between gut microbes and the immune system, playing a key role in maintaining a sustainable host-microbial relationship. Even though it is an antibody, it helps bacteria at times in so doing, showing the importance of the microbiome to its host.

long-term carbon cycle The carbon cycle that involves the exchange of carbon between rocks and the Earth's surface; it operates over millions of years. *See* carbon cycle, short-term carbon cycle.

ibid. An abbreviation for the Latin word ibīdem, meaning "in the same place," used in a scholarly reference to refer to the reference immediately cited beforehand. "It was shown (Smith, 2002) and it was also shown (ibid.)," means: "It was shown (Smith, 2002) and it was also shown (Smith, 2002)."

intron The part of a gene that is not coded into a protein.

invertebrate An animal without a backbone. This includes insects, worms, sea urchins, and many other animals.

kingdom A group of related phyla. The taxon between phylum and domain.

kin selection Natural selection for altruism in which an individual passes on its genes by helping its kin, which share genes with it. To occur, the relatedness of the altruist to the recipient of the altruism times the benefit received by the recipient from the altruism must exceed the cost of the act of altruism to the altruist.

lymphoid tissues Cells and organs that make up the lymphatic system, such as white blood cells, bone marrow, and the thymus, spleen, and lymph nodes. They are organized structures that support immune responses.

lysis (the verb is **to lyse**) The bursting open of a cell, killing it. Bacteriophage often do this to bacteria.

lysogeny (adjective is **lysogenic**) The integration of the viral DNA into the host DNA, making the DNA of virus and host one continuous strand of DNA. In this state, the virus does not hurt the host, and replicates its DNA when the host replicates the host's DNA. The lysogenic DNA of the virus can leave the host DNA and start reproducing in the host again, in which case it is harmful to the host. It can also leave the host DNA and stay lysogenic, integrating into a new host, staying harmless to the new host. When it does this, it can take some host DNA with it. It can also give that DNA to its new host when it integrates into the DNA of its new host. This is called transduction. This usually happens to another organism of the same species, but can occur between species. Transduction is a source of variability for natural selection to act on, and can be a mechanism of evolution of new adaptive traits, of evolutionary innovation. Note that lysis and lysogeny are similar words, but have very different meanings; be careful not to confuse them with each other.

lytic Describes a phage that uses its bacterial host cell to reproduce and make large numbers, and then bursts and kills the cell. Thus, the phage causes lysis of its host cell. *See* lysis.

marine snow A continuous shower of debris falling through the sea, originating in its upper layers, that resembles falling snow, and is made up of parts or all of dead or dying animals, seaweeds, plankton, bacteria, archaea, fecal matter, sand, soot, and other inorganic dust. Most of it is made up of and generated by organisms. Some of it gets buried in the seafloor, sequestering large amounts of carbon. It delivers nutrients that are used by organisms at the bottom of the sea.

Medea A Greek goddess who killed her children, so was a bad mother. Used to name the Medea Hypothesis.

Medea Hypothesis The hypothesis of Peter Ward that proposes that life is suicidal and tends to hurt life. It disagrees with the Autocatalytic Biodiversity Hypothesis/Pachamama Hypothesis and the Gaia Hypothesis.

metabolism The set of life-sustaining chemical reactions in organisms, including obtaining energy from food, building molecules like proteins, and elimination of wastes.

meter The basic unit of length in the metric system, equal to just over three feet.

mineralization The process by which chemicals present in organic matter are decomposed or oxidized into easily available forms for plants. This is done by decomposers, including bacteria, soil mesofauna, soil macrofauna, and other groups.

mitochondrion (plural: **mitochondria**) The organelle of the cell that carries out cellular respiration, using oxygen to produce energy. Mitochondria have their own genomes. They are only in eukaryotes; prokaryotes do not have them.

mollusc (also spelled **mollusk**) An invertebrate of a large phylum which includes snails, slugs, mussels, clams, and octopuses. They have a soft, unsegmented body and live in aquatic or damp habitats, and most kinds have an external shell containing calcium carbonate.

morphology The physical form and structure of an organism. Feathers, wings, and colors are aspects of a bird's morphology.

mucosa The moist, inner lining of some organs and body cavities, such as the nose, mouth, lungs, and stomach. Digestive system bacteria help create the immune system. In vertebrates, they promote the early development of the digestive system's mucosal immune system and aid its operation later in life, and symbiotic bacteria stimulate and organize the lymphoid tissue associated with the digestive system's mucosa. The digestive system's mucosa can then produce antibodies to pathogenic bacteria.

natural selection The process by which organisms better adapted to their environment tend to survive with higher probability and produce more offspring than less well-adapted ones. It results from differences in the genotype and phenotype. It is a key mechanism of evolution. It can result in extraordinary adaptations. It is a major mechanism by which the Autocatalytic Biodiversity Hypothesis operates. It often causes an increase in biodiversity.

negative feedback A process in which an effect decreases the magnitude of its cause. It is stabilizing. In negative feedback, one factor, call it factor A, causes another factor B to increase, and factor B causes factor A to decrease. There can be more factors in the loop. So if A increases B, which increases C, which comes back and decreases A, there is negative feedback. Since it makes a circle, one often uses the term negative feedback loop to refer to the process. An example of negative feedback is: Predators eat more of their prey when there is more of it, causing less prey; this means there is less food for the predators to eat, so the number of predators goes down, and they thus eat less prey, allowing prey numbers to increase, and so on. *See* feedback and positive feedback.

neoteny The sexual maturity of an animal while it is still in a larval state, or the retention of juvenile features in the adult animal. Larvae that underwent neoteny can reproduce. This can lead to macroevolutionary breakthroughs. The vertebrate line was started by neoteny in the sea squirt. The adult sea squirt cannot move and has no nervous system, but its swimming larva has a nervous system. This larva underwent a mutation that made it able to reproduce while still a larva; this started the vertebrate line.

niche, also called **ecological niche** The role of a species in an ecosystem. For example, one aspect of the niche of the wolf is that it is a predator that eats deer, rabbits, and other animals.

nitrogen fixation The process by which molecular nitrogen in the atmosphere (N_2) is converted to a form of nitrogen such as ammonia (NH_3) that can be used by plants. Nitrogen fixation can be carried out by lightning, but most is carried out by life, mostly by nitrogen-fixing bacteria, some of which live freely in the soil and some of which live in plant roots. The nitrogen-fixing bacteria in plant roots are symbiotic with plants. The plants give them a home and carbohydrates, and they provide the plant with nitrogen in a form that the plant can use.

nitrogen-fixing bacteria Bacteria that fix nitrogen. *See* nitrogen fixation.

nucleobase A molecule that is one of the basic building blocks of DNA and RNA. There are four nucleobases in DNA and four in RNA. Three of the nucleobases in DNA are in RNA. One of the nucleobases in DNA is not in RNA. One of the nucleobases in RNA is not in DNA.

order A group of related families. The taxon between family and class.

organelle A specialized structure in a cell that carries out a specific function for the cell. It is analogous to an organ in the body. The mitochondrion is an organelle.

organic carbon Carbon found in nature that comes from or is in living organisms. *See* carbon.

oxygenic photosynthesis The chemical reaction that uses the energy of sunlight to convert carbon dioxide and water into oxygen and carbohydrates (food). Oxygenic photosynthesis is carried out mainly by green plants, phytoplankton, and cyanobacteria. It produces the carbohydrate that feeds a great deal of the biosphere. It is called *oxygenic* because also produces oxygen, the energy source for a tremendous number of species. Oxygenic photosynthesis is the opposite reaction of respiration. *See* photosynthesis.

Pachamama The ever-present mother goddess of Earth and time who has her own self-sufficient and creative power to sustain life on Earth, revered by indigenous peoples of the Andes, such as the Quechua and Aymara. Pachamama is usually translated as Mother Earth, but a more literal translation is World Mother in Quechua and Aymara. The Pachamama Hypothesis is named for this goddess.

Pachamama Hypothesis *See* Autocatalytic Biodiversity Hypothesis.

paralogous genes Genes at different locations in the genome in the same organism with a similar structure, indicating that they diverged from a common ancestral gene.

pathogen (the adjective is **pathogenic**) An organism that causes disease.

phenotype Characteristics and traits of an organism that can be seen. The term covers the organism's morphology, which is its physical form and structure. The phenotype also includes developmental processes, biochemical and physiological properties, and the organism's behavior. The phenotype includes the outward appearance (shape, structure, coloration, pattern, size, weight), and the form and structure of the internal parts, like bones and organs. Phenotype contrasts with genotype. *See* genotype.

photic zone The top layer, nearest the surface of the ocean or a lake, where enough light penetrates the water to allow photosynthesis to occur. Photosynthetic phytoplankton and bacteria live in this zone.

photosynthesis The chemical reaction that uses the energy of sunlight to convert carbon dioxide to carbohydrate (food). Oxygenic photosynthesis produces oxygen. Anoxygenic photosynthesis produces sulfur. *See* anoxygenic photosynthesis, oxygenic photosynthesis.

phylum (plural: **phyla**) A group of related classes. The taxon between class and kingdom.

phytolith A microscopic silica grain formed by plants in their leaves and stems. Plants that form them include all grasses, and some herbs, shrubs, and trees. A small amount of carbon becomes encapsulated in each grain. This sequesters carbon, effectively removing it from the atmosphere for millennia. Phytoliths are an effective mechanism by which life sequesters carbon while keeping it available to the biosphere.

phytoplankton Small, often unicellular, organisms in the sea and freshwater ecosystems that carry out photosynthesis. They are very important to the biosphere and the Autocatalytic Biodiversity Hypothesis because they are the base of aquatic food webs, produce much of the Earth's oxygen, and sequester a great deal of carbon. Their sequestering of carbon helps regulate Earth's temperature and the acidity of aquatic ecosystems.

placental mammal a mammal that carries the fetus in the uterus of the mother to a relatively late stage of development. Placental mammals contain the vast majority of living mammals. They are the most complex mammals, and include wolves, deer, lions, monkeys, and humans.

plasmids Small circular pieces of DNA in cells that are separate from the chromosome. They replicate independently. They contain genes, often ones that benefit the organism. They can transfer to other cells. They can integrate into the cell's chromosome (bacteria only have one chromosome). They are mostly in bacteria.

point mutation A mutation in only one base in the DNA.

positive feedback A process in which an effect increases the magnitude of its cause. It is destabilizing. Factor A causes factor B to increase and B causes A to increase. There can be more than two factors. If A increases B, which increases C, which loops back and increases A, there is positive feedback. Since it loops back into a circle, it is called a positive feedback loop. For example, people have caused increased CO_2 in the air, warming the temperature, and causing the ice over the sea on the coast of Greenland to melt. This exposes the sea, which is darker than the ice that has melted and disappeared. Darker surfaces absorb more heat than lighter surfaces. So melting the ice and exposing the darker sea causes more heat to be absorbed, temperatures get hotter, and more ice melts. Positive feedback can lead to catastrophe. *See* feedback and negative feedback.

power of ten (or powers of ten) A shorthand way of writing a large number in which a number between 1 and 9 is multiplied by 10 raised to an exponent, or power. For example, 5×10^6 is 5 times 10 to the 6th power, which is 5 multiplied by 10 six times, or 5 with 6 zeroes after it, or 5,000,000. And 6.7×10^3 is 6.7 times 10 to 3rd power, or 6,700.

Precambrian supereon Time in Earth's history spanning from just after the formation of the Earth about 4.6 bya to about 541 mya. Life began in this supereon. Complex animals appeared right after it ended.

predation An ecological interaction where one species, the predator, kills and eats another species, its prey. It is a form of symbiosis if the population of the prey is considered, rather than individual prey organisms.

predator An animal that eats other animals, its prey. Predators increase biodiversity. They regulate prey populations. They selectively eat better competitors among their prey species, preventing the species that is the better competitor from driving the poorer competitor locally extinct, thus maintaining high diversity. Removal of predators can result in a catastrophic trophic cascade with the collapse of the ecosystem. *See* trophic cascade.

prey An animal that is eaten by another animal, its predator.

primordial soup The liquid, probably water, containing various chemicals where life first evolved via chemical evolution.

prokaryote An organism whose cell lacks a nucleus, having its DNA existing throughout its cell. Prokaryotes have no organelles, such as mitochondria. They are generally unicellular, and consist of two domains, the bacteria and archaea.

protein Large biological molecules made up of amino acids that are used by organisms either for structure or to run chemical reactions. Proteins that run chemical reactions are called enzymes. Proteins are major molecules of life and necessary for life to exist. The order of amino acids in proteins determines their nature and function, and is coded by DNA and RNA. RNA codes for proteins directly. Since DNA codes for RNA, DNA codes for proteins indirectly.

protist A member of a large group of eukaryotes that are mainly unicellular and include amoebas, diatoms, and slime molds. They are not a natural group; rather, the biological category protist is used for convenience.

protocell A self-organized sphere made of lipids that has some similarities to, but is simpler than, a living cell. Protocells are proposed as stepping-stones toward the evolution and origin of cells.

protozoa An informal term for a group of unicellular eukaryotes. They are not a true taxon. They include the *Amoeba, Paramecium*, and *Euglena*.

pseudogene A DNA sequence that resembles a gene and was formerly a gene, but has been mutated into an inactive form with no function over the course of evolution. Pseudogenes that are paired with a functioning gene that they were duplicated from can evolve freely without harming the organism they are in, and can potentially mutate into a gene with a new adaptive function.

reciprocal altruism Altruism and cooperation by which two individuals in a population who are "friends" or allies help each other.

recombination, also called **sexual recombination** The exchange or trading of segments of DNA between chromosomes in sexual reproduction. It can increase DNA content and genetic variability while maintaining the compatibility of the different parts of the genome. The author of this book has proposed that this is the principal reason why sexual reproduction evolved and is usually favored by natural selection.

reduced carbon Carbon that is not combined with oxygen. The carbon in carbon dioxide, CO_2, is combined with oxygen, so it is not reduced carbon. The carbon in methane, CH_4, is not combined with oxygen, so is reduced carbon. Sequestration of reduced carbon increases atmospheric oxygen levels because it decreases the amount of carbon that can combine with and hence deplete the free molecular oxygen in the atmosphere, and it does so without any loss of oxygen from the atmosphere. (It also lowers the temperature). Sequestration of carbon that is not reduced lowers the temperature, but does not increase atmospheric oxygen levels because it buries oxygen with the carbon.

regulatory gene A gene that regulates one or more other genes, turning them on or off.

repetitive sequences Segments of DNA that occur in multiple copies throughout the genome. The DNA bases repeat a number of times. There are a number of different types of these. They can consist of repeated sequences in tandem, or of repeats dispersed throughout the genome.

respiration, also called **cellular respiration** A reaction that converts oxygen and carbohydrate to carbon dioxide and water, producing energy. In eukaryotic cells, it takes place in the mitochondrion. It produces the energy needed for complex organisms to exist and thrive. Respiration is the opposite reaction to oxygenic photosynthesis.

retrotransposon A transposable element, which is also called a transposon, that codes DNA to RNA and then back from RNA to DNA. Usually in nature, DNA codes for RNA, which then codes for a protein. Usually RNA does not code for DNA. When RNA codes for DNA, it is the reverse of the normal process, hence the term retro, and the name retrotransposon. *See* transposable element.

RNA Biological molecule that can act as an enzyme or, in some viruses, as the genetic material. Normally DNA codes for it, and it codes for proteins. Sometimes, it can code for DNA. Some viruses have RNA as their genetic material; some have DNA. It is thought that RNA was the first molecule in the chemical evolution of life, since it can act as both genetic material and an enzyme.

sequence, also called **DNA sequence** *Noun*: The order of the four chemical bases in a DNA molecule of an organism. The bases are abbreviated A, C, G, and T. *Verb*: To determine the order of the four chemical bases in the DNA of an organism by the use of technology and chemistry.

sexual reproduction The production of new organisms by the combination of genetic information of two separate organisms. It can generate variability while maintaining the compatibility of the parts of the genotype and phenotype.

short-term carbon cycle The carbon cycle involving the short-term carbon reservoir, in which carbon is stored in the atmosphere, oceans, and biosphere, with the ocean containing the largest amount of carbon. It takes months to centuries to recycle carbon through the short-term reservoir. *See* carbon cycle and long-term carbon cycle.

species A group of organisms with similar characteristics. A group of organisms that can breed with each other and produce fertile offspring in nature.

structural gene A gene that codes for any RNA or protein product other than a product that regulates other genes.

super organism A group or colony of highly social organisms that act collectively as if they were one organism. The intelligence of the group can exceed that of any individual. Examples are highly social bees, ants, naked mole-rats, and humans.

symbiosis A relationship between two (or more) species in which each benefits the other. An example is pollination, where an animal such as a bee obtains pollen and nectar for a nutritious meal from the flower and the plant is pollinated by the bee and so can reproduce. Symbiosis can also mean a relationship between two species in which one benefits the other; this includes mutualism, commensalism, and parasitism.

symbiotic coevolution (term coined by the author) A form of coevolution between two species that benefits both of them. It can involve horizontal gene transfer between the two species, in one or both directions. It can lead to large adaptive evolutionary breakthroughs in one or both species. Pollination of plants by animals and dispersal of seeds by animals both came about by symbiotic coevolution. There is also symbiotic genetic coevolution, which always involves the transfer of genes between species. A few other authors have used this term before the author of this book, but not with the meaning used in this book. *See* symbiotic genetic coevolution.

symbiotic genetic coevolution (term coined by the author) Symbiotic coevolution between two species in which at least one of the species has its variability enhanced by the transfer of genes from one species to the other.

taxon (plural: **taxa**) A group of organisms used for classification of any rank, such as a species, family, class, phylum, or domain. Variety, subspecies, and species are the lowest ranks, and domain is the highest rank.

taxonomic Concerning the classification of biological organisms.

transcription factor A protein that regulates the production of RNA by DNA, by binding to DNA, usually enhancing RNA production.

transduction The transfer of DNA from one host to another by a lysogenic virus. The virus is lysogenic in the first host. It leaves the host and takes some DNA of the host with it. Then it becomes lysogenic with a second host, as its DNA integrates into the second host, which gains the DNA the virus brought from the first host. This can lead to new adaptive evolutionary changes in the host that receives the DNA. Transduction is usually between hosts of the same species, but can be between hosts of different species.

trophic This refers to the feeding and nutrition relationships of different species in a food web.

trophic level The position of a species in a food web. A food web starts at trophic level 1 with primary producers such as plants that obtain their energy from sunlight via photosynthesis and nutrients from the soil. Level 2 is herbivores that eat plants, level 3 is carnivores that eat herbivores, and level 4 is carnivores that eat other carnivores and perhaps herbivores. The last level consists of the decomposers, such as vultures, fungi, and bacteria, which recycle the nutrients back into the soil, making them available to plants again. Ecological communities with higher biodiversity form more complex and branched trophic paths and food webs. *See* food web.

trophic cascade A catastrophic loss of species in an ecosystem as a result of the removal a key species in the food web, such as a key predator like the gray wolf (*Canis lupus*), or a key primary producer such as a species of phytoplankton.

vertebrate An animal with a backbone. This includes fish, amphibians, reptiles, birds, and mammals.

virulence The harmfulness of a pathogenic organism, such as a disease-causing virus, bacterium, or fungus.

virus A submicroscopic infectious agent that replicates only inside the living cells of an organism. Viruses cannot live on their own. Viruses infect all life forms. They have a capsid of protein surrounding their genetic material. Their genetic material can be either DNA or RNA. They are the most diverse life forms, and have greatly enhanced the evolution and increased the diversity of cellular life forms. There is controversy over whether they are living organisms. I define life as that which can undergo natural selection, so consider viruses to be living organisms.

viral shunt The constant attacking and killing of prokaryotes by viruses in the sea. It recycles nutrients and greatly fertilizes the sea. Marine phages kill 20 to 40% of ocean bacteria every day.

weathering The breaking down of rocks, soil, minerals, wood, or even artificial materials through contact with the Earth's atmosphere, water, and biological organisms. In it, CO_2 is combined with minerals in chemical reactions. The resulting material is then carried via rivers and creeks to the sea, where it falls to the ocean bottom and is buried. Weathering is a key step in sequestering carbon, removing it from the biosphere. Weathering is the main way carbon is removed from the atmosphere nonbiologically. Weathering can be done biologically too, which also sequesters carbon.

Index

Note: Page numbers in *italics* indicate figures.